Guidebook to the
Homeobox Genes

Other books by Sambrook & Tooze Publications

Guidebook to the Cytoskeletal and Motor
  Proteins
Edited by Thomas Kreis and Ronald Vale

Guidebook to the Extracellular Matrix and
  Adhesion Proteins
Edited by Thomas Kreis and Ronald Vale

Guidebook to the Homeobox Genes
Edited by Denis Duboule

Guidebook to the Secretory Pathway
*Edited by Jonathan Rothblatt, Peter Novick,
and Tom Stevens*

Guidebook to the
Homeobox Genes

Edited by
Denis Duboule

University of Geneva
Geneva, Switzerland

A SAMBROOK & TOOZE PUBLICATION
AT OXFORD UNIVERSITY PRESS

Oxford University Press, Walton Street, Oxford OX2 6DP

Oxford New York
Athens Auckland Bangkok Bombay
Calcutta Cape Town Dar es Salaam Delhi
Florence Hong Kong Istanbul Karachi
Kuala Lumpur Madras Madrid Melbourne
Mexico City Nairobi Paris Singapore
Taipei Tokyo Toronto

and associated companies in
Berlin Ibadan

Oxford is a trade mark of Oxford University Press

Published in the United States
by Oxford University Press Inc., New York

© Sambrook and Tooze Publishing Partnership, 1994
Reprinted 1995

A catalogue record for this book is available from the British Library

Library of Congress Cataloging-in-Publication-Data
Guidebook to the homeobox genes / edited by Denis Duboule.
(Guidebook series)
Includes bibliographical references and index.
1. Homeobox genes.   I. Duboule, Denis.
II. Series: Guidebook series (Oxford, England)
QH447.8H65G85   1994   574.87'322—dc20   93—47573

ISBN 0-19-859939-0 (hbk)
ISBN 0-19-859940-4 (pbk)

Printed in Great Britain by
The Bath Press, Avon

Preface

Ten years have passed since the homeobox was simultaneously discovered in Switzerland and in the USA. Hundreds of genes containing this DNA-binding motif have now been isolated from an increasing number of animals and plants. The high number of homeobox genes, and their involvement in a large variety of biological processes as well as the lack of coordination in the use of nomenclature and classification systems have generated some confusion, not only among newcomers in the field, but also among specialists. The identification of homeobox sequences which are more and more divergent from the original prototype has raised problems of both classification and phylogeny. Moreover, while some genes are currently known and used under different names, others have been renamed several times, so that searches in data libraries are often the beginning of a long and difficult process.

This book aims to provide an overview of homeobox genes. Three different aspects have been considered: in the first two chapters, a brief historical outline is given as well as an introduction to some developmental processes in which homeobox genes exert their functions, the third chapter proposes a comprehensive classification and structural analysis of homeobox-containing genes. Finally, a list of short descriptions of individual homeobox genes is provided in alphabetical order. As newly described genes appear every month in the literature, this list could not possibly be exhaustive. In addition, while some genes escaped inclusion, it also proved difficult to obtain information from quite a number of sources. Thus we strongly urge those people who cannot find their favourite gene, or who would like to modify or complete a particular entry, to send their information to the editor. This will certainly improve further editions of the book.

The short descriptions have been provided by many different colleagues to whom I am very grateful. In particular, I should like to thank M. Affolter, M. Akam, G. Aisemberg, M. Baltzinger, F. Banuett, D. Bittner, B. Blumberg, R. Bodmer, E. Boncinelli, J. Brockes, P. Brulet, T. Bürglin, D. Cribbs, P. Dollé, T. Chouard, M. Cleary, S. Cohen, C. Desplan, E. DeRobertis, R. Di Lauro, W. Driever, A. Fainsod, M. Featherstone, M. Fortini, R. Finkelstein, A. Fjøse, M. Frasch, M. Frohman, B. Galliot, W. Gehring, T. Gerster, C. Goridis, S. Hake, R. Harvey, W. Herr, R. Hill, P. Holland, J. Hirsh, T. Humphreys, J.-C. Izpisúa-Belmonte, B. Jegalian, A. Joyner, K. Jagla, Y. Yan, F. Karch, J. Kehrl, M. Kessel, D. Kimelman, T. Kornberg, R. Krumlauf, A. Kuroiwa, G. Lemke, A. Leutz, P. Lonai, K. Mahon, W. McGinnis, F. Meijlink, M. Mlodzik, T. Morinaga, E. Morita, E. Neufeld, M. Noll, W. Odenwald, G. Oliver, E. Olson, N. Patel, R. Peterson, S. Poole, B. Robert, J. Rubenstein, J. Runstadler, A. Ruiz i Altaba, C. Rushlow, T. Saiga, K. Saigo, E. Salo, C. Scheidereit, M. Schena, B. Schierwater, H. Schöler, P. Sharpe, C. Shashikant, A. Shenk, H. Sive, P. Vos, U. Walldorf, C. Walther, P. Webster, M. Wilkinson, C. Wright, Y. Xu, Z. Xue, M. Yaniv as well as those who contributed but are not cited.

Special thanks are due to Thomas Bürglin for his help and to Christina Kjaer for her very efficient secretarial assistance.

Geneva
August 1993

D.D.

Contents

A History of the Homeobox

A History of the Homeobox

Walter J. Gehring

■ The discovery of the homeobox

Almost ten years have passed since the discovery of the homeobox and it seems appropriate to tell the "homeobox story", how it happened and what advances were made in this field over the past decade. Of course, such accounts are quite subjective and the same events are seen differently by different observers. The discovery of the homeobox was only possible in the congenial atmosphere of a laboratory in which a group of creative and hardworking young scientists made a concerted effort to elucidate the mechanisms of development.

A key to the advancement of our understanding of the genetic control of development was provided by the **homeotic genes.** From the pioneering work of Jacob and Monod (1961) on prokaryotes it was known that there exists a special class of **regulator genes** whose function it is to regulate the activity of groups of target genes, for example, the structural genes within the lac operon or the entire genome of bacteriophage lambda. The first regulator genes that were identified encoded repressors, but subsequently activators were also found in prokaryotes indicating that regulation could be either positive or negative. The isolation of the lac repressor by Gilbert and Müller-Hill (1966) and of the lambda repressor (Ptashne 1967a,b) settled the debate whether the repressors were RNA or protein molecules and showed that they are sequence-specific DNA binding proteins. The repressor proteins can reversibly bind to the operator DNA and thereby exert their gene regulatory function.

In higher organisms the identification of regulator genes proved to be much more difficult and no operons were found in eukaryotes. However, there was a class of mutations discovered by *Drosophila* geneticists, called **homeotic mutations,** which were likely to represent regulator genes involved in the genetic control of development. The first homeotic mutant, *bithorax,* was discovered by Bridges in 1915 (see Bridges & Morgan, 1923) and later shown by Lewis to be part of a cluster of genes called the Bithorax-Complex (see Lewis, 1978). Homeotic mutations lead to partial or complete segmental transformations, suggesting that they are involved in the genetic control of the body plan. For example, bithorax mutations lead to a transformation of the third thoracic segment carrying halteres towards a second thoracic segment with wings, and can generate four-winged flies. Furthermore, Lewis recognized their important role in evolution, since single mutations can induce major changes of the body plan.

My interest in homeotic genes dates back to the time when I was a graduate student and discovered a spontaneous *Drosophila* mutation which transforms the antennae on the head of the fly into legs. I named the mutant gene Nasobemia (Gehring, 1966) since the existence of such an animal had been preconceived by the poet Christian Morgenstern (1932), who had written a poem on the Nasobem, an imaginary animal which can walk on its nose. This mutation, which later was physically mapped to the *Antennapedia* gene, provided the entry point for studying the fundamental problem of how the body plan, the "architecture" of the fly, is genetically controlled. At that time, most molecular biologists considered such "morphological" mutants as much too complex to ever be understood at the molecular level and most classical biologists thought that the molecular approach was useless and besides the point. Nevertheless, when I was a postdoctoral fellow at Yale, Alan Garen and I began to analyze DNA binding proteins in *Drosophila.* However, at that time the technical obstacles to identify homeotic gene products were insurmountable.

This situation changed dramatically when methods for gene cloning were developed, and it became clear to me that this was the way to proceed. The development of a method for "walking along the chromosome" by the group of David Hogness (Bender et al., 1983) allows the cloning of essentially any *Drosophila* gene for which some mutations are available. Therefore, I decided to attempt the cloning of the *Antennapedia* gene, and two of my postdoctoral fellows, Richard Garber and Atsushi Kuroiwa, embarked upon a long and strenuous chromosomal walk which lasted for three and a half years. Although prominent molecular biologists told us that we were probably going to clone the gene for thymidine kinase or any other good old enzyme, I was convinced that *Antennapedia* was a developmental control gene probably encoding a gene regulatory protein. At some point along the walk we almost gave up, but due to the perseverance of my collaborators and my continued encouragement the walk was completed and the *Antennapedia* locus, as defined by deletions, was cloned along with some adjacent genes (Garber et al., 1983). In order to map the transcription unit more precisely, a number of cDNA clones complementary to the chromosomal DNA of the *Antennapedia* region were isolated. The *Antennapedia* gene turned out to be extremely large, spanning more than 100 kb, with huge introns. When hybridizing the cDNA clones to the chromosomal DNA from the walk, in order to map the exon sequences, Richard Garber found that there was crosshybridization between the cDNA and chromosomal DNA sequences outside the *Antennapedia* gene. This was the first sign of the homeobox. The band of crosshybridization was found in the last lane, right at the edge of the gel, but I was immediately convinced that this was the most important band on the gel. In fact, I was looking for such crosshomologies, since Ed Lewis had postulated much earlier, that homeotic genes like *bithorax* and *Antennapedia* might have arisen by tandem duplication. Therefore, we were expecting to find crosshybridization between different homeotic genes.

The crosshybridizing gene turned out to be *fushi tarazu*, a segmentation gene located next to *Antennapedia*, which was first described by Barbara Wakimoto who gave it its Japanese name meaning "not enough segments", since it lacks alternate body segments (Wakimoto and Kaufman, 1981). Of course, I asked a postdoctoral fellow from Japan, Atsushi Kuroiwa, to clone *fushi tarazu*. It turned out that in both genes, *Antennapedia* and *fushi tarazu*, the crosshybridizing DNA sequences were confined to the last (3' most) exon. Shortly before that time a new collaborator, William McGinnis, had joined my laboratory and he was enthusiastic about my proposal to define the homologous sequences between *Antennapedia* and *fushi tarazu* more precisely. After working out suitable conditions for low-stringency hybridization, he scanned the entire *Drosophila* genome for sequences crosshybridizing with the various segments of the *Antennapedia* and *fushi tarazu* transcripts. Soon, he found a ladder of bands hybridizing to these 3' exon sequences from *Antennapedia* and *fushi tarazu*, representing at least a dozen genes, most of which were subsequently cloned (see below). In order to find out whether other homeotic genes also contained this crosshybridizing DNA sequence, I called Pierre Spierer who had just come back to Switzerland from David Hogness' laboratory where he and Welcome Bender had done an extensive "chromosome walk" in the *bithorax* region. I asked him just for two clones, those covering the first and the last exon of the homeotic *Ultrabithorax* gene. The Hogness group kindly provided the two clones for testing our hypothesis. In a short period of time, Bill McGinnis in close collaboration with Michael Levine and Ernst Hafen showed that the last exon of *Ultrabithorax* also crosshybridizes with those of *Antennapedia* and *fushi tarazu*. **Eureka**, at that point, we knew that we had discovered something important and the name **homeobox** (McGinnis et al., 1984a) was coined. Subsequent sequencing of the crosshybridizing DNA segments ruled out the possibility that the crosshybridization was due to a simple repeated sequence and defined the homeobox as 180 bp DNA segment with 75-77% sequence identity between *Antennapedia*, *fushi tarazu* and *Ultrabithorax*. Furthermore, the three homeoboxes share the same open reading frame and many of the nucleotide changes preserve the same amino acid in this frame, indicating that the homeobox encodes a polypeptide segment which we designated as the **homeodomain** (McGinnis et al., 1984b). Independently, the sequence homology between *Antennapedia*, *fushi tarazu* and *Ultrabithorax* was discovered by Matthew Scott and Amy Weiner (1984) who were working in Thomas Kaufman's laboratory in Bloomington, Indiana. In parallel, they had carried out a chromosome walk across the Antennapedia Complex (Scott et al., 1983) and found

that *Antennapedia* cDNA clones crosshybridized with the neighbouring *fushi tarazu* gene and also with *Ultrabithorax* sequences obtained from David Hogness. Their sequencing data paralleled ours.

At that time Ernst Hafen and Michael Levine, in my laboratory, with some helpful advice from Michael Akam in Cambridge, had worked out the technique of *in situ* hybridization of labelled DNA probes to RNA transcripts in frozen tissue sections. As soon as *Antennapedia* was cloned, they succeeded in localizing its transcripts in developing embryos and imaginal discs (Hafen et al., 1983; Levine et al., 1983). Their data indicated that these transcripts were first localized in a single stripe or belt in the thoracic region and accumulated strongly in the ganglion of the second thoracic segment. This finding was consistent with the notion that the normal *Antennapedia* gene specifies the second thoracic segment with second legs and not the antennae. Thus, *in situ* hybridization allowed us to identify to some extent homeotic genes on the basis of their expression pattern. Using the *Antennapedia* and *fushi tarazu* (or *Ultrabithorax*) homeoboxes as duplicate probes, several homeobox-containing genes were isolated. On the basis of their chromosomal localization and the location of their transcripts, two of the clones were tentatively identified as *Deformed* and *abdominal-A,* representing additional homeotic genes from the Antennapedia- and the Bithorax-complex, respectively (McGinnis et al., 1984a). This proved that the homeobox can be used to isolate other homeotic genes and justified its name.

Surprisingly, in an evolutionary survey using DNA from various species Bill McGinnis showed that the homeobox is not confined to insects but that it is also found in vertebrates, including chicken, mice and man (McGinnis et al., 1984b). In collaboration with Eddy De Robertis and Andres Carrasco, the first vertebrate homeobox was cloned from the frog, *Xenopus laevis*. It turned out to be remarkably similar in sequence to the *Drosophila* homeobox (Carrasco et al., 1984). The first mouse homeobox genes were cloned in collaboration with Frank Ruddle who happened to be on sabbatical leave in my laboratory at that time (McGinnis et al., 1984c). Again the degree of sequence similarity to the *Drosophila* homeobox was astounding. This raised the possibility that the genetic control of development might be much more universal than previously assumed.

The development of a technique of *in situ* hybridization led immediately to another perhaps equally important discovery. When Atsushi Kuroiwa had cloned the segmentation gene *fushi tarazu* (Kuroiwa et al., 1984), he wanted to examine, in collaboration with Ernst Hafen, the expression pattern of the *fushi tarazu* transcripts in the normal embryo. Atsushi expected the transcripts to be localized in those alternate body segments that are missing in the mutant embryo. A few weeks later, I shall never forget that moment, Ernst and Atsushi called me to the microscope and showed me the beautiful pattern of seven stripes on the embryos (Hafen et al., 1984a). This was a breakthrough in the understanding of the genetic control of segmentation.

Stripes in every segment marked the discovery of the *engrailed* gene; Anders Fjose and Bill McGinnis cloned *engrailed* in just a few months by crosshybridization on the basis of homeobox homology (Fjose et al., 1985), while Tom Kornberg and his collaborators carried out a long chromosome walk to find *engrailed* (Poole et al., 1985), as we had done previously for *Antennapedia*. *In situ* hybridization yielded another important finding; Marek Mlodzik, while studying the homeobox-containing *caudal* gene, discovered that the *caudal* transcripts form a concentration gradient in the early embryo prior to cellularization (Mlodzik et al., 1985). Embryologists had speculated for a long time that gradients may be involved in specifying positional information, but this was the first physical demonstration of such a gradient. Thus, the homeobox was not only found in most homeotic genes, but also in some of the genes specifying molecular gradients and segmentation patterns at other levels of the regulatory hierarchy. It clearly provided a key to the understanding of development.

■ Structure and function of the homeodomain

The first hint at the possible function of the homeodomain came from computer searches through protein data bases which revealed a small but significant degree of sequence similarity to the yeast mating-type proteins MAT *a1* and *α2* (Shepherd et al., 1984). The MAT genes have a key role in determining cell type identity in yeast, i.e. whether a cell of the *a* or *α* mating type is formed or an *a/α* diploid cell (Strathern et al., 1981). This observation is particularly important, since *α2* was shown to be a repressor of *a*-specific genes (Johnson & Herskowitz, 1985) and *a1* also appeared to be a transcriptional regulator. Therefore, it seemed likely that homeotic genes might also encode **gene regulatory proteins**. Sequence similarities between the homeodomain and gene regulatory proteins of prokaryotes also suggested that the homeodomain might contain a **helix-turn-helix DNA binding motif** (Laughon and Scott, 1984; Shepherd et al., 1984) similar to the one found by X-ray crystallographic studies in several prokaryotic gene regulatory proteins (see Pabo & Sauer, 1984). However, the statistical significance of these sequence comparisons is very doubtful (Brennan & Matthews, 1989), and a detailed structural analysis was required to prove this point (see below). Nevertheless, these considerations soon led to the working hypothesis, that homeotic genes encode gene regulatory proteins acting as transcription factors which bind to specific *cis*-regulatory regions in their target genes, and that the homeodomain represents the DNA-binding domain of these proteins.

In following years a considerable amount of data has confirmed this working hypothesis. *In vitro* binding studies with protein extracts first indicated that the homeodomain proteins can bind sequence-specifically to DNA (Johnson and Herskowitz, 1985; Desplan et al., 1985, 1988; Fainsod et al., 1986; Hoey & Levine, 1988; Cho et al., 1988; Müller et al., 1988; Beachy et al., 1988; Laughon et al.,

1988). Using the isolated purified homeodomain polypeptide of *Antennapedia,* we demonstrated that the homeodomain is indeed sufficient for binding to specific oligonucleotide sequences (Müller et al., 1988). The homeodomain was shown to bind as a monomer with high affinity (K_D in the range of 10^{-9} to 10^{-10} M) to its DNA binding site and the halflife of the homeodomain-DNA complex was estimated to be approximately 90 min (Affolter et al., 1990). *In vitro* transcription assays and transfection experiments in cultured cells indicated that homeodomain proteins can function as activators and/or repressors of transcription (Jaynes and O'Farrell, 1988; Thali et al., 1988; Biggin and Tjian, 1989; Han et al., 1989; Krasnow 1989; Winslow et al., 1989). Independent evidence came from the analysis of mammalian transcription factors of the Pou protein family, which were isolated as transcription factors and subsequently found to contain a homeodomain in addition to a Pou-specific domain (Herr et al., 1988). *In vivo*, i.e. in the embryo, transcriptional regulation was first demonstrated for the homeodomain protein *bicoid* interacting with its target gene *hunchback* (Driever & Nüsslein-Volhard, 1989; Driever et al., 1989; Struhl et al., 1989) and most clearly proven for the *fushi tarazu* protein which directly interacts with the *fushi tarazu* autoregulatory enhancer (Schier & Gehring, 1992; see below). Therefore, we can conclude that homeodomain proteins can function as transcription factors.

The **threedimensional structure** of the homeodomain in solution was first determined for *Antennapedia* by NMR spectroscopy in a fruitful collaboration with Kurt Wüthrich and his group in Zürich (Qian et al., 1989). For this purpose, my student Martin Müller purified large amounts of the *Antennapedia* homeodomain polypeptide to homogeneity and we brought it to Zurich for structural analysis. The homeodomain contains three well-defined *α*-helices and a more flexible fourth helix. Helices two and three form a helix-turn-helix motif virtually identical to those observed in various prokaryotic repressors, except that the "recognition" helix 3 is elongated considerably by helix 4. Subsequently, the structure of the homeodomain-DNA complex in solution was also solved by NMR spectroscopy (Otting et al., 1990). The recognition helix was found to contact the DNA in the major groove as in prokaryotic repressors, but there are additional contacts between the loop immediately preceding the helix-turn-helix motif and the DNA backbone, and the flexible amino terminal arm which establishes contacts with bases in the minor groove of the DNA. Methylation and ethylation interference studies using wildtype and mutant polypeptides provided independent evidence for this model (Percival-Smith et al., 1990). The structure of the *engrailed* homeodomain and its complex with DNA was also solved by X-ray crystallography (Kissinger et al., 1990) using the model based on the NMR data of the *Antennapedia* homeodomain. The structures of the two homeodomains are quite similar, although the resolution is not yet high enough to detect any minor differences. Even the homeodomain of MAT*α2* (Wolberger et al., 1991) has a very similar structure to that of *Antennapedia* despite a three amino acid insertion in the loop between helices 1 and 2. This clearly shows that yeast indeed has homeodomain proteins, even though the sequence identity is only 28% with the homeodomain of *Antennapedia* for example.

The putative target sequences identified for several homeodomain proteins are very similar to each other. In fact, several distantly related homeodomain proteins apparently recognize the same sequence elements *in vitro* (Hoey and Levine, 1988; Desplan et al., 1988). However, in the case of the *bicoid* and *paired* homeodomain proteins it was shown in a gene activation assay in yeast and by *in vitro* binding experiments that the amino acid at position 9 of the recognition helix (amino acid 50 in the homeodomain) has a strong influence on the binding specificity (Hanes and Brent, 1989; Treisman et al., 1989). The *bicoid* homeodomain has a lysine residue at this position (amino acid 50) which in *Antennapedia* and *fushi tarazu* is replaced by glutamine. Markus Affolter in my laboratory first realized that the major difference between the consensus binding sites of *bicoid* (GGATTA) and that of *Antennapedia* and *fushi tarazu* (C C/A ATTA) resides in the two GG residues preceding the ATTA core motif which are replaced by CC or CA in *Antennapedia* and *fushi tarazu*. This led to the hypothesis that glutamine 50 contacts these two C residues in the *fushi tarazu* binding site. This was confirmed by DNA binding studies which showed that the wildtype *fushi tarazu* protein with glutamine at position 50 (*ftzQ50*) preferentially binds to CCATTA, whereas the mutant protein with lysine at this position (*ftzQ50K*) has much higher affinity for GGATTA (Percival-Smith et al., 1990). The NMR structure analysis showed that indeed glutamine 50 contacts these two C residues in the *Antennapedia* homeodomain-DNA complex (Otting et al., 1990; Billeter et al., 1993). These *in vitro* data were subsequently used to test the **DNA binding specificity *in vivo***. For this purpose, we analyzed the upstream enhancer of the *fushi tarazu* gene which functions as an autoregulatory element. Earlier work by Yasushi Hiromi had shown that the upstream enhancer element can direct the expression of a reporter gene (β-galactosidase) with a minimal promoter in a pattern of seven stripes (Hiromi & Gehring, 1987). However, in a *fushi tarazu*-mutant embryo no stripes are formed, indicating that *fushi tarazu* protein is required for the expression in stripes. The simplest hypothesis to explain these results assumes that the *fushi tarazu* protein interacts directly with its own enhancer and enhances the expression of its own gene in a positive autocatalytic feedback loop. However, these experiments do not rule out an indirect mechanism of autoregulation, in which *fushi tarazu* activates another protein which in turn binds to the *fushi tarazu* enhancer. In a series of elegant experiments, my student Alexander Schier showed that the interaction between the *fushi tarazu* protein and the autoregulatory enhancer is direct (Schier & Gehring, 1992). The *fushi tarazu* enhancer contains multiple *fushi tarazu in vitro* binding sites (Pick et al., 1990). Mutating three of these binding sites to GGATTA strongly reduces enhancer activity *in vivo*. The effect of this down-mutation can be suppressed by introducing a single mutation in the *fushi tarazu* gene substituting glutamine by lysine. The FTZ Q50->K mutant recognizes the GGATTA motif and expression in the seven stripes is fully restored. This experiment provides definitive evidence for a direct positive autoregulatory interaction *in vivo*. The *fushi tarazu* protein binds directly to its target sites in the enhancer and enhances transcription of its own gene.

■ Homeobox genes in the genetic hierarchy controlling *Drosophila* embryogenesis

The principles of the genetic control of *Drosophila* embryogenesis have been worked out by a combined genetic, embryological and molecular approach (see Ingham, 1988 for review). This analysis was based on a large collection of maternal-effect and embryonic lethal mutants which were systematically isolated by Christiane Nüsslein-Volhard, Eric Wieschaus and their collaborators. On the top of the regulatory hierarchy are the **coordinate genes** which are first expressed maternally during oogenesis. They determine both the antero-posterior and the dorso-ventral axes of the future embryo (see Nüsslein-Volhard, 1991). The antero-posterior axis is mainly determined by the mRNAs for *bicoid* and *nanos* which are maternally expressed and prelocalized at the anterior and posterior pole of the egg respectively. After fertilization these mRNAs are translated into the respective proteins which form concentration gradients with opposite polarity. The *bicoid* protein can be considered the major anterior cytoplasmic determinant which exerts a long range effect on the antero-posterior pattern. It behaves as a **morphogen**, i.e. different concentration levels determine different morphological structures. *Nanos* appears to be the primary posterior determinant (Wang and Lehmann, 1991) whereas the RNA and protein gradient of *caudal* (mentioned above) seems to form secondarily, in response to the *bicoid* protein gradient. The dorso-ventral axis is established by a different mechanism; although the *dorsal* protein is distributed uniformly throughout the syncytial embryo it subsequently forms a nuclear concentration gradient (see Govind and Steward, 1991) with its highest concentration on the ventral side of the embryo.

The gradient patterns along the antero-posterior axis are subsequently converted into a periodic or repetitive pattern of body segments through the action of the **segmentation genes**. These genes can be subdivided into (i) **gap genes,** which are required in regions that are several segments wide, (ii) the **pair-rule genes**, which affect alternate segments and (iii) the **segment polarity genes**, which are active in parts of every segment. The repetitive pattern of similar segments is converted into a sequential pattern of individually different segments by the action of the **homeotic genes** which give each one of the segments its unique identity. These genes are also called homeotic selector genes. They are clustered in the Antennapedia Complex and the Bithorax Complex respectively, and they all contain homeoboxes. However, there are some homeobox-containing genes at all levels of the developmental hierarchy, e.g. *bicoid* and *caudal* are maternally expressed genes involved in the establishment of morphogenetic gradients; *orthodenticle* and *empty spiracles* are gap genes involved in the formation of the head and posterior abdominal structures; *fushi tarazu* and *even-skipped* are segmentation genes of the pair-rule type; *engrailed* is a segment polarity gene. There are also homeobox genes involved in dorso-ventral patterning, as for example *zerknüllt,* and in the formation of fine grain patterns like *rough* in the differentiation of ommatidial cells in the compound eyes, where *rough* plays an important role. In each case studied so far, *Drosophila* homeobox genes are involved in determining either segmental identity, regionalization or cell identity, i.e. cell fate.

■ The function of homeotic genes

As first pointed out by E.B. Lewis, there are two types of homeotic mutations with opposite effects: Loss-of-function mutants which are generally recessive and due to a deletion or inactivation of the gene, and dominant gain-of-function mutants in which the gene product apparently is not altered but expressed ectopically. Loss-of-function mutants of the *Ultrabithorax* gene lead to a transformation of the third thoracic segment (T3) to a second thoracic segment (T2) and the corresponding transformation of the halteres (small balancers) into a second pair of wings. In contrast, dominant gain-of-function mutants of *Ultrabithorax* lead to the opposite transformation of T2 to T3 associated with the transformation of the wings into a second pair of halteres. This allows the "construction" of flies with either four wings or four halteres.

The nature of the dominant gain-of-function mutants remained unclear until a molecular analysis of such mutants became possible for the *Antennapedia* gene. Recessive loss-of-function mutants of *Antennapedia* lead to a transformation of T2 to T1 (Wakimoto and Kaufman, 1981), whereas the dominant gain-of-function mutants result in a transformation in the opposite direction, i.e. T1 and the head segments towards T2. In the adult fly, the latter transformation leads to the conversion of the antennae into second legs (T2). Since the loss-of-function mutants are missing the second thoracic segment (which is replaced by another T1 segment) and the gain-of-function mutants have one or more additional T2 segments, it can tentatively be concluded that *Antennapedia* specifies T2 in the body plan. The dominant gain-of-function mutations raised an apparent paradox, since most of them are caused by inversions with one breakpoint in the *Antennapedia* locus and the other in an unrelated gene. This raises the intriguing question of how a mutation which breaks a gene apart can give rise to a gain-of-function. The molecular analysis of several of these gain-of-function mutants revealed that they are due to gene fusions in which the protein-coding region is left intact and fused to a foreign enhancer and promoter which leads to the ectopic expression of the *Antennapedia* protein (Frischer et al., 1986; Schneuwly et al., 1987b).

If ectopic expression of *Antennapedia* indeed leads to the dominant phenotypes and *Antennapedia* indeed specifies the second thoracic segment, one should be able to construct an artificial gain-of-function mutant and induce T2 structures ectopically. Such a mutant was constructed by inserting the *Antennapedia* cDNA into a heat-shock vector containing a heat inducible promoter and introduction of this construct into the germline of transgenic flies (Schneuwly et al., 1987a). By application of a brief heatshock the expression of *Antennapedia* protein can be induced at defined developmental stages in all cells of the animal. As predicted, the application of a heatshock at early embryonic stages leads to a transformation of T1 and the head segments to T2, and induction of *Antennapedia* expression in early third instar larvae results in the dramatic conversion of the antennae into second legs (Schneuwly et al., 1987a; Gibson and Gehring, 1988). This was the first successful attempt at redesigning the fruitfly.

However, much remains to be learnt about the mode of action of homeotic genes. For example, the timing of heat induction is of crucial importance. The target cells seem to be sensitive only during critical developmental stages when the determination of the respective structures normally takes place. Once the developmental decision is made, the state of determination is quite stable and the ectopic overexpression of a single master control gene like *Antennapedia* is no longer effective in changing cell fate. Also the transformation into T2 segments is only effective in the anterior region of the body, which can probably be explained by the fact that the genes of the Bithorax Complex are epistatic over *Antennapedia* in the posterior segments (Gonzales-Reyes et al., 1990). Deletion of the Bithorax Complex leads to the expression of *Antennapedia* in the posterior thoracic and the abdominal segments and to their transformation into T2 (Hafen et al., 1984b).

It should be added that the rule that loss-of-function mutations lead to transformations in the anterior direction, whereas gain-of-function mutations result in posterior transformations applies only to the genes of the Bithorax complex and *Antennapedia,* but not to the more anteriorly expressed genes like *Sex combs reduced* and *Deformed* which behave in the opposite way. In accordance with these observations Lewis has proposed that loss-of-function mutants generally lead towards the evolutionary groundstate which he considers to be T2, and gain-of-function mutants lead away from it. But at the present time, we do not know enough about the evolution of homeotic genes to evaluate critically this attractive hypothesis.

■ The phylogeny of the homeobox

Our initial evolutionary survey indicated that the homeobox is found in *Drosophila,* in other arthropods and in annelids (protostomes) as well as in vertebrates (deuterostomes), but using low stringency hybridization, which requires at least 40-50% sequence identity, we could not detect homeobox sequences in animals such as the nematode *Ascaris,* which lack overt segmentation (McGinnis et al., 1984b). The apparent correlation between segmentation and the presence of the homeobox vanished when more homeoboxes were cloned and when more sensitive methods, using degenerate oligonucleotide probes (Bürglin et al., 1989), and more recently amplification by PCR (Schierwater et al., 1991) were used. Homeoboxes have now been found in fungi, plants and animals. In **fungi** which have a limited capacity for cell differentiation, homeobox-containing genes occupy a key position in the determination of the mating-type in both budding and in a fission yeast and in the determination of filamentous versus yeast-like growth forms in *Ustilago* (Schultz et al., 1991).

In **plants** homeoboxes have been identified only recently in *Arabidopsis* (Ruberti et al., 1991) and in the *Knotted* gene of maize (Vollbrecht et al., 1991). In animals, homeobox sequences have been reported from most groups of metazoa ranging from cnidarians (Schierwater et al., 1991) and planarians (Garcia-Fernandez et al., 1991) to man, but no data on protozoa are available.

As mentioned above, homeotic selector genes are clustered in *Drosophila* and organized into the *Antennapedia* and Bithorax Complex which presumably arose by tandem duplication and subsequent divergence. In other insects a similar gene arrangement is found in the silkmoth *Bombyx mori* (Ueno et al., 1992), whereas in *Tribolium,* the red flour beetle, a single homeotic gene cluster was found (Beeman, 1987; Stuart et al., 1991). Interestingly, homeobox genes with sequence similarity to *Drosophila* homeotic genes are also clustered in vertebrates. Both in mice and man four homeobox gene clusters have been mapped on different chromosomes, which presumably arose by duplication of a single ancestral gene cluster. As first suggested by Lewis (1978), for the genes of the Bithorax Complex, and later also found for the *Antennapedia* Complex, the genes are arranged in the same order along the chromosome as they are expressed along the antero-posterior axis of the embryo, i.e. the genes that are located most 5′ in the cluster are expressed most posteriorly, whereas the more 3′ located genes are progressively expressed in more anterior regions (Gaunt et al., 1988; Duboule and Dollé, 1989; Graham et al., 1989). Therefore, the evolutionary conservation not only concerns the homeobox, but also the homeobox gene organization. In the ancestors of both insects and vertebrates the homeobox genes presumably formed a single cluster, as in *Tribolium*, which subsequently was split into two partial complexes both in diptera *(Drosophila)* and in lepidoptera (silkmoth). There may also be clustering of homeobox genes in the nematode *C. elegans* (Bürglin et al., 1991; Kenyon and Wang, 1991). During vertebrate evolution, the ancestral cluster presumably was duplicated at least twice to form four homologous clusters. Both the DNA sequences and the expression patterns were conserved between the corresponding genes (paralogs) of the different clusters, which underlines the importance of these genes in development and evolution.

The reasons for the antero-posterior gene order in the clusters and its evolutionary conservation are only par-

tially understood. One of the selective forces keeping the genes together stems from the fact that adjacent genes share common *cis*-regulatory elements (Peifer et al., 1987). Therefore, adjacent genes have to remain closely linked, since translocation of one of the partners would deprive one or the other gene of its *cis*-regulatory elements. Furthermore, intergenic boundary regions have been discovered, which presumably form higher order chromatin structures involved in the regulation of those genes located between two such boundaries (Gyurkovicz et al., 1990). However, more work is needed in order to understand these evolutionary constraints.

The homeobox genes located in the Hox clusters of the mouse are expressed in posterior segments up to the level of the hindbrain (see Hunt and Krumlauf, 1991), but no genes have been found to be expressed in the forebrain except Quox 1 in the quail (Xue et al., 1991). In *Drosophila* two homeobox-containing genes, *empty spiracles* (Dalton et al., 1989; Walldorf and Gehring, 1992), and *orthodenticle* (Finkelstein and Perrimon, 1991), are known to be expressed in the anterior head region. Using *empty spiracles* as a probe, two partially homologous human cDNA clones were isolated (EMX1 and 2), the murine homologues of which are expressed in the cerebral cortex of mouse embryos (Simeone et al., 1992). This indicates that the spatial expression patterns are also conserved for certain homeobox genes located outside of the *HOX* clusters.

The question of whether vertebrate homeobox genes are also functionally homologous to homeotic selector genes of *Drosophila* has been approached by constructing both dominant gain-of-function and recessive loss-of-function mutants in transgenic mice. A gain-of-function mutant was constructed by fusing the *Hoxa-7* gene to an ubiquitously expressed actin promoter. The transgenic mice in which *Hoxa-7* is ectopically expressed have an extra cervical vertebra, a proatlas (Kessel and Gruss, 1991). This is an important morphological change since all mammals (with two or three exceptions) have seven cervical vertebrae. In the transgenic mutant mice the occipital segment is transformed into a vertebra, corresponding to a proatlas as it is found in some reptiles. Therefore, the gain-of-function mutation causes a posterior transformation. A loss-of-function mutation was generated by disruption of the *Hoxc-8* gene (Le Mouellic et al., 1992). In this case, anterior transformations are induced, e.g. the transformation of a lumbar vertebra into a thoracic vertebra with an additional rib. These transformations in opposite directions are as expected from the corresponding mutations in *Drosophila* and provide evidence that vertebrates possess true homeotic genes which are involved in specifying the antero-posterior axis in much the same way as in *Drosophila.* This interpretation is further supported by gene transfer experiments between mouse and *Drosophila,* which demonstrate that the *Antennapedia*-like *Hoxb-6* gene from the mouse can induce antennal legs when it is ectopically expressed in *Drosophila*, similar to the *Drosophila Antennapedia* gene (Malicki et al., 1990).

In conclusion, the mechanisms of the genetic control of development and of the body plan in particular are much more **universal** than anticipated. The homeobox has identified a class of master control genes and provided a key to the understanding of how the body plan is established not only in *Drosophila* but also in higher vertebrates including man.

■ References

Affolter, M., Percival-Smith, A., Müller, M., Leupin, W. and Gehring, W.J. (1990). Proc. Natl. Acad. Sci. U.S.A. 87, 4093.

Beachy, P.A., Krasnow, M.A., Gavis, E.R. and Hogness, D.S. (1988). Cell 55, 1069.

Beeman, R.W. (1987). Nature 327, 247.

Bender W., Spierer, P. and Hogness, D.S. (1983). J. Mol. Biol. 168, 17.

Biggin, M.D. and Tjian, R. (1989). Cell 58, 433.

Billeter, M., Qian, Y.Q., Otting, G., Müller, M., Gehring, W.J. and Wüthrich, K. (1993). J. Mol. Biol. 234, 1084.

Brennan, R.G. and Matthews, B.W. (1989). J. Biol. Chem. 264, 1903.

Bridges, C. and Morgan, T.H. (1923). Carnegie Inst. Wash. Publ. 327, 1-251.

Bürglin, T.R., Finney, M., Coulson, A. and Ruvkun, G. (1989). Nature 341, 239.

Bürglin, T.R., Ruvkun, G., Coulson, A., Hawkins, N.C., McGhee, J.D., Schaller, D., Wittmann, C., Müller, F. and Waterston, R.H. (1991). Nature 351, 703.

Carrasco, A.E., McGinnis, W., Gehring, W.J. and De Robertis, E.M. (1984). Cell 37, 409.

Cho, K.W.Y., Goetz, J., Wright, V.E., Fritz, A., Hardwicke, J. and De Robertis, E.M. (1988). EMBO J. 7, 2139.

Dalton, D., Chadwick, R. and McGinnis, W. (1989). Genes Devel. 3, 1940.

Desplan, C., Theis, J. and O'Farrell, P.H. (1985). Nature 318, 630.

Desplan, C., Theis, J. and O'Farrell, P.H. (1988). Cell 54, 1081.

Driever, W. and Nüsslein-Volhard, Ch. (1989). Nature 337, 138.

Driever, W., Thoma, G. and Nüsslein-Volhard, C. (1989). Nature 340, 363.

Duboule, D. and Dollé, P. (1989). EMBO J. 8, 1497.

Fainsod, A., Bogorad, L.D., Ruusala, T., Lubin, M., Crothers, D.M. and Ruddle, F.H. (1986). Proc. Natl. Acad. Sci. U.S.A. 83, 9532.

Finkelstein, R. and Perrimon, N. (1991). Development 112, 899.

Fjose, A., McGinnis, W. and Gehring, W.J. (1985). Nature 313, 284.

Frischer, L.E., Hagen, F.S. and Garber, R.L. (1986). Cell 47, 1017.

Garber, R.L., Kuroiwa, A. and Gehring, W.J. (1983). EMBO J. 2, 2027.

Garcia-Fernandez, J., Baguna, J. and Salo, E. (1991). Proc. Natl. Acad. Sci. U.S.A. 88, 7338.

Gaunt, S.J., Sharpe, P.T. and Duboule, D. (1988). Development (Supplement) 104, 169.

Gehring, W. (1966). Jul. Klaus, Arch. 41, 44.

Gibson, G. and Gehring, W.J. (1988). Development 102, 657.

Gilbert, W. and Müller-Hill, B. (1966). Proc. Natl. Acad. Sci. U.S.A. 56, 1891.

Gonzales, Reyes, A., Urquia, N., Gehring, W.J., Struhl, G. and Morata, G. (1990). Nature 344, 78.

Govind, S. and Steward, R. (1991). Trends Genet. 7, 119.

Graham, A., Papalopulu, N. and Krumlauf, R. (1989). Cell 57, 367.

Gyurkovics, H., Gausz, J., Kummer, J. and Karch, F. (1990). EMBO J. 9, 2579.

Hafen, E., Kuroiwa, A. and Gehring, W.J. (1984a). Cell 37, 833.

Hafen, E., Levine, M. and Gehring, W.J. (1984b) Nature 307, 287.

Hafen, E., Levine, M., Garber, R.L. and Gehring, W.J. (1983). EMBO J. 2, 617.

Han, K., Levine, M.S. and Manley, J.L. (1989). Cell 56, 573.

Hanes, S.D. and Brent, R. (1989). Cell 57, 1275.

Herr, W., Sturm, R.G., Clerc, L.M., Corcoran, L.M., Baltimore, D., Sharp, P.A., Ingraham, H.A., Rosenfeld, M.G., Finney, M., Ruvkun, G. and Horvitz, H.R. (1988). Genes Devel. 2, 1513.

Hiromi, Y. and Gehring, W.J. (1987). Cell 50, 963.

Hoey, T. and Levine, M. (1988). Nature 332, 858.

Hunt, P. and Krumlauf, R. (1991). Cell 66, 1075.

Ingham, P. (1988). Nature 335, 25.

Jacob, F. and Monod, J. (1961). J. Mol. Biol. 3, 318.

Jaynes, J.B. and O'Farrell, P.H. (1988). Nature 336, 744.

Johnson, A. and Herskowitz, I. (1985). Cell 42, 237.

Kenyon, C. and Wang, B. (1991). Science 253, 516.

Kessel, M. and Gruss, P. (1991). Cell 67, 89.

Kissinger, C.R., Liu, B., Martin-Blanco, E., Kornberg, T.B. and Pabo, C.O. (1990). Cell 63, 579.

Krasnow, M.A., Saffman, E.E., Kornfeld, K. and Hogness, D.S. (1989). Cell 57, 1031.

Kuroiwa, A., Hafen, E. and Gehring, W.J. (1984). Cell 37, 825.

Laughon, A. and Scott, M.P. (1984). Nature 310, 25.

Laughon, A., Howell, W. and Scott, M.P. (1988). Development 104 (Supplement), 75.

Le Mouellic, H., Lallemand, Y. and Brûlet, P. (1992). Cell 69, 251.

Levine, M., Hafen, E., Garber, R.L. and Gehring, W.J. (1983). EMBO J. 2, 2037.

Lewis, E.B. (1978). Nature 276, 565.

Malicki, J., Schughart, K. and McGinnis, W. (1990). Cell 63, 961.

McGinnis, W., Garber, R.L., Wirz, J., Kuroiwa, A. and Gehring, W.J. (1984b). Cell 37, 403.

McGinnis, W., Hart, C.P., Gehring, W.J. and Ruddle, F.H. (1984c). Cell 38, 675.

McGinnis, W., Levine, M.S., Hafen, E., Kuroiwa, A. and Gehring, W.J. (1984a). Nature 308, 428.

Mlodzik, M., Fjose, A. and Gehring, W.J. (1985). EMBO J. 4, 2961.

Morgenstern, Ch. (1932). Alle Galgenlieder, Inselverlag Wiesbaden.

Müller, M., Affolter, M., Leupin, W., Otting, G., Wüthrich, K. and Gehring, W.J. (1988). EMBO J. 7, 4299.

Nüsslein-Volhard, Ch. (1991). Development (Supplement 1), 1.

Otting, G., Qian, Y.Q., Billeter, M., Müller, M., Affolter, M., Gehring, W.J. and Wüthrich, K. (1990). EMBO J. 9, 3085.

Pabo, C.O. and Sauer, R.T. (1984). Ann. Rev. Biochem. 53, 293.

Peifer, M., Karch, F. and Bender, W. (1987). Genes Devel. 1, 891.

Percival-Smith, A., Müller, M., Affolter, M. and Gehring, W.J. (1990). EMBO J. 9, 3967.

Pick. L., Schier, A., Affolter, M., Schmidt-Glenewinkel, T. and Gehring, W.J. (1990). Genes Dev. 4, 1224.

Poole, S.J., Kauvar, L.M., Drees, B. and Kornberg, T. (1985). Cell 40, 37.

Ptashne, M. (1967a). Proc. Natl. Acad. Sci. U.S.A. 57, 306.

Ptashne, M. (1967b). Nature 214, 232.

Qian, Y.Q., Billeter, M., Otting, G., Müller, M., Gehring, W.J. and Wüthrich, K. (1989). Cell 59, 573.

Ruberti, I., Sessa, G., Lucchetti, S. and Morelli, G. (1991). EMBO J. 10, 1787.

Schier, A.F. and Gehring, W.J. (1992). Nature 356, 804.

Schierwater, B., Murtha, M., Dick, M., Ruddle, F.H. and Buss, L. (1991). J. Exp. Zool. 260, 413.

Schneuwly, S., Klemenz, R. and Gehring, W.J. (1987a). Nature 325, 816.

Schneuwly, S., Kuroiwa, R. and Gehring, W.J. (1987b). EMBO J. 6, 201.

Scott, M.P. and Weiner, A.J. (1984). Proc. Natl. Acad. Sci. U.S.A. 81, 4115.

Scott, M.P., Weiner, A.J., Polisky, B.A., Hazelrigg, T.I., Pirrotta, V., Scalenghe, F. and Kaufman, T.C. (1983). Cell 35, 763.

Shepherd, J., McGinnis, W., Carrasco, A.E., De Robertis, E.M. and

Gehring, W.J. (1984). Nature 310, 70.

Simeone, A., Gulisano, M., Acampora, D., Rambaldi, M. and Boncinelli, E. (1992). EMBO J. 11, 2541.

Strathern, J., Hicks, J. and Herskowitz, I. (1981). J. Mol. Biol. 147, 357.

Struhl, G., Struhl, K. and Macdonald, P.M. (1989). Cell 57, 1259.

Stuart, J., Brown, S.J., Beeman, R.W. and Denell, R.E. (1991). Nature 350, 72.

Thali, M., Müller, M.M., DeLorenzi, M., Matthias, P. and Bienz, M. (1988). Nature 336, 598.

Treisman, J., Gonczy, P., Vashishtha, M., Harris, E. and Desplan, C. (1989). Cell 59, 553.

Ueno, K., Hui, C.-C., Fukuta, M. and Suzuki, Y. (1992). Development 114, 555.

Vollbrecht, E., Veit, B., Sinha, N. and Hake, S. (1991). Nature 350, 241.

Wakimoto, B.T. and Kaufman, T.C. (1981). Dev. Biol. 81, 51.

Walldorf, U. and Gehring, W.J. (1992). EMBO J. 11, 2247.

Wang, C. and Lehmann, R. (1991). Cell 66, 637.

Winslow, G.M., Hayashi, S., Krasnow, M., Hogness, D.S. and Scott, M.P. (1989). Cell 57, 1017.

Wolberger, C.A., Vershon, K., Liu, B., Johnson, A.D. and Pabo, C.O. (1991). Cell 67, 517.

Xue, Z.-G., Gehring, W.J. and LeDouarin, N.M. (1991). Proc. Natl. Acad. Sci. U.S.A. 88, 2427.

■ *Walter J. Gehring:*
Biozentrum,
University of Basel,
CH-4056 Basel,
Switzerland

2

The Homeobox in Cell Differentiation and Evolution

The Homeobox in Cell Differentiation and Evolution

E. M. De Robertis

The purpose of this overview is to provide an introduction to the considerable diversity of homeodomain protein sequences and functions recorded in this guidebook. It is intended for readers that might be unaware of the origin of terms such as Hoxa, Hoxb, Pax, POU, LIM, Msx, Cdx, Nkx, Pbx, Dlx and gsc. The profound changes that the discovery of a seemingly universal homeobox gene network controlling body form is causing on current views on animal evolution are discussed at the end of the chapter.

With the benefit of hindsight, it perhaps should not have been surprising that metazoans share similar mechanisms to specify the position of their cells with respect to the rest of the organism. After all, eukaryotic cells, which provided the building blocks for starting the evolutionary adventure leading to the present variety of life forms on earth, have a plethora of other common components (e.g., structural proteins such as histones, biochemical pathways, membranes, organelles, etc.). However, when the simple experiment of taking a Drosophila homeobox probe and isolating a homologous gene from another species was carried out (Carrasco et al., 1984), initially from the vertebrate Xenopus laevis, it is safe to say that no one would have predicted the degree of conservation in the molecular mechanisms that control development.

The discovery of homeobox genes opened the door for understanding the development of those organisms, such as vertebrates, that lacked the sophisticated genetics available in Drosophila. Soon the monumental effort of cloning all the mammalian homeobox genes was under way. Because the probes were all derived from the original Antennapedia (Antp) clone (reviewed by Gehring, preceding chapter), most of the mouse and human genes isolated in the early years were of the Antp class of homeobox genes. This was surely fortunate because with so many other homeoboxes lurking in the genome, the confusion caused by the isolation of so many new genes would have been even greater than what we were about to experience.

■ The *Antp*-type complexes

It was also fortunate that, almost from the beginning, the efforts of those working on the mouse were dedicated to showing that homeobox genes were organized in complexes in genomic DNA, and to mapping on which chromosome they were located (Rabin et al., 1985). This strategy was indeed the correct one. After an extraordinary amount of work the human and mouse *Antp*-type complexes were mapped to four gene complexes, containing about ten homeobox genes each, as shown in Figure 1. The name *Hox* was coined for genes in the mammalian *Antp* complexes. Within the *Hox-1* (now *HOXA*) through *Hox-4* (now *HOXD*) complexes, each gene was numbered according to the order in which it was discovered (Martin et al., 1987). This created some confusion because the numbering did not follow a linear order (Fig.1), but was much better than having no systematic nomenclature at all. By 1992 it was clear that the genes within each vertebrate *HOX* complex could be aligned into 13 groups of genes (called paralogues) that share distinct similarity in the homeobox region (Scott, 1992). At the third Homeobox Workshop, which took place in Ascona, Switzerland, in 1992, it was collectively agreed to rename the *HOX* complexes as A, B, C and D. Within each complex the most 5′ group (i.e., *labial*-related) is designated as 1 and the most 5′ paralogous group (related to the *Abdominal-B* genes) is designated as 13. Once the initial shock is over (e.g. what used to be *Hox-2.1* now corresponds to *Hoxb-5*, see Fig.1) this nomenclature is expected to be very beneficial, because it

will apply not only to the mouse, but to vertebrates in general, all of which share a common organization of the *Antp*-type genes, and for which the name *Hox* genes is now reserved. The non-*Hox* homeobox genes receive different names, usually with a 3-letter abbreviation, and the origin of their names will be explained later on this chapter. In this brief overview we place some emphasis in the nomenclature of homeobox genes because its anarchic nature, together with the large numbers of genes involved, has been a major obstacle for non-aficionados attempting to follow the literature.

When the sequence of the mammalian *Antp*-type homeobox was compared to that of their *Drosophila* counterparts (Boncinelli et al., 1988), a most remarkable fact became apparent: homeoboxes at the 3′ end of the mammalian *HOX* complexes were similar in nucleotide sequence to those at the 3′ end of the *Drosophila* homeotic complexes (corresponding to the gene called *labial*), and those toward the 5′ end of the mammalian complex resembled *Abdominal-B*, maintaining the orientation present in the *Drosophila* homeotic complexes, as illustrated in Figure 2 for homeoboxes of the *Hoxb* complex. Within each *Hox* complex every mammalian gene is transcribed in the same direction (with the exception of *Evx 1* and *Evx 2*, Boncinelli et al., 1991, see Fig.1). This rule is not followed in the *Drosophila Antp* complex (Kaufman et al., 1990, see Fig.1). In addition, fruit flies also have divergent, non-homeotic, homeobox genes interspersed within the *Antp* complex (i.e., *fushi tarazu, zerknüllt, bicoid*), which are absent from vertebrate Hox complexes.

More recently, a cluster of *Antp*-type homeobox genes

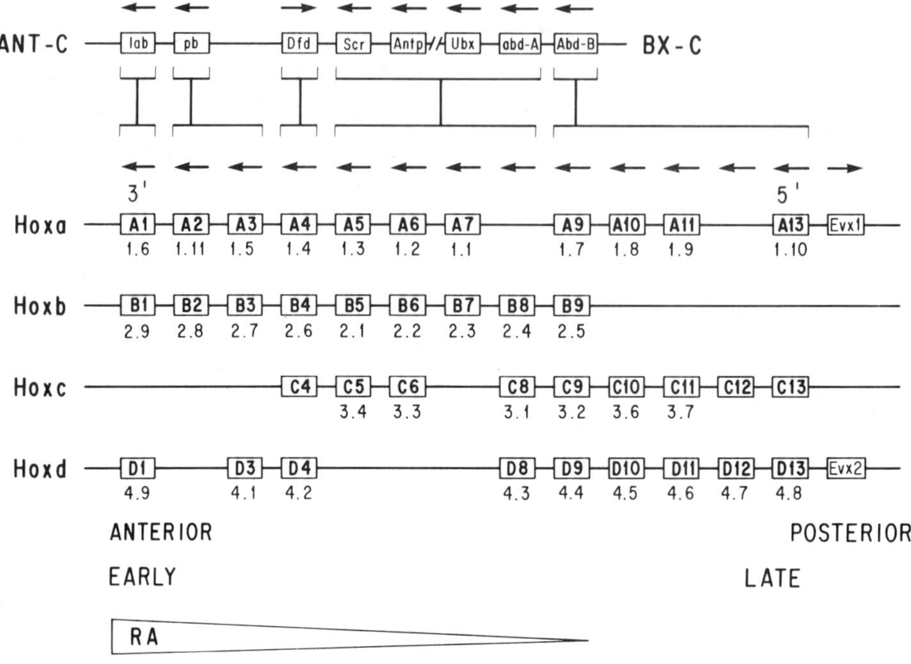

Figure 1. Nomenclature of vertebrate *Hox* genes (see also the chapter by T. Bürglin). The Ascona Workshop nomenclature (Scott, 1992) is shown within the boxed regions (*Hoxa, Hoxb, Hoxc, Hoxd*), and the old numerical nomenclature for the mouse (*Hox-1, Hox-2*, etc.) is shown underneath each gene. *Drosophila* genes of the *Antennapedia (ANT-C)* and *Bithorax (BX-C)* complexes are shown at the top. Direction of transcription is indicated by the arrows. Note that genes on the left (3' end of the vertebrate complexes) are expressed more anteriorly and earlier than genes on the right. These genes are also more sensitive to stimulation by retinoic acid (RA).

has been identified in the nematode *C. elegans* (Bürglin et al., 1991; Kenyon and Wang, 1991). As indicated in Figure 2, this cluster also contains homeoboxes that are related to the *Drosophila* genes *labial, Deformed, Antp* and *Abdominal-B*. Even hydra, which is probably one of the earliest metazoans (cnidarians are diploblastic, *i.e.*, lack mesoderm; they also lack a through-gut, *i.e.*, have a sole opening that serves both as mouth and anus) has *labial, proboscipedia, Deformed* and *Antp*-like homeoboxes (Murtha et al., 1991; Schierwater et al., 1991; Schummer et al., 1992).

The resemblance of *Hox* genes and *Drosophila*

```
Consensus  RKRGRTTYTRYQTLELEKEFHFNRYLTRRRRIEIAHALCLTERQIKIWFQNRRMKWKKEN

Labial     NNS---NF-NK-LT--------------A-------NT-Q-N-T-V----------Q--RV
Hoxb 1     PGGL--NF-TR-LT---------K--S-A--V---AT-G-N-T-V----------Q--RE
Ceh-13     NGTN--NF-TH-LT-------TAK-YN-T--T---SN-K-Q-A-V----------E--RE

Deformed   P--Q--A---H-I--------Y---------------T-V-S--------------D-
Hoxb 4     P--S--A---Q-V--------Y---------V----------S--------------DH
Ceh-15     E--Q-TA---N-V--------THK----K----V--S-M-----V----------H----

Antp       -----Q------------------------------------------------------
Hoxb 7     -----Q---------------------Y--------------T------------------
mab-5      S--T-Q--S-S----------YHK----K--Q--SET-H-----V----------H---A

Abd-B      VRKK-KP-SKF---------L--A-VSKQK-W-L-RN-Q-----V----------N--NS
Hoxb 9     SRKK-CP--K----------L--M----D--H-V-RL-N-S---V----------M--L-
Ceh-11     S-K--Q--Q----SV--AK-QQSS-VSKKQ-E-LRLQTQ--D------------A--K
```

Figure 2. Homeodomains within the *Drosophila* complexes share sequence similarities in very different organisms. *labial, Deformed, Antp* and *Abd-B* are *Drosophila* genes, *Hoxb-1, -4, -7* and *-9* are mouse genes, while *Ceh-13, Ceh-15, mab-5* and *Ceh-11* are homeodomain sequences from the nematode *C. elegans*. The hyphens indicate amino acid identity to a consensus *Antp*-type homeobox. Note that the four groups of sequences share similarities in amino acid changes.

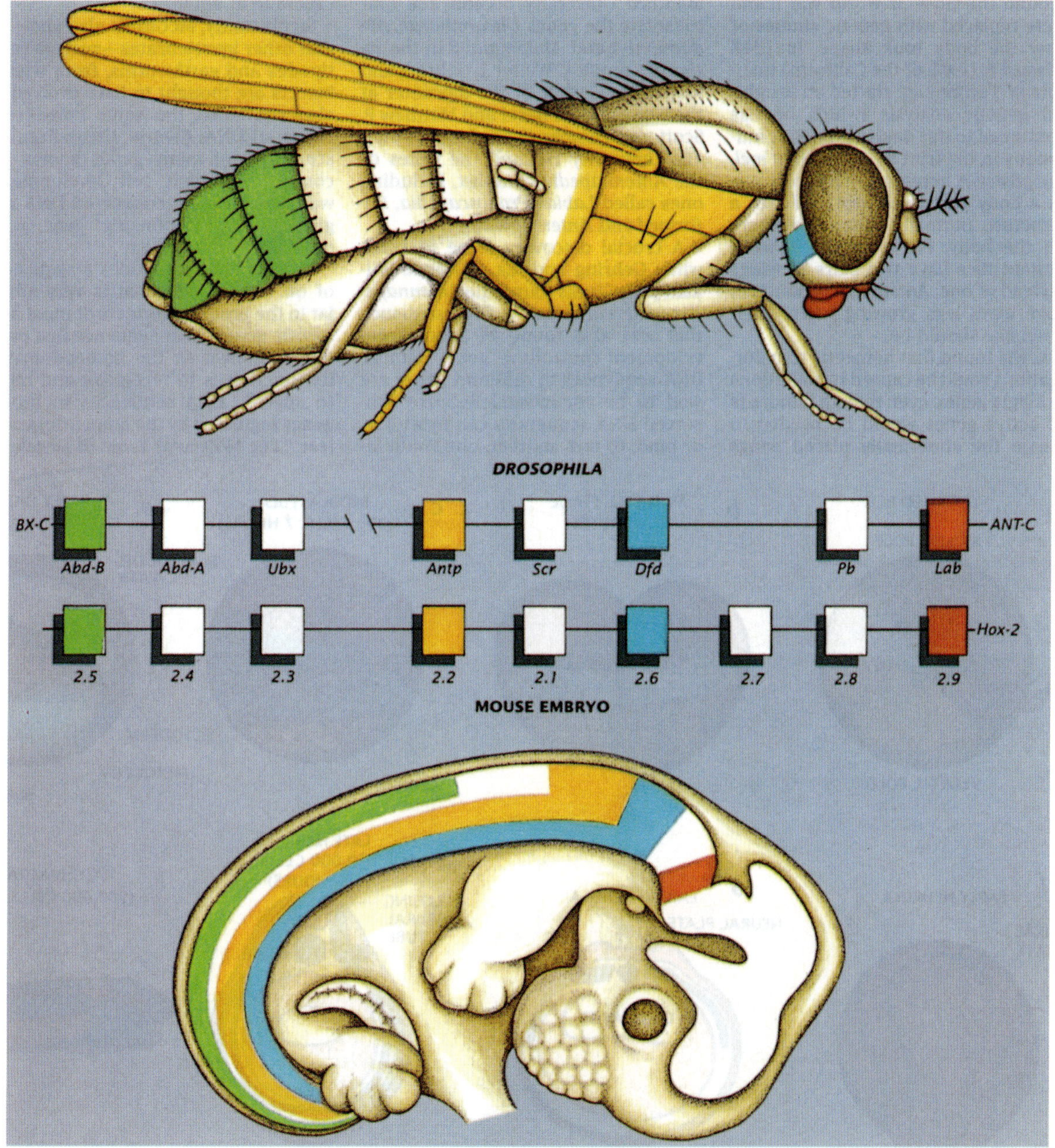

Figure 3. Homeobox genes of the *Antp*-type are expressed in the body in same order as they occupy in genomic DNA. For example, genes of the *labial* class are expressed anteriorly and genes of the *Abdominal-B* class posteriorly. This extraordinary conservation suggests that the gene system that controls antero-posterior cell specification may have arisen very early, perhaps, only once, during metazoan evolution. Reproduced, with permission, from E. De Robertis, G. Oliver and C.V.E. Wright, Scientific American, July 1990, pp. 46-52.

homeotic genes is not only structural, but also functional. *In situ* hybridization studies revealed that the genomic organization of the *Hox* genes is colinear with their expression in the various body regions (Gaunt et al., 1988; Graham et al., 1989, Duboule and Dollé, 1989). In other words, as is the case in *Drosophila*, genes at the 3' end of the complex are expressed anteriorly, and those at the 5' end are expressed more posteriorly, as depicted in Figure 3. This rule applies for all four mammalian *Hox* complexes.

The discovery of colinearity meant that the model pro-

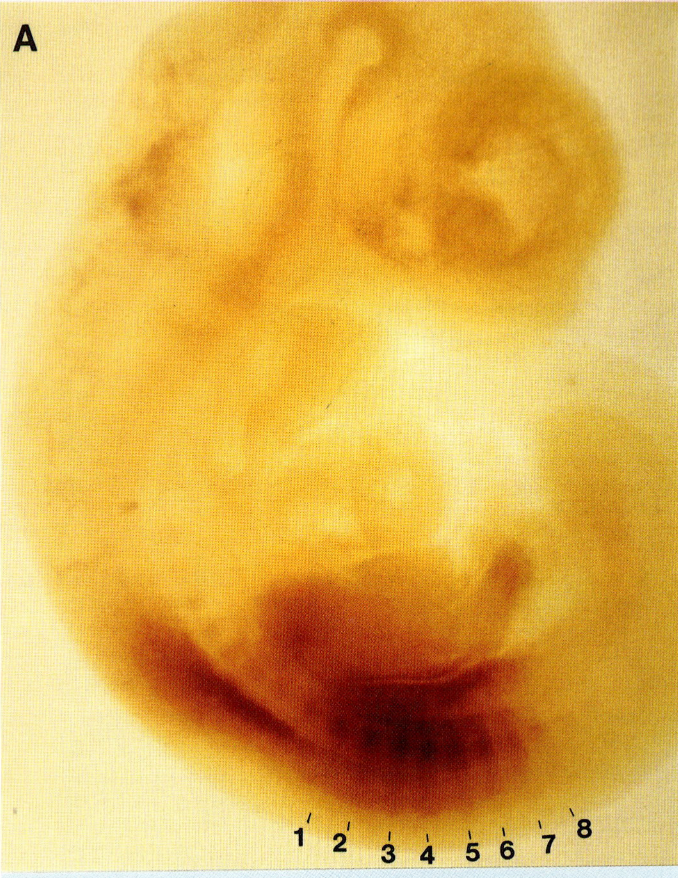

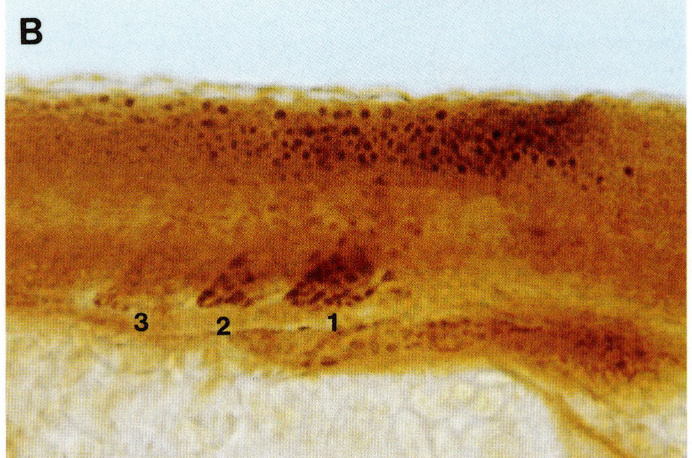

Figure 4. Whole-mount staining of mouse and zebrafish embryos with an anti *Hoxc-6* antibody. A) Mouse embryo, 10.5 days. Note that in the mesoderm the *Hoxc-6* protein is detectable in eight segments. B) Zebrafish embryo after 22 hours of development. Note that in fish mesoderm the *Hoxc-6* protein is detectable in only three somites (numbered), decreasing towards the posterior. In both embryos, the spinal cord also shows staining, which is out-of-register with the mesodermal expression. Mouse embryo prepared by Guillermo Oliver, fish embryo by Anders Molven.

posed by Edward Lewis (1978) for *Drosophila* applied to the mouse as well (Gaunt et al., 1988). Thus, the molecular mechanisms that determine the antero-posterior (A-P) axis had been conserved in evolution to a degree beyond anyone's wildest expectations (Graham et al., 1989; Duboule and Dollé, 1989; reviewed by De Robertis et al., 1991; McGinnis and Krumlauf, 1992). Within the hindbrain, for example, the anterior borders of the *Hox* genes demarcate a series of segmental structures in the hindbrain, the rhombomeres, that had previously been mostly ignored. With the discovery of colinearity in mammals, the study of homeobox genes suddenly became an accepted, even fashionable, scientific pursuit.

■ The function of Vertebrate *Hox* genes

Remarkably localized expression patterns suggested early on that *Hox* genes would be involved in regional specification (Awgulewitsch et al., 1986). Figure 4A shows a mouse embryo stained with an antibody directed against *Hoxc-6* protein, which can be considered a run-of-the-mill *Hox* protein. *Hoxc-6* is expressed in a band, or belt, encompassing the anterior spinal cord and eight mesodermal somites. Thus *Hox* genes demarcate regions of the embryo, rather than particular tissues or cell types. A main conclusion derived from such localization studies is that the vertebrate embryo is first subdivided into regions of *Hox* gene activity before particular tissues or organs are formed.

Early loss-of-function (involving microinjection of antibodies into *Xenopus* embryos, Wright et al., 1989) and gain-of-function (by misexpression of *Hoxa-7* driven by an actin promoter, Kessel et al., 1990) experiments suggested that vertebrate *Hox* genes, like their *Drosophila* counterparts, provide identity to body regions, and can lead to homeotic transformations. The mammalian HOX proteins are functional in transgenic fruit flies and can, remarkably, alleviate the effect of homeotic mutations (Malicki et al., 1990, McGinnis et al., 1990). The initial *Hox* gene knockouts in mouse failed to reveal homeotic transformations (Chisaka and Capecchi, 1991; Lufkin et al., 1991; Chisaka et al., 1992). This negative result was probably due to the use of genes expressed anteriorly (*Hoxa-1* and *Hoxa-3*), which contribute extensively to the neural crest cells that form most of the vertebrate head and neck regions. The complexity of the morphogenesis of these anterior regions facilitates the identification of anatomical deficiencies while making true homeotic transformations more difficult to detect. Interestingly, knockout of *Hoxc-8*, a gene expressed in the thoracic region, produces anterior homeotic transformations in the vertebral column of the mouse (Le Moulleic et al., 1992). Overexpression of *Hoxd-11* in the chick limb bud using retroviral vectors leads to homeotic transformations of anterior foot digits into more posterior ones (Morgan et al, 1992).

When homeodomain proteins are misexpressed in the posterior of the embryo, anterior homeotic transformations were found. For example, a transgenic mouse line misexpressing *Hoxc-6* leads to the formation of extra pairs of ribs in lumbar vertebrae (Jegalian and De Robertis, 1992). Misexpression of *Hoxc-8* also results in anterior transformations in thoracic segments T11, T12 and lumbar 1 (Pollock et al. 1992). These results were not predicted by the *Drosophila* studies. One interpretation is that misexpression of a gene normally expressed in the thorax will impart thoracic characteristics (i.e., ribs) to lumbar segments in which these homeodomain proteins are not expressed (*Hoxc-6* and *Hoxc-8* proteins are not detectable by antibody staining posterior to T9 and T10 respectively; Jegalian and De Robertis, 1992; Awgulewitsch and Jacobs, 1990).

While much remains to be done, the general conclusion that is emerging from the various functional studies is that the A-P axis is specified by similar arrays of homeobox genes in fruit flies and mammals. This has had considerable impact on the way we view the evolution of metazoans.

On the other hand, we know very little about the molecular mechanisms by which homeobox genes affect the development of pattern, either in *Drosophila* or vertebrates. One interesting finding is that homeodomain proteins, which are transcription factors, regulate the activity of target genes encoding cell adhesion molecules. *Hoxb-9*, *Hoxb-8* and *Hoxc-6* have been shown to regulate the promoter of N-CAM in co-transfection experiments (Jones et al., 1992 and 1993), and *Ultrabithorax* target genes include *scabrous* (a gene involved in intercellular communication during neurogenesis; Graba et al. 1992) and *connectin* (a cell adhesion molecule involved in the establishment of neuromuscular synapses; Gould and White, 1992). Another finding is that in *C. elegans* the *Antp*-like gene *mab-5* (Fig.2) acts as a developmental switch that controls cell migration of a particular neuronal precursor cell (Salser and Kenyon, 1992). This relationship between homeobox genes and cell movements is also found in the case of the *Xenopus* gene *goosecoid*, which directs dorso-anterior cell migration during gastrulation (Niehrs et al., 1993). The hypothesis that homeodomain proteins may affect pattern formation by regulating cell adhesion and migration seems to merit further exploration.

■ A great variety of homeobox genes

Up to this moment we have only discussed homeobox genes of the *Antennapedia*-type of which there are about 40 in the mammal. But this is only the tip of the iceberg. The vast majority of homeobox genes correspond to the non-*Antp*-class (Scott et al., 1989). *C. elegans*, for example, contains scores of homeobox genes, which by one estimate may account for up to 0.3% of the total number of genes in the nematode (Bürglin et al., 1989). Some of the non-*Hox* homeobox genes have colorful names; we now turn to explaining the origin of some of them in order to facilitate the use of this guidebook.

■ The *Pax/Prd* family

When the *Drosophila paired* gene was cloned, a domain of 182 amino acids upstream of the homeobox was found to be conserved in the segmentation genes *gooseberry*-proximal and *gooseberry*-distal (Bopp et al., 1986). In mouse, at least 8 genes contain this paired-box, and have been designated *Pax* genes. *Pax* genes can have a homeobox or may contain only the paired-box (Gruss and Walther, 1992). The paired-box also acts as a DNA-binding domain (Treisman et al., 1991). Those *Pax* genes that contain a paired box and a complete homeobox, therefore, have two distinct DNA-binding domains. *Pax-2* and *Pax-8* contain part of the homeobox, but lack the last two thirds of it, while having an intact Pax domain (Gruss and Walther, 1992). One interesting aspect of *Pax* genes is that out of 8, three have been found to correspond to previously known mutations in the mouse (*undulated* affects *Pax-1*, *small-eyes* is *Pax-3*, and *splotch* is a mutation in *Pax-6*). This is rather surprising, because of 40 *Antp*-type *Hox* genes found in the mouse, not a single one corresponds as yet to a known mutation. It is therefore not unlikely that this difference in the rate of spontaneous mutation may reflect a difference in the mechanism of action, perhaps in the DNA-binding affinity, of *Pax* and *Hox* proteins (reviewed by Gruss and Walther, 1992).

■ The POU genes

In 1988 various transcription factors were purified and cloned. Some of them, called *Pit-1* (pituitary-1) and *Oct 1* and *2* (for octamer-binding factors) were found to contain homeodomains. These studies were based entirely on biochemical purification and, therefore, it follows that had the homeobox not been discovered by genetic means, it would still have been identified biochemically. A development-controlling gene from *C. elegans*, called *Unc-86* (short for *Uncoordinated*-86), was cloned at about the same time and found to have a similar primary structure. The name POU was coined to designate a family of related transcription factors that includes *Pit-1*, *Oct 1* and *2*, and *Unc 86* (Herr et al., 1988). POU proteins contain, upstream of the homeodomain, an additional conserved domain which varies from 147 to 156 amino acids in length, which was designated the POU-specific domain. The POU-specific domain functions as a second DNA-binding domain (Ruvkun and Finney, 1991).

POU proteins have a very rich biology, which has been reviewed recently (Rosenfeld, 1991). The POU family is very large, and many of its members are expressed in the CNS. One of the founding members of the family, *Pit-1* (also known as *GHF-1*), is expressed only in the pituitary at mid-embryogenesis, providing a striking example of a homeobox gene that is expressed in an exclusively tissue-specific way, in contrast to *Antp*-type genes that are expressed in regions containing many tissues. Mutations in the *Pit-1* gene cause dwarfism in mice due to lack of growth hormone. An inhibitory (dominant-negative) POU protein, i-POU, which is able to dimerize with a POU activator protein in the *Drosophila* nervous system, but unable to bind to DNA, has been described (Treacy et al., 1991).

The Oct proteins are an important component of the basic transcription apparatus; they interact with an octamer motif (ATGCAAAT) found in the promoter/enhancer regions of a great many genes. The promoter specificity of *Oct-2* is regulated by alternative spicing of a protein domain located downstream of the homeobox, so that one form can activate the U2 snRNA promoter while another form activates preferentially mRNA promoters (Tanaka et al., 1992).

■ The LIM proteins

These proteins contain, in addition to a diverged homeobox, two tandemly arranged cysteine/histidine-rich regions, whose function is as yet unknown. The members of the LIM family received this name because the first three genes to be isolated were *lin-11* (a nematode cell lineage gene, Freyd et al., 1990), *Isl-1* (a factor required for insulin gene transcription in islet of Langerhans cells,

Karlsson et al., 1990), and *mec-3* (a gene involved in the development of mechano-sensory cells in *C. elegans*, Way and Chalfie, 1988). In *Xenopus*, several members of this family have been isolated and one of them, *XLIM-1*, has an interesting pattern of expression, which includes the dorsal blastopore lip and the notochord (Taira et al., 1992).

■ The Msx group

This is a group of homeobox genes isolated using *msh* (a *Drosophila* muscle-specific homeobox isolated in W. Gehring's laboratory) as a probe. *Msx-1* and *Msx-2* were previously known as *Hox-7* and *Hox-8*. *Msx-1* and *Msx-2* are expressed in the mesodermal progress zone (tip) of the limb bud and of the mandibular and maxillary arches (Robert et al., 1991; Davidson et al., 1991). This corresponds to the zone of these structures in which cell proliferation is most active. Msx expression is thought to be under the control of growth factors released by the apical ectodermal ridge of the limb bud.

Hox-7 was a designation that exemplified the difficulties with the previous, frequently arbitrary, homeobox nomenclature. *Msx-1* is not part of an *Antp*-type homeobox complex and therefore should not have been designated *Hox-7* to begin with. In addition, there are only four *Hox* complexes, so *Hox-5* and *Hox-6* were non-existent (this was due to historical reasons; a *Hox 4* gene was assigned to an incorrect chromosome, as was also the case for the mouse *Hox 3.3* gene which for a period was called *Hox 6.1*). The case of the *Msx* genes underscores why the *Hox* designation, which has proven so useful, should be reserved for *Antp*-type gene clusters.

■ HNF-1-*like factors*

This family of transcription factors was initially defined by an hepatocyte-derived nuclear factor required for the transcription of albumin, fibrinogen and other liver-specific promoters (Frain et al., 1989). Its homeobox is an "extra large" one, containing an insertion of 21 amino acids which loops out between two of the homeobox helices. *HNF-1* also has a stretch of weak homology to the POU domain upstream of the homeobox. Unlike other homeobox genes, *HNF-1* can dimerize in solution (Chouard et al., 1990) but, interestingly, this reaction is greatly facilitated by a protein cofactor (Mendel et al., 1991). Expression of *HNF-1* in the rat 15.5 day embryo is restricted to hepatocytes, gut and stomach epithelia, and the developing kidney (Lazzaro et al., 1992), suggesting that it functions mainly as a tissue-specific transcription factor.

■ The endoboxes, *Cdx* and *XlHbox 8*

Most of the *Antp*-type genes are expressed in neural and mesodermal tissues. Only a few homeobox genes, such as *HNF-1*, seem to be predominantly expressed in endoderm. One is *Cdx-1*, a mouse homologue of *Drosophila caudal* that is expressed in posterior gut epithelium at late stages of mouse embryogenesis (Duprey et al., 1988). Another *caudal* homologue, *CHox-cad*, is expressed early on in the chick primitive streak and later in gut epithelium (Frumkin et al., 1991). A gene that is expressed exclusively in a narrow band of the gut epithelium (duodenum, pancreatic anlagen and interconnecting ducts), *XlHbox 8*, has been described (Wright et al., 1988). *XlHbox 8* encodes a divergent homeobox and may represent the first member of a family of homeobox genes involved exclusively in endodermal differentiation.

■ The *NK/Nkx* group

The *NK* homeoboxes were described by Niremberg and Kim in *Drosophila*, where three members of these divergent homeoboxes are clustered in the genome (Kim and Niremberg, 1989). The function of this homeobox fruit fly gene cluster is not known as yet. Interestingly, the *NK*-type homeoboxes are highly conserved in evolution, from mouse to planarians. The mouse contains at least four *Nkx* genes, which are most similar to *Drosophila NK-2* (Price et al., 1992). The presence of *NK* homeoboxes in Planaria (Garcia-Fernandez et al., 1991) suggests that this group of homeobox genes is a very ancient one, a finding which certainly merits further investigation.

■ The plant homeoboxes

The gene encoding the *knotted* mutation in maize was cloned and found to contain a divergent homeobox (Vollbrecht et al., 1991). Two additional related maize genes have also been isolated using *knotted* as a probe. The *knotted* homeobox is most similar (35% identity) to the homeobox of *Pbx*, a gene that is translocated in certain human pre-B acute lymphoblastic leukemias (Kamps et al., 1990). The *knotted* homeobox has a 3 amino acid insertion at the turn between helices 1 and 2 (Vollbrecht et al., 1991). Interestingly, this insertion is also present in the yeast homeodomain protein *MAT alpha2* and in human *Pbx*. In *Arabidopsis* two homeobox genes have been isolated using an oligonucleotide directed against the highly conserved helix 3 (Ruberti et al., 1991). Their highly divergent homeodomains lack the three amino acid insertion, are related to each other, and are followed by a leucine zipper motif that may be involved in dimerization (Ruberti et al., 1991).

■ The head homeoboxes: *Dlx, Otx, Emx,* and *gsc*

The *Antp*-type homeobox genes are all involved in the development of the vertebrate trunk and tail. The most anteriorly-expressed gene of the *Hox* complexes has its anterior-most boundary at the border between rhombomeres 2 and 3 of the hindbrain (Hunt and Krumlauf, 1991). Until recently we had no candidates for homeobox genes specifying position in the head (and brain) region, except for *engrailed-2* (*en-2*), which is expressed in a narrow band at the hindbrain-midbrain junction (Davis and Joyner, 1988). In an almost incredible conservation of gene functions, it has been recently found that genes involved in *Drosophila* head development are also expressed in a region-specific way in mammalian mid- and forebrain. In the fruit fly *Distalless* is a gene involved in the development of appendages. Its expression is most abundant in appendages deriving from the head segments, such as antennae and maxillae (Cohen, 1990). Two murine *Distalless* (*Dlx*) homologues, are expressed in specific regions of the mouse embryo forebrain (Price et al., 1991; Porteus et al., 1991). Homologues of other homeobox genes involved in *Drosophila* head development, *orthodenticle* and *empty spiracles* (Finkelstein and Perrimon, 1991), designated *Otx 1* and *2* and *Emx 1* and *2*, are expressed in more anterior regions of the forebrain, in a nested pattern starting at the olfactory lobe, so that the regions of expression are contained within the others in the sequence *Emx 1 < Emx 2 < Otx 1 < Otx 2* (Simeone et al., 1992). Two murine NK-type genes, *Nkx-2.1* (also called *TTF-1*) and *Nkx-2.2* are also expressed in distinct subregions of the forebrain (Lazzaro et al., 1991; Price et al., 1992). Many of the boundaries of expression of these brain-specific homeobox genes coincide with anatomical landmarks which had been suggested (with little support from mainstream neurobiologists) to represent segment-like regions in the developing midbrain and forebrain (Puelles et al., 1987). The study of the brain homeoboxes is so exciting and important because it may reveal the basic units of organization of the mammalian brain, which is arguably the most complex anatomical structure in the animal kingdom.

Another gene that may be involved in the generation of pattern in the head region, but at an earlier stage, is *goosecoid* (*gsc*). This gene was isolated from the *Xenopus* dorsal lip (the region also known as Spemann's organizer) and has the ability to induce twinned body axes after microinjection into ventral cells of the *Xenopus* embryo (Cho et al., 1991). This gene has a DNA-binding specificity similar to that of the *Drosophila* head determinant *bicoid* (Driever and Nüsslein-Volhard, 1988) and has similarities in the first two-thirds of the homeobox to *gooseberry*; for this reason it was named *goosecoid* (Blumberg et al., 1991). In both *Xenopus* and mouse, *gsc* is expressed in the most anterior gastrulating cells, that give rise to the prechordal plate (Cho et al., 1991; Blum et al., 1992). The fate of the *goosecoid*-expressing cells is to become head mesoderm and pharyngeal endoderm. It is hoped that the study of *goosecoid*, and of other genes expressed in the early gastrula (Blumberg et al., 1991; Taira et al., 1992; Dirksen and Jamrich, 1992), will provide insights into the initial steps in the formation of the vertebrate body axis.

■ Homeobox genes and evolution

Even from this very incomplete and brief overview it should be clear that many types of homeodomain proteins exist, and that they can have very different functions. As is recorded in the pages that follow, hundreds of homeobox genes have been isolated from a wide variety of organisms, ranging from plants to man. It is fair to ask at this point in time whether general principles are emerging from this potpourri of genes and sequences. One such principle is certainly the universal nature of the molecular mechanisms that regulate embryonic development. Another general principle that is emerging is related to our understanding of evolution.

Evolution is the central question for many of those interested in the nature of life on earth. Evolutionary and embryological studies have intertwined to various degrees throughout their history (for an excellent review of the pre-homeobox days see Raff and Kaufman, 1983). Naturalists like William Bateson were interested in how new body structures are generated in the course of evolution, and to this end analyzed malformations that occurred spontaneously in the wild. The problem they were trying to resolve was that new species seemed not to arise through multiple small changes but rather by abrupt, discontinuous, changes. In the words of Bateson (1894):

> *"Upon the accepted view it is held that the Discontinuity of Species has been brought about by a Natural Selection...(acting upon)...a continuous series of variations. Of the difficulties besetting this doctrine enough was said in the introductory pages. These difficulties have oppressed all who have thought upon these matters...and they have caused some anxiety even to the faithful.""*

But there was a way round this problem if one could demonstrate that mutations sometimes affect large regions of the body, leading to extensive changes in body form:

> *"The case of the modification of the antenna of an insect into a foot, of the eye of a crustacean into an antenna, of a petal into a stamen, and the like, are examples of the same kind. It is desirable and indeed necessary that such variations, which consist in the assumption by one member of a meristic series, of the form or characters proper to other members of the series, should be recognized as constituting a distinct group of phenomena...I therefore propose...the term Homeosis...for the essential phenomenon is not that there merely has been a change, but that something has been changed into the likeness of something else."*
> *"The discontinuity of species results from the discontinuity of variation. Discontinuity results from the fact that bodies of living things are made of repeated parts. Meristic variation in numbers of parts is often integral, and thus discontinuous. ... (A structure) may suddenly appear in the likeness of some other member of the series, assuming at one step the condition to which the member copied attained presumably by a long course of Evolution."*

Thus the homeobox saga starts with natural history, continues with the isolation of the initial *bithorax* mutation by Calvin Bridges and Thomas H. Morgan (1923) and the extensive genetic analysis of this complex locus by Edward Lewis (starting in 1948 and continuing to the present day), and entered the molecular era in 1983 with the discovery of the homeobox (reviewed by Gehring, preceding chapter). It seems fitting that the circle should now be closed with the molecular analysis of the evolution of form.

We have emphasized how homeobox genes have been so extensively conserved throughout evolution. Probably the "homeobox record" of organisms, which can be determined for example by PCR analysis of genomic DNA (e.g., Murtha et al., 1991) will prove one of the best methods of establishing phylogenetic relationships between organisms. From plants to hydra, from Planarians to *Drosophila*, from *Drosophila* to vertebrates, homeobox genes seem to orchestrate pattern formation. Many *Hox* and non-*Antp* homeobox genes (*Dlx, Msx, Otx, Emx,* etc.) have persisted through evolution, presumably because they work in a network and cannot be easily eliminated individually.

It has been found that the genomic organization of, say, *Hox* genes in frogs and mice are quite similar. The number of *Hoxb* genes is the same, and their distances in genomic DNA are comparable (Leroy and De Robertis, 1992; Dekker et al., 1992). Frogs and mice, however, have very obvious anatomical differences. The evolutionary conservation in overall homeobox gene structure therefore presents us with a paradox. Are we to conclude that *Hox* genes are not involved in the evolution of the body plan because they share similarities in different species?

Presumably the differences will lie in the regulatory elements in DNA that control where homeobox genes are expressed. For example, Figure 4A and 4B show the expression of *Hoxc-6* protein in mouse and zebrafish embryos. While the distribution of this protein is restricted to an A-P band in trunk mesoderm and spinal cord in both animals, differences are readily seen in the mesoderm. In mouse mesoderm *Hoxc-6* spans eight somites, compared to only three in zebrafish mesoderm. Such seemingly small changes could have profound effects on body shape. Sequence comparisons of the intergenic regions of *Hox* genes in different species should permit the identification of the DNA regions responsible for such changes. A good example of how the role of homeobox genes in evolution might be approached experimentally is provided by the work of Lufkin et al. (1992). They performed a "promoter swap" experiment in which the *Hoxd-4* gene product was expressed under the control of the *Hoxa-1* promoter in transgenic mice. This swap caused a gene normally expressed in cervical vertebrae to be expressed in more anterior regions, including the region of the occipital bone. The resulting mice had a homeotic transformation of the occipital bone which gave rise to up to four vertebral neural arches. This is very interesting, because it has long been known that the first four somites of higher vertebrates do not give rise to vertebrae, but instead fuse to each other and give rise to the "neocranium". The most ancient vertebrates, the agnathans or jawless fishes (presently represented by the lampreys, but which played in the past a more prominent role as documented by the fossil record), have four additional vertebrae and the brain is encased only by the "paleocranium". Thus, misexpression of a *Hox* gene caused the reappearance of atavistic characters that had been absent for many millions of years in the evolution of the mammalian lineage (Lufkin et al., 1992). This experiment vividly illustrates how changes in the expression patterns of *Hox* genes might lead to changes in the body plan during the course of evolution. (Renucci et al. 1992).

With four *Hox* complexes in the genome, evolution could have a fertile substratum to derive a large variety of vertebrate body shapes using rather similar transcription factors. The position of cells in all metazoan organisms seems to be specified by very similar, perhaps universal, molecular mechanisms. Therefore, it is quite possible that the driving force of evolution may have been natural selection acting upon mutations of master regulatory genes guiding embryonic development, such as for example homeotic genes, as suggested initially by Bateson.

Less than ten years after the discovery of the homeobox, we are presently at a stage in which one can isolate a gene and compare its homologues in very distant organisms. What we have learnt indicates that the molecular mechanisms that control the development of body form were already present in the earliest metazoans. We can now examine the entire history of animal life through the narrow prism of an individual regulatory gene.

This guidebook records the herculean task of characterizing hundreds of homeobox genes in many different species. Many of the hundreds of investigators involved in this effort must have wondered, or might still wonder, whether they were ever going to make sense out of this jumble of genes. Others, particularly those working in organisms with poor genetics, may have felt the stigma of participating in a presumed unintellectual "fishing expedition". Even to the faithful, the importance of sequencing yet another gene must have seemed very small at times. The revolution in biological thought that this work was bringing about, however, was not lost on those interested in the study of evolution. In the words of Stephen Jay Gould (1991):

"The discovery of homeoboxes allowed molecular geneticists to 'go fishing' for vertebrate genes that might be related to the homeotics of Drosophila - the key to segmental architecture in insects. Forget all the folk wisdom about big ones that got away; this has been one of the finest fishing expeditions in human history."

■ References

Awgulewitsch, A., Utset, M.F., Hart, C.P., McGinnis, W. and Ruddle, F.H. (1986). Spatial restriction in expression of a mouse homeo box locus within the central nervous system. Nature 320, 328-335.

Awgulewitsch, A. and Jacobs, D. (1990). Differential expression of Hox 3.1 protein in subregions of the embryonic and adult spinal cord. Development 108, 411-420.

Bateson, W. (1894). Materials for the study of variation treated with especial regard to discontinuity in the origin of species. (London: McMillan and Co.).

Blum, M., Gaunt, S., Cho, K. W. Y., Steinbeisser, H., Bittner, D., Blumberg, B. and De Robertis, E. M. (1992). Gastrulation in the mouse: the role of the homeobox gene goosecoid. Cell 69, 1097-1106.

Blumberg, B., Wright, C.V.E., De Robertis, E.M. and Cho, K.W.Y. (1991). Organizer-specific homeobox genes in *Xenopus laevis* embryos. Science 253, 194-196.

Boncinelli, E., Somma, R., Acampora, D., Pannese, M., D'Esposito, M., Faiella, A. and Simeone, A. (1988). Organization of human homeobox genes. Hum. Reprod. 3, 880-886.

Boncinelli, E., Simeone, A., Acampora, D. and Mavilio, F. (1991). HOX gene activation by retinoic acid. Trends Genet. 7, 329-334.

Bopp, D., Burri, M., Baumgartner, S., Frigerio, G. and Noll, M. (1986). Conservation of a large protein domain in the segmentation gene paired and in functionally related genes of *Drosophila*. Cell 47, 1033-1040.

Bridges, C.B. and Morgan, T.H. (1923). The third-chromosome group of mutant characters of Drosophila melanogaster. (Washington: Carnegie Institution of Washington), pp. 137-139.

Bürglin, T.R., Finney, M., Coulson, A. and Ruvkun, G. (1989). *C. elegans* has scores of homeobox-containing genes. Nature 341, 239-243.

Bürglin, T.R., Ruvkun, G., Coulson, A., Hawkins, N.C., McGhee, J.D., Schaller, D., Wittman, C., Müller, F. and Waterston, R.H. (1991). Nematode homeobox cluster. Nature 351, 703.

Carrasco, A.E., McGinnis, W., Gehring, W.J. and De Robertis, E.M. (1984). Cloning of an X. laevis gene expressed during early embryogenesis coding for a peptide region homologous to Drosophila homeotic genes. Cell 37, 409-414.

Chisaka, O. and Capecchi, M. (1991). Regionally restricted developmental defects resulting from targetted disruption of the mouse homeobox gene *Hox-1.5*. Nature 350, 473-479.

Chisaka, O., Musci, T.S. and Capecchi, M.R. (1992). Developmental defects of the ear, cranial nerves and hindbrain resulting from targeted disruption of the mouse homeobox gene *Hox-1.6*. Nature 355, 516-520.

Cho, K.W.Y., Blumberg, B., Steinbeisser, H. and De Robertis, E.M. (1991). Molecular nature of Spemann's organizer: the role of the *Xenopus* homeobox gene *goosecoid*. Cell 67, 1111-1120.

Chouard, T., Blumenfeld, M., Bach, I., Vandekerckhove, J., Cereghini, S. and Yaniv, M. (1990). A distal dimerization domain is essential for DNA-binding by the atypical HNF-1 homeodomain. Nucl. Acids Res. 18, 5853-5863.

Cohen, S.M. (1990). Specification of limb development in the *Drosophila* embryo by positional cues from segmentation genes. Nature 343, 173-177.

Davidson, D.R., Crawley, A., Hill, R.E. and Tickle, C. (1991). Position-dependent expression of two related homeobox genes in developing vertebrate limbs. Nature 352, 429-431.

Davis, C.A. and Joyner, A.L. (1988). Expression patterns of the homeobox-containing genes *En-1* and *En-2* and the proto-oncogene *int-1* diverge during mouse development. Genes Dev. 2, 1736-1744.

Dekker, E.J., Pannese, M., Houtzager, E., Boncinelli, E. and Durston, A. (1992). Hox-2 genes are expressed sequentially during the development of Xenopus laevis and are differentially inducible by retinoic acid. Development Supplement, 195–202.

De Robertis, E.M., Morita, E.A. and Cho, K.W.Y. (1991). Gradient fields and homeobox genes. Development 112, 669-678.

Dirksen, M.L. and Jamrich, M. (1992). A novel, activin-inducible, blastopore lip-specific gene of *Xenopus laevis* contains a *fork head* DNA-binding domain. Genes Dev. 6, 599-608.

Driever, W. and Nüsslein-Volhard, C. (1988). The *bicoid* protein determines position in the *Drosophila* embryo in a concentration-dependent manner. Cell 54, 95-104.

Duboule, D. and Dollé, P. (1989). The structural and functional organization of the murine HOX gene family resembles that of *Drosophila* homeotic genes. EMBO J. 8, 1497-1505.

Duprey, P., Chowdhury, K., Dressler, G.R., Balling, R., Simon, D., Guénet, J.-L. and Gruss, P. (1988). A mouse gene homologous to the *Drosophila* gene *caudal* is expressed in the epithelial cells from the embryonic intestine. Genes Dev. 2, 1647-1654.

Finkelstein, R. and Perrimon, N. (1991). The molecular genetics of head development in *Drosophila melanogaster*. Development 112, 899-912.

Frain, M., Swart, G., Monaci, P., Nicosia, A., Stämpfli, S., Frank, R. and Cortese, R. (1989). The liver-specific transcription factor LFB1 contains a highly diverged homeobox DNA binding domain. Cell 59, 145-157.

Freyd, G., Kim, S.K. and Horvitz, H.R. (1990). Novel cystein-rich motif and homeodomain in the product of the *Caenorhabditis elegans* cell lineage gene *lin-11*. Nature 344, 876-879.

Frumkin, A., Rangini, A., Ben-Yehuda, A., Gruenbaum, Y. and Fainsod, A. (1991). A chicken *caudal* homologue, *CHox-cad*, is expressed in the epiblast with posterior localization and in the early endodermal lineage. Development 112, 207-219.

García-Fernández, J., Baguñá, J. and Saló, E. (1991). Planarian homeobox genes: Cloning, sequence analysis, and expression. Proc. Natl. Acad. Sci. U.S.A. 88, 7338-7342.

Gaunt, S.J., Sharpe, P.T. and Duboule, D. (1988). Spatially restricted domains of homeo-gene transcripts in mouse embryos: relation to a segmented body plan. Development Supplement 104, 169-179.

Gehring, W. (1994). Preceding chapter.

Gould, A.O. and White, R.A.H. (1992). Development 116, 1163–1174.

Gould, S.J. (1991). Of mice and mosquitoes. Natural History July 1991, 12-20.

Graba, Y., Aragnol, D., Laurenti, P., Garzino, V., Charmot, D., Berenger, H. and Pradel, J. (1992). Homeotic control in *Drosophila*; the *scabrous* is an *in vivo* target of *Ultrabithorax* proteins. EMBO J. 11, 3375-3385.

Graham, A., Papalopulu, N. and Krumlauf, R. (1989). The murine and *Drosophila* homeobox gene complexes have common features of organization and expression. (1989). Cell 57, 367-378.

Gruss, P. and Walther, C. (1992). Pax in development. Cell 69, 719-722.

Herr, W., Sturm, R.A., Clerc, R.G., Corcoran, L.M., Baltimore, D., Sharp, P.A., Ingraham, H.A., Rosenfeld, M.G., Finney, M., Ruvkun, G. and Horvitz, H.R. (1988). The POU domain: A large conserved region in the mammalian *pit*-1, *oct*-1, *oct*-2, and *Caenorhabiditis elegans* unc-86 gene products. Genes Dev. 2, 1513-1516.

Hunt, P. and Krumlauf, R. (1991). Deciphering the Hox code: Clues to patterning branchial regions of the head. Cell 66, 1075-1078.

Jegalian, B. and De Robertis, E.M. (1992). Homeotic transformations in the mouse induced by overexpression of a human *Hox 3.3* transgene. Cell 71, 901–910.

Jones, F.S., Prediger, E.A., Bittner, D.A., De Robertis, E.M. and Edelman, G.M. (1992). Cell adhesion molecules as targets for Hox genes: N-CAM promoter activity is modulated by cotransfection with Hox 2.5 and 2.4. Proc. Natl. Acad. Sci. U.S.A. 89, 2086-2090.

Jones, F.S., Holst, B.D., Minowa, O., De Robertis E.M. and Edelman, G.M. (1993). Binding and transcriptional activation of the promoter for the neural cell adhesion molecule by HoxC6 (Hox-3.3). Proc. Natl. Acad. Sci. U.S.A. 90, 6557–6561.

Kamps, M.P., Murre, C., Sun, X. and Baltimore, D. (1990). A new homeobox gene contributes the DNA binding domain of the t(1;19) translocation protein in Pre-B ALL. Cell 60, 547-555.

Karlsson, O., Thor, S., Norberg, T., Ohlsson, H. and Edlund, T. (1990). Insulin gene enhancer binding protein Isl-1 is a member of a novel class of proteins containing both a homeo-and a cys-his domain. Nature 344, 879-882.

Kaufman, T.C., Seeger, M.A. and Olsen, G. (1990). Molecular and genetic organization of the *Antennapedia* gene complex of *Drosophila melanogaster*. Adv. Genet. 27, 309-362.

Kenyon, C. and Wang, B. (1991). A cluster of *Antennapedia*-class homeobox genes in a nonsegmented animal. Science 253, 516-517.

Kessel, M., Balling, R. and Gruss, P. (1990). Variations of cervical vertebrae after expression of a *Hox 1.1* transgene in mice. Cell 61, 301-308.

Kim, Y. and Niremberg, M. (1989). *Drosophila* NK-homeobox genes. Proc. Natl. Acad. Sci. U.S.A. 86, 7716-7720.

Lazzaro, D., Price, M., De Felice, M. and Di Lauro, R. (1991). The transcription factor TTF-1 is expressed at the onset of thyroid and lung morphogenesis and in restricted regions of the foetal brain. Development 113, 1093-1104.

Lazzaro, D., De Simeone, V., De Magistris, L., Lehtonen, E. and Cortese, R. (1992). LFBI and LFB3 homeoproteins are sequentially expressed during kidney development. Development 114, 469-479.

Le Mouellic, H., Lallemand, Y. and Brûlet, P. (1992). Homeosis in the mouse induced by a null mutation in the *Hox-3.1* gene. Cell 69, 251-264.

Leroy, P. and De Robertis, E.M. (1992). Effects of lithium chloride and retinoic acid on the expression of genes from the *Xenopus laevis* Hox 2 complex. Dev. Dynamics 194, 21-32.

Lewis, E.B. (1978). A gene complex controlling segmentation in Drosophila. Nature 276, 565-570.

Lufkin, T., Dierich, L., LeMeur, M., Mark, M. and Chambon, P. (1991). Disruption of the *Hox-1.6* homeobox gene results in defects in a region corresponding to its rostral domain of expression. Cell 66, 1105-1119.

Lufkin, T., Mark, M., Hart, C., Dolle, P., LeMeur, M. and Chambon, P. (1992). Homeotic transformation of the occipital bone of the skull by ectopic expression of a homeobox gene. Nature 359, 835–841.

Malicki, J., Schughart, K. and McGinnis, W. (1990). Mouse *Hox-2.2* specifies thoracic segmental identity in *Drosophila* embryos and larvae. Cell 63, 961-967.

Martin, G.R., Boncinelli, E., Duboule, D., Gruss, P., Jackson, I., Krumlauf, R., Lonai, P., McGinnis, W., Ruddle, F. and Wolgemuth, D. (1987). Nomenclature for homoeo-box-containing genes. Nature 325, 21-22.

Mendel, D.B., Khavari, P.A., Conley, P.B., Graves, M.K., Hansen, L.P., Admon, A. and Crabtree, G.R. (1991). Characterization of a cofactor that regulates dimerization of a mammalian home-odomain protein. Science 254, 1762-1767.

McGinnis, N., Kuziora, M.A. and McGinnis, W. (1990). Human *Hox-4.2* and *Drosophila deformed* encode similar regulatory specificities in *Drosophila* embryos and larvae. Cell 63, 969-976.

McGinnis, W. and Krumlauf, R. (1992). Homeobox genes and axial patterning. Cell 68, 283-302.

Morgan, B.A., Izpisúa-Belmonte, J.C., Duboule, D. and Tabin, C.J. (1992). Targeted misexpression of Hox 4.6 in the avian limb bud causes apparent homeotic transformations. Nature 358, 236–239.

Murtha, M.T., Leckman, J.F. and Ruddle, F.H. (1991). Detection of homeobox genes in development and evolution. Proc. Natl. Acad. Sci. U.S.A. 88, 10711-10715.

Niehrs, C., Keller, R., Cho, K.W.Y. and De Robertis, E. (1993). The homeobox gene *goosecoid* controls cell migration in *Xenopus* embryos. Cell 72, 491–503.

Pollock, R.A., Jay, G. and Bieberich, C.J. (1992). Altering the boundaries of *Hox 3.1* expression: Evidence for antipodal gene regulation. Cell 71, 911–923.

Porteus, M.H., Bulfone, A., Ciaranello, R.D. and Rubenstein, J.L.R. (1991). Isolation and characterization of a novel cDNA clone encoding a homeodomain that is developmentally regulated in the ventral forebrain. Neuron 7, 221-229.

Price, M., Lemaistre, M., Pischetola, M., Di Lauro, R. and Duboule, D. (1991). A mouse gene related to *Distal-less* shows a restricted expression in the developing forebrain. Nature 351, 748-751.

Price, M., Lazzaro, D., Pohl, T., Mattei, M.G., Rüther, U., Olivo, J.-C., Duboule, D. and Di Lauro, R. (1992). Regional expression of the homeobox gene *Nkx-2.2* in the developing mammalian forebrain. Neuron 8, 241-255.

Puelles, L., Amat, J.A. and Martinez de la Torre, M. (1987). Segment-related, mosaic neurogenetic pattern in the forebrain and mesencephalon of early chick embryos: I. Topography of AChE-positive neuroblasts up to stage HH18. J. Comp. Neurol. 266, 247-268.

Rabin, M., Hart, C.P., Ferguson-Smith, A., McGinnis, W., Levine, M. and Ruddle, F.H. (1985). Two homeobox loci mapped in evolutionary related mouse and human chromosomes. Nature 314, 175.

Raff, R.A. and Kaufman, T.C. (1983). Embryos, genes and evolution. MacMillan Publishing Company, New York.

Renucci, A., Zappavigna, V., Zàkàny, J., Izpisúa-Belmonte, J.-C., Bürki, K. and Duboule, D. (1992). Comparison of mouse and human HOX-4 complexes defines conserved sequences involved in the regulation of *Hox-4.4*. EMBO J. 11, 1459-1468.

Robert, B., Lyons, G., Simandl, B.K., Kuroiwa, A. and Buckingham, M. (1991). The apical ectodermal ridge regulates *Hox-7* and *Hox-8* gene expression in developing chick limb buds. Genes Dev. 5, 2363-2374.

Rosenfeld, M.G. (1991). POU-domain transcription factors: pou-er-ful developmental regulators. Genes Dev. 5, 897-907.

Ruberti, I., Sessa, G., Lucchetti, S. and Morelli, G. (1991). A novel class of plant proteins containing a homeodomain with a closely linked leucine zipper motif. EMBO J. 10, 1787-1791.

Ruvkun, G. and Finney, M. (1991). Regulation of transcription and cell identity by POU-domain proteins. Cell 64, 475-478.

Salser, S.J. and Kenyon, C. (1992). Activation of a *C. elegans* *Antennapedia* homologue in migrating cells controls their direction of migration. Nature 355, 255-258.

Schierwater, B., Murtha, M., Dick, M., Ruddle, F.H. and Buss, L.W. (1991). Homeoboxes in Cnidarians. J. Exp. Zool. 260, 413-416.

Schummer, M., Scheurlen, I., Schaller, C. and Galliot, B. (1992). HOM/HOX homeobox genes are present in hydra *(Chlorohydra viridissima)* and are differentially expressed during regeneration. EMBO J. 11, 1815-1823.

Scott, M.P. (1992). Homeobox nomenclature. Cell, in press.

Scott, M.P., Tamkun, J.W. and Hartzell, G.W., III (1989). The structure and function of the homeodomain. Biochim. Biophys. Acta 989, 25-48.

Simeone, A., Acampora, D., Guliano, M., Sornaiuolo, A. and Boncinelli, E. (1992). Nested expression domains of four homeobox genes in developing rostral brain. Nature 356, 687-690.

Taira, M., Jamrich, M., Good, P.J. and Dawid, I.B. (1992). The LIM domain-containing homeo box gene *Xlim-1* is expressed specifically in the organizer region of *Xenopus* gastrula embryos. Genes Dev. 6, 356-366.

Tanaka, M., Lai, J.-S. and Herr, W. (1992). Promoter-selective activation domains in Oct-1 and Oct-2 direct differential activation of an sRNA and mRNA promoter. Cell 68, 755-767.

Treacy, M.N., He, X. and Rosenfeld, M.G. (1991). I-POU: a novel POU-domain protein inhibiting neuron-specific gene activation. Nature 350, 577-584.

Treisman, J., Harris, E. and Desplan, C. (1991). The paired box encodes a second DNA-binding domain in the paired homeodomain protein. Genes Dev. 5, 594-604.

Vollbrecht, E., Veit, B., Sinha, N. and Hake, S. (1991). The developmental gene *Knotted-1* is a member of a maize homeobox gene family. Nature 350, 241-243.

Way, J.C. and Chalfie, M. (1988). *mec-3*, a homeobox-containing gene that specifies differentiation of the touch receptor neurons in *C. elegans*. Cell 54, 5-16.

Wright, C.V.E., Schnegelsberg, P. and De Robertis, E.M. (1988). XlHbox 8: a novel *Xenopus* homeo protein restricted to a narrow band of endoderm. Development 104, 787-795.

Wright, C.V.E., Cho, K.W.Y., Hardwicke, J., Collins, R.H. and De Robertis, E.M. (1989). Interference with function of a homeobox gene in *Xenopus* embryos produces malformations of the anterior spinal cord. Cell 59, 81-93.

■ *E.M. De Robertis:*
Molecular Biology Institute and Department
of Biological Chemistry,
University of California,
Los Angeles,
CA 90024–1737,
USA

3

A Comprehensive Classification of Homeobox Genes

A Comprehensive Classification of Homeobox Genes

Thomas R. Bürglin

■ Contents

■ Definition of the homeodomain

Consensus sequence

The homeodomain was originally described as a highly conserved protein motif of 60 to 63 amino acids, encoded by a short DNA fragment, the homeobox, that was found in *Drosophila* developmental control genes (McGinnis et al., 1984; Scott and Weiner, 1984). As more and more homeobox-containing genes were isolated it became obvious that for closely related homeodomains the sequence similarity could be extended upstream and/or downstream of the core 60 amino acids (for a previous compilation, see Scott et al., 1989). In contrast, highly divergent homeodomain sequences have been identified for which the sequence similarity is shorter than the originally defined 60 amino acids. The definition of the homeodomain used here is based on the canonical 60 amino acid core as it is most often used.

Figure 1A shows a consensus sequence based on a compilation of 346 homeodomain sequences. The consensus is biased towards *Antennapedia* class sequences, which constitute a major proportion of the known sequences. The numbering scheme beneath this consensus is that used in all the published descriptions of the homeodomain structure (Qian et al., 1989; Kissinger et al., 1990; Wolberger et al., 1991). The three α-helices identified in the 3-D structure of *engrailed* (Kissinger et al., 1990) and *MATα2* (Wolberger et al., 1991) are represented as shaded cylinders above the consensus, and the numbering for these three helices is shown on top. The NMR study of *Antennapedia* (Qian et al., 1989) identified a kink in the third helix, at position 52/53, such that this helix was considered as two separate helices, helix 3 and helix 4.

NMR and X-ray analysis of DNA-protein complexes for *Antp*, *en*, and *MATα2* allowed the identification of the residues critical for sequence specific contacts and contacts to the DNA backbone (Kissinger et al., 1990; Otting et al., 1990; Wolberger et al., 1991). These studies have been complemented by *in vivo* studies on the DNA-binding properties of homeodomains, which confirm the importance of some of the DNA-amino acid contacts of the structural studies (e.g. Hanes and Brent, 1991; Furukubo-Tokunaga et al., 1992; Schier and Gehring, 1992). A very critical residue that determines sequence specificity is found at position 9 of helix 3 (for review see Treisman et al., 1992).

Figure 1B gives a schematic summary of various contacts made by the amino acids of the homeodomain as deduced from structural and genetic data. Not every homeodomain makes each interaction shown; in particular some DNA contacts are slightly different in *MATα2*. Residues marked "H" form the hydrophobic core in which the highly conserved residues in helix 1 interact with the extremely conserved residues tryptophan and phenylalanine of helix 3. The main residues providing base-specific contacts are found in helix 3, which sits in the major groove of the DNA. However, the amino-terminus of the homeodomain, as well as the beginning of helix 2 contribute DNA contacts, albeit mainly to the backbone.

An analysis of the frequency of occurrence of each amino acid at each position (a "spectrum" of amino acids) of the homeodomain indicates that certain positions are significantly more conserved than others. Indeed, consistent with the structural studies, the highly conserved positions form part of the hydrophobic core, or are residues critical for homeodomain-DNA interactions. A summary of the conserved amino acids is given in Fig.1C where the amino acids are listed in decreasing order of their frequency at a particular position within the homeodomain. Amino acids found in fewer than five sequences are omitted. In this representation, conserved positions have been arbitrarily divided into three groups, corresponding to degrees of conservation. There are seven extremely highly conserved positions, two in helix 1, and five in helix 3. Only the tryptophan residue at position 48 and the asparagine residue at position 51 are invariant, thus far (two recently identified plant genes not included in the 346 sequences, *Zmhbx1a* and Lp *hbx7* have the asparagine residue replaced). Three other positions predominantly contain one amino acid, but occasionally permit substitutions: position 16, normally a leucine, can accommodate a methionine; positions 20 and 49, normally phenylalanine residues, can be substituted by tyrosine. Position 53 is almost invariably an arginine, with the only exceptions being the first (highly divergent) homeodomains in *zfh-2* and *ATBF1*. Position 57 accommodates either an arginine or a lysine residue. A second group of highly conserved positions has fewer than six different amino acids at a particular position. Six such positions are found: 5, 26, 40, 45, 47, and 57. Thirteen positions fall into the third group of moderately conserved positions, which can have up to 9 different amino acids at a particular position.

These positions, especially the more conserved ones, define the homeodomain and any new sequence should match most of these positions in order to be considered a homeodomain. Residues in the DNA-binding region of helix 3 are especially conserved and constitute a 'trademark' for the homeodomain. Figure 2 provides a spectrum of amino acids for each position based on 346 sequences, and can be used to assess how well a new homeodomain matches. Any unusual amino acids, particularly in more conserved positions, should be subject to particular scrutiny, as they might be of special functional importance for that homeodomain.

Classifying homeodomains

The purpose of comparing and classifying the sequences of homeobox-containing genes is to determine evolutionary relationships between these different genes, to assist structural and functional studies, and to provide tools, such as degenerate oligonucleotides, for further gene hunts. In order to arrange the homeodomain sequences into logical groups, the terms "superclass", "class" and "family" will be used. Their use reflects the hierarchy of relationships, such that a superclass encompasses several classes, and a class can be subdivided further into particular families.

Figure 3 shows a comparative tree based on homeodomain sequences of selected example genes of all the different classes and some of the families. Such a tree pro-

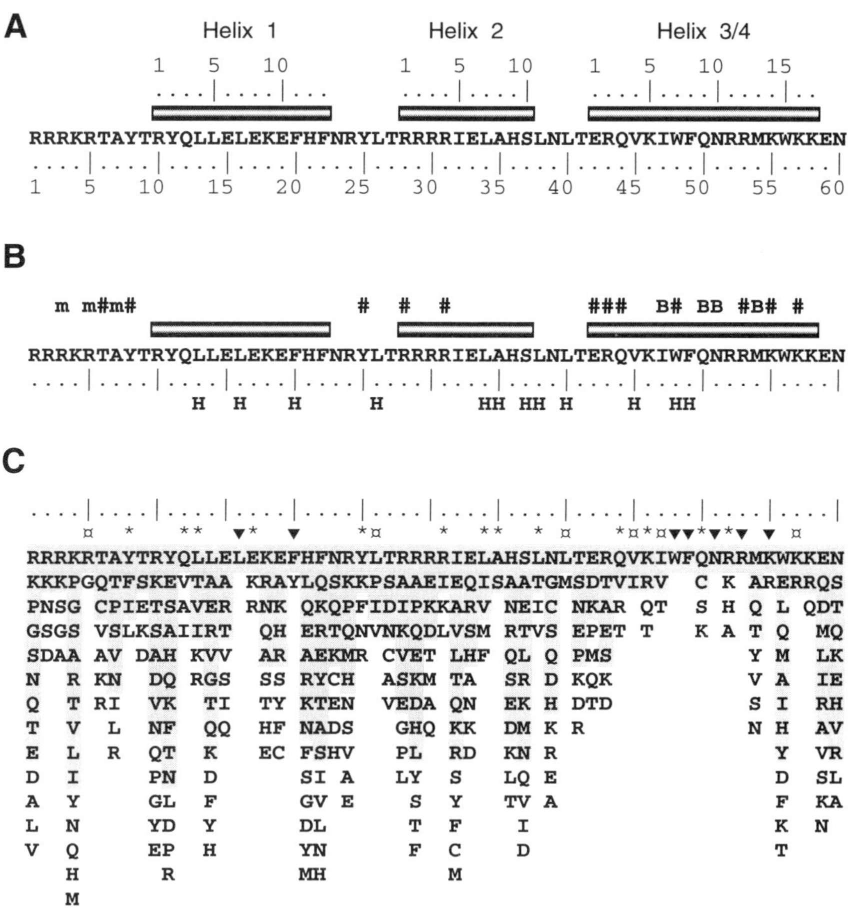

Figure 1: Homeodomain consensus sequence.
A) Amino acid consensus sequence based on 346 homeodomain sequences. Standard numbering scheme is shown beneath the consensus. The three α-helices (a composite derived from the structures of the *Antp*, *en* and *MATα2* homeodomains) and their numbering scheme is shown above the consensus.
B) Schematic representation of intramolecular and protein-DNA contacts made by individual amino acid of the homeodomain. This is a summary of the contacts observed for the *Antp*, *en* and *MATα2* homeodomains. Not all these contacts are made by each of the homeodomains, but this simplified overview allows estimating which contacts could be made by a homeodomain of undetermined structure. Amino acids designated "H" contribute to the hydrophobic core that is responsible for the tertiary structure of the homeodomain. Residues designated "B" contact bases in the major groove and are responsible for sequence specific DNA contacts. Residues designated "m" make contacts in the minor groove, and "#" indicates residues that contact the sugar-phosphate backbone.
C) Amino acids encountered at a given position in the homeodomain (based on 346 sequences). For each position the amino acid most frequently encountered is listed at the top, while other amino acids are listed beneath in decreasing order of their frequency of occurrence. The group of shaded amino acids are found in at least 85% of the 346 homeodomain sequences. Amino acids occurring less than 5 times (1.5%) are not shown. The symbols on top indicate highly conserved positions: ▼ are the most highly conserved positions, with only 1 or 2 amino acids found at that position, ¤ indicates highly conserved positions with 3 to 5 different amino acids found at a particular position, and * indicates conserved positions, with not more than 9 different amino acids.

vides an easy means of grouping homeodomain sequences into classes and families (see methods for information on interpretation of trees). However, other criteria need to be taken into account for classification purposes. Members of different classes have in general less than 55-57% identity in the homeodomain with homeodomains of genes in other classes. Figure 4 shows the percent identity between selected members of different classes of homeodomains. Information derived from the analysis of sequence conservation is one criterion for

		I	V	L	M	F	Y	W	C	A	G	S	T	P	N	Q	H	K	R	D	E
1	R	3	7	7	4	1	0	0	1	9	38	35	19	39	26	20	1	46	65	10	12
2	R	1	1	4	0	2	0	0	2	0	3	7	2	2	8	3	0	122	181	6	0
3	R	0	0	1	0	0	0	4	0	5	6	9	3	1	2	0	0	123	189	0	2
4	K	12	12	12	6	1	10	0	2	28	49	29	15	54	10	8	8	71	17	0	2
5	R	0	0	0	0	0	0	0	0	0	5	1	0	0	0	0	0	2	338	0	0
6	T	1	21	1	1	0	0	0	31	7	0	0	219	1	0	53	0	6	5	0	0
7	A	15	17	10	0	1	0	0	1	88	4	47	79	57	16	1	1	1	8	0	0
8	Y	44	0	6	1	134	156	0	0	1	0	0	0	0	0	0	4	0	0	0	0
9	T	0	0	1	0	0	0	0	0	4	0	86	210	0	4	1	1	7	1	6	25
10	R	3	17	1	1	0	5	0	0	22	9	35	41	12	16	13	2	47	97	20	5
11	Y	4	4	13	1	14	79	1	1	32	3	33	13	6	13	21	26	20	5	8	49
12	Q	5	20	1	0	0	1	0	0	9	0	1	0	0	1	304	0	0	0	0	4
13	L	28	30	140	0	0	0	0	0	1	0	1	98	0	0	1	0	25	22	0	0
14	L	2	19	154	2	8	7	1	2	24	16	3	15	1	4	15	6	13	20	12	22
15	E	8	14	3	0	1	0	0	1	51	4	8	18	0	0	5	0	3	30	0	200
16	L	0	0	343	3	0	0	0	0	0	0	0	0	0	0	0	0	0	0	0	0
17	E	0	0	0	0	0	2	0	0	0	0	1	0	0	3	4	0	28	5	1	302
18	K	1	2	2	1	1	0	0	1	13	2	13	10	0	21	16	8	183	63	1	8
19	E	1	2	4	3	5	10	1	5	29	0	12	3	0	1	2	23	24	19	0	202
20	F	0	0	0	0	332	14	0	0	0	0	0	0	0	0	0	0	0	0	0	0
21	H	0	0	60	5	9	6	0	1	22	7	8	1	0	11	29	117	18	20	6	26
22	F	9	8	7	1	127	22	0	3	13	1	10	19	0	7	33	5	29	25	3	24
23	N	1	0	0	0	0	1	1	14	2	2	30	26	0	199	28	5	24	1	5	7
24	R	0	9	0	27	0	0	0	1	6	1	10	0	44	11	35	26	54	116	0	6
25	Y	0	0	0	0	9	279	0	0	0	0	2	3	0	8	0	0	38	7	0	0
26	L	35	20	211	0	0	0	0	0	0	0	0	0	80	0	0	0	0	0	0	0
27	T	0	7	0	1	0	1	0	15	15	3	89	165	0	19	0	0	2	0	28	1
28	R	30	22	8	2	0	1	0	1	31	12	22	1	8	0	0	1	23	163	2	19
29	R	1	2	10	0	7	10	0	3	46	4	9	7	41	2	34	11	21	84	20	34
30	R	3	4	0	8	0	0	0	2	5	0	2	13	0	0	5	1	22	178	14	89
31	R	35	0	8	2	2	0	0	1	2	0	0	0	0	0	0	0	14	282	0	0
32	I	88	31	31	5	7	9	2	5	38	0	10	23	0	0	18	2	17	10	2	48
33	E	1	3	3	4	1	1	0	1	8	4	17	1	0	8	48	11	7	18	6	204
34	L	130	25	154	21	15	1	0	0	0	0	0	0	0	0	0	0	0	0	0	0
35	A	0	2	0	0	0	0	0	0	256	1	83	0	0	1	0	0	0	3	0	0
36	H	2	2	8	4	0	0	0	0	46	4	26	6	0	41	27	72	22	41	22	23
37	S	6	9	31	18	0	3	1	2	63	3	67	31	0	14	13	2	22	24	5	32
38	L	20	16	278	0	0	0	0	0	2	0	2	28	0	0	0	0	0	0	0	0
39	N	0	1	1	1	0	3	0	60	6	66	29	1	0	91	24	15	11	10	20	7
40	L	2	1	327	16	0	0	0	0	0	0	0	0	0	0	0	0	0	0	0	0
41	T	0	0	2	0	0	0	0	1	3	2	68	161	19	34	3	0	11	6	8	28
42	E	0	0	0	8	0	0	0	0	2	0	1	6	16	1	6	0	38	1	41	226
43	R	0	3	0	0	0	1	0	3	31	0	17	67	1	4	4	0	13	167	7	28
44	Q	0	53	0	0	0	0	0	0	0	0	0	7	0	1	260	2	4	19	0	0
45	V	134	209	2	0	1	0	0	0	0	0	0	0	0	0	0	0	0	0	0	0
46	K	2	2	0	0	0	0	0	0	1	0	0	10	0	0	32	0	257	38	0	4
47	I	234	99	0	0	0	0	0	0	0	0	0	9	0	2	0	2	0	0	0	0
48	W	0	0	0	0	0	0	346	0	0	0	0	0	0	0	0	0	0	0	0	0
49	F	0	0	0	0	342	4	0	0	0	0	0	0	0	0	0	0	0	0	0	0
50	Q	0	0	0	0	0	0	0	34	0	0	9	0	0	0	291	2	8	2	0	0
51	N	0	0	0	0	0	0	0	0	0	0	0	0	0	346	0	0	0	0	0	0
52	R	0	0	0	0	0	1	0	0	5	0	1	3	0	1	3	9	21	302	0	0
53	R	0	0	2	0	0	0	0	0	0	0	0	0	0	0	0	0	0	344	0	0
54	M	1	10	0	167	2	11	0	2	78	0	10	20	0	6	36	0	0	3	0	0
55	K	0	0	0	0	0	0	0	0	0	0	0	0	0	0	0	0	322	24	0	0
56	W	15	1	27	19	9	15	113	3	17	4	3	5	0	2	19	15	9	0	13	57
57	K	0	0	0	0	0	0	0	0	0	4	0	0	0	0	0	0	289	53	0	0
58	K	2	0	1	0	0	0	0	0	1	1	1	2	0	2	5	3	213	115	0	0
59	E	25	14	30	37	0	0	0	0	16	3	14	3	0	5	49	3	9	24	47	63
60	N	4	13	12	4	2	3	1	0	10	4	38	34	0	81	34	27	30	12	4	29

(The consensus at left, read vertically beside the rows, spells **Homeodomain**.)

Figure 2: Number of amino acids at each position of the homeodomain.
At left the homeodomain consensus is shown vertically. The top row indicates the individual amino acid for each column.

determining whether a homeobox gene belongs to a new class. A second criterion is whether a particular homeobox gene is more related to a homeobox gene found in a species of a different phylum than to any other homeobox gene. For example in a comparison against 348 homeodomain sequences, the homeodomains of *C. elegans*

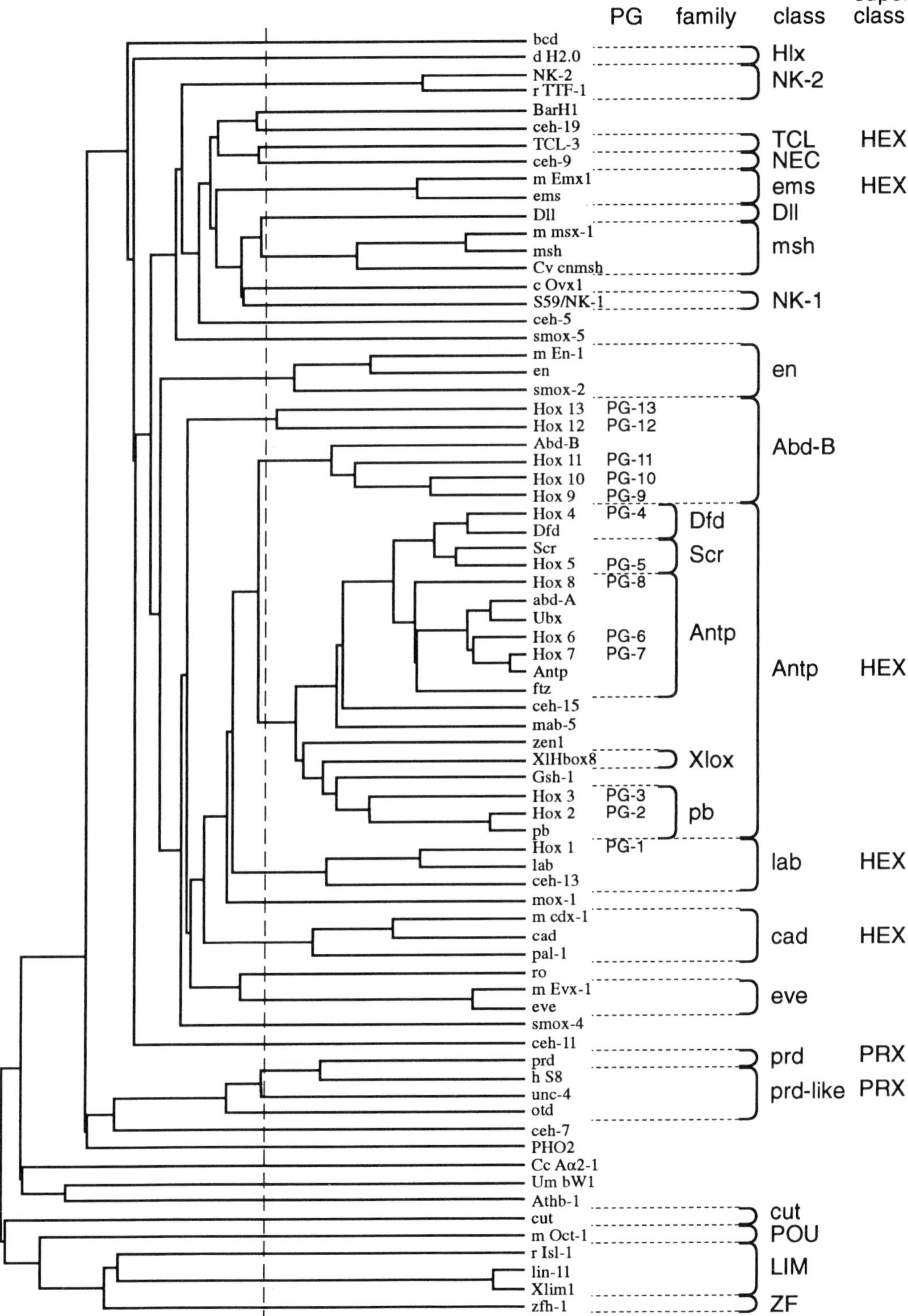

Figure 3: Comparative tree showing the different families and classes.
This tree is an adaptation of the large tree in Figure 6 giving only examples of genes for each of the families and classes. The paralog groups of the vertebrate clusters are indicated on the right, and parentheses mark the families and classes. The more similar two sequences are two each other, the shorter the horizontal distance is from the branch point to the endpoint which indicates a particular gene. The vertical dashes line marks approximately the distance that separates different classes from each other.

A Comprehensive Classification of Homeobox Genes 31

	lab	pb	Antp	Abd-B	cad	ems	msh	bcd	eve	en	H2.0	prd	otd	lin-11	ceh-7	ceh-9	S59	Oct-1	cut	PHO2	Athb-1
lab	100	65	67	53	48	47	53	43	50	50	40	35	38	33	42	52	52	30	28	32	33
pb	65	100	67	57	47	53	53	48	62	45	45	40	42	37	42	48	55	33	27	30	37
Antp	67	67	100	58	53	42	45	42	50	50	40	37	37	33	40	52	50	33	28	27	35
Abd-B	53	57	58	100	50	47	50	38	48	40	42	37	32	28	40	52	47	32	30	32	33
cad	48	47	53	50	100	38	43	37	40	40	40	33	35	28	40	42	43	28	25	23	38
ems	47	53	42	47	38	100	52	40	50	45	42	47	38	35	48	48	53	33	30	33	33
msh	53	53	45	50	43	52	100	42	47	47	35	38	40	30	45	52	58	35	23	38	35
bcd	43	48	42	38	37	40	42	100	42	45	38	38	38	27	42	38	47	25	28	22	33
eve	50	62	50	48	40	50	47	42	100	45	43	43	50	38	43	42	50	35	32	28	35
en	50	45	50	40	40	45	47	45	45	100	35	38	32	38	38	42	48	25	33	37	32
H2.0	40	45	40	42	40	42	35	38	43	35	100	38	33	33	38	48	40	28	27	23	30
prd	35	40	37	37	33	47	38	38	43	38	38	100	60	45	42	42	42	28	37	37	27
otd	38	42	37	32	35	38	40	38	50	32	33	60	100	38	40	33	45	27	27	38	27
lin-11	33	37	33	28	28	35	30	27	38	38	33	45	38	100	35	32	33	37	35	33	28
ceh-7	42	42	40	40	40	48	45	42	43	38	38	42	40	35	100	40	50	25	30	30	35
ceh-9	52	48	52	52	42	48	52	38	42	42	48	42	33	32	40	100	55	32	28	32	37
S59	52	55	50	47	43	53	58	47	50	48	40	42	45	33	50	55	100	30	30	37	37
Oct-1	30	33	33	32	28	33	35	25	35	25	28	28	27	37	25	32	30	100	25	32	32
cut	28	27	28	30	25	30	23	28	32	33	27	37	27	35	30	28	30	25	100	20	27
PHO2	32	30	27	32	23	33	38	22	28	37	23	37	38	33	30	32	37	32	20	100	28
Athb-1	33	37	35	33	38	33	35	33	35	32	30	27	27	28	35	37	37	32	27	28	100

Figure 4: Percent identity between representatives of different families and classes of homeodomains (see Fig. 7 for sequences).

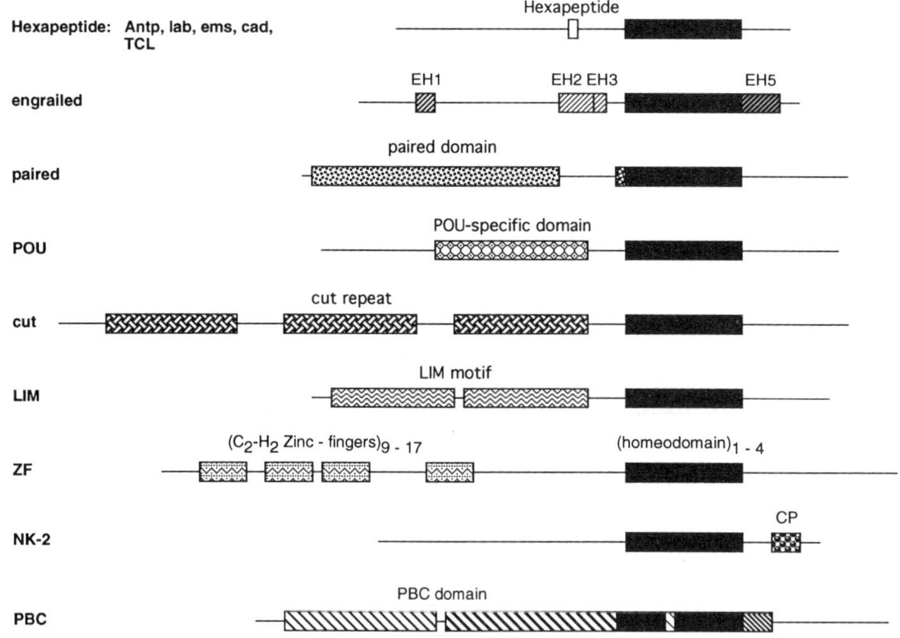

Figure 5: Schematic representation of classes of homeobox genes encoding conserved motifs outside of the homeodomain. On the left are the names given to the different classes. The black box represents the homeodomain, and the other boxes represent conserved motifs specific to individual classes. The length of the boxes is approximately proportional to the size of the domains. The connecting linker regions (black lines) can be substantially different between genes, and do not reflect true length. The Hexapeptide is found in several different classes, and thus this group is referred to as the Hexapeptide superclass. For the ZF class, the number of zinc-fingers (9–17), as well as the number of homeodomains (1–4) varies, and zinc-fingers can be found between homeodomains.

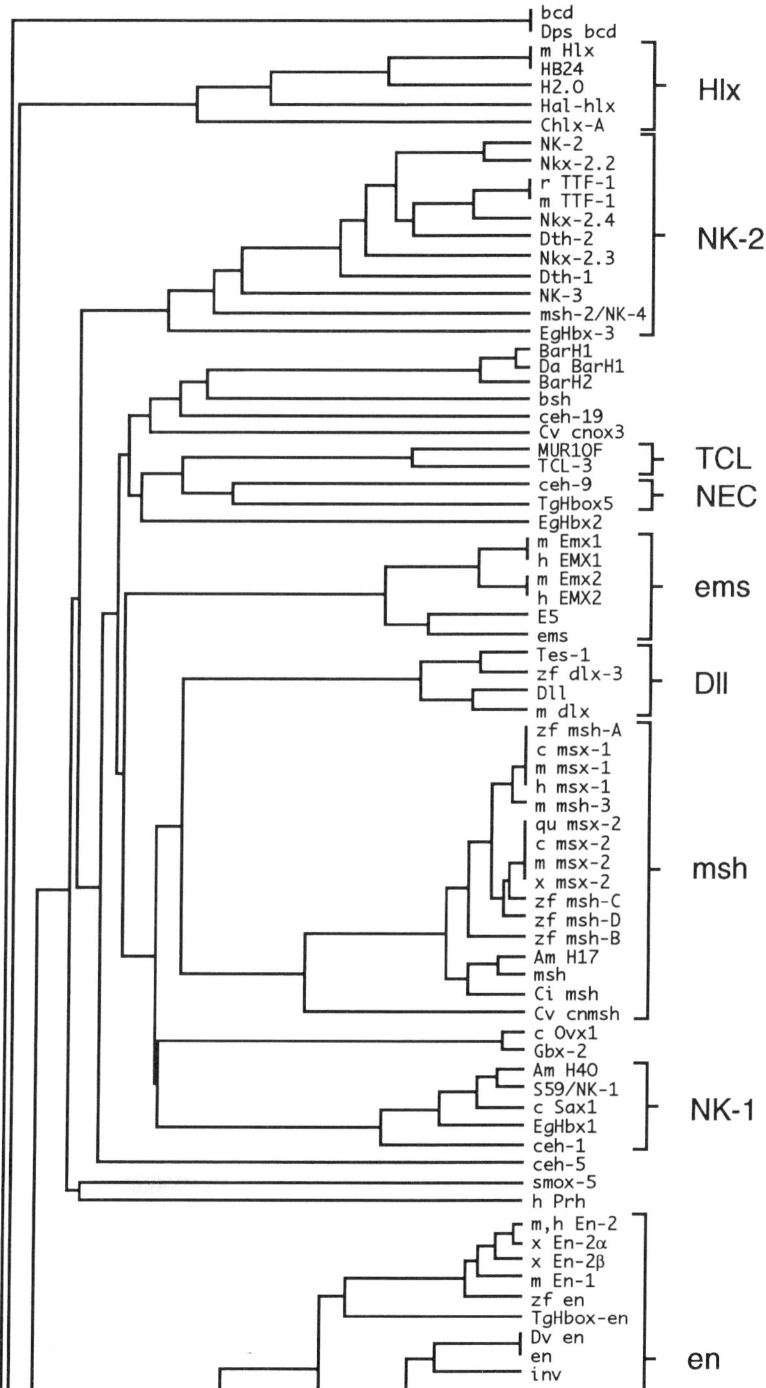

Figure 6: Comparative tree of homeodomain sequences.
The more similar two sequences are two each other, the shorter the horizontal distance is from the branch point to the endpoint that is marked by a gene. If two sequences are identical, then there is no horizontal branch at all (e.g. *bcd* and Dps *bcd*). The details of the tree generation are described in the methods section. Related groups (families or classes) of homeodomains are marked by square brackets and the name of that class or family is indicated on the right. The brackets with roman numerals inside the LIM class indicate families.

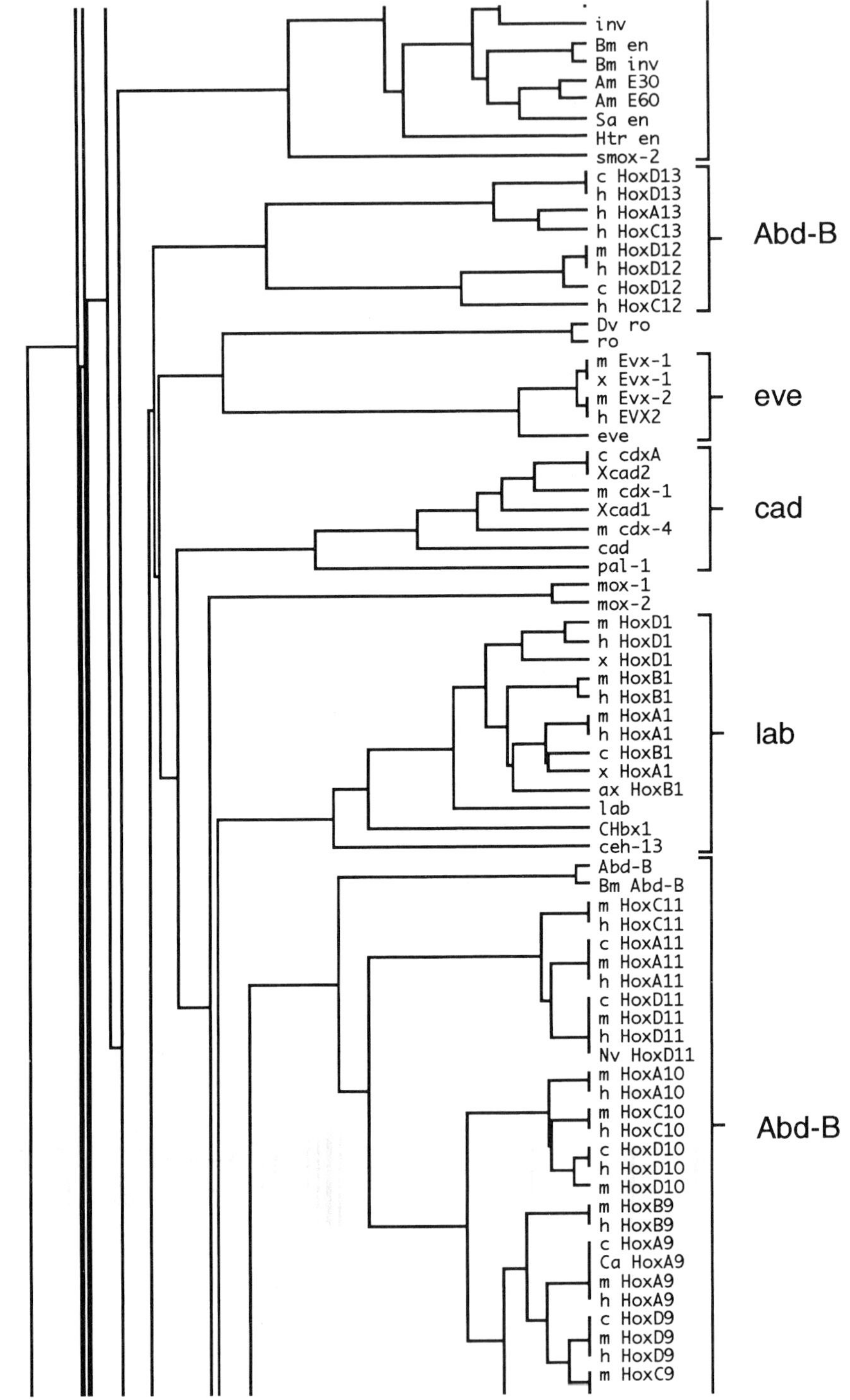

ceh-9 and sea urchin *TgHbox5* are 65% identical, while the next best matches are separated by 5% (i.e. *NK-3, Nkx-2.3*, mouse *MUR10F* are 60% identical to *TgHbox5, ceh-9* is 58% identical to *NK-3* and *MUR10F*). The next best matches for these two genes include members of various classes (for *TgHbox5*: 2 matches at 58%, 14 matches at 56%, 15 matches at 55%, etc., for *ceh-9*: 5 matches at 56%, 15 matches at 55% etc.) and are thus clearly not especially related to *ceh-9/TgHbox5*.

Often, individual homeodomain classes have telltale signs of sequence similarity outside of the homeodomain that further support the distinction as a separate class. For

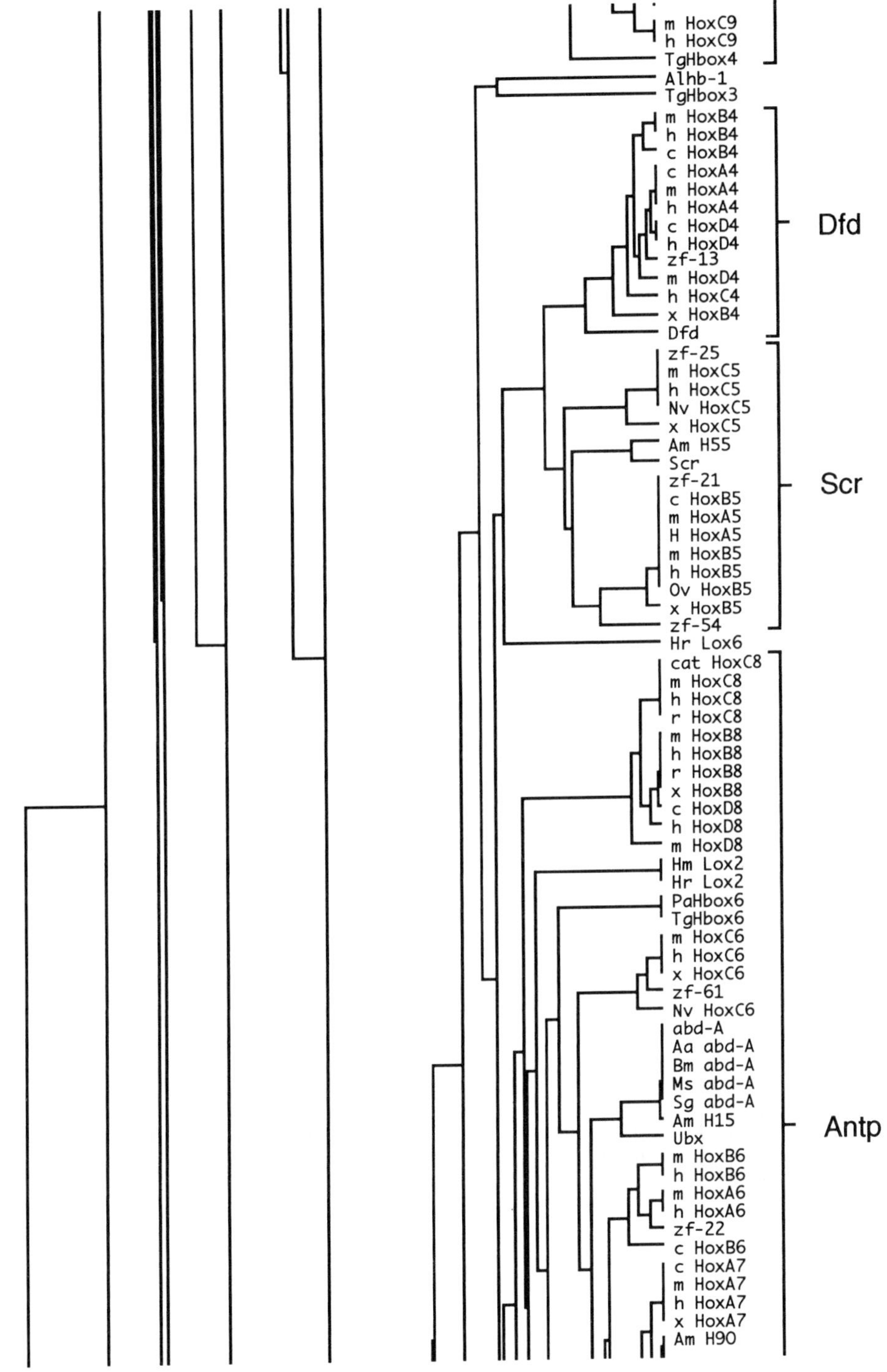

example, *TgHbox5* and *ceh-9* have extra amino acids conserved just upstream of the homeodomain. Figure 5 shows a schematic diagram of those homeobox gene classes that contain special conserved domains outside of the homeodomain. Several classes of homeobox genes that contain a "Hexapeptide" motif (see below), have been grouped together into the Hexapeptide (HEX) superclass (Figs. 3, 5).

The criteria for a family are less well defined. In general any subgroupings within a class into particular distinct groups are considered families, the only requirement being that they are also conserved over large evolutionary distances (such as flies to vertebrates). While the classifi-

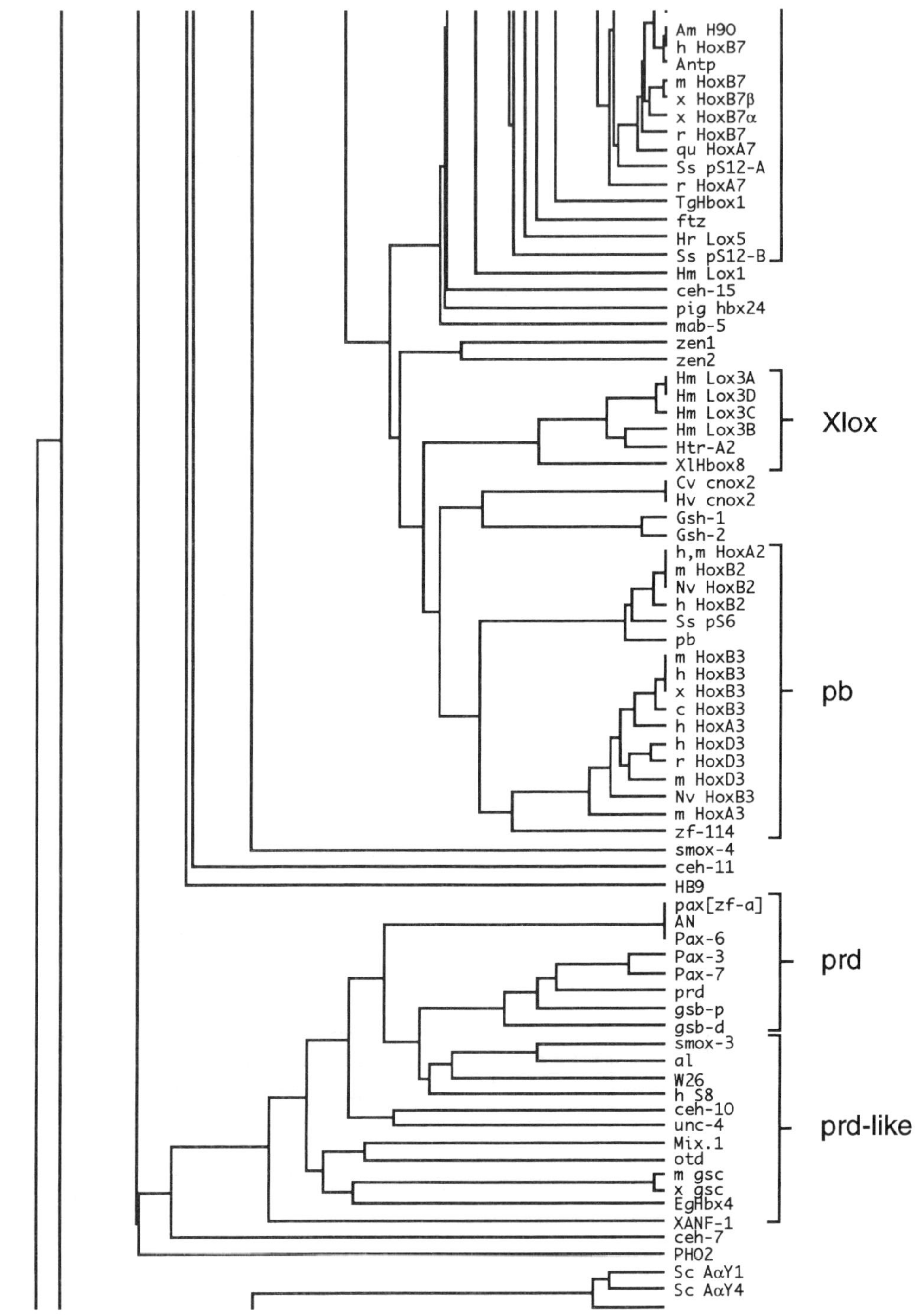

cation scheme presented here goes somewhat towards a "taxonomy" for homeobox genes, the current dataset is quite incomplete, and much further data is needed for an understanding of how homeobox genes and species coevolved during phylogeny. Furthermore, these classification attempts are limited by necessity, because the emergence of new genes happens continually during evolution, and thus somewhat arbitrary decisions have to

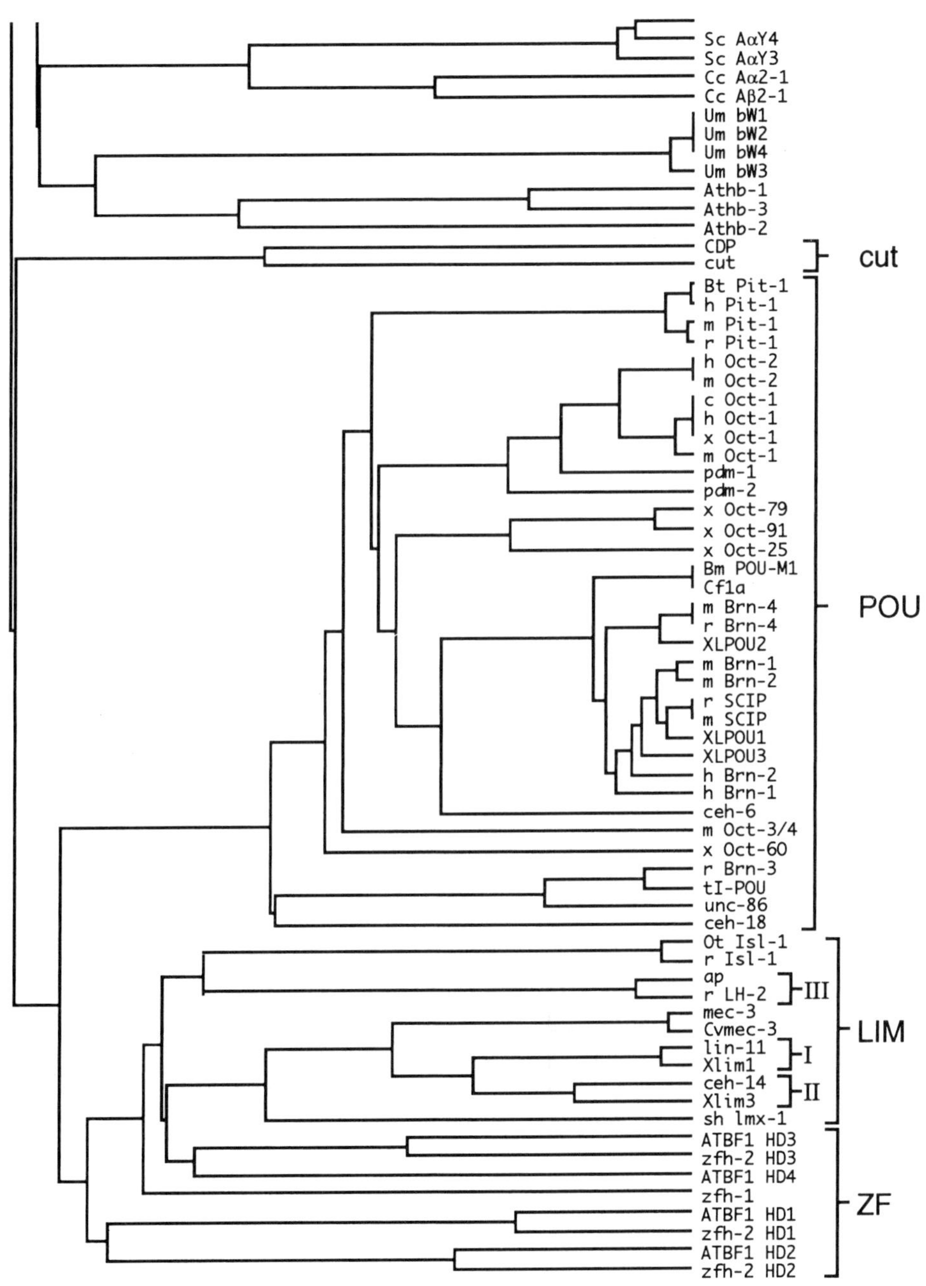

be taken to define classes and families.

The following sections introduce the various classes of homeobox genes. Figure 6 presents a comparative tree with most of the known homeobox genes, and Figure 7 is a fairly complete compilation of the homeodomain sequences of these genes (some recently described genes are shown only in Fig. 6). The subsequent comparisons of genes and classes are based on these homeodomain sequences and occasional additional data such as homeo-domain flanking sequences.

```
    ....|...10....|...20....|...30....|...40....|...50....|...60
    RRRKRTAYTRYQLLELEKEFHFNRYLTRRRRIELAHSLNLTERQVKIWFQNRRMKWKKEN      Consensus

    P..T..TF.SS.IA...QH.LQG....AP.LAD.SAK.A.GTA......K...RRH.IQS      bcd
    P..T..TF.SS.IA...QH.LQG....AP.LAD.SAK.A.GTA......K...RRH.IQS      Dps bcd

    .SWS.AVFSNL.RKG..IQ.QQQK.I.KPD.RK..AR....DA...V.........RHTR      H2.0
    .SWS.AVFSNL.RKG...R.EIQK.V.KPD.KQ..AM.G..DA...V.........RHSK      m H1x
    .SWS.AVFSNL.RKG...R.EIQK.V.KPD.KQ..AM.G..DA...V.........RHSK      HB24
    .KWN.AVFRLM.RRG...S.QSQK.VAKPE.RK..DA.S..DA...........RQ.I        Hal-hlx
    GMLR.AVFSDV.RKA...T.QKQK.ISKPD.KK..SK.G.KDS...........RNSK        m Dbx
    GILR.AVFSED.RKA...M.QKQK.ISKTD.KK..IN.G.K.S...........RNSK        Chlx-A

    K.KR.VLF.KA.TY...RR.RQQ...SAPE.EH..SLIR..PT........H.Y.T.RAQ      NK-2
    K.KR.VLFSKA.TY...RR.RQQ...SAPE.EH..SLIR..PT........H.Y.M.RAR      Nkx-2.2
    ..KR.VLFSQA.VY...RR.KQQK..SAPE.EH..SMIH..PT........H.Y.M.RQA      r TTF-1
    ..KR.VLFSQA.VY...RR.KQQK..SAPE.EH..SMIH..PT........H.Y.M.RQA      m TTF-1
       R.VLFSQA.VY...RR.KQQK..SAPE.EH..SMIH..PT........H.Y.M.RQA      Nkx-2.4
    ..KR.ILFSQA.IY...RR.KQQK..SAPE.EH..NLI...PT........H.Y.C.RSQ      Dth-2
    ..KP.VLFSQA.VF...RR.KQQ...SAPE.EH..S..K..ST.........Y.C.RQR       Nkx-2.3
    K.KR.VLFSKK.I....RH.RQKK..SAPE.EH..NLIG.SPT........H.Y.M.RAH      Dth-1
    KK.S.A.FSHA.VF...RR.AQQ...SGPE.S.M.K..R...T.........Y.T.RKQ       NK-3
    K.KP.VLFSQA.V....CR.RLKK...GAE.EII.QK...SAT.........Y.S.RGD       d msh-2/NK-4
    QSKR.VLFNKF.ISQ...R.RKQ....AQE.Q....TIG..PT........HAY.M.RLF      EgHbx3

    Q.KA...F.DH..QT...S.ERQK..SVQE.Q....K.D.SDC...T.Y....T...RQT      BarH1
    Q.KA...F.DH..QT...S.ERQK..SVQE.Q..S.K.D.SDC...T.Y....T...RQT      Da BarH1
    Q.KA...F.DH..QT...S.ERQK..SVQD.M...NK.E.SDC...T.Y....T...RQT      BarH2
    ..KA..VFSDP..SG...R.EGQ...SPPE.V...TA.G.S.T...T........H..QL      bsh
    E.KP.Q..SAR..DR..T..QTDK..SVNK..Q.SQT.....T.I.T......T....QL      ceh-19
    C.KP..VFSDL..MV..R..NNRK..STPQ.TN..DR.G.NQT...T.Y..........T      Cv cnox3

    KKKP..SF..L.IC....R..RQK..ASAE.AA..KA.KM.DA...T......T..RRQT      TCL-3
    .KKP..SFS.S.V....RR.LRQK..ASAE.AA..KA.RM.DA...T......T..RRQT      MUR10F

    .KKA..TFSGK.VF....Q.EAKK..SSSD.S...KR.DV..T.........T....IE       ceh-9
    KKKT..VFS.S.VFQ..ST.EVK...SSSE.AG..AN.H...T..........N...RQM      TgHbox5
    QK.A.VSFSSS.VHV..ER.DRQK..SSAE.A.MSRD.G.S.T..........Y.T..RA      EgHbx2

    PK.I...FSPS...K..HA.ES.Q.VVGAE.KA..QN...S.T...V......T.H.RMQ      ems
    PK.I...FSPS...R..RA.EK.H.VVGAE.KQ..G..S.S.T...V......T.Y.RQK      m Emx1
    PK.I...FSPS...R..RA.EK.H.VVGAE.KQ..G..S.S.T...V......T.Y.RQK      h EMX1
    PK.I...FSPS...R..HA.EK.H.VVGAE.KQ.....S...T...V......T.F.RQK      m Emx2
    PK.I...FSPS...R..HA.EK.H.VVGAE.KQ.....S...T...V......T.F.RQK      h EMX2
    PK.V...FSPT...K..HA.EG.H.VVGAE.KQ..QG.S...T...V......T.H.RMQ      E5

    M.KP..I.SSL..QQ.NRR.QRTQ..ALPE.A...A..G..QT..........S.Y..MM      Dll
    V.KP..I.SSF..AA.QRR.QKTQ..ALPE.A...A..G..QT..........S.F..MW      Tes-1
    I.KP..I.SS...AA.QRR.QKAQ..ALPE.A...AQ.G..QT..........S.F..LY      zf dlx-3
    I.KP..I.SSL..QA.NRR.QQTQ..ALPE.A...A..G..QT........K.S.F..LM      m dlx
    I.KP..I.SSL..QA.NHR.QQTQ..ALPE.A...A..GV.QT........K.S.Y..LI      Xdll

    N.KP..PF.TQ...S...K.REKQ..SIAE.A.FSS..R...T..........A.A.RLQ      msh
    N.KP..PF.TQ...S...K.REKQ..IAE.A.FSS..H...T...........A.A.RLQ      Am H17
    N.KP..PF.TA...A..RK.RQKQ..SIAE.A.FSS..S...T..........A.A.RLQ      zf msh-A
    N.KP..PF.TA...A..RK.RQKQ..SIAE.A.FSS..S...T..........A.A.RLQ      c msx-1
    N.KP..PF.TA...A..RK.RQKQ..SIAE.A.FSS..S...T..........A.A.RLQ      m msx-1
    N.KP..PF.TA...A..RK.RQKQ..SIAE.A.FSS..S...T..........A.A.RLQ      h msx-1
    N.KP..PF.TS...A..RK.RQKQ..SIAE.A.FSS......T..........A.A.RLQ      x msx-1
    N.KP..PF.TA...A..RK..QKQ..SIAE.A.FSS..S...T..........A.A.RLQ      m msh-3
    N.KP..PF.TS...A..RK.RQKQ..SIAE.A.FSS......T..........A.A.RLQ      qu msx-2
7Λ  N.KP..PF.TS...A..RK.RQKQ..SIAE.A.FSS......T..........A.A.RLQ      c msx 2
```

Figure 7: Compilation of homeodomain sequences.

In this figure and all subsequent sequence alignments the following conventions apply: A consensus (based on the most frequently encountered amino acid at a particular position) is shown at the top, and dots indicate identities to the consensus. Dashes indicated a gap that was introduced for optimal alignment. In cases where sequences are incomplete or unavailable, empty spaces have been left; a "*" marks a stop codon. Sequences are essentially presented in the same order as in the tree in Figure 6, except for some sequences or groups which were rearranged to reflect the sequence relationships based on *HOX* cluster comparisons. Underlined amino acids of *r HOXC4* are probably not part of the homeobox gene, since they are very different from corresponding amino acids in *Hoxc-4* genes

```
....|...10....|...20....|...30....|...40....|...50....|...60
RRRKRTAYTRYQLLELEKEFHFNRYLTRRRRIELAHSLNLTERQVKIWFQNRRMKWKKEN   Consensus

N.KP..PF.TS...A..RK.RQKQ..SIAE.A.FSS......T.........A.A.RLQ   m msx-2
N.KP..PF.TS...A..RK.RQKQ..SIAE.A.FSS......T.........A.A.RLQ   x msx-2
N.KP..PF.TS...A..RK.RQKQ..SIAE.A.FSN......T.........A.A.RLQ   zf msh-C
N.KP..PF.TS...A..RK.RQKQ..SIAE.A.FSS..T...T.........A.A.RLQ   zf msh-D
N.KP..PFSTS...S..RK.RQKQ..SIAE.A.FSN......T.........A.A.RLQ   zf msh-B
N.KP..PF.TQ..MS..K.REKQ..SIAE.A.FSN..S...T.........A.S.RLQ   Ci msh
N.KP..PFSVN...T..QK.KRKQ..SISE.A..SEL.R...T.I........A.Q.RSK   Cv cnmsh

S..R...F.SE.........CKK..SLTE.SQI..A.K.S.V.........A...RIK   c Ovx1

P..A...F.YE..VS..NK.KTT...SVCE.LN..L..S...T.........T....Q.   S59/NK-1
A..A...F.YE..VA..NK.KTT...SVCE.LN..L..S...T.........T....Q.   Am H40
P..A...F.YE..VA..NK.RAT...SVCE.LN..L..S...T.........T....QH   c Sax1
...A...F.YE..VT..NK.QST...SVYE.LN..L......T.........T....Q.   EgHbx1
M..A...F.YE..VA..NK.KTS...SVVE.LN..IQ.Q.S.T.........T....H.   ceh-1

PK.P..VF.DE..EK..ES.NTSG..SGST.AK..E..G.SDN...V......T.Q..ID   ceh-5
..KT..TFSNC..N...NN.NRQ....PTD.DRI.KH.G..NT..IT......A.L.R.A   smox-5
.KGGQVRFSNE.TI....K.ETQK..SPPE.KR..KL.Q.S.....T......A..RRLK   c Prh
.KGGQVRFSND.TI....K.ETQK..SPPE.KR..KM.Q.S.....T......A..RRLK   h Prh

EK.P...FSSE..AR.KR..NE.....E...QQ.SSE.G.N.A.I......K.A.I..ST   en
EK.P...FSSE..AR.KR..NE.....E...QQ.SSE.G.N.A.I......K.A.I..ST   Dv en
DK.P...FSGT..AR.KH..NE.....EK..QQ.SGE.G.N.A.I......K.A.L..SS   inv
EK.P...FSAE..AR.KR..AE.....E...QQ.SRD.G...A.I......K.A.I..AS   Am E30
EK.P...FSGE..AR.KR..AE.....E...QQ.SRD.G...A.I......K.A.I..AS   Am E60
EK.P...FSGA..AR.KH..AE.....E...QS..AE.G.A.A.I......K.A.I..AS   Bm en
EK.P...FSGP..AR.KH..AE.....E...QS..AE.G.A.A.I......K.A.I..AS   Bm inv
EK.P...FSGE..AR.KH..TE.....E...Q...RE.G.N.A.I......K.A.I..AS   Sa en
DK.P...F.AE..QR.KA..QT.....EQ..QS..QE.S.N.S.I......K.A.I..AT   h En-2
DK.P...F.AE..QR.KA..QT.....EQ..QS..QE.S.N.S.I......K.A.I..AT   m En-2
DK.P...F.AD..QR.KA..QT.....EQ..QS..QE.S.N.S.I......K.A.I..AT   x En2α
DK.P...F.AE..QR.KA..QT.....EQ..QS..QE.G.N.S.I......K.A.I..ST   x En2β
DK.P...F.AE..QR.KA..QA...I.EQ..QT..QE.S.N.S.I......K.A.I..AT   m En-1
DK.P...F.AE..QR.KA..QT.....EQ.AQS..QE.G.N.S.I......K.A.I..AS   zf en
EK.P...FSAS..QR.KQ..QQSN...EQ..RS..KE.T.S.S.I......K.A.I..AS   TgHbox-en
EK.P...F.GD..AR.KR..SE.K...EQ..TC..KE...N.S.I......K.A.M..AS   Htr en
LK.P..SF.VP..KR.SQ..EK....DEL..KK..TE.D.R.S.........K.A.T..AS   smox-2

Q..Q..TFSTE.T.R..V...R.E.IS.S..F...ET.R...T.I........A.D.RIE   ro
Q..Q..TFSTE.T.R..V...R.E.IS.S..F...ET.R.S.T.I........A.D.RIE   Dv ro

KDKY.VV..DF.R......YCTS..I.I..KS...QT.S.S...........A.ERTS.   cad
KDKY.VV..DH.R........YS..I.I..KA...AA.G............A.ER.V.   c cdxA
KDKY.VV..DH.R........YS..I.I..KA...AA.G............A.ER.V.   Xcad2
KDKS.VV..DH.R........YS..I.I..KS...AN.G............A.ER.V.   m cdx-1
KDKY.VV..DH.R........YS..I.I..KA...AT.G.S..........A.ER.IN   sh cdx-3
KDKY.VV..DQ.R........YS..I.I..KA...VN.G.S.T.........A.ER.I.   Xcad1
KEKY.VV..DH.R..........I.I..KS...VN.G.S..........A.ER.LI   zf cad1
ADKY.MV.SD..R.......TSPFI.SD.KSQ.STM.S.....I........A.DRRDK   pal-1

NGTN..NF.TH..T.......TAK.VN.T..T.I.SN.K.Q.A...........E..RE   ceh-13
NNSG..NF.NK..T...............A....I.NT.Q.N.T.........Q..RV   lab
AFSL..SFSTR..T........S...S.A..L.V.R..R.RDA...V........Q..RE   CHbx-1
PSAI..NFSTK..T........K....A....I.NC.Q.NDT.........Q..RE   m HoxD1
SSAI..NFSTK..T........K....A....I.NC.H.NDT.........Q..RE   h HoxD1
PCNV..NF.TK..T........K....A....I.N..Q.NDT.........Q..RE   x HoxD1
PGGL..NF.TR..T........K.S.A..V.I.AT.E.N.T..........Q..RE   m HoxB1
PSGL..NF.TR..T........K.S.A..V.I.AT.E.N.T..........Q..RE   h HoxB1
```

7B

in other species such as mouse. References or accession numbers for some recently published and unpublished genes: *PaHbox6* (X54494); *XANF-1* (X60099); *HES-1* (Thomas and Rathjen, 1992); *PHBX1* (M95929); *Xb HoxA9* (Stickland et al., 1992); *Zmhbx1a* & *Zmhbx1b* (Bellmann and Werr, 1992), *Ot Isl-1* (X64884); *al* (*aristaless*, Schneitz et al., 1992); *Nv HoxC10* (X68975); *Nv HoxD10* (X68976); *m HoxC4* (Geada et al., 1992); *bsh* (L06475), *Mhbx/K-2* (Cserjesi et al., 1992; Kern et al. 1992); *sh cdx-3*, *sh lmx-1* (German et al., 1992); *c, h Prh* (Crompton, 1992); *AmphiHox3* (Holland et al., 1992); *EgHbx1, 2, 3,* and *4* (Oliver et al., 1992); *zf zgsc* (L03395).

```
        ....|...10....|...20....|...30....|...40....|...50....|...60
        RRRKRTAYTRYQLLELEKEFHFNRYLTRRRRIELAHSLNLTERQVKIWFQNRRMKWKKEN      Consensus

        PNAV..NF.TK..T.........K...A..V.I.A..Q.N.T............Q..RE      m HoxA1
        PNAV..NF.TK..T.........K...A..V.I.A..Q.N.T............Q..RE      h HoxA1
        PNTA..NF.TK..T.........K...A..V.I.AA.Q.N.T............Q..RE      x HoxA1
        PNTI..NF.TK..T.........K...A..V.I.AT.E.N.T............Q..RE      c HoxB1
        QNSI..NF.TK..S.............A..V.I.AT.E.N.T............Q..RE      ax HoxB1

        V..Y...F..D..GR.....YKEN.VS.P..C...AQ...P.STI.V........D.RQR     eve
        M..Y...F..E.IAR.....YREN.VS.P..C...AA...P.TTI.V........D.RQR     m Evx-1
        M..Y...F..E.IAR.....YREN.VS.P..C...AA...P.TTI.V........D.RQR     x Evx-1
        V..Y...F..E.IAR.....YREN.VS.P..C...AA...P.TTI.V........D.RQR     m Evx-2
        V..Y...F..E.IAR.....YREN.VS.P..C...AA...P.TTI.V........D.RQR     h EVX2
        T..Y...F..E..SR.....LREN.VS.T..S...SM...S.TTI.........A.RRR      Af eveC

        SKKG.QT.Q...TSV..AK.QQSS.VSKKQ.E..RLQTQ..D..I..........A...K      ceh-11

        V.K..KP.SKF.T.......L..A.VSKQK.W...RN.Q................N..NS      Abd-B
        V.K..KP.SKF.T.......L..A.VSKQK.W...RN.................N..NS      Bm Abd-B

        G.K..VP..KL..K...N.YAI.KFINKDK.RRISAAT..S....T.......V.D..IV     c HoxD13
        G.K..VP..KL..K...N.YAI.KFINKDK.RRISAAT..S....T.......V.D..IV     h HoxD13
        G.K..VP..KV..K...R.YAT.KFI.KDK.RRISATT..S....T.......V.E..VI     h HoxA13
        G.K..VP..KV..K.....YAASKFI.KEK.RRISATT..S....T.......V.E..VV     h HoxC13
        A.K..KP..KQ.IA...N..LV.EFIN.QK.K..SNR...SDQ...........K.RVV      m HoxD12
        A.K..KP..KQ.IA...N..LV.EFIN.QK.K..SNR...SDQ...........K.RVV      h HoxD12
        S.K..KP..KQ.IA...N..LL.EFIN.QK.K..SNR...SDQ...........K.RVV      c HoxD12
        S.K..KP.SKL..A...G..LV.EFI..Q..R..SDR...SDQ...........K.RLL      h HoxC12

        T.K..CP.SKF.IR...R..F..V.INKEK.LQ.SRM....D.............E..LS     m HoxC11
        T.K..CP.SKF.IR...R..F..V.INKEK.LQ.SRM....D.............E..LS     h HoxC11
        T.K..CP..K..IR...R..F.SV.INKEK.LQ.SRM....D.............E..I.     c HoxA11
        T.K..CP..K..IR...R..F.SV.INKEK.LQ.SRM....D.............E..I.     m HoxA11
        T.K..CP..K..IR...R..F.SV.INKEK.LQ.SRM....D.............E..I.     h HoxA11
        S.K..CP..K..IR...R..F..V.INKEK.LQ.SRM....D.............E..L.     c HoxD11
        S.K..CP..K..IR...R..F..V.INKEK.LQ.SRM....D.............E..L.     m HoxD11
        S.K..CP..K..IR...R..F..V.INKEK.LQ.SRM....D.............E..L.     h HoxD11
        S.K..CP..K..IR...R..F..V.INKEK.LQ.SRM....D.............E..L.     Nv HoxD11

        G.K..CP..KH.T.......L.M....E..L.ISR.VH..D.............L..M.      m HoxA10
        G.K..CP..KH.T.......L.M....E..L.ISR.VH..D.............L..M.      h HoxA10
        G.K..CP..KH.T.......L.M....E..L.ISKTI...D.............L..M.      m HoxC10
        G.K..CP..KH.T.......L.M....E..L.ISKTI...D.............L..M.      h HoxC10
        G.K..CP..KH.T.......L.M....E..L.ISK.I..D.............L..M.       Nv HoxC10
        G.K..CP..KH.T.......L.M....E..L.ISK.V...D.............L..MS      c HoxD10
        G.K..CP..KH.T.......L.M....E..L.ISK.V...D.............L..MS      h HoxD10
        G.E..CP..KH.T.......L.M....E..L.ISK.V...D.............L..MS      m HoxD10
        G.K..CP..KH.T.......L.M....E..L.ISK.V...D.............L..MS      Nv HoxD10

        S.K..CP..K..T.......L.M....D..H.V.RL...S..............M..M.      m HoxB9
        S.K..CP..K..T.......L.M....D..H.V.RL...S..............M..M.      h HoxB9
        S.K..CP.SK..T.......L.M....D..H.V.RL...S..............M..L.      x HoxB9
        T.K..CP..KH.T.......L.M....D..Y.V.RL.................M..I.       c HoxA9
        T.K..CP..KH.T.......L.M....D..Y.V.RL.................M..I.       Ca HoxA9
        T.K..CP..KH.T.......L.M....D..Y.V.RL.................M..I.       m HoxA9
        T.K..CP..KH.T.......L.M....D..Y.V.RL.................M..I.       h HoxA9
        T.K..CP..KH.T.......L.M....D..Y.V.RL.................M..I.       Xb HoxA9
        T.K..CP..K..T.......L.M....D..Y.V.RI.................M..MS       c HoxD9
        T.K..CP..K..T.......L.M....D..Y.V.RI.................M..MS       m HoxD9
        T.K..CP..K..T.......L.M....D..Y.V.RI.................M..MS       h HoxD9
        T.K..CP..K..T.......L.M....D..Y.V.RV.................M..M.       m HoxC9
```

7C

```
        ....|...10....|...20....|...30....|...40....|...50....|...60
        RRRKRTAYTRYQLLELEKEFHFNRYLTRRRRIELAHSLNLTERQVKIWFQNRRMKWKKEN     Consensus

        T.K..CP..K..T.......L..M....D..Y.V.RV.................M..M.      h HoxC9
        G.K..CP..KF.T.......L..M....D..L.I.RL.S...............M..Q.      TgHbox4

        PK.Q......H.I........Y.............I..T.V.S...I.............D.    Dfd
        PK.S......Q.V........Y.........V.I..A.C.S...I.............DH     m HoxB4
        PK.S......Q.V........Y.........V.I..A.C.S...I.............DH     h HoxB4
        PK.S......Q.V........Y.........V.I....C.S...I.............DH     c HoxB4
        AK.S......Q.V........Y.........V.I..T.R.S...I.............DH     x HoxB4
        PK.S......Q.V........Y.........V.I..T.C.S...I.............DH     zf-13
        PK.S......Q.V..................I..T.C.S.............DH           c HoxA4
        PK.S......Q.V..................I..T.C.S.............DH           m HoxA4
        PK.S......Q.V..................I..T.C.S.............DH           h HoxA4
        PK.S......Q.V..................I..T.C.S...I.........DH           c HoxD4
        PK.S......Q.V..................I..T.C.S...I.........DH           h HoxD4
        PK.S......Q.V..................I..T.C.P..I.........DH            m HoxD4
        PK.S.A....Q.V.........Y........I....C.S...I.........DH           h HoxC4
        PK.S......Q.V.........Y........I....C.S...I.........DH           m HoxC4
        GPARGV.NC.Q.V.........Y........I....C.S...I.........DH           r HoxC4

        GK.G.QT...Q.T..........S..V.....F.I.Q..G.S...I...........R.H     TgHbox3
        EK.Q......N.V........THK....K...V....M...............H....       ceh-15
        DK.A..S.S...T............NG.....I...G....I.............D.        Hr Lox6

        TK.Q..S....T...............I..A.C....I.........L...H             Scr
        VK.Q..S....T...............I..A.C....I.............H             Am H55
        GK.S..S....T...............I.NN.C.N...I.............DS           zf-25
        GK.S..S....T...............I.NN.C.N...I.............DS           m HoxC5
        GK.S..S....T...............I.NN.C.N...I.............DS           h HoxC5
        GK.S..S....T...............I.NN.C.N...I.............DS           Nv HoxC5
        GK.S..S....T..............D...I.NN.C.N...I.............DT        x HoxC5
        GK.A.......T...............I..A.C.S...I.............D.           zf-21
        GK.A.......T...............I..A.C.S...I.............D.           c HoxB5
        GK.A.......T...............I..A.C.S...I.............D.           m HoxA5
        GK.A.......T...............I..A.C.S...I.............D.           h HoxA5
        GK.A.......T...............I..A.C.S...I.............D.           m HoxB5
        GK.A.......T...............I..A.C.S...I.............D.           h HoxB5
        GK.A.......T...............I..A.C.S...I.............D.           Ov HoxB5
        GK.A.......T...............I..T.C.S...I.............D.           x HoxB5
        EN.A.......A...............IR.A.C.S...I.............D.           zf-54
        G.GRGQT.....T.............M..A.C.S...I.............D.            Ss pS12-B

        ..SG.QT.S...T.......L..P....K....VS.A.G.................         cat HoxC8
        ..SG.QT.S...T.......L..P....K....VS.A.G.................         m HoxC8
        ..SG.QT.S...T.......L..P....K....VS.A.G.................         h HoxC8
        ..SG.QT.S...T.......L..P....K....VS.A.G.................         r HoxC8
        ...G.QT.S...T.......L..P....K....VS.A.G.................         m HoxB8
        ...G.QT.S...T.......L..P....K....VS.A.G.................         h HoxB8
        ...G.QT.S...T.......L..P....K....VS.A.G.................         r HoxB8
        ...G.QT.S...T.......L..P....K....VS.A.G.................         x HoxB8
        ...G.QT.S.F.T.......L..P....K....VS.A.G.................         c HoxD8
        ...G.QT.S.F.T.......L..P....K....VS.A.A.................         h HoxD8
        ...G.QT.S.F.T.......L..P....K....VS.T.A.................         m HoxD8
        H..GLQTCS...T........QC.P...CK.W..VS.A.G....I.........RE         pig Hbx24
        .K.C.QT.....T.......................S.L.G....I.........Y..S      TgHbox1

        .K.G.QT...A.T........Y.....K....I.QAVC.S...I.............R        PaHbox6
        .K.G.QT...A.T........Y.....K....I.QAVC.S...I.............R        TgHbox6
        ...G.QI.S...T..............I.NA.C....I.............S             m HoxC6
        ...G.QI.S...T..............I.NA.C....I.............S             h HoxC6
```

7D

```
         ....|...10....|...20....|...30....|...40....|...50....|...60
         RRRKRTAYTRYQLLELEKEFHFNRYLTRRRRIELAHSLNLTERQVKIWFQNRRMKWKKEN      Consensus

         ...G.QI.S...T....................I.NA.C.....I..............S      x HoxC6
         ...G.QI.S...T....................I.NASC.....I..............S      Nv HoxC6
         ...G.QI.S...T....................I.NA.C.....I..............T      zf-61
         G..G.QT.....T.......Y............I..A.C.....I..............S      m HoxB6
         G..G.QT.....T.......Y............I..A.C.....I..............S      h HoxB6
         A..G.QT.....T....................I...C.....I..............      c HoxB6
         G..G.QT.....T....................I.NA.C.....I..............      m HoxA6
         G..G.QT.....T....................I.NA.C.....I..............      h HoxA6
         G..G.QT.....T....................I..A.C.....I..............      zf-22

         .K.G.QT.....T....................I..A.C.....I.............H      c HoxA7
         .K.G.QT.....T....................I..A.C.....I.............H      m HoxA7
         .K.G.QT.....T....................I..A.C.....I.............H      h HoxA7
         .K.G.QT.....T....................I..A.C.....I.............H      x HoxA7
         .K.G.QT.....T....................Y..A.C.....I.............H      qu HoxA7
         GK.G.QT.....T...............AV.I..A.C.....I.............H      r HoxA7
         .K.G.QT.....T.........Y..........I..T.C.....I.............      m HoxB7
         .K.G.QT.....T.........Y..........I..A.C.....I.............      h HoxB7
         .K.G.QT.....T....................I..T.C.....I.............      x HoxB7β
         .K.G.QT.....T....................I..V.C.....I.............      x HoxB7α
         .K.G.QT.....T.........Y..........I..GVC.....I.............      r HoxB7
         .K.GSQT.....T...............V.I..V.C.....I.............DH      Ss pS12-A

         .K.G.QT.....T....................I..A.C.....I.............      Antp
         .K.G.QT.....T.........Y..........I..A.C.....I.............      Am H90
         ...G.QT...F.T..........H.........I..A.C.....I.........L...L      abd-A
         ...G.QT...F.T..........H.........I..A.C.....I.........L...L      Aa abd-A
         ...G.QT...F.T..........H.........I..A.C.....I.........L...L      Bm abd-A
         ...G.QT...F.T..........H.........I..A.C.....I.........L...L      Ms abd-A
         ...G.QT...F.T..........H.........I..A.C.....I.........L...L      Sg abd-A
         ...G.QT...F.T........Y.H.........I..A.C.....I.........L...L      Am H15
         ...G.QT.....T.........T.H........M..A.C.....I.........L...I      Ubx
         ...G.QT.....T.........T.H........M..A.C.....I.........L...I      Bm Ubx
         ...G.QT.....T.........K...........S.T.Y...I.........E...V      Hm Lox2
         ...G.QT.....T.........K...........S.T.Y...I.........E...V      Hr Lox2
         .K.T.QT.....T.........YS...........I....A.S...I.............      Hr Lox5
         SK.T.QT.....T........I......DI.NA.S.S...I.........S..DR      ftz
         QK.T.QT.....T.........K......I..T.T....I......      smox-1
         KK.G.QT.S.H.T.........Y.K..........NR..G.S...I.........K...D      Alhb-1
         SK.T.QT.S.S.T.......YHK....K..Q.ISET.H...............H...A      mab-5
         SK.I.....SI........QN....S.L...QI.AI.D...K...........V....DK      Cv cnox2
         SK.I.....SI........QN....S.L...QI.AI.D...K...........V....DK      Hv cnox2
            CSFGHRKII...R..KY......D..L.F.RN.D.S.S.I.V........Q...Q      Cv cnox1

         SK.M...F.ST......R..AS.M..S.L....I.TY...S.K..........V.H...G      Gsh-1
         GK.M...F.ST......R..SS.M..S.L....I.TY...S.K..........V.H...G      Gsh-2
         NK.T...S.S..F........DK.IS.P..V...S.......HI..............ME      Htr-A2
         NK.T......A......L..K.IS.P..V...VM.......HI..............E      Xlhbox8
         LK.S...F.SV..V...N..KS.M..Y.T....I.QR.S.C..............F..DI      zen1
         SK.S...FSSL..I...R...L.K..A.T....ISQR.A................L..ST      zen2

         P..L.....NT..............K..C.P....I.A..D........V........H.RQT      pb
         S..L.....NT..............K..C.P..V.I.AL.D........V........H.RQT      h HoxA2
         S..L.....NT..............K..C.P..V.I.AL.D........V........H.RQT      m HoxA2
         S..L.....NT..............K..C.P..V.I.AL.D........V........H.RQT      m HoxB2
         S..L.....NT..............K..C.P..V.I.AL.D........V........H.RQT      Nv HoxB2
         A..L.....NT..............K..C.P..V.I.AL.D........V........H.RQT      h HoxB2
         PG.L.....NT..............K..C.P..V.I.AL.D........V........H.RQT      Ss pS6
 7E      SK.A.....SA..V..............C.P..V.M.NL...S...I..........Y..DQ      m HoxB3
```

```
     ....|...10....|...20....|...30....|...40....|...50....|...60
     RRRKRTAYTRYQLLELEKEFHFNRYLTRRRRIELAHSLNLTERQVKIWFQNRRMKWKKEN   Consensus

     SK.A.....SA..V............C.P..V.M.NL...S...I.........Y..DQ    h HoxB3
     SK.A.....SA..V............C.P..V.M.NL...S...I.........Y..DQ    x HoxB3
     SK.A.....SA..V............C.P..V.M.NL...S...I.........Y..DE    c HoxB3
     SK.AA....SA..V............C.P..V.M.NL.S.....I.........Y..DQ    Nv HoxB3
     SK.A.....SA..V............V.P..V.M.NL......I.........Y..DQ     h HoxA3
     SK.G.....P..V............M.P..V.M.NL......I.........Y..DQ      m HoxA3
     SK.V.....SA..V............C.P..V.M.NL......I.........Y..DQ     h HoxD3
     SK.V.....SA..V............C....V.M.NL......I.........Y..DQ     r HoxD3
     SKSV..T..SA..V............C.P..V.M.NL......I.........Y..DQ     m HoxD3
     SK.A.V.F.SS..........SA..C.N..L.M.EL.K..D..I.........Y..DH     zf-114
     GK.A.....SA..V............C.P..V.M.AM.......I.........Y...Q    AmphiHox3

     SF.N...F.D...IC..R..SHIQ..S.ID..H..QN.....K..........VR.R.R.   smox-4
     GKCR.SRTAFTSQQL..L..K..K..S.PK.F.V.T..M...T.............RSK    HB9

     Q..C..TFSAS..D...RA.ERTQ.PDIYT.E...QRT....ARIQV..S...ARLR.QH   prd
     Q..S..TF.AE..EA..RA.SRTQ.PDVYT.E...QTTA...ARIQV..S...ARLR.HS   gsb-p
     Q..S..TFSND.IDA..RI.ARTQ.PDVYT.E...Q.TG...AR.QV..S...ARLR.QL   gsb-d
     LQ.N..SF.QE.IEA.....ERTH.PDVFA.ER..AKID.P.ARIQV..S...A..RR.E   pax[zf-a]
     LQ.N..SF.QE.IEA.....ERTH.PDVFA.ER..AKID.P.ARIQV..S...A..RR.E   AN
     LQ.N..SF.QE.IEA.....ERTH.PDVFA.ER..AKID.P.ARIQV..S...A..RR.E   Pax-6
     Q..S..TF.AE..E...RA.ERTH.PDIYT.E...QRAK...AR.QV..S...AR.R.QA   Pax-3
     Q..S..TF.AE..E....A.ERTH.PDIYT.E...QRTK...ARFQV..S...AR.R.QA   Pax-7
     Q..N..TFNSS..QA..RV.ERTH.PDAFV.E...RRV..S.AR.QV......A.FRRNE   m S8
     Q..N..TFNSS..QA..RV.ERTH.PDAFV.ED..RRV....AR.QV......A.FRRNE   m MHbx/K-2
     Q..N..TFNSS..QA..RV.ERTH.PDAFV.ED..RRV....AR.QV......A.FRRNE   h PHBX1
     Q..Y..TF.SF..E....A.SRTH.PDVFT.E...MKIG...ARIQV......A..R.QE   al
     Q..I..TF.SL..K...RA.QETH.PDIYT.ED..LRID...AR.QV......A.FR.TE   smox-3
     K..H..IF.Q..ID....A.QDSH.PDIYA.EV..GKTE.Q.DRIQV......A..R.TE   ceh-10
     ...T..NFSGW..E...SA.EASH.PDVFM.EA..MR.D.L.SR.QV......A..R.RE   unc-4
     QK.H..RF.PA..N...RC.SKTH.PDIFM.E.I.MRIG...SR.QV......A....RK   W26
     Q.....FF.QA..DI..QF.QT.M.PDIHH.E...RHIYIP.SRIQV......A.VRRQG   Mix.1
     Q..E..TF..A..DV..AL.GKT..PDIFM.E.V.LKI..P.SR.QV..K...A.CRQQL   otd
     Q..E..TF..S..DV..AL.AKT..PDIFM.E.V.LKI..P.SR.QV..K...A.CRQQQ   m Otx1
     Q..E..TF..A..DV..AL.AKT..PDIFM.E.V.LKI..P.SR.QV..K...A.CRQQQ   m Otx2
     K..H..IF.DE..EA..NL.QETK.PDVGT.EQ..RKVH.R.EK.EV..K...A..RRQK   m gsc
     K..H..IF.DE..EA..NL.QETK.PDVGT.EQ..RRVH.R.EK.EV..K...A..RRQK   x gsc
     K..H..IF.DE..EA..NL.QETK.PDVGT.EQ..RKVH.R.EK.EV..K...A..RRQK   zf zgsc
     S..E..I..PE..EAM.EV.GV...PDVSM.E...SR.GIN.SKIQV..K...A.LRNLE   EgHbx4
     G..P...F..S.IEI..NV.RV.S.PGIDV.E...SK.A.D.DRIQ.......A.L.RSH   XANF-1
     G..P...F.QN.VEV..NV.RV.C.PGIDI.ED..QK...E.DRIQ.......A.M.RSR   m HES-1

     IP.R..TF.VE..YL..MY.AQSQ.VGCDE.ER..RI.S.D.Y..........IRMRR.A   ceh-7
     Q.P...RAKGEA.DV.KRK.EI.PTPSLVE.KKISDLIGMP.KN.R.......A.LR.KQ   PHO2
     SKKP.PKFHSEYTPL..LY.R..A.P.YAD.RV..EKTGMLT..ITV....H.RRA.GPL   Sc AαY1
     GK.S.PKFHSEYTPV..LY....A.P.YAD.RI..EKTGMLT..ITV....H.RRA.GPL   Sc AαY4
     YKKP.PKFHSEYTPL..LY....A.P.FAD.RM..EKTGMQT..ITV....H.RRA.GPL   Sc AαY3
     PPS.KQ.FNVHYIPV...Y.EY.A.P.AQD.AL..RKSMMSA..IEV....H.ARAR..G   Cc Aα2-1
     LEV...PFNSEYTPL...Y.EY.A.PSA.D.EW..RKTMMSV..IEV....H.RRAR..G   Cc Aβ2-1
     PLKTGRGHDSEAVRI..QA.KHSPNI.PAEKFR.SEVTG.KPK..T.......NRKG.K.   Um bW1
     PLKTGRGHDSEAVRI..QA.KHSPNI.PAEKFR.SEVTG.KPK..T.......NRKG.K.   Um bW2
     PLKTGRGHDSEAVRI..QA.KHSPNI.PAEKFR.SEVTG.KPK..T.......NRKG.K.   Um bW4
     PLKTGRGHDSEAVQI..EA.KHSPNI.PAEKFR.SEVTG.KPK..T.......NRKG.K.   Um bW3
     LPE.KRRL.TE.VHL...S.ETENK.EPE.KTQ..KK.G.QP...AV......AR..TKQ   Athb-1
     LGE.KKRLNLE.VRA...S.ELGNK.EPE.KMQ..KA.G.QP..IA......AR..TKQ    Athb-3
     NS..KLRLSKD.SAI..ET.KDHST.NPKQKQA..KQ.G.RA...EV......ART.LKQ   Athb-2
     STARKGHFGPVINQK.HEH.KTQP.PS.SVKES..EE.G..F...NK..ET..HSARVAS   Zmhbx1a
     IKDRKGHFGPVISQK.HEH.KTQP.PS.SLKES..EE.G..FH..NR..E...HFARLAS   Zmhbx1b
     KKP..QMK.PF..ET..RVYAMET.PSEAI.A..SEK.G..D..LQM..CH..L.D.NTS  Lp hbx7
  7F SPKGKSSISPQARAF..QV.RRKQS.NSKEKE.V.KKCGI.PL..RV..I.K..RS.*    MATa1
```

```
         ....|....10....|....20....|....30....|....40....|....50....|....60
         RRRKRTAYTRYQLLELEKEFHFNRYLTRRRRIELAHSLNLTERQVKIWFQNRRMKWKKEN    Consensus

         K.KR..TISIAAKDA..RH.GEQNKPSSQEILRM.EE...EKEV.RV...C...QRE.RVK    Bt Pit-1
         K.KR..TISIAAKDA..RH.GEQNKPSSQEIMRM.EE...EKEV.RV...C...QRE.RVK    h Pit-1
         K.KR..TISVAAKDA..RH.GEHSKPSSQEIMRM.EE...EKEV.RV...C...QRE.RVK    m Pit-1
         K.KR..TISIAAKDA..RH.GEHSKPSSQEIMRM.EE...EKEV.RV...C...QRE.RVK    r Pit-1
         ..K...SIETNVRFA...S.LA.QKP.SEEILLI.EQ.HMEKEVIRV...C...Q.E.RI.    h Oct-2
         ..K...SIETNVRFA...S.LA.QKP.SEEILLI.EQ.HMEKEVIRV...C...Q.E.RI.    m Oct-2
         ..K...SIETNIRVA...S.LE.QKP.SEEITMI.DQ..MEKEVIRV...C...Q.E.RI.    c Oct-1
         ..K...SIETNIRVA...S.LE.QKP.SEEITMI.DQ..MEKEVIRV...C...Q.E.RI.    h Oct-1
         ..K...SIETNIRVA...S.LE.QKP.SEEITMI.DQ..MEKEVIRV...C...Q.E.RI.    x Oct-1
         ..K...SIETNIRVA...S.ME.QKP.SEDITLI.EQ..MEKEVIRV...C...Q.E.RI.    m Oct-1
         ..K...SIETTIRGA...A.LA.QKP.SEEITQ..DR.SMEKEV.RV...C...Q.E.RI.    pdm-1
         ..K...SIETTVRTT...A.LM.CKP.SEEISQ.SER..MDKEVIRV...C...Q.E.RI.    pdm-2
         K.KH..SIEDNVRHT..NY.MQCSKPSAQEIAQI.RE..MEKDV.RV...C...Q.G.RQV    x Oct-79
         K.KH..SIENNVKCT..NY.MQCSKPSAQEIAQI.RE..MEKDV.RV...C...Q.G.RQV    x Oct-91
         K.KR..NIENIVKGT..SY.MKCPKPGAQEMVQI.KE..MDKDV.RV...C...Q.G.RQG    x Oct-25
         K.K...SIEVSVKGA..QH..KQPKPSAQEITS..D..Q.EKEV.RV...C...Q.E.RMT    Bm POU-M1
         K.K...SIEVSVKGA..QH..KQPKPSAQEITS..D..Q.EKEV.RV...C...Q.E.RMT    Cf1a
         K.K...SIEVSVKGV..TH.LKCPKPAAQEISS..D..Q.EKEV.RV...C...Q.E.RMT    m Brn-4
         K.K...SIEVSVKGV..TH.LKCPKPAAQEISS..D..Q.EKEV.RV...C...Q.E.RMT    r Brn-4
         K.K...SIEVSVKGV..TH.LKCPKPAALEITS..D..Q.EKEV.RV...C...Q.E.RMT    X1POU2
         K.K...SIEVSVKGA..SH.LKCPKPSAQEITN..D..Q.EKEV.RV...C...Q.E.RMT    m Brn-1
         K.K...SIEVSVKGA..SH.LKCPKPSAQEITS..D..Q.EKEV.RV...C...Q.E.RMT    m Brn-2
         K.K...SIEVGVKGA..SH.LKCPKPSAHEITG..D..Q.EKEV.RV...C...Q.E.RMT    r SCIP
         K.K...SIEVGVKGA..SH.LKCPKPSAHEITG..D..Q.EKEV.RV...C...Q.E.RMT    m SCIP
         K.K...SIEVGVKGA..NH.LKCPKPSAHEITS..D..Q.EKEV.RV...C...Q.E.RMT    XLPOU1
         K.K...SIEVSVKGA..SH.LKCPKPSAPEITS..D..Q.EKEV.RV.CC...Q.E.RMT    XLPOU3
         K.K...SIEVSVKGA..SH.LKCPKPSAQEITS..D..Q.EKEV.RV...C...Q.E.R     h Brn-2
         K.K...SIEVSVKGA..SH.LKCPKPSSQEITN..D..Q.EKEV.RV...C...Q.E.R     h Brn-1
         K.K...SIEVNVKSR..FH.QS.QKPNAQEITQV.ME.Q.EKEV.RV...C...Q.E.RIA    ceh-6
         K.K...SIEANVKSI..SS.MKLSKPSAQDISS..EK.S.EKEV.RV...C...Q.E.RIT    Djpou1
         .K....SIENRVRWS..TM.LKCPKPSLQQITHI.NQ.G.EKDV.RV...C...Q.G.RSS    m Oct-3/4
         K.KM..CFDTVLKGQ..GH.MC.QKPGA.ELT.I.KE.S.EKDV.RV...C...Q.E.SKF    x Oct-60
         KK....SIAAPEKRS..AY.AVQPRPSSEKIAAI.EK.D.KKNV.RV...C.Q.Q.Q.R      r Brn-3
         KK....SIAAPEKRS..AY.AVQPRPSGEKIAAI.EK.D.KKNV.RV...C.Q.Q.Q.RIV    tI-POU
         KK....SIAAPEKR...QF.KQQPRPSGE.IASI.DR.D.KKNV.RV...C.Q.Q.Q.RDF    unc-86

         SKKQ.VLFSEE.KEA.RLA.ALDP.PNVGTIEF..NE.G.AT.TITN..H.H..RL.QQV    cut
         LKKP.VVLAPEEKEA.KRAYQQKP.PSPKTIED..TQ...KTST.IN..H.Y.SRIRR.L    CDP
         LKKP.VVLAPEEKEA.KRAYQQKP.PSPKTIE...TQ...KTST.IN..H.Y.SRIRR.L    Clox

         TT.V..VLNEK..HT.RTCYAA.PRPDALMKEQ.VEMTG.SP.VIRV....K.C.D..RS    r Isl-1
         TT.V..VLNEK..HT.RTCYNA.PRPDALMKEQ.VEMTG.SP.VIRV....K.C.D..RT    Ot Isl-1
         PK.P..IL.TQ.RRAFKAS.EVSSKPC.KV.ET..AETG.SV.V.QV....Q.A.M..LA    sh lmx-1
         TK.M..SFKHH..RTMKSY.AI.HNPDAKDLKQ.SQKTG.PK.VLQV....A.A..RRMM    ap
         TK.M..SFKHH..RTMKSY.AI.HNPDAKDLKQ..QKTG..K.VLQV....A.A.FRRNL    r LH-2
         ..GP..TIKQN..DV.NEM.SNTPKPSKHA.AK..LETG.SM.VIQV......S.ERRLK    mec-3
         ..GP..TIKQN..DV.NEM.SNTPKPSKHA.AKK.LETG.SM.VIQV......S.ERRLK    Cvmec-3
         ..GP..TIKAK..ET.KNA.AATPKP..HI.EQ..AETG.NM.VIQV......S.ERRMK    lin-11
         ..GP..TIKAK..ET.KAA.AATPKP..HI.EQ..QETG.NM.VIQV......S.ERRMK    Xlim1
         NK.P..TISAKS.ET.KQAYQTSSKPA.HV.EQ..SETG.DM.V.QV......A.E.RLK    ceh-14
         .AK..ET.KNAYDNSPKPA.HV.EQ.SSETG.DM.V.QV                         Xlim3

         DK.L..TI.PE..EI.YQKYLLDSNP..KMLDHI..EVG.KK.V.QV....T.ARER.GQ    ATBF1 HD3
         NK.L..TILPE..NF.YECYQSESNPS.KMLE.ISKKV..KK.V.QV....S.A.D..SR    zfh-2 HD3
         QK.F..QM.NL..KV.KSC.NDY.TP.MLECEV.GNDIG.PK.V.QV....A.A.E..SK    ATBF1 HD4
         KV.V...INEE.QQQ.KQHYSL.ARPS.DEFRMI.AR.Q.DP.V.QV....N.SRER.MQ    zfh-1
         NK.P..RI.DD..RV.RQY.DI.NSPSEEQIK.M.DKSG.PQKVI.H..R.TLF.ERQR.    ATBF1 HD1
         QK.A..RI.DD..KI.RAH.DI.NSPSEESIM.MSQKA..PMKV..H..R.TLF.ERQR.    zfh-2 HD1
         K.SS..RF.D...RV.QDF.DA.A.PKDDEFEQ.SNL...PT.VIVV....A.Q.AR.NY    ATBF1 HD2
    7G   K.AN..RF.D..IKV.QEF.EN.S.PKDSDLEY.SKL.L.SP.VIVV....A.Q.QR.IY    zfh-2 HD2
```

■ The HOM-C/HOX cluster and related homeobox genes

The HOX cluster and the Hexapeptide superclass

The first, and also best known homeobox genes are the homeotic selector genes of *Drosophila melanogaster*. These genes are organized into two complexes, the Antennapedia complex (ANT-C) and the Bithorax complex (BX-C), summarily referred to as the homeotic gene complex HOM-C (Akam, 1989). Four evolutionarily related complexes, the *HOX* clusters, have been found in vertebrates (see e.g. Akam, 1989; Boncinelli et al., 1991; McGinnis and Krumlauf, 1992). The role of the genes in the HOM-C/HOX cluster in pattern formation along the anterior-posterior axis of animals has been discussed in

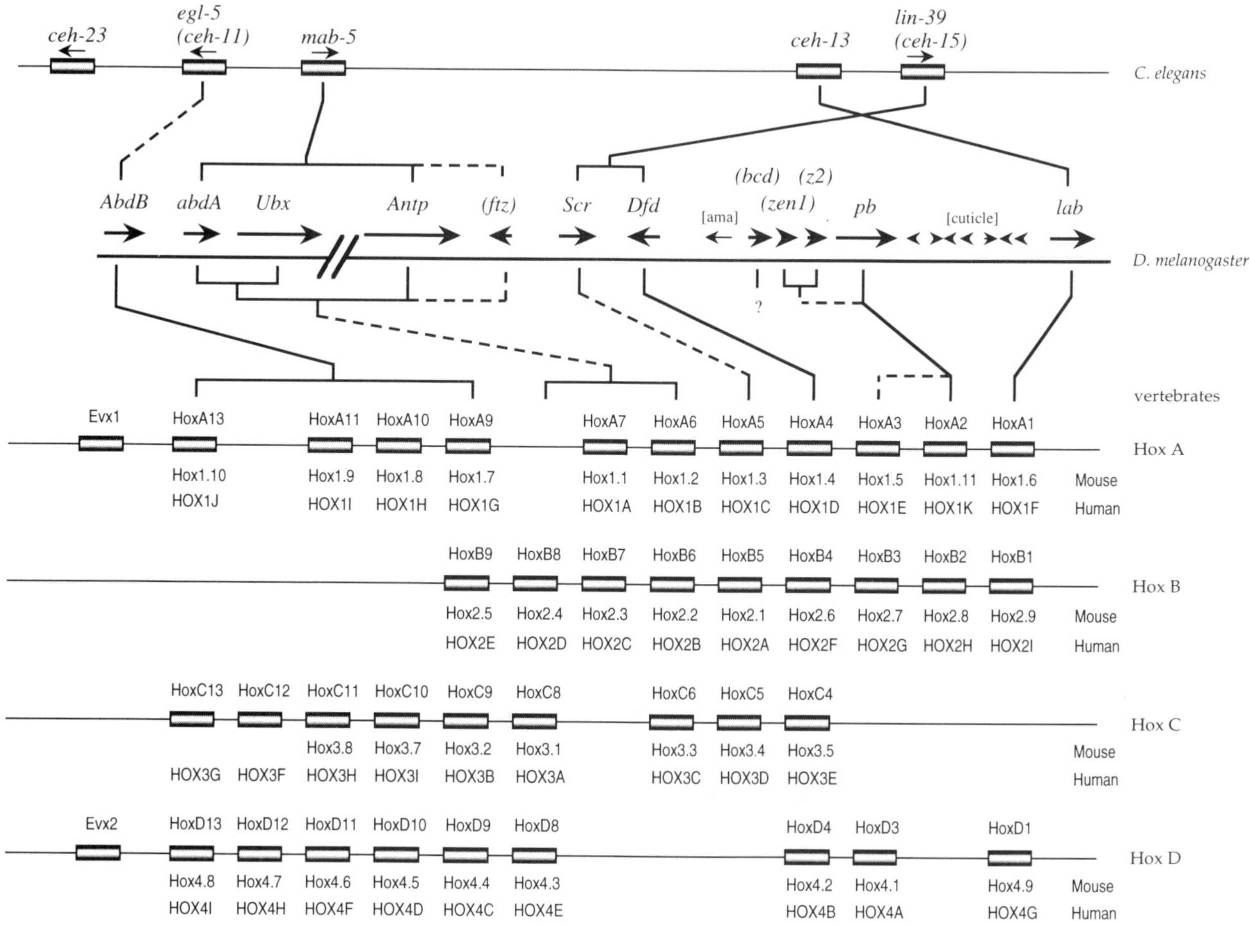

Figure 8: Organization of the *C. elegans* cluster, the *Drosophila melanogaster* HOM-C cluster, and the vertebrate *HOX* cluster. The organization of the 4 mammalian *HOX* clusters is shown in the lower part of the figure. The new nomenclature (Scott, 1992) is shown above the schematic representation of the homeobox genes (cylinders), while the old nomenclature is shown beneath each cluster for human and mouse. Above the vertebrate clusters is a representation of the *Drosophila* ANT-C and BX-C clusters. Large arrows indicate the individual transcription units of homeobox genes, while small arrows represent non-homeobox genes and their names are in square brackets. Gene names of non-homeotic homeobox genes are in parentheses. At the top is a schematic representation of the homeobox cluster in *C. elegans*. The distance between *ceh-13* and *mab-5* is about 250 kb, and since *C. elegans* genes are relatively small and closely placed (Sulston et al., 1992), there are probably many non-homeobox genes in that interval. Black lines and brackets between the *C. elegans*, *D. melanogaster* and mammalian clusters mark related homeodomain sequences. Dashed lines indicate less certain sequence relationships.

```
consensus    IYPWMK                    consensus    IYPWMK

cad         SKPPYFDWMKKPAYPA          Antp        MPSPLYPWMRSQFGKC
m cdx-1     PRRTPYEWMRRSVAAA          Ubx         SNHTFYPWMAIAGECP
sh cdx-3    QRRNLCEWMRKPAQPS          Dps Ubx     SNHTFYPWMAIAGECP
c cdxA      QRRTPYEWMRRSIPST          Df Ubx      GNHTFYPWMAIAGEST
Xcad1       HRRHHYEWLRKPVPQA          abd-A       ADLPRYPWMTLTDWMG
Xcad2       QRRNPYEWMRRTGVPT          m HoxC6     ASIQIYPWMQRMNSHS
pal-1       YQQSVWPFMDYQQFQG          h HoxC6     ASIQIYPWMQRMNSHS
                                      x HoxC6     GSIQIYPWMQRMNSHS
ems         DSYQLYPWLLSRHGRI          Nv HoxC6    TSIQIYPWMQRMNSHS
h EMX1      DPLHFYPWVLRNRFFG          m HoxB6     CSTPVYPWMQRMNSCN
h EMX2      DPSTFYPWLIHRYRYL          h HoxB6     CSTPVYPWMQRMNSCN
                                      zf-22       PSAPVYPWMQRMNSCN
lab         SSIPTYKWMQLKRNVP          x HoxB7α    ANLRIYPWMRSAGADR
x HoxD1     SYVSTFDWMKVKRNPP          x HoxB7β    ANLRIYPWMRSAGSDR
m HoxA1     SPAQTFDWMKVKRNPP          h HoxB7     SNFRIYPWMRSSGTDR
x HoxA1     GPTQTFDWMKVKRNPP          m HoxB7     SNFRIYPWMRSSGPDR
c HoxB1     SRARTFDWMKVKRNPP          x HoxA7     SHFRIYPWMRSSGPDR
h HoxB1     PTARTFDWMKVKRNPP          c HoxA7     ANFRIYPWMRSSGPDR
m HoxB1     LTPRTFDWMKVKRNPP          m HoxA7     ASFRIYPWMRSSGPDR
                                      qu HoxA7    GQFRIYPWMRSSGPDR
XlHbox8     RTLLPFPWMKSTKSH           c HoxB8     SPTQLFPWMRPQAAAG
                                      m HoxB8     SPTQLFPWMRPQAAAG
pb          DSVPEYPWMKEKKTSR          x HoxB8     SPTQLFPWMRPQAAGR
h HoxB2     PPAPEFPWMKEKKSAK          c HoxD8     SPAQMFPWMRPQAAPG
m HoxB3     LTKQIFPWMKESRQTS          m HoxD8     SPSQMFPWMRPQAAPG
h HoxB3     LTKQIFPWMKESRQTS          m HoxC8     SPSLMFPWMRPHAPGR
h HoxD3     ISKQIFPWMKESRQNS
zf-114      LSTMKYPWMRETHAPT          TgHbox1     KSTSGYPWMPVAGPNV
AmphiHox3   ANTKIYPWMKESRQNS          mab-5       ISQPVFPWMKMGGAKG
                                      smox-1      SNVVVYPWMNPKGTDI
Dfd         GERIIYPWMKKIHVAG          ftz         NGAGDFNWSHIEETLA
c HoxD4     QPAVVYPWMKKVHVNS
m HoxD4     QPAVVYPWMKKVHVNS          TCL-3       LTGLTFPWMESNRRYT
h HoxD4     QPAVVYPWMKKVHVNS          MUR10F      LAGLTFPWMDSGRRFA
m HoxA4     KEPVVYPWMKKIHVSA
h HoxA4     KEPVVYPWMKKIHVSA          -------------------------
c HoxA4     KEPVVYPWMKKIHVST
c HoxB4     KEPVVYPWMKKVHVST          m msx-1     EKLDRTPWMQSP-RFSP
m HoxB4     KEPVVYPWMRKVHVST          h MSX1      EKPERTPWMQSP-RFSP
h HoxC4     KQPIVYPWMKKIHVST          c msx-1        GRSPWMQSP-RFSP
m HoxC4     KQPIVYPWMKKIHVST          x msx-1     ESPERSSWIQSP-SFSP
x HoxB4     QDPVVYPWMKKAHISK          x msx-2     NSEDGTSWSKDGGSYSP
                                      m msx-2     NSEDGAPWIQEPGRYSP
Scr         NPPQIYPWMKRVHLGT          qu msx-2    NSEDGTSWIQEAGRYSP
m HoxA5     AQPQIYPWMRKLHISH          c msx-2     NSEDGTSWIQEAGRYSP
h HoxA5     AQPQIYPWMRKLHISH          zf msh-c    SDKGQSGWFQTT-SYTS
m HoxB5     QTPQIFPWMRKLHISH          zf msh-d    EPDDCTSWVMNS-RYSQ
h HoxB5     QTPQIFPWMRKLHISH
x HoxB5     QSPQIFPWMRKLHINH          MATa1       NRLEIYHHIKKEKSPK
zf-21       QAPQIFPWMRKLHISH          Cv cnox2    ERSRLYPITSPLHGQR a
h HoxC5     APPQIYPWMTKLHMSH          Hv cnox2    DRSRLFPILSPLHGQR a
                                      Cv cnox2    GASSQYPYTREPHEHS b
                                      Hv cnox2    GTSSQYPYTREPHEHT b
```

Figure 9: Compilation of Hexapeptide sequences.
The top line shows a consensus for the Hexapeptide. The *C. elegans* gene *pal-1* is a member of the *cad* class of homeobox genes, and thus a Hexapeptide sequence might be expected upstream of the homeodomain. The best fit with the Hexapeptide that can be found in the appropriate position upstream of the homeodomain is shown. If this sequence does perform the function of the Hexapeptide, it would appear that phenylalanine and tryptophan have exchanged positions; this might be feasible, given that both are large hydrophobic amino acids. A remnant of the Hexapeptide can be found in the appropriate place in the *Drosophila* gene *ftz*. Sequences upstream of the homeodomain with weak similarity to the Hexapeptide can be found in genes of the *msh* class, *MATa1*, and *cnox2* (marked with "a"), but the significance is dubious. In *cnox2* a second region a little further upstream shows also some vague similarities to the Hexapeptide (marked "b").

other places (e.g. Kaufman, 1990; Gehring, this book; De Robertis, this book). The evolution of these clusters has received wide interest, and has been subject to detailed analysis (e.g. Kappen et al., 1989; Krumlauf, 1992; Holland, 1992; Duboule, 1992; Schubert et al., 1993). Figure 8 shows the *Drosophila* and vertebrate clusters, including the new nomenclature for the HOX genes (Scott, 1992). The genes in the four vertebrate clusters can be aligned such that 13 groups are formed that are highly related by sequence. These groups are referred to as paralog groups (PG-1 to PG-13). Each cluster appears to be missing genes for some of the paralog groups, suggesting that the vertebrate cluster duplication took place after all 13 groups had formed, and that during or after the duplication events some genes were lost. The lines and brackets between the fly and vertebrate clusters indicate how the genes are related between the two clusters. A cluster containing five homeobox genes has also been found in the nematode *C. elegans* (Bürglin and Ruvkun, 1993; Kenyon and Wang, 1991). Possible correspondences to the fly cluster are indicated in Figure 8. Several searches for homeobox genes have been performed that had been designed to detect *Antp* class genes (Kambs et al., 1989; Bürglin et al., 1989; Hawkins and McGhee, 1990; Schaller et al., 1990; Naito et al., 1992). Thus it becomes increasingly less likely that many additional *Antp* class genes in that region will be detected. This is in agreement with the notion that the common ancestor of vertebrates and arthropods contained only about five to six homeobox genes (Kappen et al., 1989; Schubert et al., 1993).

The genes in the HOM-C/HOX cluster have been grouped into three classes: *Antp*, *Abd-B* and *lab*. Genes of the *Antp* and *lab* classes encode a short conserved peptide motif upstream of the homeodomain. Various names have been given to this sequence, originally referred to as the pentapeptide (Mavilio et al., 1986), but I will refer to it as the "Hexapeptide", since the main conserved core is 6 amino acids long (Fig. 9). Other classes, whose genes are not localized to the cluster, also encode the Hexapeptide motif. Classes found in this Hexapeptide superclass are: *Antp*, *lab*, *cad*, *ems* and *TCL*. The Hexapeptide is usually separated by an intron from the homeodomain.

Antp class

The homeodomains of the genes in the center of the HOM-C and HOX clusters are very similar to each other and are grouped into the *Antennapedia* (*Antp*) class of genes. One can subdivide the *Antp* class into several families, which often possess additional sequence conservation outside of the homeodomain. In particular, the amino terminal region of several families is conserved even between different families (see e.g. De Robertis et al., 1989; Scott et al., 1989; Krumlauf, 1992).

Dfd family: This family is formed by *Deformed* (*Dfd*) and the vertebrate PG-4 group. The homeodomain of the *C. elegans* gene *ceh-15* displays marginally more sequence similarity to the *Dfd* family (max. 77% identity) than to

the other families. However, it is not yet clear whether its most recent common ancestor amongst the HOM-C cluster sequences is indeed with *Dfd*, since *Dfd* is more similar to *Antp* (82%), and *ceh-15* contains an intron at position 44/45 like *pb* and *lab*, while the *Dfd* family has lost the intron.

Scr family: Examination of Figure 6 shows the homeodomain of *Scr* clusters with those of the PG-5 genes. However, other types of analysis do not clearly group *Scr* with the PG-5 genes (Schubert et al., 1993) and it is not yet certain if a common ancestor existed before vertebrates and arthropods separated. In the meantime, I will consider these genes a separate family here.

Antp family: *Antennapedia* (*Antp*), *Ultrabithorax* (*Ubx*), *abdominal-A* (*abd-A*), *fushi tarazu* (*ftz*), PG-6, PG-7, and PG-8. This family of genes encodes homeodomains that are extremely conserved between flies and vertebrates (up to 98%). It is assumed that there existed only one common ancestral gene, and that independent duplication events gave rise to the PG-6, PG-7 and PG-8 genes in vertebrates and to the *Antp*, *Ubx*, *abd-A* and *ftz* genes in flies (Akam, 1989; Kappen et al., 1989; Krumlauf, 1992; Schubert et al., 1993). This is supported to some extent by the fact that the leech gene *lox2* seems to be equally related to *abd-A* and *Ubx* (Wysocka-Diller et al., 1989; Shankland et al., 1991).

pb family: The vertebrate paralog group PG-2 is closely related to the *Drosophila* gene *pb*. A second vertebrate group, PG-3, appears to have arisen as a duplication of PG-2.

Xlox family: The *Xenopus* gene *Xlhbox8* is most closely related to several leech genes (*lox*) thus justifying the grouping of these genes into a new family, called "*Xlox*". The homeodomains encoded by these genes belong to the *Antp* class, and *Xlhbox8* contains a Hexapeptide upstream of the homeodomain (Fig. 9). While the chromosomal location of these genes is not known with respect to their respective HOM-C/HOX clusters, it is likely that these genes are not located in the cluster because none of the known genes in the cluster belong to this family.

Gsh genes: Recently, two mouse genes (*Gsh-1*, *Gsh-2*) have been isolated that do not map to the HOX cluster (Singh et al., 1991), yet they do belong to the *Antp* class of genes. These genes might possibly be derived from paralog group 2 or 3, or might even be the descendents of *HoxC2* or *HoxC3* (which seem to be missing in the HOX cluster). Since no related genes from other phyla are known currently, I am not yet considering them a separate family.

Abd-B class

A second class within the HOM-C/HOX cluster is the *Abdominal-B* (*Abd-B*) group of genes. Since the *Abd-B* class genes are more divergent (see Figure 3) than other cluster genes and since they do not have a Hexapeptide, it appears justified to treat these genes as a different class (see also Akam, 1989; Izpisúa-Belmonte et al., 1991; McGinnis and Krumlauf, 1992). A remnant of the Hexapeptide may exist however, as all genes from the PG-9, 10 and 12, as well as *Abd-B* itself have a Trp residue at position -6 or -7 from the homeodomain (Izpisúa-Belmonte et al., 1991). While *Drosophila* has only one gene, *Abd-B*, there are 5 paralog groups, PG-9 to PG-13, in vertebrates. The *Abd-B* homeodomain is 52 to 75% identical to the various human and mouse *Abd-B* class homeodomains. The fly and mammalian homeodomains in turn, are 43 to 72% identical to *Antp* and various other *Antp* class genes. The further removed from the center of the HOX cluster the paralog groups are (PG-9 to PG-13), the more divergent their homeodomain sequences are. In *C. elegans* the one gene which possibly could correspond to the *Abd-B* class genes based on its position is *ceh-11*. It is highly divergent, but so are the PG-12 and PG-13 genes. There are a few amino acids in common between *ceh-11* and *Abd-B* (Kenyon and Wang, 1991), although these residues are not conserved in the vertebrate *Abd-B* class genes. A few of these residues are present in the *eve* class genes, and since the *eve* class genes, at least in vertebrates, are next to the *Abd-B* genes in the HOX cluster, this might be an indication that the *eve* class genes were indeed derived from the *Abd-B* class (see also *eve* section).

lab class

The *labial* (*lab*) group of genes, at the other end of the cluster, are about equally divergent from central cluster genes as are the genes of the *Adb-B* class and are sufficiently distinct to form their own class (Akam, 1989). The *lab* homeodomain is 80 to 85% identical to the PG-1 genes, and the homeodomains of the *lab* class are 55-67% identical to *Antp* class genes. The *lab* class genes have distinct Hexapeptides, which are more similar to the *cad* class Hexapeptides than to *Antp* class Hexapeptides (Fig. 9). The chicken *lab* class gene *CHbx1* does not seem to match any of the known mouse or human genes of PG-1 and might thus not be localized to the cluster and/or might be the "missing" *HoxC1* gene.

HOM-C cluster curios

A few genes in the *Drosophila* HOM-C cluster are not involved in axial patterning (i.e. *ftz, zen1, zen2, bcd*). The *ftz* homeodomain is similar enough to *Antp, Ubx* and *abd-A* to be considered part of the *Antp* family. It seems clear, that *ftz* must have arisen from one of the *Antp* family genes by a duplication event. Concomitant with a change in function, the gene seems to be undergoing or has undergone more rapid changes in sequence than other genes in the cluster, since only a very cryptic Hexapeptide (Fig. 9) is found upstream of the homeodomain. Furthermore, a *ftz* homologue might not even be present in honeybees (Walldorf et al., 1989), suggesting a relative recent emergence of this gene.

zen1 and *zen2* are a pair of genes also located in the cluster that appear to have taken on novel roles during development. While it seems clear that they are derived from the *pb* family (based on position in the cluster, Fig. 8, and sequence similarity, Figs. 3 and 6), it is not apparent whether they correspond to the ancestor of the PG-3 group, or if they arose by an independent duplication event from the PG-2 group. *bcd* is the most divergent gene in the HOM-C cluster and it is not clear from which other cluster gene it might have arisen. Given that *bcd* has a very special role during early embryogenesis in *Drosophila* (see *bcd* entry in this book), it is not surprising that it may have undergone rapid evolutionary divergence for this specialized function, such that its homeodomain is only about 45% identical to other genes in the HOM-C/HOX cluster. One cannot completely exclude though that *bcd* has "jumped" into the cluster and is not derived from the HOM-C cluster.

HOX cluster related genes in *Cnidarians*

In hydra, several homeobox genes have been isolated, some of which are most closely related to the genes of the HOM-C and HOX cluster (Schierwater et al., 1991; Murtha et al., 1991; Schummer, 1992; Shenk et al., 1993). However, since only partial sequence information is available for some of the genes, and because of the large evo-

```
m HoxB4    EPVVYPWMRKV--------------HVSTVNPNYAGGEPKRSRTAYTRQQVLELEKEF
            :  | |                       |   :||||||:| ||:|||   ||
h HB9      AGMILPKMPDF---------------NSQAQSNLLGKCRRSRTAFTSQQLLEL--EF
            | | | :                      |        :| |||:|| ||:||   ||
h HOXB3    SKQIFPWMKESRQNSKQKNSCATAGESCEDKSPPGPASKRVRTAYTSAQLVELEKEF


m HoxB4    HYNRYLTRRRRVEIAHALCLSERQIKIWFQNRRMKWKKDHKLPN
            :|:||:| :| |:|  | |:| |:||||||||||||:   |
h HB9      KFNKYLSRPKRFEVATSLMLTETQVKIWFQNRRMKWKRSKKAKE
            ||:||:||:| |:|   |  ||| |:|||||||||| |:   |||
h HOXB3    HFNRYLCRPRRVEMANLLNLTERQIKIWFQNRRMKYKKDQKAKG
```

Figure 10: Alignment of human *HB9* with mouse *Hoxb-4* and human *HOXB3*
HB9 is a gene containing a homeodomain with not more than 56% identity to any other homeodomain. An improved alignment can be obtained by introducing a 2 amino acid gap in the human *HB9* gene, increasing the identity to 68% (compared to *pb*). Also apparent is a cryptic Hexapeptide upstream of the homeodomain. The homeodomain of human *HOXB3* is underlined. Vertical lines indicate identities and colons indicate conservative changes between sequences. Conserved residues are grouped as follows: I,V,L,M; K,R; E,D; T,S; F,Y,W.

lutionary distance between hydra and flies, the assignments of these genes to particular classes and families in the HOM-C/HOX cluster should be regarded only as tentative. For example, in this analysis *cnox2* is more similar to the *Gsh-1/Gsh-2* genes than to the *Dfd* family. Further information about these genes and the hydra cluster (if there is a cluster) is needed to corroborate any assignments, in particular since some hydra genes might not belong to a cluster, like the *Gsh-1/Gsh-2* genes in vertebrates.

A region of similarity to the Hexapeptide in *cnox2* (Schummer et al., 1992) is not absolutely conserved between *Hydra vulgaris* and *Chlorohydra viridissima* (Fig. 9, marked "a"), although this gene is otherwise highly conserved between the two species. A somewhat better match to the Hexapeptide can be found a little further upstream (Fig. 9, marked "b"), but additional evidence is necessary to confirm if any of these regions are indeed the functional equivalent of the Hexapeptide. The presence of *Antp* class genes in *Cnidarians*, and of *eve* and *msh* class genes (see below), suggests that members of other classes remain to be found.

eve class

The vertebrate *even-skipped* (*eve*) class genes are located at the posterior end of the HOX cluster (Fig. 8), suggesting that this class of genes arose also from genes in the cluster. In *Drosophila* though, the *eve* gene is not linked to the HOM-C cluster. Like the *Abd-B* class, these genes do not contain a Hexapeptide upstream of the homeodomain (nor do they contain any other substantial regions of conservation outside of the homeodomain). Their homeodomain sequences are clearly very different from the *Antp* and *Abd-B* class homeodomains, and in addition, the intron position (see Fig. 22) is in a different location than in other genes in the cluster. Thus if the *eve* class is indeed evolutionarily most closely related to the genes in the HOM-C/HOX cluster, it has clearly undergone substantial changes. An *eve* class gene has been isolated from corals (Af *eveC*, Miles and Miller, 1992) indicating that this class is very ancient.

An evolving homeobox gene?

A highly divergent human homeobox gene, *HB9*, was described by Deguchi and Kehrl (1991) that was only about 50% identical to other genes (Fig. 10). A much improved alignment can be obtained by introducing a 2 amino acid gap into *HB9*. While it is possible that the two missing amino acids are due to alternative splicing, as in the case of *tI-POU* and *I-POU*, (Treacy et al., 1992), none of the *Antp* class genes have an intron at that position. The introduction of the gap increases the identity to 68% compared to *pb*. In addition, a cryptic Hexapeptide appears to be present upstream of the homeodomain. Consequently, this homeobox gene appears to have been derived from some *Antp* class gene, and possibly might even be one of the "missing" genes of the HOX cluster (e.g. *Hoxd-5* or *Hoxd-6*, Fig. 8) that is in the process of evolving rapidly.

■ Non-cluster members of the Hexapeptide superclass

cad class

The *caudal* (*cad*) class of homeobox genes encodes a Hexapeptide upstream of the homeodomain, but is not located in the HOM-C/HOX cluster. The function in *Drosophila* and the expression patterns in vertebrates indicates that these selector genes are important in axial pattern formation (for review see McGinnis and Krumlauf, 1992). The Hexapeptides are quite similar to the Hexapeptides of the *lab* class, possibly indicating that the *cad* genes were derived from the *lab*/PG-1 end of the cluster. So far 4 genes of this class have been identified in vertebrates (Figs. 6,7). The *C. elegans* gene *pal-1*, while clearly a member of the *cad* class, does not have a readily recognizable Hexapeptide upstream of the homeodomain (see Fig. 9 and legend).

ems class

Like the *cad* class, the genes of the *empty-spiracles* (*ems*) class have a Hexapeptide and they function in axial pattern formation mainly in the head region. There are at least two genes in mammals and two in flies, which appear to have arisen from independent duplication events (see Fig. 6).

TCL class

The human gene *TCL-3* is one of the better known cases where a homeobox-containing gene has been implicated in oncogenesis (Hatano et al., 1991; Kennedy et al., 1991; Lu et al., 1991). A highly related gene has been isolated from *Drosophila* (M. Frasch and W. McGinnis, personal communication). This group of genes is referred to as the "*TCL*" class. A similar gene has also been isolated from mouse (*MUR10F*), but since that homeodomain is less similar to *TCL-3* than the *Drosophila* gene, there might be at least two different homeobox genes of the *TCL* class in vertebrates. *TCL-3* and *MUR10F* encode Hexapeptides that are somewhat divergent from Hexapeptides of other classes (Fig. 9).

■ *PRX* superclass

prd class

The *paired* (*prd*) class of homeobox genes is characterized by the fact that the homeodomains of these genes contain a serine residue at position 9 of helix 3. In addition, a highly conserved domain of about 130 amino acids, the "*paired*" domain (Bopp et al., 1986), is found at the amino termini of these genes. The *prd* domain has been shown to bind DNA (Treisman et al., 1991), and recently it has been shown that the transcription factor BSAP is encoded by *Pax-5* (Adams et al., 1992). This domain has been found to occur without an associated homeodomain (Bopp et al., 1989; Chalepakis et al., 1991). In contrast, thus far all homeodomain proteins of this class (containing a serine residue at position 9 of helix 3) contain also *prd* domains. The *prd* class homeodomains show in addi-

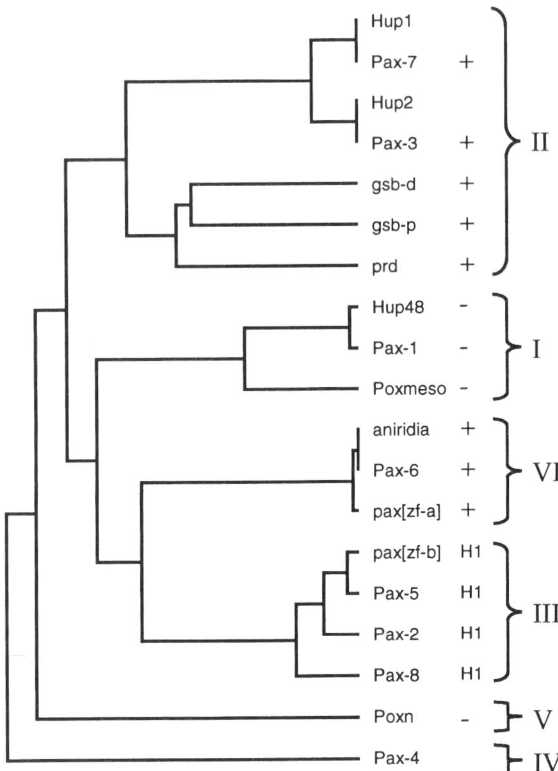

Figure 11: Comparative tree of *paired* domain sequences. The comparative tree is based exclusively on the *paired* domain. Several families can be distinguished, and they have been numbered I-VI (Burri et al., 1989, Walther et al., 1991). The "+" indicates genes that encode a homeodomain, the "-" indicates genes that do not encode a homeodomain, and the "H1" indicates genes that encode a short region of similarity to the helix 1 of the homeodomain.

tion some extended sequence conservation just upstream of the homeodomain (Bopp et al., 1986; Goulding et al., 1991). Figure 11 shows a comparative tree based solely on the *prd* domain and includes genes that do not encode a homeodomain. The aligned sequences of the *prd* domains are shown in Figure 12. The *prd* class genes can be grouped into 6 families based on this domain (Burri et al., 1989, Walther et al., 1991), but for several of these families no homologues in *Drosophila* have been found yet, and it is thus not clear when these families have emerged during evolution.

Family I: The genes of this family, formed by mouse *Pax-1* and *Drosophila Pox meso*, do not encode homeodomains.

Family II: This includes the *Drosophila* genes *prd*, *gsb-d*, *gsb-p* and the mouse genes *Pax-3* and *Pax-7*. Apparently independent duplication events gave rise to the multiple genes in vertebrates and flies. This family contains a small conserved region (the Octapeptide) between the homeodomain and the *prd* domain (Burri et al., 1989)

Family III: So far defined only by vertebrate genes (*Pax-2*, *Pax-5*, *Pax-8*). Interestingly, these genes encode a remnant of a homeodomain (Krauss et al., 1991b; Walther and Gruss, 1991). This suggests that the homeodomain was only recently lost in evolution. Like in family II, these genes also encode an octapeptide between the *prd* domain and the homeodomain remnant.

Family IV: Defined solely by *Pax-4*.

Family V: Only defined by *Pox neuro*.

Family VI: Defined by several vertebrate genes containing a homeobox.

prd-like class

There are a number of genes whose homeodomain sequences are similar to the *prd* class homeodomains, but they do not encode a *paired* domain, nor do the homeodomain sequences have a serine residue at position 9 of helix 3. Given that these homeodomains are in average about 55-75% identical to the *prd* class homeodomains, these genes have been grouped into a "*prd*-like" class. The *prd*-like class and the *prd* class are grouped together into the "*PRX*" superclass. While *prd*-like class genes have been isolated from *Drosophila*, vertebrates and *C. elegans*, almost none of them can be identified as homologues and grouped into families. The one clear exception so far is the *otd* family, since recently genes closely related to *Drosophila otd* have been cloned from mouse and humans (Simeone et al., 1992). Another recently described *Drosophila* gene, *al* (Schneitz et al., 1992), might be the homologue of the *Schistosoma* gene *smox-3* (Webster and Mansour, 1992). It seems quite feasible that a number of these genes remain to be identified in various organisms. While most of the *prd*-like genes encode a glutamine residues at position 9 of helix 3, several genes (e.g. *gooscoid*, *otd*) encode a lysine residues at that position, but it is not clear if these genes are evolutionarily related or if independent events gave rise to that lysine residue. A feature quite common to *prd*-like class genes appears to be the presence of an intron between amino acids 46 and 47 (Fig. 20).

■ Other classes

en class

The *engrailed (en)* class of genes is characterized by a series of small conserved domains outside of the homeodomain (Fig. 13). These domains were named EH1, EH2, EH3, EH4 (the homeodomain) and EH5 (Joyner and Hanks, 1991). EH1 is a small conserved region (~13 amino acids) towards to amino terminus of en class genes, EH2 and EH3 are adjacent conserved regions of about 20 and 8 amino acids just upstream of the homeodomain. EH2 contains residues that are somewhat reminiscent of the Hexapeptide. EH5 is a carboxy-terminal extension of the

```
Consensus   MPLGQGGVNQLGGVFVNGRPLPNHIRQKIVELAHHGIRPCDISRQLRVSHGCVSKILGRYYETGSIRPGA
            ◇••••• •◇•••••••    •   •• ◇•  • • ••• ••• • •• ••••••• •◇  •• ◇ •

HuP1        T.....R.........I.........H....M.........V.................C..Q.........
m Pax-7     T.....R.........I.........H....M.........V.................C..Q.........
HuP2        T.....R.........I.........H....M.........V.................C..Q.........
m Pax-3     T.....R.........I.........H....M.........V.................C..Q.........
gsb-d       PFQ...R.........I.........RQ...M.AA.V...V.................N.FQ.........V
gsb-p       PFQ...R.........I.........L....M.AS.V...V.................N..Q.........V
prd         .NS...R.........I........N..L....M.AD.....V...............N..Q.........V
HuP48       AEQTY.E..............A..LR.....QL.........................A..N.....L...
m Pax-1     .EQTY.E..............A..LR.....QL.........................A..N.....L...
poxmeso     QCPQY.E..............AT.MR.....RL.........................A..H.....L...
aniridia    .QNSHS..............DST.........S.A.......I.Q..N.................R.
m Pax-6     .QNSHS..............DST.........S.A.......I.Q..N.................R.
pax[zf-a]   .QNSHS..............DST.........S.A.......I.Q..N.................R.
pax[zf-b]   AMHRH...............DVV..R.....Q.V.......................K..V
m Pax-5     IRT.H...............DVV..R.....Q.V.......................K..V
h Pax-5     SRT.H...............DVV..R.....Q.V.......................K..V
m Pax-2     AMHRH...............DVV..R.....Q.V.......................K..V
m Pax-8     IRS.H..L.....A.......EVV..R..D...Q.V....................V
poxn        PHT..A..............DCV.RR..D..LC.V............L........T.F.........S
m Pax-4      LSS......L.........LDT..Q..Q..IR.M.......S.K..N..........R..VLE.KC
            .........|.........|.........|.........|.........|.........|.........|
                    10        20        30        40        50        60        70


Consensus   IGGSKPRQVATPDVVKKIAEYKRENPGMFAWEIRDRLLKEGVCDNDTVPSVSSISRILRNKFGK
            ••••• ◇ ◇ ◇• ◇    •    ◇      ◇• •••    •        •• • • •◇◇•

HuP1        ..............E...E.........S.........D.H..RS...........V..I....
m Pax-7     ..............E...E.........S.........D.H..RS...........V..I....
HuP2        ......K..T...E...E.........S....K...DA...RN..............S....
m Pax-3     ......K..T...E...E.........S....K...DA...RN..............S....
gsb-d       ......-.....IESR.E.L.QSQ..I.S....AK.IEA....KQNA.........L..GSS.S
gsb-p       ......K-.TS.EIETR.D.LRK...SI.S....EK.I...FA.---P..T.....L..GSDRG
prd         ........-I...EIENR.E....SS....S....EK.IR.....RS.A....A...LV.GRDAP
HuP48       ........-.T..N...H.RD..QGD..I..........AD....KYN.............I.S
m Pax-1     ........-.T..N...H.RD..QGD..I..........AD....KYN.............I.S
poxmeso     ........-.T..K..NY.R.L.QRD..I.........S..I..KTN..............L.S
aniridia    ........-....E.S...Q....C.SI.........S....T..NI......N.V...LASE
m Pax-6     ........-....E.S...Q....C.SI.........S....T..NI......N.V...LASE
pax[zf-a]   ........-....E..G...Q....C.SI.........S....T..NI......N.V...LASE
pax[zf-b]   ......K-....K..E......Q..T..........A..............N..I.T.VQQ
m Pax-5     ......K-....K..E......Q..T..........A.R............N..I.T.VQQ
h Pax-5     ......K-....K..E......Q..T..........A.R............N..I.T.VQQ
m Pax-2     ......K-....K..D......Q..T..........A..I...........N..I.T.VQQ
m Pax-8     ......K-....K..E..GD..Q..T..........A..............N..I.T.VQQ
poxn        .....TK.....T.....IRL.E..S........EQ.QQQR...PSS...I...N.....SGLW
m Pax-4     ........-L...A..AR..QL.D.Y.AL.....QHQ.CT..L.TQ.KA......N.V..ALQ
            .........|.........|.........|.........|.........|.........|....
                    80        90       100       110       120       130
```

Figure 12: Alignment of *paired* domain sequences.
References for *paired* box genes not containing a homeobox: *Pax-1* ([Chalepakis et al., 1991); *Pax-2* (Dressler et al., 1990; Adams et al., 1992); *Pax-4* (Walther et al., 1991); h and m *Pax-5* (Walther et al., 1991; Adams et al., 1992); *Pax-8* (Plachov et al., 1990); *pax[zf-b]* (Krauss et al., 1991a; Krauss et al., 1991b); *poxn* (*pox neuro*) and *pox meso* (Bopp et al., 1989); *HuP1, HuP2, HuP48* (Burri et al., 1989).

homeodomain of about 20 amino acids that is followed by a charged region. The current evidence suggests that one ancestral *engrailed* gene existed in the ancestor of vertebrates and arthropods, and that later independent duplication events gave rise to *en* and *inv* in flies and *En-1* and *En-2* in vertebrates (Joyner and Hanks, 1991; Holland, 1992). It is interesting to note that the two *engrailed* genes in honeybee appear more closely related to *en* than to *inv*, suggesting that a further duplication might have occurred in that species.

POU class

The POU class was originally defined based on four different genes: *Pit-1, Oct-1, Oct-2* and *unc-86* (Herr et al., 1988). The POU-specific domain is an approximately 80 amino acid long conserved domain upstream of the homeo-

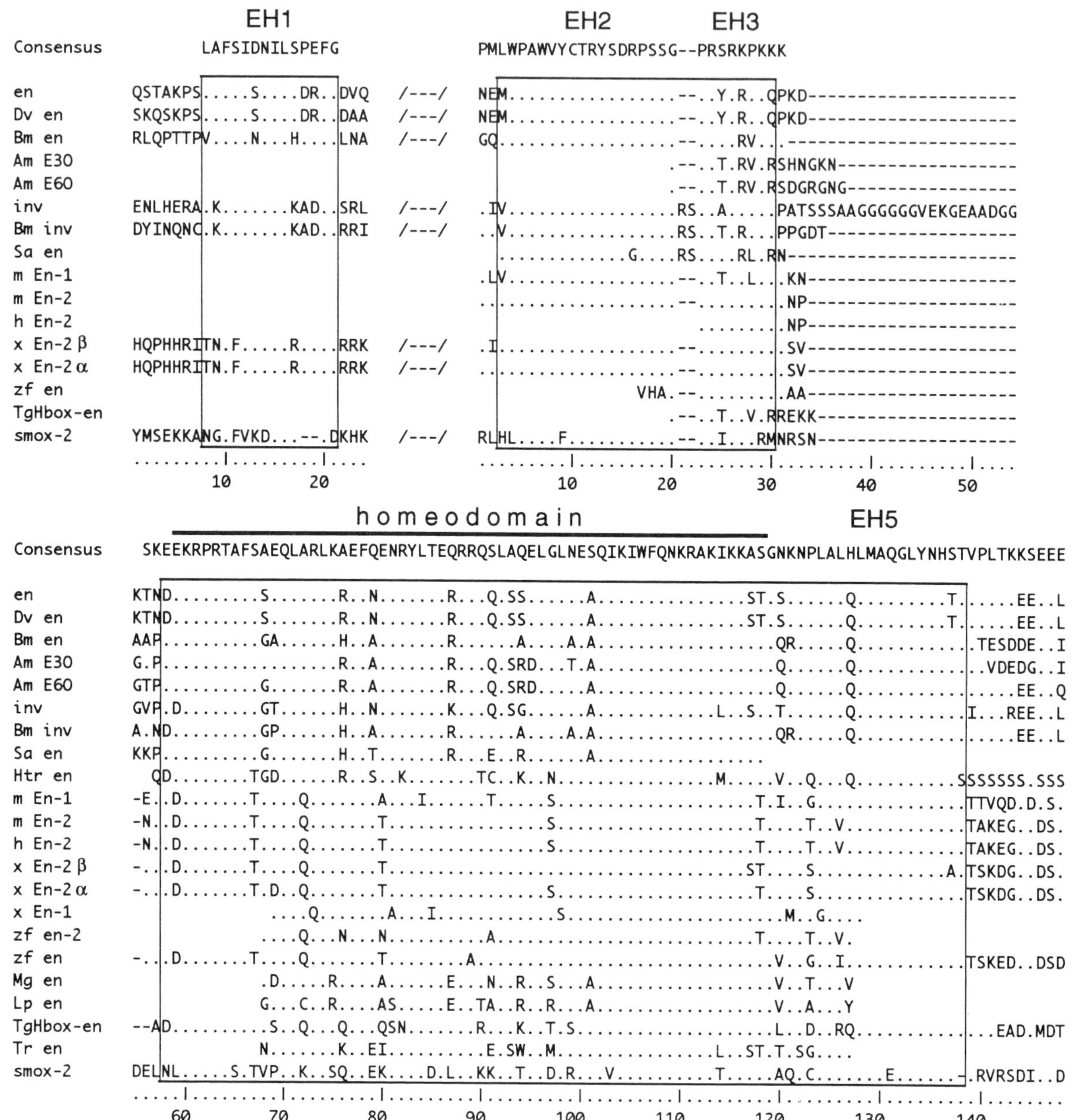

Figure 13: Sequence alignment of *en* class genes.
The black bar denotes the homeodomain, and the boxes indicate the domains that have been conserved between *Schistosoma*, vertebrates and flies. The slashes indicate a long non-conserved region of variable length. Two conserved regions can be found upstream of the homeodomain, and a third region of conservation is just downstream of the homeodomain, followed by a charged region.

domain with a variable linker in between. Several of the POU class genes were originally isolated as transcription factors, which provided the most convincing evidence that homeobox containing genes encode transcription factors (for review see Ruvkun and Finney, 1991; Schöler, 1991). The POU-specific domain is required for cooperative, high affinity DNA-binding (Sturm and Herr, 1988; LeBowitz et al., 1989; Ingraham et al., 1990) and has so far always been found in association with a POU homeodomain. The POU homeodomain is characterized by a cysteine residue in position 9 of helix 3. Figure 14 shows a comparative tree of POU class genes based on the POU-specific domain and

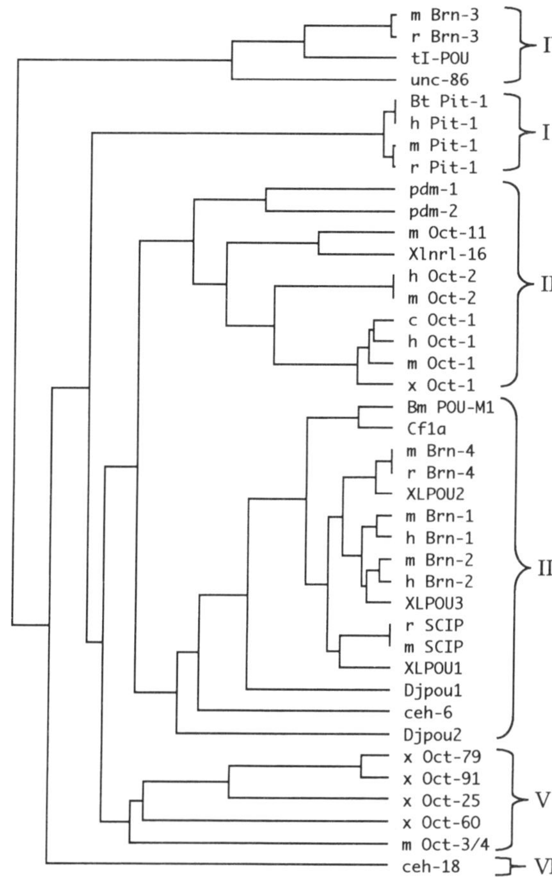

Figure 14: Comparative tree of POU domain sequences. The comparison is based on the POU-specific domain, the linker and the POU homeodomain. The sequences can be grouped into six subclasses (I-VI), five of which have been described previously (He et al., 1989; Ruvkun and Finney, 1991).

POU homeodomain, and Figure 15 shows the sequence alignment over this region. The genes have been grouped into families (He et al., 1989), and currently there are 6 families.

Family I: So far only known from mammalian species.

Family II: In addition to *Oct-1* and *Oct-2*, this family contains at least one more member in vertebrates (*Oct-11*). The *Drosophila* genes of this family, *pdm-1* and *pdm-2* appear not to be direct homologues of *Oct-1* and *Oct-2*, but instead, independent duplication events in flies and vertebrates seem to have given rise to these genes.

Family III: Currently four members are known in vertebrates, while in *Drosophila* and *C. elegans* only one gene of this family has been isolated so far. In flatworms two genes of this family have been isolated (Fig. 14).

Family IV: Currently only one gene of this family has been identified each in mammals, flies and nematodes.

Family V: A diverse assortment of genes. At least 4 members are known from *Xenopus*. These genes are quite divergent from each other, but the linker region between the POU-specific domain and the POU homeodomain shows that these genes are related to each other.

Family VI: The only gene known so far is the nematode gene *ceh-18*. This gene is very divergent from other POU genes, and has a much longer linker region than any other POU gene (D. Greenstein, S. Hird, M. Finney and G. Ruvkun, personal communication). Given this uniqueness, it seems likely that related genes exist in flies and vertebrates.

cut class

In the *Drosophila* homeobox gene *cut*, a conserved domain of about 80 amino acids, the "*cut*-repeat" (Fig. 16), was observed three times upstream of the homeodomain (Blocklinger et al., 1988). There are probably not many (maybe just one) members of this class present in *Drosophila*, since a degenerate PCR approach failed to reveal any additional genes (Bodmer et al., 1990). Recently, a human transcription factor, *CDP* (Neufeld et al., 1992), and a dog transcription factor, *Clox* (Andres et al., 1992) were isolated that contain a homeodomain related to *cut* and three *cut*-repeats upstream of the homeodomain. A partial cDNA (cm18e7) sequence stemming from the *C. elegans* random cDNA sequencing project (Waterston et al., 1992) exhibits significant similarity to the *cut*-repeat, suggesting that a *cut* class gene is also present in nematodes.

There is no evidence for the possible function of the *cut*-repeat, but the following points suggest that the *cut*-repeat might be involved in DNA-binding as are the *paired* and the POU-specific domains. Firstly, the footprint of *CDP* is quite large (Neufeld et al., 1992). Secondly, like the *prd* and the POU-specific domains, the *cut* class homeodomains do not have a glutamine residue at position 9 of helix 3, but rather a histidine. Thirdly, if human *CDP* and *Drosophila cut* are direct homologues, then the fact that they are only 48% identical in the homeodomain suggests that selective pressure on the homeodomain is not very strong, possibly because some of the DNA-binding functions are taken over by the *cut*-repeat.

Zinc-finger meets homeodomain: the *ZF* class

This is without question the most bizarre class of homeobox genes. While one *Drosophila* gene, *zfh-1*, has been described containing 9 zinc-fingers and one homeodomain, a related gene, *zfh-2*, has 16 zinc-fingers and 3 homeodomains (Fortini et al., 1991), while the human gene *ATBF1* has 17 zinc-fingers and 4 homeodomains (Morinaga et al., 1991). *ATBF1* is possibly the human homologue of the fly gene *zfh-2*, since the first (HD1), the second (HD2), and the third homeodomain (HD3) of each gene correspond to each other (see Fig. 6, Lundell and Hirsh, 1992). The fourth homeodomain of *ATBF1* may be a duplication of the third homeodomain. The homeo-

POU - specific domain

```
                              A                                              B
Consensus    P SHSDEDTPDSDELEQFAKQFKQRRIKLGFTQADVGLALGTLYGNVFGSLSQTTICRFEALQLSFKNMCKLKPLLNKWLEEAE SS--
               ◇• •• • ◇•◇ ◇• •• ◇• ◇  ◇     •••◇•• ••• ••• • •◇ •   •◇ • •◇ • •◇      ◇

unc-86     . |PT..M..-.PRQ..T..EH.........V......K..AH.KMPGV.....S......S.T..HN..VA...I.HS...K..|EA--
tI-POU     G |L.P.T..-.PR...A..ER.........V......K..AN.KLPGV.AV..S......S.T..HN..IA...I.QA......|AQ--
r Brn-3      |         R...A..ER.........V......S..AN.KIPGV.....S......S.T..HN..IA...I.QA......|GP--
m Brn-3      |                  V......S..AN.KIPGV.....S......S.T..HN..IA...I.QA......|GA--
x Oct-60   G |PGTE..GMTLE.M.E...EL..K.VA..Y..G.I.H...I...KM.---.........S...T............EQ..G...|NN--
x Oct-91   S |.D.E.EA.N.G.M.....DL.HK..TM.Y......Y...V.F.KT.---.........S..............RS..H.V.|NN--
x Oct-79   T |.D.E.EA.N.G.M.....DL.HK..TM.Y......Y...V.F.KT.---.........S..............RS..H.V.|NN--
x Oct-25   S |.DNE.EV.SES.M.....DL.HK.VS..Y......Y...V...KM.---.........S..........Q...F.ER.VV...|NN--
m Oct-3/4  E |ESQ.MKA-LQK.......LL..K..T..Y.......T..V.F.K..---.........L.....R...E..V...D|NN--
Cf1a       I |.GGE....T..D..A.............................---.........................Q.....D|.T--
Bm POU-M1  R |DQPE....T..D..A.............................---.........................Q.....D|.T--
XLPOU3     Q |D.......T..D................................---.........................D|..--
XLPOU2     Q |D....E..T...................................---.........................D|..--
r Brn-4    Q |D....E..T...................................---.........................D|..--
m Brn-4    Q |D....E..T...................................---.........................D|..--
XLPOU1     Q |D.....A.S..D................................---........................TD|.T--
r SCIP     G |E.....A.S..D................................---........................TD|..--
m SCIP     G |E.....A.S..D................................---........................TD|..--
h Brn-2      |         .D.................................---.........................D|..--
m Brn-2    D |P.......T..D................................---.........................D|..--
h Brn-1      |         .D.................................---.........................D|..--
m Brn-1    D |P.......T..D................................---.........................D|..--
Djpou1     R |.GKT..L.S..D.....M.........Y................---...............R...Q...H..D|..--
ceh-6      D |ISD.SEQTCP.D..G...........Y.....V........I.---.........................F.....D|.T--
Bt Pit-1   E |EPI.M.S.EIR...K..NE..V.....Y..TN..E..AAVH.SE.---...........N.......A....AI.S......|QV--
h Pit-1    E |EPI.M.S.EIR...K..NE..V.....Y..TN..E..AAVH.SE.---...........N.......A....AI.S......|QV--
r Pit-1    E |EPI.M.S.EIR......NE..V.....Y..TN..E..AAVH.SE.---...........N.......A....AI.S......|QV--
m Pit-1    E |EPI.M.S.EIR......NE..V.....Y..TN..E..AAVH.SE.---...........N.......A....AI.S......|QV--
pdm-2      Q |.-PE.T.-.LE.......T...........G.....M.K....D.---.....S.....N............Q....D.D|.TVA
pdm-1      . |.-PE.T.-.LE.......T...........G.D...M.K....D.---.....S.....N............Q...DD.D|RTIQ
m Oct-11     |                             .G.....M.K....D.---.....S.....N..............E...ND..|..--
Xlnrl-16     |                             .G.....M.K....D.---.....S.....N..............E...ND..|.A--
h Oct-2    . |..PE.PS-.LE......RT...........G.....M.K....D.---.....S.....N..............E...ND..|TM--
m Oct-2    . |..PE.PS-.LE......RT...........G.....M.K....D.---.....S.....N..............E...ND..|TM--
x Oct-1    . |.-LE.PS-.LE......T............G.....M.K....D.---.....S.....N..............E...ND..|NI--
c Oct-1    . |.-LE.PS-.LE......T............G.....M.K....D.---.....S.....N..............E...ND..|NL--
h Oct-1    . |.-LE.PS-.LE......T............G.....M.K....D.---.....S.....N..............E...ND..|NL--
m Oct-1    . |.-LE.PS-.LE......T............G.....M.K....D.---.....S.....N..............E...ND..|NL--
              ........|.........|.........|.........|.........|.........|.........|.........|.. ....
                     10        20        30        40        50        60        70        80
```

Figure 15: Alignment of POU domain sequences.
Boxes indicate the POU-specific domain and the POU homeodomain. The POU-specific domain can be subdivided into two subregions, called A and B. Within individual subclasses, conservation in the linker region between the POU-specific domain and the POU homeodomain can be observed. The black dots mark absolutely conserved positions, and open diamonds mark positions with conserved residues.

domains within *ATBF1* and *zfh-2* are not very similar to each other, and in particular the first homeodomain within each protein is very divergent (Fig. 6, 7). In particular, they are the only two examples so far that contain an arginine residue at position 9 of helix 3. The homeodomain of *zfh-1*, and the third homeodomains of *zfh-2* and *ATBF1* are marginally more similar to the LIM class of homeodomains than to any other class (Figs. 3 and 6). Since the LIM motif is also a metal-binding motif and contains zinc, this similarity is most likely evolutionarily relevant (see below).

The zinc-fingers are of the C_2-H_2 type, and, interestingly, when database searches are performed, some of the best matching zinc-fingers found in the database are developmental control genes like *Krüppel* and *hunchback*. Clearly this class of homeobox genes seems to provide an evolutionary link to zinc-finger genes. While both *ATBF1* and *zfh-2* have been shown to be transcription factors (Morinaga et al., 1991; Lundell and Hirsh, 1992), it will be interesting to see what role all the putative DNA-binding motifs have in these monstrous proteins.

ATBF1 was reported to have some sequence similarity to the POU-specific domain upstream of HD1. However, this similarity is not significant, since most of the con-

POU homeodomain

```
Consensus                            GR KRKKRTSIEVNVKGALEKHFLKCPKPSSQEITSIADSLQLEKEVVRVWFCNRRQKEKRMT PP
                                        ◇◇◇ ••      ◇  •• •    ◇•    ◇   ◇ ◇   • ◇ • •◇•••••• •••◇ •

unc-86      --------MKQKDTIGDINGILPNTD |.KR.....AAPE.RE..QF.KQQ.R..GER.A....R.D.K.N........Q...Q..DF| RS
tI-POU      --------AKNKRRDPDAPSVLPAGE |.KR.....AAPE.RS..AY.AVQ.R..GEK.AA..EK.D.K.N........Q...Q..IV| SS
r Brn-3     --------QREKMNKPE---LFNGGE |.KR.....AAPE.RS..AY.AVQ.R...EK.AA..EK.D.K.N........Q...Q..|
m Brn-3     --------QREKMNKPE---LFNGGE |.KR.....AAPE.RS..AY.AVQ.R...EK.AA..EK.D.K.N..|
x Oct-60    ---------DNLQEMIHKAQIEEQNR |...M..CFDTVL..Q..G..MCNQ..GAR.L.E..KE.S...D..............SKF| RM
x Oct-91    ---------KNLQEIISRGQIIPQVQ |...H.....N...CT..NY.MQ.S...A...AQ..RE.NM..D.............G..QV| Y.
x Oct-79    ---------ENLQEIISRGQIIPQVQ |...H.....D..RHT..NY.MQ.S...A...AQ..RE.NM..D.............G..QV| Y.
x Oct-25    ---------DNLQELINREQVIAQTR |...R..N..NI...T..SY.M.....GA..MVQ..KE.NMD.D.............G..QG| M.
m Oct-3/4   ---------ENLQEICKSETLV--QA |RKR......NR.RWS..TM.........L.Q..H..NQ.G...D.............G..SS| IE
Cf1a        ---------TGSPTS-IDKIAAQ-GR |..........S......Q..H.Q....A.....L...............|..
Bm POU-M1   ---------TGSPTS-IDKIAAQ-GR |..........S......Q..H.Q....A.....L...............|..
XLPOU3      ---------SGSPTS-IDKIAAQ-GR |..........S......S.......AP....L.........C.......|..
XLPOU2      ---------TGNPTS-IDKIAAQ-GR |..........S...V..T.......AAL....L................|..
r Brn-4     ---------TGSPTS-IDKIAAQ-GR |..........S...V..T.......AA...S.L................|..
m Brn-4     ---------TGSPTS-IDKIAAQ-GR |..........S...V..T.......AA...S.L................|..
XLPOU1      ---------TGSPTN-LDKIAAQ-GR |..........G......N.......AH....L................|.A
r SCIP      ---------SGSPTN-LDKIAAQ-GR |..........G......S.......AH...GL................|.A
m SCIP      ---------SGSPTN-LDKIAAQ-GR |..........G......S.......AH...GL................|.A
h Brn-2     ---------SGSPTS-IDKIAAQ-GR |..........S......S.......A.....L................|
m Brn-2     ---------SGSPTS-IDKIAAQ-GR |..........S......S.......A.....L................|..
h Brn-1     ---------TGSPTS-IDKIAAQ-GR |..........S......S.......NL................|
m Brn-1     ---------TGSPTS-IDKIAAQ-GR |..........S......S.......A....NL................|..
Djpou1      ---------SESPTN-FDKISAQ-SR |.........A...SI..SS.M.LS...A.D.S.L.EK.S................I.|.S
ceh-6       ---------TGSPNSTFEKMTGQAGR |................SR..F..QSNQ..NA....QV.ME...............IA|.N
Bt Pit-1    ------------GALYNEKVGAN-ER |...R..T.SIAA.D...R..GEQN.......LRM.EE.N.............R...VK| TS
h Pit-1     ------------GALYNEKVGAN-ER |...R..T.SIAA.D...R..GEQN.......MRM.EE.N.............R...VK| TS
r Pit-1     ------------GALYNEKVGAN-ER |...R..T.SIAA.D...R..GEHS.......MRM.EE.N.............R...VK| TS
m Pit-1     ------------GALYNEKVGAN-ER |...R..T.S.AA.D...R..GEHS.......MRM.EE.N.............R...VK| TS
dPOU-28     KSGGGVFNINTMTSTLSSTPESILGR |R.........TT.RTT...A..MNC..T.E..SQLSER.NMD...I.............IN|.S
dPOU-19     ATGG-VFDPAALQATV-STPE-IIGR |R.........TTIR.....A..ANQ..T.E...QL..R.SM.............IN|.S
m Oct-11    -PSDPSASTPSSYPTLSEVF----GR |.........T.IRLT...R.QDN.....E..SM..EQ.SM.....|
Xlnrl-16    -PSDTALTPSANYAQLSEMF----GR |.........T.IRLT...R.QDN.....E..SM..EQ.V|
h Oct-2     -SVDSSLSPSNQLSSPSLGFDGLPGR |R.........T..RF....S..ANQ..T.E..LL..EQ.HM....I.............IN|.C
m Oct-2     -SVDSSLSPSNQLSSPSLGFDGLPGR |R.........T..RF....S..ANQ..T.E..LL..EQ.HM....I.............IN|.C
x Oct-1     -TSDSTLTNQSVLNSPGHGMEGL-NR |R.........T.IRV....S..ENQ..T.E...M...Q.NM....I.............IN|..
c Oct-1     -SSDSTLSSPSALNSPGQGIEGV-NR |R.........T.IRV....S..ENQ..T.E...M...Q.NM....I.............IN|..
h Oct-1     -SSDSSLSSPSALNSP--GIEGL-SR |R.........T.IRV....S..ENQ..T.E...M...Q.NM....I.............IN|..
m Oct-1     -SSDSTASSPSALNSPGLGAEGL-NR |R.........T.IRV....S.MENQ..T.ED..L..EQ.NM....I.............IN|..
                              ...|.........|.........|.  .......|.........|.........|.........|.........|........|..
                               90       100       110      120       130       140       150       160       170
```

15B

served residues of the POU-specific domain are not present, and the region of similarity is not conserved in the related gene *zfh-2*.

LIM class

The LIM motif, a repeat of about 60 amino acids containing conserved cysteine and histidine residues (Fig. 17), was first defined in two *C. elegans* genes, *mec-3* and *lin-11*, and in the rat transcription factor *Isl-1* (Freyd et al., 1990; Karlsson et al., 1990). Two copies of the LIM motif are usually found upstream of the homeodomain, but several genes have been isolated that contain only two LIM motifs, the best known case being the putative human oncogene *rhombotin 1* (Boehm et al., 1990b; McGuire et al., 1989; Rabbitts and Boehm, 1990). Genes related to *rhombotin 1* have also been isolated from *Drosophila* and mice (Boehm et al., 1990a; Boehm et al., 1991).

The LIM motif exhibits relatively poor primary sequence conservation between different genes, and between the two repeats, although it is clear that the first and the second repeat are slightly distinct from each other. The conserved cysteine and histidine residues are readily apparent, but otherwise, amino acids between these conserved residues seem to be mainly conserved based on their biochemical properties (e.g. there are several positions which are hydrophobic). The LIM class homeodomain sequences are much better conserved than the LIM motifs themselves, and can thus be used to identify related families: *C. elegans lin-11* and *Xenopus Xlim1* form family I, *C. elegans ceh-14* and *Xlim-3* form family II, and *Drosophila apterous* and rat *LH2* form family III. Other prospective candidates for new families are the *Isl-1* genes, and *mec-3*.

Additional genes containing regions of similarity to the LIM motif have been found: *CRIP*, a cysteine-rich intestinal protein (Birkenmeier and Gordon, 1986), *ESP1*, cut repeat

```
cut rp1 ANQALDTLHISRRVRELLSVHNIGQRLFAKYILGLSQGTVSELLSKPKPWDKLTEKGRDSYRKMHAWACDDNAVMLLKSLIPKKD
        :||   | | |:| |   ||||||:|   |:||||||:|||:| :|||| ||| :|:: : || :  |:  :: |:|:   :
CDP rp1 EGEEMDTAEIARQVKEQLIKHNIGQRIFGHYVLGLSQGSVSEILARPKPWNKLTVRGKEPFHKMKQFLSDEQNILALRSIQGRQR
        ||||||||||||||||||||||||||||||||||||||||||||||||||||||||||||||||||||||||||||||||||
Clox rp1          SRQVKEQLIKHNIGQRIFGHYVLGLSQGSVSEILARPKPWNKLTVRGKEPFHKMKQFLSDEQNILALRSIQGRQR
        : | :| ||||  |  | ||:|| |||||||||::||||||| || :|||| :|| || || :  |  |
cut rp2 MFNNLNTEDIVRRVKEALSQYSISQRLFGESVLGLSQGSVSDLLARPKPWHMLTQKGREPFIRMKMFLEDENAVHKLVASQYKIA
        |: : | :: | ||||  |||||||||||:| ||||| |||||||||||||| ::| |     |
CDP rp2 MYQEVDTIELTRQVKEKLAKNGICQRIFGEKVLGLSQGSVSDMLSRPKPWSKLTQKGREPFIRMQLWLNGELGQGVLPVQGQQQG
        ||||||||||||||||||||||||||||||||||||||||||||||||||||||||||||||||||||||||
Clox rp2 MYQEVDTIELTRQVKEKLAKNGICQRIFGEKVLGLSQGSVSDMLSRPKPWSKLTQKGREPFIRMQLWLNGELGQGVLPVQGQQQG
        : |::|| ::|  :|| |  | | |:|||| |||||||||::||:|||| |: ||||||||||            |  :
cut rp3 LTQDLDTHDITTKIKEALLANNIGQKIFGEAVLGLSQGSVSELLSKPKPWHMLSIKGREPFIRMQLWLSDANNVERLQLLKNERR
        :: :|||| || ::|| | ||:||:||| :|||:|||||:| :||||| ||:|||||:||||| | |||||:|  :|     :
CDP rp3 MSPELDTYGITKRVKEVLTDNNLGQRLFGETILGLTQGSVSDLLARPKPWHKLSLKGREPFVRMQLWLNDPNNVEKLMDMKRMEK
        |||||||||||||||||||||||||||||||||||||||||||||||||||||||||||||||||||||||||||
Clox rp3 MSPELDTYGITKRVKEVLTDNNLGQRLFGETILGLTQGSVSDLLSRPKPWHKLSLKGREPFVRMQLWLNDPNNVEKLMDMKRMEK
        |    |  || :||::||||  ||||  :   ||   |  ||   ||   :   :: |
cm18e7                              QTALAEKILARSQGTLSDLLRMPKPWSVM-KNGRATFQRMSNWLGLDPDVRRALCFLPKED
        .........|.........|.........|.........|.........|.........|.........|.........|.....
                 10        20        30        40        50        60        70        80
```

Figure 16: Alignment of *cut* repeats.

The 3 *cut* repeats (rp) of *Drosophila cut* and human *CDP* are aligned underneath each other, in addition to the partial cDNA sequence (cm18e7) of a *C. elegans* gene obtained from the random cDNA sequencing project (Waterston et al., 1992). The black bar indicates the *cut* repeat as defined by this alignment. Reference for dog *Clox* (Andres et al., 1992).

LIM motif

```
mec-3     .VDHENQNKCNCCNEQIY-DRYIYRMD-NRSYHENCVKCTICESPLAE---KCFWKNG--RIYCSQ-HYYKDHS
lin-11    .ISEISGNECAACAQPIL-DRYVFTVL-GKCWHQSCLRCCDCRAPMSM---TCFSRDG--LILCKT-DFSRRYS
Xlim-1    MVHCAGCERPIL-DRFLLNVL-DRAWHVKCVQCCECKCNLTE---KCFSREG--KLYCKN-DFFRRFG
sh lmx-1  .RAVSPKSVCEGCQRVIS-DRFLLRLN-DSFWHEQCVQCASCKEPLET---TCFYRDK--KLYCKY-HYEKLFA
Isl-1     .KKKRLISLCVGCGNQIH-DQYILRVSPDLEWHAACLKCAECNQYLDE-SCTCFVRDG--KTYCKR-DYIRLYG
Ot Isl-1  .KKKSGIAMCVGCGSQIH-DQYILRVAPDLEWHAACLKCSECSQYLDE-TCTCFVRDG--KTYCKR-DYVRLFG
ap        .KITRNLDDCSGCGRQIQ-DRFYLSAV-EKRWHASCLQCYACRQPLER-ESSCYSRDG--NIYCKN-DYYSFFG
r LH-2    .ISSDRAALCAGCGGKIS-DRYYLLAV-DKQWHMRCLKCCECKLNLES-ELTCFSKDG--SIYCKE-DYYRRFS
h rhom-1  .QPKGKQKGCAGCNRKIK-DRYLLKAL-DKYWHEDCLKCACCDCRLGEVGSTLYTKAN--LILCRR-DYLRLFG
h rhom-3  .QPDTKPKGCAGCNRKIK-DRYLLKAL-DKYWHEDCLKCACCDCRLGEVGSTLYTKAN--LILCRR-DYL.
h rhom-2  .QIPPSLLTCGGCQQNIG-DRYFLKAI-DQYWHEDCLSCDLCGCRLGEVGRRLYYKLG--RKLCRR-DYL.
m rhom-2  .QIPPSLLTCGGCQQNIG-DRYFLKAI-DQYWHEDCLSCDLCGCRLGEVGRRLYYKLG--RKLCRR-DYLRLFG
d rhom        .QLCAGCGKHIQ-DRYLLRAL-DMLWHEDCLKCGCCDCRLGEVGSTLYTKGN--LMLCKR-DYLR.
```
> (brace 1 — first LIM motif)

```
mec-3     IHRCAGCKKGVSPTDMVYKLKAGLVFHVECHCCSLCGRHLSP-GEQILVDDTMMTVSCMS-HYPPQMDDNAPGAIG.
lin-11    QRCAGCDGKLEKEDLVRRAR-DKVFHIRCFCCSVCQRLLDT-GDQLYIMEG-NRFVCQS-DFQTATKTSTPTSIH.
Xlim-1    TKCAGCAQGISPSDLVRRAR-SKVFHLNCFTCMMCNKQLST-GEELYIIDE-NKFVCKE-DYLNNNNAAKENSFI.
sh lmx-1  VKCGGCFEAIAPNEFVMRAQ-KSVYHLSCFCCCVCERQLQK-GDEFVLKEG--QLLCKG-DYEKERELLSLVSPA.
Isl-1     GIKCAKCSIGFSKNDFVMRAR-SKVYHIECFRCVACSRQLIP-GDEFALRED--GLFCRA-DHDVV-ERASLGAGDP.
Ot Isl-1  IKCAKCNIGFCSSDLVMRAR-DNVYHMECFRCSVCSRQLVP-GDEFSLRDE--ELLCRA-DHGLLLERASAGSPIS.
ap        TRRCSRCLASISSNELVMRAR-NLVFHVNCFCCTVCHTPLTK-GDQYGIIDA--LIYCRT-HYSIAREGDTASSSMS.
r LH-2    VQRCARCHLGISESEMVMRAR-DLVYHLNCFTCTTCNKMLTT-GDHFGMKDS--LVYCRL-HFEALLQGEYPPHFNH.
h rhom-1  TTGNCAACSKLIPAFEMVMRAR-DNVYHLDCFACQLCNQRFCV-GDKFFLKNN--MILCQM-DYEEG-QLNGTFESQVQ*
m rhom-2  QDGLCASCDKRIRAYEMTMRVK-DKVYHLECFKCAACQKHFCV-GDRYLLINS--DIVCEQ-DIYEWTKINGII*
```
> (brace 2 — second LIM motif)

```
zyxin rp1 .VEAATSELCGFCRKPLSRTQPAVRAL-DCLFHVECFTCFKCEKQLQG--QQFYNVDE--KPFCED-CYAG
      rp2 TLEKCSVCKQTIT-DR-MLKAT-GNSYHPQCFTCVMCHTPLEG--ASFIVDQA-NQPHCVD-DYHRKYA
      rp3 PRCSVCSEPIM-ETVRVVAL-EKNFHMKCYKCEDCGRPLSI-ENGCFPLDG--HVLCMK-CHTVRAKTAC*
            MPEPGKD                           EAD

ESP1      .HVHWEPNMCPRCNKRVYFAEKVTSL--GKDWHRPCPRCERCSKTLTP--GGHAEHDG--QPYCHKPCYGILFGPKGVNTGAVGSYIYDKDPEGTVQP*
CRIP      MPKCPKCDKEVYFAERVTSL--GKDWHRPCLKCEKCGKTLTS--GGHAEHEG--KPYCNHPCYSAMFGPKGFGRGGAESHTFK*
hCRP rp1  MPNWGGGKKCGVCQKTVYFAEEVQCE--GNSFHKSCFLCMVCKKNLDS--TTVAVHGE--EIYCKS-CYGKKYGPKGYGYGQGAGTLSTDK.[26aa].
      rp2 QKIGGSERCPRCSQAVYAAEKVIGA--GKSWHKACFRCAKCGKGLES--TTLADKDG--EIYCKG-CYAKNFGPKGFGFGQGAGALVHSE*
          .........|.........|.........|.........|.........|.........|.........|.........|
                   10        20        30        40        50        60        70        80
```

Figure 17: Alignment of LIM motifs.

The two parentheses mark the first and the second LIM motif for the different genes. No sequences were omitted between the two motifs (except as indicated in *hCRP*).

References for genes without homeodomain: *ESP1* (Nalik et al., 1989); *CRIP* (Birkenmeier and Gordon, 1986); *hCRP* (Liebhaber et al., 1990); *h rhom1* (Boehm et al., 1990b), (McGuire et al., 1989); *h rhom2, h rhom3, m rhom2* (Boehm et al., 1991); *d rhom* (Boehm et al., 1990a); *LH-2* (Xu et al., 1992); *zyxin* (Sadler et al., 1992); *sh lmx-1* (German et al., 1992). Accession number for Ot *Isl-1*: X64884.

an estradiol-stimulated protein in brain (Nalik et al., 1989), and *hCRP*, a human cysteine-rich protein (Liebhaber et al., 1990). *hCRP* has two repeats which are more closely related to each other than to either LIM repeat 1 or 2; *CRIP* and *ESP1* have only one repeat. In addition, *CRIP*, *ESP1*, and *hCRP* share a conserved element downstream of the region of similarity to the LIM motif. A recently isolated gene, *zyxin*, contains three regions of similarity to the LIM motif. *zyxin* repeat 1 and 2 appear to be about equally similar to either repeat 1 or 2 of the homeobox gene LIM motifs, and repeat 3 distinguishes itself by having more residues in the repeat than any other LIM repeat (see Fig. 17). It seems clear that these genes, especially *zyxin*, are evolutionarily related to the LIM class homeobox genes, and that their cysteine-histidine motifs can be counted as LIM motifs, but they might represent special subclasses.

It has been speculated that the LIM motif might be involved in protein-protein interactions (Freyd et al., 1990; Rabbitts and Boehm, 1990). Biochemical characterization of *zyxin* indicates that it is a zinc binding protein that binds to α-actinin and is associated with the actin cytoskeleton (Sadler et al., 1992). In addition *zyxin* directly interacts with the chicken homologue of *hCRP in vitro* (Sadler et al., 1992). This suggests that the LIM motif does indeed function as a protein-protein interaction domain. *CRIP* isolated from rat intestinal mucosa is bound to zinc and it appears to function as a zinc carrier protein in the intestine (Hempe and Cousins, 1991). The LIM motifs of *lin-11* have been shown to bind zinc and iron when produced in *E. coli* (Li et al., 1991), but no iron, copper or cadmium was observed in *zyxin* isolated from avian muscle (Sadler et al., 1992). Thus zinc binding is a feature common to LIM motifs. Interestingly, the most closely - albeit only very marginally - related other homeodomain class is the ZF class, whose genes contain zinc-fingers of the C_2-H_2 type. Consequently, one might speculate that the LIM motif is derived from a "standard" zinc-finger, and that its original function might have been to bind DNA. Subsequently, as the LIM motif evolved, it might have undergone changes in its function, at least in genes like *zyxin* and *CRIP*, with the homeodomain becoming dispensable.

NK-2 class

The *NK-2* class of homeobox genes is a complex assortment of fly genes (*NK-2*, *NK-3*, *NK-4*) and vertebrate genes, and several flatworm genes have also been isolated (Figs. 6, 7). The first vertebrate gene of this class to be isolated was the transcription factor *TTF-1* (Guazzi et al., 1990). There will probably be several families in this class, one obvious family formed by *NK-2*, *Nkx-2.2*, *-2.3* and *-2.4* Several of the fly and vertebrate genes share a small conserved region of about 17 amino acids downstream of the homeodomain, referred to as the "conserved peptide" (CP) (Price et al., 1992).

msh class

The evolution of *msh* class genes has recently been discussed by Holland (1992). Suffice it to say that there are at least three *"msh"* class genes in mammals and four in zebra fish, and the exact relationships between these genes remains somewhat open until all members have been cloned. The class itself is very ancient, since a member has been cloned from hydra (Schummer et al., 1992).

A comparison of various vertebrate *msh* class genes over their complete protein sequences indicates that there are probably no major conserved domains outside of the homeodomain. The vertebrate *msx1* genes have a motif upstream of the homeodomain that is reminiscent of the Hexapeptide (Fig. 9). However, this region is very poorly conserved in other *msh* class genes (Fig. 9), while genes containing bona fide Hexapeptides show good conservation in that region. Thus it is questionable if this motif does indeed correspond to the Hexapeptide, and consequently the *msh* class has not been included in the Hex superclass.

Dll class

Currently there are at least 4 different *Distal-less* (*Dll*) class genes in vertebrates (*Dlxs*), while only one fly gene is known (Fig. 6).

NK-1 class

The homeodomains of the *NK-1* class have been highly conserved between flatworms, nematodes, flies and vertebrates (Fig. 6). It is not known if there are any conserved domains outside the homeodomain, or how many members there are in vertebrates and flies.

NEC class

The *C. elegans* gene *ceh-9* and the sea urchin gene *Tghbox5* fulfill the criteria for a new class (see section 1). A recently identified homeobox gene from vertebrates is very similar to the *Tghbox5* gene and this new class has been termed NEC (D. Deitcher and C. Cepko, personal communication). It is not known yet whether there are additional large conserved domains outside of the homeodomain.

Hlx class

Genes similar to the *Drosophila* gene *H2.0* have been isolated from vertebrates and ascidians. Chicken *Chlx-A* and mouse *Dbx* genes as well as the *Ascidia* gene *Hal-hlx* are quite different from *H2.0* and the mouse *Hlx* gene; therefore one might expect that this class, termed *Hlx*, has several distinct families in vertebrates and flies.

Orphans

Examination of the tree in Figure 6 shows that there are a series of homeobox genes for which no homologues have been found in other divergent species: *BarH1* and *BarH2*, *bsh*, *ceh-19*, *EgHbx2*, *Ovx-1*, *Gbx-2*, *ceh-5*, *smox-5*, *Prh*, *ro*, *mox-1* and *mox-2*, *Gsh-1* and *Gsh-2*, *smox-4*, and *ceh-7*. In particular the *Bar* genes and *Ovx-1/Gbx-2* are promising candidates that might define novel classes of homeodomains, and homologues for these genes might exist in other phyla. Yet one cannot discount the possibility that some genes have acquired new functions and have undergone rapid sequence divergence (such as probably hap-

```
                 ....|....|....|....|..   .|....|....|....            |....|....|....|....|
atypical hds          ▫  *    **  ▼*  ▼      *▫    *   **  *          ▫  *▫*▫▼▼*▼*▼  ▼ ▫
PBX1        ARRKRRNFNKQATEILNEYFYSHLSNPYPSEEAKEELAKKCG------------------------------------ITVSQVSNWFGNKRIRYKKNI
PBX2        ARRKRRNFSKQATEVLNEYFYSHLSNPYPSEEAKEELAKKCG------------------------------------ITVSQVSNWFGNKRIRYKKNI
PBX3        ARRKRRNFSKQATEILNEYFYSHLSNPYPSEEAKEELAKKCS------------------------------------ITVSQVSNWFGNKRIRYKKNI
ceh-20      ARRKRRNFSKQATEVLNEYFYGHLSNPYPSEEAKEDLARQCN------------------------------------ITVSQVSNWFGNKRIRYKKNM
Kn1         KKKKKGKLPKEARQQLLSWWDQHYKWPYPSETQKVALAESTG------------------------------------LDLKQINNWFINQRKRHWKPS
ZMH1        RKRRAGKLPGDTTSILKQWWQEHSKWPYPTEDDKAKLVEETG------------------------------------LQLKQINNWFINQRKRNWHNN
ZMH2        RKRRAGKLPGDTASTLKAWWQAHSKWPYPTEEDKARLVQETG------------------------------------LQLKQINNWFINQRKRNWHNN
MATα2       KPYRGHRFTKENVRILESWFAKNIENPYLDTKGLENLMKNTS------------------------------------LSRIQIKNWVSNRRRKEKTIT
Um bE1      DVGCRNLSEDLPAYHMRKHFLLTLDNPYPTQEEKETLVRLTNESTARVGQSSVNRPP---------------------LEVHHVTLWFINARRRSGWSH
Um bE2      VVGCRDLSEDLPAYHMRKHFLHTLDNPYPTQEEKEGLVRLTNESTARVGLSKANRPP---------------------LEVHQLTLWFINARRRSGWSH
Um bE3      GEGCRNLSEDLPAYHMRNHFLHTLENPYPTQEEKEGLVRLTNESTARIRPSNAIRPP---------------------LEVHQLTLWFINARRRSGWSH
Um bE4      VVGCRELSEDLPAYHMRKHFLHTLENPYPTQEEKETLVRLTNESTARVGQSIVNRPP---------------------LEVHQLTLWFINARRRSGWSH
Um bE5      AVVCRNLSEDLPAYHMRKHFLLTLDNPYPTQEEKQNLVRLTNESTVRVGSSNPARPP---------------------LEVHQLTLWFINARRRSGWSH
Um bE6      AVEYRNLSEDLPAYHMRKHFLHTLDSPYPTQEEKETLVRLTNESTARVGQSSVNRPP---------------------LEVHQLTLWFINARRRSGWSH
Um bE7      AVGCRNLSEDLPAYHMRKHFLLTLDSPYPTQEEKEGLVRLTNESTARVGLSNATRPP---------------------LEVHQLTLWFINARRRSGWSH
Sp MAT1Pi   MTTVRGQCSKCTKPHLMRWLLLHYDNPYPSNSEFYDLSAATG------------------------------------LTRTQLRNWFSNRRR*
Cc Aβ1-1    DKNEPTSPTPAYVEPCARWLKDNWYNPYPSGEVRTQIARQTR------------------------------------TSRKDIDAWFIDARRRIGWNE
Cc Aβ4-1    DASPKEPAEPAYIEPSCRWLKDNWYNPYPSPQVRSSIAKQTG------------------------------------ASRKDIDAWFIDARKRIGWNE
Sc AαZ3     EWQENMPPVPPFIGACYEWLLQHLHNPYPSKSEKQIILQLLQDESHRASEGRLTPPSSPRSACSILPSLKPALDTCKTFDDIDRWFSAARIRIGWTH
Sc AαZ4     GAELSATPLPPYIEPCYRWLVNHLDNPYPTKAIKEELLDQARQRTSPDVAQH---------------------LALGDIDNWFIAARARMGWGD
LFB1        GRRNRFKWGPASQQILFQAYERQ---KNPSKEERETLVEECNRAECIQRGVSPSQAQGLGSNL-------------VTEVRVYNWFANRRKEEAFRH
LFB3        MRRNRFKWGPASQQILYQAYDRQ---KNPSKEEREALVEECNRAECLQRGVSPSKAHGLGSNL-------------VTEVRVYNWFANRRKEEAFRQ
pros        TPLHSSTLTPMHLRKAKLMFFWV---RYPSSAVLKMYFPDIKFNK------------------------------NNTAQLVKWFSNFREFYYIQM
PEM         MPLQGSRFAQHRLRELESILQR----TNSFDVPREDLDRLMD------------------------------------ACVSRVQNWFKIRRAAARRTR

typical hds
Antp        RKRGRQTYTRYQTLELEKEFHFN---RYLTRRRRIEIAHALC------------------------------------LTERQIKIWFQNRRMKWKKEN
prd         QRRCRTTFSASQLDELERAFERT---QYPDIYTREELAQRTN------------------------------------LTEARIQVWFSNRRARLRKQH
otd         TASRTTTFTRAQLDVLEALFGKT---RYPDIFMREEVALKIN------------------------------------LPESRVQVWFKNRRAKCRQQL
cut         SKKQRVLFSEEQKEALRLAFALD---PYPNVGTIEFLANELG------------------------------------LATRTITNWFHNHRMRLKQQV
Oct-1       RRKKRTSIETNIRVALEKSFLEN---QKPTSEEITMIADQLN------------------------------------MEKEVIRVWFCNRRQKEKRIN
PHO2        QRPKRTRAKGEALDVLKRKFEIN---PTPSLVERKKISDLIG------------------------------------MPEKNVRIWFQNRRAKLRKKQ
MATa1       SPKGKSSISPQARAFLEQVFRRK---QSLNSKEKEEVAKKCG------------------------------------ITPLQVRVWFINKRMRSK*
Cc Aα2-1    PPSKKQAFNVHYIPVLEKYFEYN---AYPTAQDRALLARKSM------------------------------------MSARQIEVWFQNHRARARKEG
Cc Aβ2-1    LEVKRTPFNSEYTPLLEKYFEYN---AYPSARDREWLARKTM------------------------------------MSVRQIEVWFQNHRRRARKEG
Sc AαY1     SKKPRPKFHSEYTPLLELYFRFN---AYPTYADRRVLAEKTG------------------------------------MLTRQITVWFQNHRRRAKGPL
Sc AαY3     YKKPRPKFHSEYTPLLELYFHFN---AYPTFADRRMLAEKTG------------------------------------MQTRQITVWFQNHRRRAKGPL
Sc AαY4     GKRSRPKFHSEYTPVLELYFHFN---AYPTYADRRILAEKTG------------------------------------MLTRQITVWFQNHRRRAKGPL
Um bW1      PLKTGRGHDSEAVRILEQAFKHS---PNITPAEKFRLSEVTG------------------------------------LKPKQVTIWFQNRRNRKGKKN
Um bW3      PLKTGRGHDSEAVQILEEAFKHS---PNITPAEKFRLSEVTG------------------------------------LKPKQVTIWFQNRRNRKGKKN
zfh-2 HD1   QKRARTRITDDQLKILRAHFDIN---NSPSEESIMEMSQKAN------------------------------------LPMKVVKHWFRNTLFKERQRN
Zmhbx1a     STARKGHFGPVINQKLHEHFKTQ---PYPSRSVKESLAEELG------------------------------------LTFRQVNKWFETRRHSARVAS
Zmhbx1b     IKDRKGHFGPVISQKLHEHFKTQ---PYPSRSLKESLAEELG------------------------------------LTFHQVNRWFENRRHFARLAS

zeste       QLPLTPRFTAEEKEVLYTLFHLH  NKYSVRETWDKIVKDFNSHPHVSAM-----------------------RNIKQIQKFWLNSRLRKQYPY
                              EEVIDIKHRKKQR
```

Figure 18: Alignment of atypical homeodomain sequences.
Numbering scheme, symbols for conserved positions, and symbols for helices are adapted from Figure 1. Dashes represent the gaps that have been introduced for optimal sequence alignment. The gene *zeste* is not considered to be a homeobox gene; however, there is sequence similarity to the region of helix 1 and helix 3 of the homeodomain. References for sequences: *PBX1, PBX2 & PBX3* (Kamps et al., 1990; Monica et al., 1991; Nourse et al., 1990); *ceh-20* (Bürglin and Ruvkun, 1992); *MATα1 & MATα2* (Astell et al., 1981; Miller, 1984); *Kn1, ZMH1, ZMH2* (Vollbrecht et al., 1991); *Um bE* (Kronstad and Leong, 1990; Schulz et al., 1990); *Sp MATPi* (Kelly et al., 1988); *Cc Aβ1-1, Aβ4-1, Aα2-1 & Aβ2-1* (Kües et al., 1992, Kües & Casselton, personal communication); *Sc AαY1, AαY3, AαY4 & AαZ4* (Stankis et al., 1992); the sequence of *AαZ3* is taken from Novotny et al. (1991), since a proposed intron (Stankis et al., 1992) would shift the sequence such that two proline residues would fall into the middle of helix 2; *pros* (Chu-Lagraff et al., 1991; Matsuzaki et al., 1992; Vaessin et al., 1991); *PEM* (Sasaki et al., 1991); *Um bW* (Gillissen et al., 1992).

pened in the case of *bcd*); thus their presence might be restricted to a particular Phylum, Subphylum, or Class.

■ Atypical homeodomains

Definition of atypical homeodomains

Sequence similarity between the *Antp* class genes and the yeast mating type genes *MATα1* and *MATα2* were detected soon after the discovery of the homeobox (Shepherd et al., 1984; Laughon and Scott, 1984). As more homeobox-containing genes were cloned, and the conserved residues better defined, it became apparent that a better alignment of the helix 1 region of *MATα2* to the consensus could be achieved by looping out three extra amino acids (Hall and Johnson, 1987). X-ray crystallography of the *MATα2* homeodomain revealed that its structure is indeed very similar to *en* and *Antp* (Wolberger et al., 1991), and that 3 amino acids are in fact looped out between helix 1 and helix 2. Several genes have now been isolated that have 3 extra amino acids between helix 1 and helix 2 from a variety of organisms (Fig. 18). Extra amino acids have also been found in several genes between helix 2 and helix 3, first in the liver transcription factor *LFB1* (Finney, 1990). The homeodomains of the fungal genes *bE* and *AαZ* have additional amino acids both between helix 1 and helix 2, and between helix 2 and helix 3. Homeodomain sequences which require insertions and/or deletions within the homeodomain to obtain optimal alignment are referred to as atypical homeodomains. The only places to insert such extra amino acids without destroying the structure of the homeodomain are in the loop regions between the α-helices.

In general the sequences of atypical homeodomains are quite divergent from other classical or typical homeodomain sequences (e.g. the *Um bE1* homeodomain ranges between only 15% to maximally 28% identities to 349 classical homeodomains). The atypical homeodomains of Cc *Aβ1-1*, *Aβ4-1*, and Sc *AαZ* are very divergent at their amino termini. In particular, the highly conserved leucine residue at position 16 (see Fig. 1) is not conserved, and proline residues are found so far into helix 1 that one has to surmise that the amino terminal structure of these homeodomains is substantially different from typical homeodomains.

The *Drosophila* gene *prospero* (*pros*) is another case where the sequence similarity to the homeodomain becomes extremely marginal, such that a search of *pros* against all available sequences in the databanks using the Blast program (Altschul et al., 1990) does not reveal any significant similarity to homeobox genes. The highly conserved leucine in helix 1 is replaced with an alanine (Figs. 1, 18), and the turn/loop structure between helix 2 and helix 3 is very different from the standard turn. Yet there are a sufficient number of key residues conserved so that *pros* can be considered an atypical homeobox gene, in particular if the intracellular location of *pros* and its function are taken into account (Vaessin et al., 1991; Chu-Lagraff et al., 1991; Matsuzaki et al., 1992)

While the above genes have extra amino acids in the homeodomain, a gap of one amino acid has to be introduced into the homeodomain of PEM (Sasaki et al., 1991) in order to achieve some satisfactory alignment. Furthermore, this constitutes the only example where a proline residue is found so far inside the amino terminus of helix 2. Further experiments should indicate if this domain can bind DNA, and if related genes exist.

HNF1/LFB1 has been reported to have similarity to the POU-specific domain (Baumhueter et al., 1990; Frain et al., 1989), but this similarity is in fact not statistically significant. One proposed alignment (Baumhueter et al., 1990) is simply meaningless, because the region of similarity to POU-A overlaps substantially the beginning of the homeodomain. In the other proposed alignment (Frain et al., 1989) the purported region of similarity does not match the pattern of the most conserved residues of POU-A. In addition, comparison of *HNF1/LFB1* and *vHNF1/LFB3* reveals several conserved regions (De Simone et al., 1991), such as the atypical homeodomain, but the purported region of POU-A similarity only partially overlaps one of these conserved regions. One would expect this region of similarity to POU-A to be highly conserved between *HNF1* and *vHNF1*, if there were an evolutionary structural relationship. Thus *HNF1* and the related gene *vHNF1* do not belong to the POU class. Examination of the regions conserved between *HNF1* and *vHNF1* upstream of the atypical homeodomain reveals a 100% conserved sequence (LYTWYVR) which shows weak similarity to the Hexapeptide. While this similarity is too small to be of statistical significance, it might be an indication that this region of *HNF1* is functionally or evolutionarily related to the Hexapeptide.

<pre>
                                    P B C - A
ceh-20   THPANLSELLDAVLKINEQTLDDNDSAKKQELQCHPMRQALFDVLCETKEKTVLTVRNQVDETPEDPQLMRLDNMLVAEGVAGPDKGGS-----------
PBX2   RGKQDIGDI.QQIMT.TD.S..EA-Q...HA.N..R.KP...S....I....G.SI.SSQE.E.V...........L.......E...GSAAAAAAAAAS
PBX1   GRKQDIGDI.QQIMT.TD.S..EA-Q.R.HA.N..R.KP...N....I......SI.GAQE.E.T...........L.......E...GSAAAAAAAAAS
PBX3   GRKQDIGDI.HQIMT.TD.S..EA-Q...HA.N..R.KP...S....I....G.SI.GAQE.D.P...........L....S..E...GSAAAAAAAAAS
exd   RKQKDIG.I.QQIMS.S..S..EA-Q.R.HT.N..R.KP...S....I......SI..TQE.E.P...........I.......E...GGAAAASAAAAS

                                    P B C - B
ceh-20   ----LGS---DASGGDQADYRQKLHQIRVLYNEELRKYEEACNEFTQHVRSLLKDQSQVRPIAHKEIERMVYIIQRKFNGIQVQLKQSTCEAVMILRSRFLD
PBX2   -GGGVSP---.N.-IEHS...S..A...HI.HS..E...Q......T..MN..RE..RT..V.P..M....S..H...SA..M..............
PBX1   -GG-A..----.N.-VEHS...A..S...QI.HT..E...Q......T..MN..RE..RT...SP.......S..H...SS..M..............
PBX3   -GG--S.----.N.-IEHS...A..T...QI.HT..E...Q......T..MN..RE..RT...SP.......G..H...SS..M..............
exd   QGGS.SIDGA.NA-IEHS...A..A...QI.HQ..E...Q......T..MN..RE..RT...TP.......Q..HK...SS..M..............

                        P B C   h o m e o d o m a i n
ceh-20   ARRKRRNFSKQATEVLNEYFYGHLSNPYPSEEAKEDLARQCNITVSQVSNWFGNKRIRYKKNMAKAQEEASMYAAKKNAHVTLGGMAG
PBX2   ...........S.............E..KK.G.........................IG.F....NI..V.TAVS..Q..HSR
PBX1   .......N.....I.....S.............E..KK.G.........................IG.F....NI....TAVTA.NVSAH.
PBX3   .............I.....S.............E..KK.S.........................IG.F....NL....TAVTAAHAVA.A
exd   ............S.I.....S.............E..RK.G.........................IG......NL.....A.GASPYS...
</pre>

Figure 19: Alignment of PBC class genes.
The thick black bar indicates the homeodomain, and the additional conserved regions are marked by a thin black bar. The conserved domain upstream of the homeodomain is the PBC domain with two subregions, A and B. *exd* accession nr.: L19295

PBC class

Extensive sequence conservation outside the atypical homeodomain has been found between the three human homeobox genes *PBX1*, *PBX2*, *PBX3* and the *C. elegans* gene *ceh-20* (Bürglin and Ruvkun, 1992). This new motif has been termed the PBC domain, and this group of homeodomains is referred to as the PBC class (Fig. 19). Interestingly, this is so far the only case where a conserved motif joins directly to the homeodomain upstream of the homeodomain without any variable linker region. This domain has conserved basic residues (see Fig. 19), and it might possibly be involved in DNA-binding like the POU-specific domain or the *prd* domain.

■ Homeobox genes in plants and fungi

Currently, three homeobox genes are known in yeast. Two of these genes, *MATα1* and *MATα2* are part of the mating type locus (MAT); *MATα1* encodes a typical homeodomain and *MATα2* an atypical homeodomain. Such a dyad of a typical and an atypical homeobox gene is also found in the fungi *Ustilago maydis*, *Schizophyllum commune* and *Coprinus cinereus* (Fig. 18). In *Schizosaccharomyces pombe* though, only an atypical homeobox gene (*MAT Pi*) was found at the MAT locus. That these genes are indeed evolutionarily related is further supported by the fact that *MATα1*, *Um bW* and *Sc AαY* have an intron in the same position in the homeobox (Fig. 22). A region of similarity to the Hexapeptide upstream of the homeodomain in *MATα1* (Fig. 9, Wright et al., 1987) is probably not significant, since there is no obvious sequence conservation in that region in the other fungal genes. It has been proposed that Cc *Aβ1-1* contains a region of similarity to the POU-specific domain (Tymon et al., 1992). However, this is not significant, since this region is not conserved in any of the other fungal atypical homeodomains, and too many gaps have to be introduced for the alignment, which is incompatible with a well conserved domain such as POU.

In plants, three genes with atypical homeodomains (*Kn1*, *ZMH1*, *ZMH2*), and seven genes encoding typical homeodomains have been isolated from *Arabidopsis*, maize and *Lycopersicon* (*Athb1*, *Athb2*, *Athb3*, *HAT22*, *Zmhbx1a*, *Zmhbx1b*, Lp *hbx7*). The *Arabidopsis* genes are related to each other (53-80% identical), and interestingly, following the homeodomain is a conserved region containing a leucine zipper (Busch and Sassone-Corsi, 1990). This similarity appears significant, since other regions outside of the homeodomain are not conserved, while the leucine residues are. This provides a potential link between homeobox containing genes and other regulatory genes containing leucine zippers. *Zmhbx1a* has been reported to have similarity to the POU-specific domain (Bellmann and Werr, 1992), but this similarity appears not significant, since the pattern of the most conserved residues of the POU-B domain does not match the pattern of the purported sequence similarity.

The gene Lp *hbx7* has a cysteine residue at position 9 of helix 3. The only other genes containing a cysteine residue at that position are the POU class genes. Hence Lp *hbx7* is possibly related to the POU class, although the homeodomain as a whole does not fall into the POU class. It remains to be determined whether there is a POU-specific domain upstream of the homeodomain of Lp *hbx7*.

Because of the less complex organization of plants and fungi, it might be expected that they will contain fewer homeobox genes than *Drosophila* and vertebrates. Furthermore, taking the long evolutionary distance between these organisms into account, it will not be possible to assign a particular plant or fungal gene to a specific class of homeodomains (defined here based on vertebrate and fly genes) in many, if not all cases. In fact, plant genes such as *Athb1*, *Athb2*, *Athb3* and *HAT22* might form their own class that is restricted to plants.

■ Homeodomain relatives?

The structural conformation of helix 2 and helix 3 of typical homeodomains is very similar to prokaryotic DNA-binding proteins of the helix-turn-helix type (Pabo and Sauer, 1984; Laughon and Scott, 1984; Qian et al., 1989; Affolter et al., 1991). However, there are also differences between the helix-turn-helix motif and the homeodomain, in particular in the way these domains bind to DNA (Qian et al., 1989; Otting et al., 1990; Treisman, 1992). None of these prokaryotic genes is sufficiently similar in primary sequence to be considered a homeodomain. The fact that several of the atypical homeodomains have insertions between helix 2 and helix 3 (see section above, Fig. 18) shows that the homeodomain does not have to restrict itself during evolution to a helix-turn-helix motif, and may have given rise to variations of DNA-binding proteins that would not be recognized as homeodomains any more.

Such an example could be the gene *zeste*, which shows some similarity to helix 3 in its DNA-binding domain (Chen et al., 1992). Secondary structure predictions suggest that there might be two further α-helices upstream of the helix 3 related motif, and in the region of the first helix, a leucine and a phenylalanine can be found that have the same spacing as the conserved residues in helix 1 of the homeodomain (Fig. 18). Other residues in that region of helix 1 also exhibit good matches to the consensus matrix (Figs. 2, 18). This region of *zeste* with similarity to the homeodomain has been shown to be the DNA-binding domain, and is highly conserved between *Drosophila melanogaster* and *Drosophila virilis* (Chen et al., 1992). Thus this DNA-binding domain might indeed be structurally and evolutionarily related to the homeodomain. However, there are also substantial differences (i.e. the phenylalanine and the tryptophan are exchanged, the region of helix 2 matches poorly and it is not quite clear how the region between helix 1 and helix 3 should be aligned with respect to helix 2) such that *zeste* should not be considered a homeobox gene.

Another example of a gene encoding a DNA-binding motif that is possibly related to the homeodomain is the yeast gene *STE12*. The minimal DNA-binding domain of the *STE12* protein is located between amino acids 41 and 204, and a motif with similarity to the homeodomain is found between amino acids 57-168 (Fig. 5 in Yuan and Fields, 1991). A putative helix 3 contains the highly conserved residues tryptophan and phenylalanine, and a putative helix 1 contains the corresponding hydrophobic residues. However, the proposed alignment requires two large loops between the three helixes, and some of the highly conserved residues in homeodomains are not conserved in *STE12*. In addition, the experimental data indicate that sequences flanking the region of similarity are necessary for DNA-binding. Thus, like *zeste*, *STE12* should not be considered a homeobox gene.

It has been proposed that the DNA-binding domain of the Myb group of proteins has similarity to the homeodomain (Frampton et al., 1989). Indeed, the recently determined structure of mouse c-Myb showed that the DNA-binding domain has 3 alpha-helixes with a hydrophobic core arranged in a pattern reminiscent of the homeodomain (Ogata, 1992). This structural similarity is not better than that of some of the prokaryotic DNA-binding proteins though, nevertheless the Myb genes might be evolutionarily related to the homeobox genes.

■ Homopolymeric amino acid stretches

A sequence feature commonly found in homeobox genes are stretches of homopolymeric amino acids (HPAA). These stretches have attracted attention because they are striking to the eye, but their functional significance is still elusive. Polyglutamine repeats, named M- or *opa* repeats, were initially found in *Antp*, *Dfd*, *en*, *bcd*, but also in other developmental control genes such as *Notch* (McGinnis et al., 1984; Wharton et al., 1985; Frigerio et al., 1986). Since then many other types of repeated amino acid stretches have been found, and Figure 20 gives a summary of the different HPAA stretches found in various homeobox genes. Figure 21a summarizes these findings in a histogram that shows the number of genes containing particular HPAA repeats for each amino acid. Notably, an inspection of the genes containing HPAA repeats shows that the repeats are not well conserved even within phylogenetic classes and families (Fig. 20). For example, *BarH2* contains several types of HPAA repeats spread throughout the protein as does *BarH1*, but when the two genes are compared to each other, it turns out that most of the repeats are not conserved. The same holds true for *Drosophila cut* and human *CDP* (Neufeld et al., 1992). Other examples include polyalanine stretches which are not conserved between *D. virilis en* and *D. melanogaster en* (Kassis et al., 1986), and *ceh-20* and the human *PBX* genes (Bürglin and Ruvkun, 1992). In the mammalian *Hoxb-3* genes, the polyglycine between the Hexapeptide and the homeodomain is not conserved in paralog genes on other clusters such as *Hoxd-3*. HPAA stretches can be regarded as an extreme case of protein regions that are rich in particular amino acids: where such non-homopolymeric enrichment is found, the amino acids enriched are almost always the same as those found in HPAA stretches (Fig. 21a). These non-HPAA enriched regions are also often found in less conserved regions of homeodomain proteins. This is illustrated in Figure 21b: the upper panel

A:	*PBX1, PBX2, PBX3, c Oct-1, c HoxD8, Dth-2, Abd-B, Da BarH1, BarH1, BarH2, Cf1a, cut, H2.0, pdm-1, Df Ubx, Dps Ubx, Md Ubx, Ubx, Dps bcd, Dv en, en, ems, eve, gsb-p, otd, Pox-n, S59/NK-1, zfh-2, h Oct-1, m Brn-1, m Brn-2, h EVX2, m Evx-1, m Evx-2, m HoxD11, m SCIP, r SCIP, m TTF-1, r TTF-1, mab-5, pal-1*
C:	-
D:	*TgHbox5*
E:	*zf-22, x HoxB7, h HoxB6, h HoxB7, m HoxA7, m HoxB6, m HoxB7, r HoxA7, ceh-18*
F:	-
G:	*Da BarH1, Dfd, Df Ubx, Dps Ubx, Md Ubx, Ubx, inv, lab, otd, S59/NK-1, x HoxB5, h HoxA10, h HoxB3, h HoxC6, h HoxD3, h HoxD9, AmphiHox3, m Evx-2, ATBF1, TCL-3, h Oct-2, m Oct-2, m SCIP, r SCIP, m Brn-1, m Brn-2, m HoxB3, m HoxC6, m HoxD9, m HoxD11, Tes-1, m TTF-1, r TTF-1*
H:	*Kn-1, BarH2, Da BarH1, cad, Cf1a, pdm-1, inv, ro, m Brn-1, m HoxA1, m SCIP, r SCIP, Tes-1*
I:	-
K:	*Kn-1, TgHbox5*
L:	-
M:	-
N:	*smox-5, Abd-B, BarH1, BarH2, cad, cut, Dfd, NK-2*
P:	*c HoxA4, cut, Da BarH1, Dps bcd, Xcad1, h HoxA10, h HoxB2, h HoxC5, h HoxD3, ATBF1, HB24, m Brn-1, m Brn-2, Hlx, m HoxB4, m SCIP, r SCIP, lin-11*
Q:	*PHO2, Sg abd-A, abd-A, Abd-B, Antp, Da BarH1, BarH2, Dps bcd, bcd, cut, Dfd, pdm-2, Dv ro, ro, ems, en, lab, msh-2, otd, pb, W26, zfh-1, zfh-2, ATBF1, HB24, m Brn-2, m cdx-1*
R:	*cad*
S:	*Htren, smox-2, TgHbox1, TgHbox5, cut, Dv en, en, zen1, zfh-2, Xcad2, CDP, h HoxD4, m HoxD4, m HoxD9, Pax-8*
T:	*smox-2, Xcad2*
V:	-
W:	-
Y:	-

Figure 20: List of genes containing homopolymeric amino acid (HPAA) stretches.
Homeobox genes were examined for the presence of stretches of HPAAs that were at least 5 residues long. For each individual amino acid (single letter code) genes are given that contain at least one stretch of that particular amino acid.

shows the frequency of the more "frequent" HPAA amino acids along the length of the *Barh2*, while the lower panel shows the regions of conservation between *BarH2* and *BarH1*. As can be seen, there is quite a good correlation between "HPAA amino acids" and conservation. Another form of "enriched regions" are repeats composed of different amino acids. The repeat composed of alternating histidine and proline residues has been termed the *prd*-repeat and has been found in genes such as *prd* and *bcd* (Frigerio et al., 1986).

The role of these repeats and enriched regions is unclear. Some of the HPAAs are the ones proposed to be involved in rapid protein degradation in the PEST hypothesis (Rogers et al., 1986). Other stretches enriched in particular amino acids are thought to be involved in transcription activation (acidic regions, glutamine-rich regions, and proline-rich regions; Ptashne, 1988; Mitchell, 1989; Struhl, 1989). On the other hand, since these hydrophilic HPAA stretches are often less well conserved, they might simply be flexible structural linker regions (that do not have to be especially conserved) between functional protein domains.

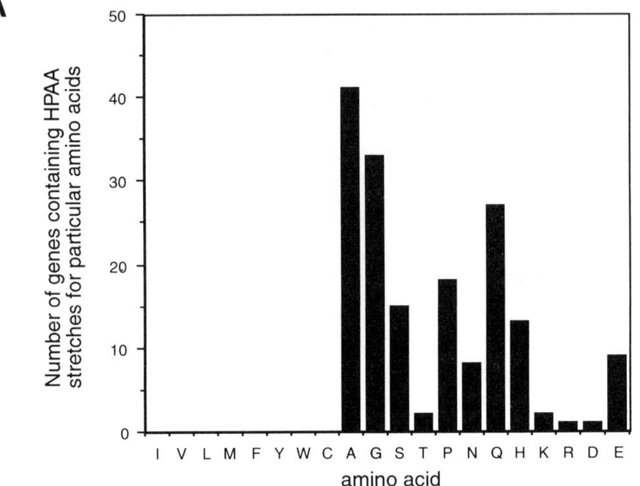

A

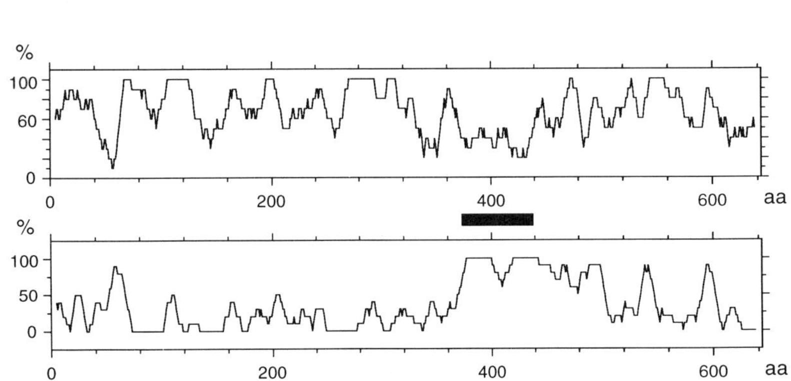

B

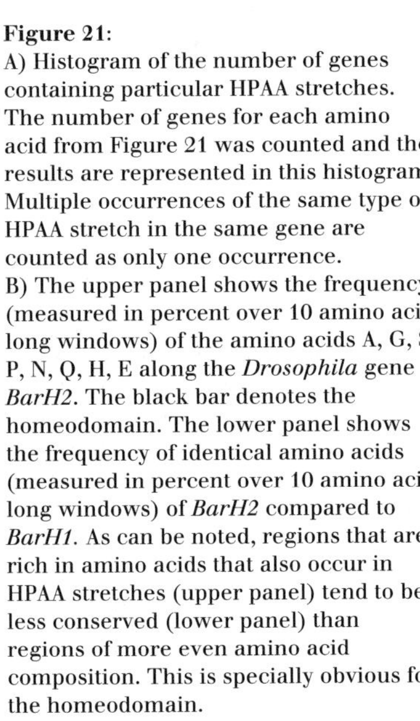

Figure 21:
A) Histogram of the number of genes containing particular HPAA stretches. The number of genes for each amino acid from Figure 21 was counted and the results are represented in this histogram. Multiple occurrences of the same type of HPAA stretch in the same gene are counted as only one occurrence.
B) The upper panel shows the frequency (measured in percent over 10 amino acid long windows) of the amino acids A, G, S, P, N, Q, H, E along the *Drosophila* gene *BarH2*. The black bar denotes the homeodomain. The lower panel shows the frequency of identical amino acids (measured in percent over 10 amino acid long windows) of *BarH2* compared to *BarH1*. As can be noted, regions that are rich in amino acids that also occur in HPAA stretches (upper panel) tend to be less conserved (lower panel) than regions of more even amino acid composition. This is specially obvious for the homeodomain.

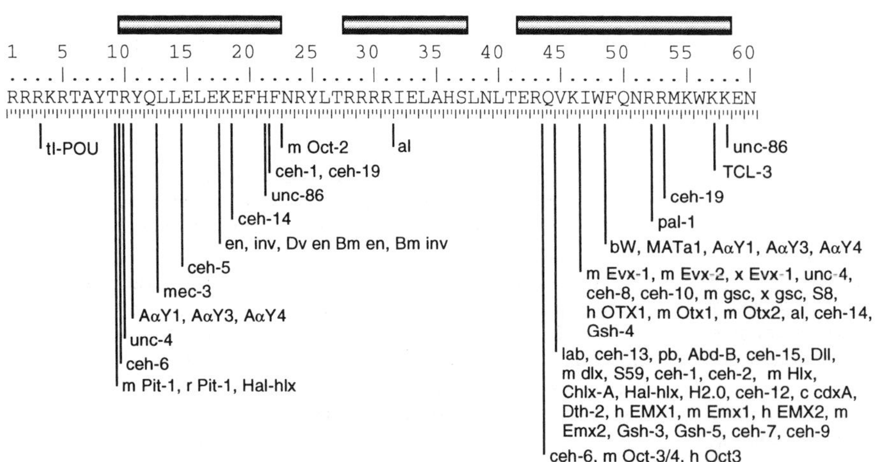

Figure 22: Compilation of intron positions.
The homeodomain consensus is shown at the top, and intron positions for particular genes are indicated underneath.

■ Introns

Introns can provide additional information to help determine relationships between different classes and families of homeodomains. Figure 22 shows a compilation of intron sites within the homeodomain. As is readily apparent, intron sites can be found in many different positions, and some genes contain even two introns within the homeodomain. Splice sites do not fall on "functional" boundaries, such as beginnings or ends of α-helices. Likewise, it is inconceivable that all introns were present in some ancestral homeobox and were subsequently lost (as some theories of intron/exon evolution propose). Clearly, loss and gain, or possibly "jumping" of introns, must have occurred many times during evolution. Splice donors and acceptors have slight sequence preferences within the exons adjacent to the splice sites, which causes certain amino acids to be found more frequently at the splice site in a given frame (Fichant, 1992). Particularly noteworthy are arginine codons, if the splice site is within the codon between base 2 and 3, and valine codons that follow the splice site, if the splice site falls between codons. Such preferences can be observed in here in the homeobox genes.

The following observations support the notion that introns have been lost and gained over time in homeobox genes, and that within related families or classes, introns can be found in the same position:

-Some classes of homeodomains have an intron in the same position, for example *Drosophila DII* and m *Dlx*, members of the *Hlx* class, or the many of the *prd*-like genes such as *gsc, ceh-10, unc-4, Otx1*.

-Some members of the HOM-C cluster (e.g. *pb, Abd-B, lab*) and non-cluster members of the Hexapeptide superclass (e.g. m *Emx1, c cdxA*) have an intron at position 44/45. This suggests that an ancestral gene that gave rise to the *lab, Antp, Abd-B, ems* and *cad* classes contained an intron at that particular position, but that this intron appears to been lost subsequently in many of the branches. For example, it appears that the mouse and human HOX clusters contain no introns in their homeodomains. On the other hand, mouse *Emx* and chicken *cdxA* (cad class) genes have retained that intron, while it was lost in *Drosophila ems* and *cad*.

-The *pal-1* gene of *C. elegans*, a *cad* class member, contains an intron in position 52, unlike chicken *cdxA*. This suggests that this intron has been inserted *de novo* at that position, or possibly "jumped" from one location to another. This is supported by the fact that this intron in this position is the only example known thus far.

-Apparently totally unrelated classes of homeodomains can have an intron in equivalent positions. For example the LIM class gene *ceh-14* has an intron in the same position as some *prd*-like genes and the mouse *Evx* genes. Thus in this case the position of the intron does not correlate with the relationship between these homeobox genes.

-The genes of the POU class have introns in many different places, and few of them are in the same position. For example *unc-86* and *tI-POU* (both members of the POU

family IV) have introns in different places.

In conclusion, while common positions of introns can be a useful marker to identify relationships amongst different homeobox genes, the large number of insertion/deletion events requires one to be cautious. For example, it is not yet clear whether members of the *Dll, Hlx* and *NK-1* classes contain an intron at position 44/45 because they are related to the ancestral Hexapeptide progenitor, or because they have acquired an intron independently at that position.

■ Appendix

A. Notes on sequences

References to sequences are in general not included in this overview (with some exceptions), since they should be present in the accompanying compilation. For any inadvertent omissions, errors or inaccuracies I wish to apologize.

The analysis of the sequences is based on the assumption that no errors are present. While every effort has been made to prevent errors in the sequences during transfer from the computer into the figures, or during the manual entering of sequences, some mistakes might have occurred. However, most mistakes present in any of the sequences are more likely to have occurred at some earlier point. During the course of assembling this homeobox sequence database the following problems have been encountered:

a) Sequencing errors, i.e. frameshifts, misreading of lanes (e.g. reading Cs instead of Gs) etc. are particularly obvious when several independent submissions are present in the database that have differences in their sequence. In such cases the consensus was taken; however, in some cases these differences may represent true allelic variants. b) Errors during translation, i.e. amino acid sequence and nucleotide sequence do not match in a figure. c) Errors in the databank, i.e. nucleotide sequences in the databank were translated in the wrong frame or spliced wrong to produce erroneous protein sequences. d) cDNA synthesis errors, for example the cysteine residue at position 49 instead of the phenylalanine in *XLPOU3* is extremely unlikely, as also noted by the authors (Baltzinger et al., 1992). e) Cloning artifacts, where apparently unrelated sequences were joined to homeobox-containing gene sequences, e.g. r *Hoxc-4*.

In cases where errors were obvious, they were corrected, if possible. Frameshifts in r *Hoxd-3* and r *Hoxb-8* were corrected based on sequence comparison to the mouse homologues. Examples of problematic residues can be found in Figure 7 by careful examination of the sequences.

The yeast gene *ARD1* was reported to have sequence similarity to homeodomains, yet the sequence clearly does not match the pattern of conserved residues (as defined in Fig. 1C), and recent results indicate that the *ARD1* protein is part of a complex with N-terminal acetyltransferase activity (Park and Szostak, 1992). The *Arabidopsis* gene *HAT24* (Schena and Davis, 1992), cloned using degenerate oligonucleotides, only fortuitously resembles the helix 3 region of the homeodomain, since in actuality it exhibits significant sequence similarity to yeast, fly and vertebrate synaptobrevins.

B. Nomenclature

The new nomenclature for the *HOX* clusters (Scott, 1992), prohibits the use of the species codes as part of the gene name. However, since in compilations such as these sequences from many different organisms are compared to each other, it is essential to have tags that identify the species associated with a particular homeobox gene. Figure 23 shows a list of the species abbreviations used throughout this compilation. Some of the abbreviations are actually integral parts of gene names, but have been listed here nonetheless for identification purposes. The new cluster nomenclature requires that the names of many genes from many organisms need to be adapted. Figure 24 shows many of the old names for mouse, rat, human, guinea pig, sheep, chicken, quail, Xenopus, newt, and axolotl genes. Additional old nomenclature of vertebrate clusters can be found in the following references: Martin et al., 1987, Baron et al., 1987, Duboule et al., 1990, Scott, 1992. The identification of some of these genes as a particular member of the cluster is based solely on sequence comparison, and they have not been mapped to the genome. Thus their assignment is at present somewhat tentative. While some zebrafish genes can unambiguously be identified as particular homologues of mouse *Hox* genes, none have been included here yet, because a few of the genes are not easily assigned. Mouse and human cluster genes have been given generic *Hox* names in this compilation. The precise way of capitalizing and hyphenating for the human (e.g. *HOXA1*) and mouse (e.g. *Hoxa-1*) genes is given in Scott (1992) and is used in the rest of the book.

As a corollary of the new *HOX* nomenclature, the use of '*Hox*' is prohibited for any gene outside the cluster. Figure 25 shows a list of new and old names for non cluster genes. A few of these names are my own provisional names. Two genes are somewhat problematic: pig *Hbx24* might be the homologue of mouse *Hoxb-8*, but there are so many inconsistent sequence differences with the *Hoxb-8* group that at present either the accuracy of the sequence or the assignment has to be questioned. Chicken *CHbx-1* is clearly most closely related to the *labial* family, yet it does not match any of the vertebrate genes well enough to warrant precise assignment.

The current rules of nomenclature for genes in the *HOX* cluster and for vertebrate homologues of fly genes pose some problems. In order to assign a new gene name accurately, knowledge of the complete organization of the cluster is required, even before all genes have been cloned and the cluster structure resolved. This problem becomes especially acute as cluster genes from lower chordates are cloned, since one can safely assume that at some point only one or two clusters will be present in these organisms, and that at that point no exact correspondence with *HOX* cluster genes in higher vertebrates is possible. For genes outside the cluster, the problem with nomenclature is exemplified by the *msh* class. Homologues were cloned from different species before it was realized that there was more than one gene of the *msh* class in vertebrates. Consequently, *Xhox7.1* and *Quox7* got their names because of their similarity to the first published gene *Hox-7.1*, while in fact they are more similar to *Hox-8*. At this point it is not even clear how many *msh* class genes there are in the different vertebrate branches, and consequently, an accurate nomenclature can only determined once all members have been cloned.

Possible solutions include giving temporary names for genes, as has been done in zebrafish (e.g. *zf-22*), or giving unique gene names, as has been done in *Drosophila*. Alternatively, one can simply give a specific name to all homeobox genes in a particular organism, and then number them sequentially (e.g. *lox* for leech homeobox genes). This approach is certainly well suited to lower organisms (e.g. *ceh*, *Eghbx*, *smox*, *Dth*), since genes in these organisms might not have precise homologues of particular vertebrate or fly genes. Such gene names do not contain any information about chromosomal location or affiliation with specific classes, and therefore the names would not have to be constantly changed as more information becomes available.

C. Methods

Most incomplete sequences (i.e. short PCR fragments or just helix 3 regions) were not included in the comparative trees or compilations. Comparative trees were generated using the program Pileup which is part of the GCG package (Devereux et al., 1984). Pileup uses a modification of the Feng and Doolittle algorithm (Feng and Doolittle, 1987), and in addition, I implemented my own version of this algorithm for comparative purposes. Additional small utility programs in C were written to support the maintenance of the homeobox gene database. Sequence alignments were performed using "mse" (provided by W. Gilbert), and transferred to a Macintosh™ computer for final editing.

The interpretation of the "comparative" trees is to be taken with a grain of salt. While the programs used are designed to reveal evolutionary relationships between related genes that have the same function, it should be noted that the homeodomain sequences that are compared here have biologically very different functions, although they are basically all DNA-binding domains. Thus the trees should merely be used to examine how closely related the homeodomain sequences are. However, in many instances, the trees do indeed reflect evolutionary relationships, especially if genes within classes are examined. The longer the branch lengths along the horizontal axis are, the more divergent two sequences are and the more statistically unreliable and insignificant the branching patterns are. Thus, the way *smox-5*, *ceh-5*, the *NK-2* class, the *NK-1* class etc. branch from each other should not be considered significant. For example, when the new *Prh* homeodomain sequences were added to the tree, the *smox-5* and the h/c *Prh* homeodomain sequences branch off the very bottom of the *Hlx* class branch. (This is not shown in Fig. 6, since *Prh* was added manually.) Only if the distance from one branch point to the next distant

```
h      human (Homo sapiens)
m      mouse (Mus musculus)
r      rat (Rattus norvegicus)
pig    Sus suum
cat    Felis catus
Ca     Cavia sp. (guinea pig)
Ov     Ovis sp. (sheep)
Bt     Bos taurus (cow)
sh     Syrian hamster
c      chicken (Gallus gallus)
qu     quail (Coturnix coturnix)
x      Xenopus laevis (clawed frog)
Xb     Xenopus borealis
ax     axolotl (Ambystoma mexicanum)
Nv     Notophthalmus viridescens (newt)
zf     zebra fish (Brachydanio rerio)
Ot     Oncorhynchus tschawytscha (Salmon)
Ss     Salmo salar (Atlantic salmon)
Lp     Lampetra planeri (brook lamprey)
Mg     Myxine glutinosa (Atlantic hagfish)
Ci     Ciona intestinalis (ascidia)
Hal    Halocynthia roretzi (ascidia)
Tr     Terebratulina retusa (sea urchin)
Tg     Tripneustes gratilla (sea urchin)
Pa     Parechinus angulosus (sea urchin)
Htr    Helobdella triserialis (leech)
Hr     Helobdella robusta (leech)
Hm     Hirudo medicinalis (leech)
Dt     Dugesia tigrina (flatworm)
Dj     Dugesia japonica (flatworm)
Eg     Echinococcus granulosus (parasitic flatworm)
Cv     Caenorhabditis vulgarensis (roundworm)
ce     Caenorhabditis elegans (roundworm)
Al     Ascaris lumbricoides  (parasitic roundworm)
Hv     Hydra vulgaris (hydra) (in conjunction with cnox)
Cv     Chlorohydra viridissima (hydra) (in conjunction with cnox)
Af     Acropora formosa (coral)
Bm     Bombyx mori (silk worm)
Ms     Manduca sexta (moth)
Sa     Schistocerca americana (grasshopper)
Sg     Schistocerca gregaria (grasshopper)
Aa     Aedes aegypti (mosquito)
Am     Apis mellis  (honeybee)
d      Drosophila melanogaster
Dv     Drosophila virilis
Dps    Drosophila pseudoobscura
Da     Drosophila ananassae
Df     Drosophila funebris
Md     Musca domestica (house fly)
At     Arabidopsis thaliana (plant)
Lp     Lycopersicon peruvianum (plant)
Um     Ustilago maydis (smut fungus)
Cc     Coprinus cinereus (fungus)
Sc     Schizophyllum commune (bracket fungus)
Sp     Schizosaccharomyces pombe (fission yeast)
```

Figure 23: Species codes as used in the nomenclature of gene names.

branch point is clearly set apart, should it be considered real. The fact that the organization of the very distant branch points (as viewed from the end point) are not very significant is exemplified by that fact that some branching patterns appear quite different when the tree generated with Pileup is compared with the unweighted pair-group tree based on the Feng and Doolittle method. Simply adding new sequences to the tree can already change the branching pattern for many sequences. Likewise, other methods such as parsimony produce different trees yet again (e.g. Kappen et al., 1989; Schubert et al. 1993).

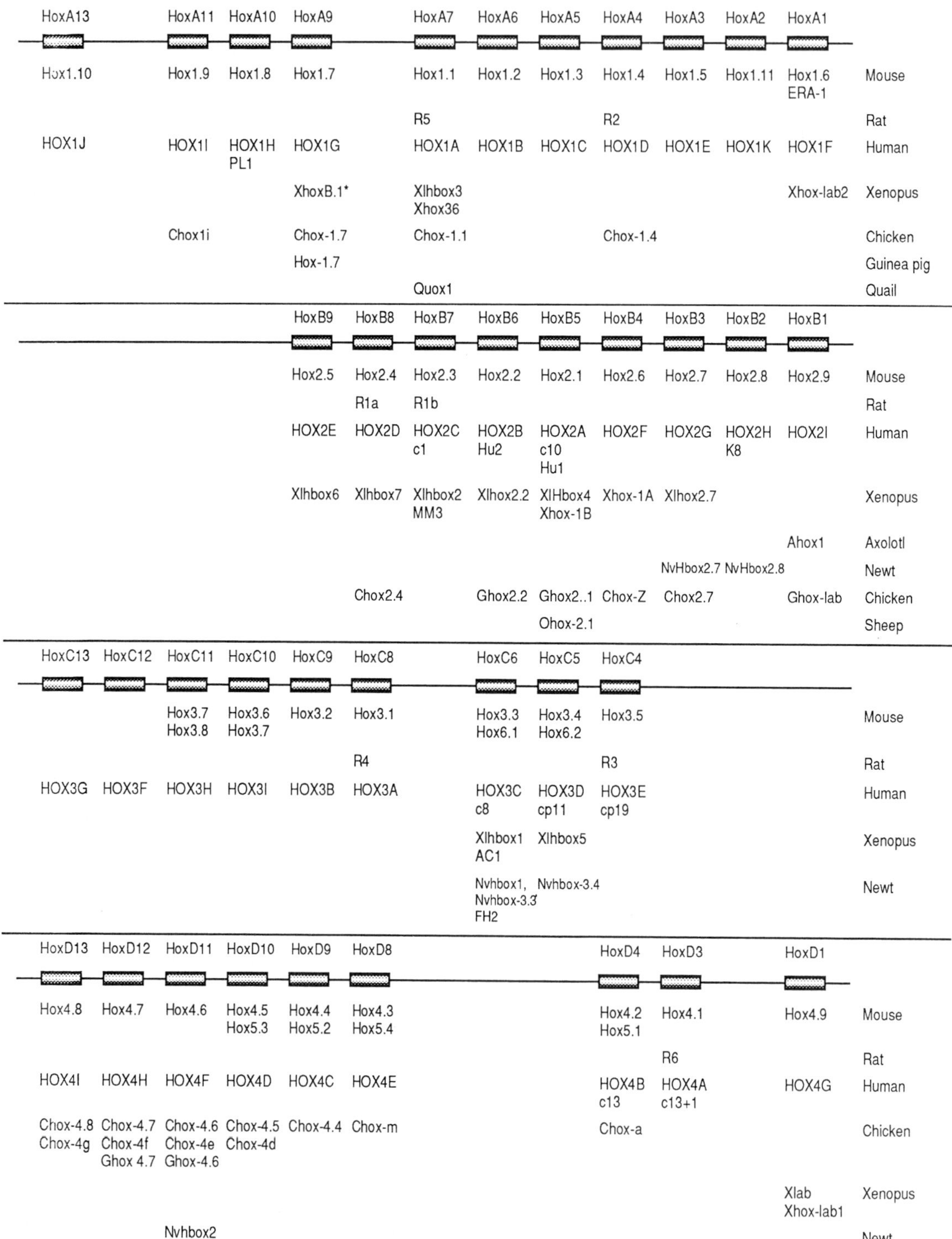

Figure 24: Cluster homeobox genes in old and new nomenclature.
For some sequences the identification is solely based on sequence comparison, and consequently, some assignments might be wrong. *XhoxB.1* (Stickland et al., 1992) is from *Xenopus borealis*, the other *Xenopus* genes from *Xenopus laevis*.

TCL-3	:	h Hox-11
MUR10F	:	m Hox-11
Hal-hlx	:	Hal Ahox1
Chlx-A	:	Chox E
cdxA	:	Chox-cad
Sax-1	:	Chox3
Ovx-1	:	Chox7
x evx-1	:	Xhox-3
c msx-1	:	Ghox7
h MSX1	:	h Hox7, h Hox7.1
m msx-1	:	m Hox7.1
x msx-1	:	Xhox-7.1
m msx-2	:	m Hox8
x msx-2	:	Xhox-7.1'
qu msx-2	:	Quox7
c msx-2	:	Ghox8 = chicken Msx-1
h PHBX1	:	PHOX1
m MHbx	:	MHox
Zmhbx1a	:	Zmhox1a
Zmhbx1b	:	Zmhox1b
Lp hbx7	:	Lp-hox7
CHbx-1	:	CHox 1 = c HoxD1 ??
pig Hbx24	:	pig Hox-2.4 = pig HoxB8 ??

Figure 25: New nomenclature for genes not in the cluster.
Genes that are not part of the cluster are now not to be named
'*Hox*' (Scott, 1992). This list gives the new names for genes that
contained '*Hox*' within their name. In some cases the
nomenclature is simply my provisional naming. *CHbx-1* and pig
Hbx24 might be paralogs or homologues of genes in the cluster,
but exhibit significant sequence differences, which make an
assignment problematic. The assignment of *Xhox-7.1* as *msx-2*
and *Xhox-7.1'* as *x msx-1* is somewhat tentative and based on
flanking sequences (the homeodomains are identical).

■ Acknowledgements

I would like to thank everyone who provided their unpublished sequences as I generated the homeobox database
during the past few years. In particular, I would like to
thank the following people for providing unpublished
sequences or other information for inclusion in this compilation: H. Bode, R. Cortese, D. Deitcher and C. Cepko, J.
Diller and E. Macagno, M. Frasch and W. McGinnis, M.
Frohman, Y. Gruenbaum, J. Hirsh, U. Kües and L.
Casselton, Y. Kurosawa, A. Lakshminarayanan, G. Oliver, I.
Sadler and M.C. Beckerle, K. Schneitz and M. Noll, F.
Schubert and P. Gruss, A. Simeone, P.Q. Thomas and P.D.
Rathjen, R. Ullrich, P. Webster, M. Affolter, U. Walldorf
and W.J. Gehring, C.V.E. Wright, and Y. Xu. I am grateful
to G. Ruvkun for continued support, to M. Cherry for suf-
fering my computer questions, and to E. Golemis and T.
Barnes for critical reading of the manuscript. Thanks goes
to the Ruvkun and Brent lab members for helpful discus-
sions and comments.

■ References

References for most sequences will be found in the
accompanying entries of this book.

Affolter, M., Percival-Smith, A., Müller, M., Billeter, M., Qian,
Y.Q., Otting, G., Wüthrich, K. and Gehring, W.J. (1991).
Similarities between the homeodomain and the Hin
recombinase DNA-binding domain. Cell 64, 879-880.

Adams, B., Dörfler, P., Aguzzi, A., Kozmik, Z., Urbánek, P.,
Maurer-Fogy, I. and Busslinger, M. (1992). *Pax-5* encodes the
transcription factor BSAP and is expressed in B lymphocytes,
the developing CNS, and adult testis. Genes Dev. 6, 1589-1607.

Akam, M. (1989). Hox and HOM: homologous gene clusters in
insects and vertebrates. Cell 57, 347-349.

Andres, V., Nadal-Ginard, B. and Mahdavi, V. (1992). *Clox*, a
mammalian homeobox gene related to *Drosophila cut*,
encodes DNA-binding regulatory proteins differentially
expresssed during development. Development 116, 321-332.

Astell, C.R., Ahlstrom-Jonasson, L., Smith, M., Tatchell, K.,
Nasmyth, K.A. and Hall, B.D. (1981). The sequence of the DNAs
coding for the mating-type loci of *Saccharomyces cerevisiae*.
Cell 27, 15-23.

Baltzinger, M., Payen, E. and Remy, P. (1992). Nucleotide
sequence of *XLPOU3* cDNA, a member of the POU domain
gene family expressed in *Xenopus laevis* embryos. Nucl. Acids
Res. 20, 1993.

Baron, A., Featherstone, M.S., Hill, R.E., Hall, A., Galliot, B. and
Duboule, D. (1987). Hox-1.6: a mouse homeo-box-containing
gene member of the Hox-1 complex. EMBO J. 6, 2977-2986.

Baumhueter, S., Mendel, D.B., Conley, P.B., Kuo, C.J., Turk, C.,
Graves, M.K., Edwards, C.A., Courtois, G. and Crabtree, G.R.
(1990). HNF-1 shares three sequence motifs with the POU
domain proteins and is identical to LF-B1 and APF. Genes Dev.
4, 372-379.

Bellmann, R. and Werr, W. (1992). Zmhox1a, the product of a
novel maize homeobox gene, interacts with the *Shrunken* 26
bp *feedback* control element. EMBO J. 11, 3367-3374.

Birkenmeier, E.H. and Gordon, J.I. (1986). Developmental
regulation of a gene that encodes a cysteine-rich intestinal
protein and maps near the murine immunoglobulin heavy
chain locus. Proc. Natl. Acad. Sci. U.S.A. 83, 2516-2520.

Blochlinger, K., Bodmer, R., Jack, J., Jan, L.Y. and Jan, Y.N. (1988).
Primary structure and expression of a product from *cut*, a locus
involved in specifying sensory organ identity in *Drosophila*.
Nature 333, 629-635.

Bodmer, R., Jan, L.Y. and Jan, Y.N. (1990). A new homeobox-
containing gene, *msh-2*, is transiently expressed early during
mesoderm formation of *Drosophila*. Development 110, 661-
669.

Boehm, T., Foroni, L., Kaneko, Y., Perutz, M.F. and Rabbitts, T.H.
(1991). The rhombotin family of cysteine-rich LIM-domain
oncogenes: distinct members are involved in T-cell
translocations to human chromosomes 11p15 and 11p13. Proc.
Natl., Acad. Sci. U.S.A. 88, 4367-4371.

Boehm, T., Foroni, L., Kennedy, M. and Rabbitts, T.H. (1990a). The
rhombotin gene belongs to a class of transcriptional regulators
with a potential novel protein dimerisation motif. Oncogene 5,
1103-1105.

Boehm, T., Greenberg, J.M., Buluwela, L., Lavenir, I., Forster, A.
and Rabbitts, T.H. (1990b). An unusual structure of a putative T
cell oncogene which allows production of similar proteins from
distinct mRNAs. EMBO J. 9, 857-868.

Boncinelli, E., Simeone, A., Acampora, D. and Mavilio, F. (1991).
HOX gene activation by retinoic acid. TIG 7, 329-334.

Bopp, D., Burri, M., Baumgartner, S., Frigerio, G. and Noll, M.
(1986). Conservation of a large protein domain in the
segmentation gene *paired* and in functionally related genes of
Drosophila. Cell 47, 1033-1040.

Bopp, D., Jamet, E., Baumgartner, S., Burri, M. and Noll, M.
(1989). Isolation of two tissue-specific *Drosophila* paired box
genes, *Pox meso* and *Pox neuro*. EMBO J. 8, 3447-3457.

Bürglin, T.R., Finney, M., Coulson, A. and Ruvkun, G. (1989).
Caenorhabditis elegans has scores of homoeobox-containing
genes. Nature 341, 239-243.

Bürglin, T.R. and Ruvkun, G. (1992). New motif in PBX genes. Nature Genet. 1, 319-320.

Bürglin, T.R. and Ruvkun, G. (1993). The *Caenorhabditis elegans* homeobox gene cluster. Curr. Opin. Genet. Dev. 3, 615–620.

Burri, M., Tromvoukis, Y., Bopp, D., Frigerio, G. and Noll, M. (1989). Conservation of the paired domain in metazoans and its structure in three isolated human genes. EMBO J. 8, 1183-1190.

Busch, S.J. and Sassone-Corsi, P. (1990). Dimers, leucine zippers and DNA-binding domains. Trends Genet. 6, 36-40.

Chalepakis, G., Fritsch, R., Fickenscher, H., Deutsch, U., Goulding, M. and Gruss, P. (1991). The molecular basis of the *undulated/Pax-1* mutation. Cell 66, 873-884.

Chen, J.D., Chan, C.S. and Pirrotta, V. (1992). Conserved DNA binding and self-association domains of the *Drosophila zeste* protein. Mol. Cell. Biol. 12, 598-608.

Chu-Lagraff, Q., Wright, D.M., McNeil, L.K. and Doe, C.Q. (1991). The *prospero* gene encodes a divergent homeodomain protein that controls neuronal identity in *Drosophila*. Development Supplement 2, 79-85.

Crompton, M.R., Bartlett, T.J., MacGregor, A.D., Manfioletti, G., Buratti, E., Giancotti, V. and Goodwin, G.H. (1992). Identification of a novel vertebrate homeobox gene expressed in haematopoietic cells. Nucl. Acids Res. 20, 5661-5667.

Cserjesi, P., Lilly, B., Bryson, L., Wang, Y., Sassoon, D.A. and Olson, E.N. (1992). MHox: a mesodermally restricted homeodomain protein that binds an essential site in the muscle creatine kinase enhancer. Development 115, 1087-1101.

De Robertis, E.M., Bürglin, T.R., Fritz, A., Wright, C.V.E., Jegalian, B., Schnegelsberg, P., Bittner, D., Morita, E., Oliver, G. and Cho, K.W.Y. (1989). Families of vertebrate homeodomain proteins. In DNA-protein interactions in transcription, J. Gralla, ed. (NY.: Alan R. Liss, Inc.), pp 107-115.

De Simone, V., De Magistris, L., Lazzaro, D., Gerstner, J., Monaci, P., Nicosia, A., and Cortese, R. (1991). LFB3, a heterodimer-forming homeoprotein of the LFB1 family, is expressed in specialized epithelia. EMBO J. 10, 1435-1443.

Deguchi, Y. and Kehrl, J.H. (1991). Nucleotide sequence of a novel diverged human homeobox gene encodes a DNA binding protein. Nucl. Acids Res. 19, 3742.

Devereux, J., Haeberli, P. and Smithies, O. (1984). A comprehensive set of sequence analysis programs for the VAX. Nucl. Acids Res. 12, 387-395.

Dressler, G.R., Deutsch, U., Chowdhury, K., Nornes, H.O. and Gruss, P. (1990). *Pax2*, a new murine paired-box-containing gene and its expression in the developing excretory system. Development 109, 787-795.

Duboule, D. (1992). The vertebrate limb: A model system to study the HOX/HOM gene network during Development and Evolution. BioEssays 14, 375-384.

Duboule, D., Boncinelli, E., De Robertis, E., Featherstone, M., Lonai, P., Oliver, G. and Ruddle, R.H. (1990). An update of mouse and human HOX gene nomenclature. Genomics 7, 458-459.

Feng, D.-F. and Doolittle, R.F. (1987). Progressive sequence alignment as a prerequisite to correct phylogenetic trees. J. Mol. Evol. 25, 351-360.

Fichant, G.A. (1992). Constraints acting on the exon positions of the splice site sequences and local amino acid composition of the protein. Human Molecular Genetics 1, 259-267.

Finney, M. (1990). The homeodomain of the transcription factor LF-B1 has a 21 amino acid loop between helix 2 and helix 3. Cell 59, 5-6.

Fortini, M.E., Lai, Z. and Rubin, G.M. (1991). The *Drosophila zfh-1* and *zfh-2* genes encode novel proteins containing both zinc-finger and homeodomain motifs. Mech. Dev. 34, 113-122.

Frain, M., Swart, G., Monaci, P., Nicosia, A., Stämpfli, S., Frank, R. and Cortese, R. (1989). The liver-specific transcription factor LF-B1 contains a highly diverged homeobox DNA binding domain. Cell 59, 145-157.

Frampton, J., Leutz, A., Gibson, T.J. and Graf, T. (1989). DNA-binding ancestry. Nature 342, 134.

Freyd, G., Kim, S. and Horvitz, R.H. (1990). Novel cysteine-rich motif and homeodomain in the product of the *Caenorhabditis elegans* cell lineage gene *lin-11*. Nature 344, 876-879.

Frigerio, G., Burri, M., Bopp, D., Baumgartner, S. and Noll, M. (1986). Structure of the Segmentation gene *paired* and the *Drosophila* PRD gene set as part of a gene network. Cell 47, 735-746.

Furukubo-Tokunaga, K., Müller, M., Affolter, M., Pick, L., Kloter, U. and Gehring, W.J. (1992). *In vivo* analysis of the helix-turn-helix motif of the *fushi tarazu* homeo domain of *Drosophila melanogaster*. Genes Dev. 6, 1082-1096.

German, M.S., Wang, J., Chadwick, R.B. and Rutter, W.J. (1992). Synergistic activation of the insulin gene by a LIM-homeo domain protein and a basic helix-loop-helix protein: building a functional insulin minienhancer complex. Genes Dev. 6, 2165-2176.

Gillissen, B., Bergemann, J., Sandmann, C., Schroeer, B., Bölker, M. and Kahmann, R. (1992). A two-component regulatory system for self/non-self recognition in *Ustilago maydis*. Cell 68, 647-657.

Goulding, M.D., Chalepakis, G., Deutsch, U., Erselius, J.R. and Gruss, P. (1991). Pax-3, a novel murine DNA binding protein expressed during early neurogenesis. EMBO J. 10, 1135-1147.

Guazzi, S., Price, M., De Felice, M., Damante, G., Mattei, M.-G. and Di Lauro, R. (1990). Thyroid nuclear factor (TTF-1) contains a homeodomain and displays a novel DNA binding specificty. EMBO J. 9, 3631-3639.

Hall, M.N. and Johnson, A.D. (1987). Homeo domain of the yeast repressor α2 is a sequence-specific DNA-binding domain but is not sufficient for repression. Science 237, 1007-1012.

Hanes, S.D. and Brent, R. (1991). A genetic model for interaction of the homeodomain recognition helix with DNA. Science 251, 426-430.

Hatano, M., Roberts, C.W.M., Minden, M., Crist, W.M. and Korsmeyer, S.J. (1991). Deregulation of a homeobox gene, HOX11, by the t(10;14) in T cell leukemia. Science 253, 79-82.

Hawkins, N.C. and McGhee, J.D. (1990). Homeobox containing genes in the nematode *Caenorhabditis elegans*. Nucl. Acids Res. 18, 6101-6106.

He, X., Treacy, M.N., Simmons, D.M., Ingraham, H.A., Swanson, L.W., and Rosenfeld, M.G. (1989). Expression of a large family of POU-domain regulatory genes in mammalian brain development. Nature 340, 35-42.

Hempe, J.M. and Cousins, R.J. (1991). Cysteine-rich intestinal protein binds zinc during transmucosal zinc transport. Proc. Natl. Acad. Sci. 88, 9671-9674.

Herr, W., Sturm, R.A., Clerc, R.G., Corcoran, L.M., Baltimore, D., Sharp, P.A., Ingraham, H.A., Rosenfeld, M.G., Finney, M., Ruvkun, G. and Horvitz, H.R. (1988). The POU domain: a large conserved region in the mammalian pit-1, oct-1, oct-2, and *Caenorhabditis elegans unc-86* gene products. Genes Dev. 2, 1513-1516.

Holland, P. (1992). Homeobox genes in vertebrate evolution. BioEssays 14, 267-273.

Holland, P.W.H., Holland, L.Z., Williams, N.A. and Holland, N.D. (1992). An amphioxus homeobox gene: sequence conservation, spatial expression during development and insights into vertebrate evolution. Development 116, 653-661.

Ingraham, H.A., Flynn, S.E., Voss, J.W., Albert, V.R., Kapiloff, M.S., Wilson, L. and Rosenfeld, M.G. (1990). The POU-specific domain of Pit-1 is essential for sequence-specific, high affinity DNA binding and DNA-dependent Pit-1-Pit-1 interactions. Cell 61, 1021-1033.

Izpisúa-Belmonte, J.-C., Falkenstein, H., Dollé, P., Renucci, A. and Duboule D. (1991). Murine genes related to the *Drosophila AbdB* homeotic gene are sequentially expressed during development of the posterior part of the body. EMBO J. 10, 2279-2289.

Joyner, A.L. and Hanks, M. (1991). The engrailed genes: evolution of function. Seminars in Developmental Biology 2, 435-445.

Kamb, A., Weir, M., Rudy, B., Varmus, H. and Kenyon, C. (1989). Identification of genes from pattern formation, tyrosine kinase, and pottassium channel families by DNA amplification. Proc. Natl. Acad. Sci. U.S.A. 86, 4372-4376.

Kamps, M.P., Murre, C., Sun, X.-H. and Baltimore D. (1990). A new homeobox gene contributes the DNA binding domain of the t(1;19) translocation protein in pre-B ALL. Cell 60, 547-555.

Kappen, C., Schughart, K. and Ruddle, F.H. (1989). Two steps in the evolution of *Antennapedia*-class vertebrate homeobox genes. Proc. Natl. Acad. Sci. U.S.A. 86, 5459-5463.

Karlsson, O., Thor, S., Norberg, T., Ohlsson, H. and Edlund, T. (1990). Insulin gene enhancer binding protein Isl-1 is a member of a novel class of proteins containing both a homeo- and a Cys-His domain. Nature 344, 879-882.

Kassis, J.A., Poole, S.J., Wright, D.K. and O'Farrell, P.H. (1986). Sequence conservation in the protein coding and intron regions of the *engrailed* transcription unit. EMBO J. 5, 3583-3589.

Kaufman, T.C., Seeger, M.A. and Olsen, G. (1990). Molecular and genetic organization of the Antennapedia gene complex of Drosophila melanogaster. Advances in Genetics 27, 309-362.

Kelly, M., Burke, J., Smith, M., Klar, A. and Beach, D. (1988). Four mating-type genes control sexual differentiation in the fission yeast. EMBO J. 7, 1537-1547.

Kennedy, M.A., Gonzalez-Sarmiento, R., Kees, U.R., Lampert, F., Dear, N., Boehm, T. and Rabbitts, T.H. (1991). *HOX11*, a homeobox-containing T-cell oncogene on human chromosome 10q24. Proc. Natl. Acad. Sci. U.S.A. 88, 8900-8904.

Kenyon, C. and Wang, B. (1991). A cluster of Antennapedia-class homeobox genes in a nonsegmented animal. Science 253, 516-517.

Kern, M.J., Witte, D.P., Valerius, M.T., Aronow, B.J. and Potter, S.S. (1992). A novel murine homeobox gene isolated by a tissue specific PCR cloning strategy. Nucl. Acids Res. 20, 5189-5195.

Kissinger, C.R., Liu, B., Martin-Blanco, E., Kornberg, T.B. and Pabo, C.O. (1990). Crystal structure of an engrailed homeodomain-DNA complex at 2.8 Å resolution: a framework for understanding homeodomain-DNA interactions. Cell 63, 579-590.

Krauss, S., Johansen, T., Korzh, V. and Fjose, A. (1991a). Expression of the zebrafish paired box gene pax[zf-b] during early neurogenesis. Development 113, 1193-1206.

Krauss, S., Johansen, T., Korzh, V. and Fjose, A. (1991b). Expression pattern of zebrafish *pax* genes suggests a role in early brain regionalization. Nature 353, 267-270.

Kronstad, J.W. and Leong, S.A. (1990). The *b* mating-type locus of *Ustilago maydis* contains variable and constant regions. Genes Dev. 4, 1384-1395.

Krumlauf, R. (1992). Evolution of the vertebrate Hox homeobox genes. BioEssays 14, 245-252.

Kües, U., Richardson,W.V.J., Tymon, A.M., Mutasa, E.S., Göttgens, B., Gaubatz, S., Gregoriades, A. and Casselton, L.A. (1992). The combination of dissimilar alleles of the *Aα* and *Aβ* gene complexes, whose proteins contain homeo domain motifs, determines sexual development in the mushroom *Coprinus cinereus*. Genes Dev. 6, 568-577.

Laughon, A. and Scott, M.P. (1984). Sequence of a *Drosophila* segmentation gene: protein structure homology with DNA-binding proteins. Nature 310, 25-31.

LeBowitz, J.H., Clerc, R.G., Brenowitz, M. and Sharp, P.A. (1989). The Oct-2 protein binds cooperatively to adjacent octamer sites. Genes Dev. 3, 1625-1638.

Li, P.M., Reichert, J., Freyd, G., Horvitz, H.R. and Walsh, C.T. (1991). The LIM region of a presumptive *Caenorhabditis elegans* transcription factor is an iron-sulfur- and zinc-containing metallodomain. Proc. Natl. Acad. Sci. U.S.A. 88, 9210-9213.

Liebhaber, S.A., Emery, J.G., Urbanek, M., Wang, X. and Cooke, N.E. (1990). Characterization of a human cDNA encoding a widely expressed and highly conserved cysteine-rich protein with an unusual zinc-finger motif. Nucl. Acids Res. 18, 3871-3879.

Lu, M., Gong, Z., Shen, W. and Ho, A. (1991). The *tcl-3* proto-oncogene altered by chromosomal translocation in T-cell leukemia codes for a homeobox protein. EMBO J. 10, 2905-2910.

Lundell, M.J. and Hirsh, J. (1992). The *zfh-2* gene product is a potential regulator of neuron-specific DOPA decarboxylase gene expression in *Drosophila*. Dev. Biology 154, 84-94.

Martin, G.R., Boncinelli, E., Duboule, D., Gruss, P., Jackson, I., Krumlauf, R., Lonai, P., McGinnis, W., Ruddle, F. and Wolgemuth, D. (1987). Nomenclature for homoeobox-containing genes. Nature 325, 21-22.

Matsuzaki, F., Koizumi, K., Hama, C., Yoshioka, T. and Nabeshima, Y. (1992). Cloning of the *Drosophila prospero* gene and its expression in ganglion mother cells. Biochem. Biophys. Res. Commun. 182, 1326-1332.

Mavilio, F., Simeone, A., Giampaolo, A., Faiella, A., Zappavigna, V., Acampora, D., Poiana, G., Russo, G., Peschle, C. and Boncinelli, E. (1986). Differential and stage-related expression in embryonic tissues of a new human homoeobox gene. Nature 324, 664-668.

McGinnis, W., Garber, R.L., Wirz, J., Kuroiwa, A. and Gehring, W.J. (1984). A homologous protein-coding sequence in *Drosophila* homeotic genes and its conservation in other metazoans. Cell 37, 403-408.

McGinnis, W. and Krumlauf, R. (1992). Homeobox genes and axial patterning. Cell 68, 283-302.

McGinnis, W., Levine, M.S., Hafen, E., Kuroiwa, A. and Gehring, W.J. (1984). A conserved DNA sequence in homoeotic genes of the *Drosophila* Antennapedia and bithorax complexes. Nature 308, 428-433.

McGuire, E.A., Hockett, R.D., Pollock, K.M., Bartholdi, M.F., O'Brien, S.J. and Korsmeyer, S.J. (1989). The t(11;14)(p15;q11) in a T-cell acute lymphoblastic leukemia cell line activates multiple transcripts, including Ttg-1, a gene encoding a potential zinc finger protein. Mol. Cell. Biol. 9, 2124-2132.

Miles, A. and Miller, D.J. (1992). Genomes of diploblastic organisms contain homeoboxes: sequence of *eveC*, an *even-skipped* homologue from the cnidarian *Acropora formosa*. Proc. R. Soc. Lond B 248, 159-161.

Miller, A.M. (1984). The yeast MATa1 gene contains two introns. EMBO J. 3, 1061-1065.

Mitchell, P.J. and Tjian, R. (1989). Transcriptional regulation in mammalian cells by sequence-specific DNA binding proteins. Science 245, 371-378.

Monica, K., Galili, N., Nourse, J., Saltman, D. and Cleary, M.L. (1991). *PBX2* and *PBX3*, new homeobox genes with extensive homology to the human proto-oncogene *PBX1*. Mol. Cell. Biol. 11, 6149-6157.

Morinaga, T., Yasuda, H., Hashimoto, T., Higashio, K. and Tamaoki, T.(1991). A human alpha-fetoprotein enhancer-binding protein, ATBF1, contains four homeodomains and seventeen zinc fingers. Mol. Cell. Biol. 11, 6041-6049.

Murtha, M.T., Leckman, J.F. and Ruddle, F.H. (1991). Detection of homeobox genes in development and evolution. Proc. Natl. Acad. Sci. U.S.A. 88, 10711-10715.

Naito, M., Kohara, Y. and Kurosawa, Y. (1992). Identification of a homeobox-containing gene located between *lin-45* and *unc-24* on chromosome IV in the nematode *Caenorhabditis elegans*. Nucl. Acids Res. 20, 2967-2969.

Nalik, P., Panayotova-Heiermann, M. and Pongs, O. (1989). Characterization of an estradiol-stimulated mRNA in the brain of adult male rats. Mol. Cell. Endocrinology 62, 235-242.

Neufeld, E.J., Skalnik, D.G., Lievens, P.M.-J. and Orkin, S.H. (1992). Human CCAAT displacement protein is homologous to the *Drosophila* homeoprotein *cut*. Nature Genet. 1, 50-55.

Nourse, J., Mellentin, J.D., Galili, N., Wilkinson, J., Stanbridge, E., Smith, S.D. and Cleary, M.L. (1990). Chromosomal translocation t(1;19) results in the synthesis of a homeobox fusion mRNA that codes for a potential chimeric transcription factor. Cell 60, 535-545.

Novotny, C.P., Stankis, M.M., Specht, C.A., Yang, H., Ullrich, R.C. and Giasson, L. (1991). The $A\alpha$ mating type locus of *Schizophyllum commune*. In: More Gene Manipulations in Fungi, J. W. a. L. Bennett L.L., ed. Academic Press, Inc.), pp 234-257.

Ogata, K., Hironobu, H., Aimoto, S., Nakai, T., Nakamura, H., Sarai, A., Ishii, S. and Nishimura, Y. (1992). Solution structure of a DNA-binding unit of Myb: a helix-turn-helix-related motif with conserved tryptophans forming a hydrophobic core. Proc. Natl. Acad. Sci. U.S.A. 89, 6428-6432.

Oliver, G., Vispo, M., Maihlos, A., Martinez, C., Sosa, B., Fielitz, W. and Ehrlich, R. (1992). Homeobox-containing genes in flatworms. Gene 121, 337-342.

Otting, G., Qian, Y.Q., Billeter, M., Müller, M., Affolter, M., Gehring, W.J. and Wüthrich, K. (1990). Protein-DNA contacts in the structure of a homeodomain - DNA complex determined by nuclear magnetic resonance spectroscopy in solution. EMBO J. 9, 3085-3092.

Pabo, C.O. and Sauer, R.T. (1984). Protein-DNA recognition. Ann. Rev. Biochem. 53, 293-321.

Park, E.-C. and Szostak, J.W. (1992). ARD1 and NAT1 proteins form a complex that has N-terminal acetyltransferase activity. EMBO J. 11, 2087-2093.

Plachov, D., Chowdhury, K., Walther, C., Simon, D., Guenet, J.-L. and Gruss, P. (1990). Pax8, a murine paired box gene expressed in the developing excretory system and thyroid gland. Development 110, 643-651.

Price, M., Lazzaro, D., Pohl, T., Mattei, M.-G., Rüther, U., Olivo, J.-C., Duboule, D. and Di Lauro, R. (1992). Regional expression of the homeobox gene Nkx-2.2 in the developing mammalian forebrain. Neuron 8, 241-255.

Ptashne, M. (1988). How eukaryotic transcriptional regulators work. Nature 335, 683-689.

Qian, Y.Q., Billeter, M., Otting, G., Müller, M., Gehring, W.J. and Wüthrich, K. (1989). The structure of the Antennapedia homeodomain determined by NMR spectroscopy in solution: comparison with prokaryotic repressors. Cell 59, 573-580.

Rabbitts, T.H. and Boehm, T. (1990). LIM domains. Nature 346, 418.

Rogers, S., Wells, R. and Rechsteiner, M. (1986). Amino acid sequences common to rapidly degraded proteins: the PEST hypothesis. Science 234, 364-368.

Ruvkun, G. and Finney, M. (1991). Regulation of transcription and cell identity by POU domain proteins. Cell 64, 475-478.

Sadler, I., Crawford, A.W., Michelsen, J.W. and Beckerle, M.C. (1992). Zyxin and cCRP: two interactive LIM domain proteins associated with the cytoskeleton. J. Cell. Biol. 119, 1573–1587.

Sasaki, A.W., Doskow, J., MacLeod, C.L., Rogers, M.B., Gudas, L.J. and Wilkinson, M.F. (1991). The oncofetal gene Pem encodes a homeodomain and is regulated in primordial and pre-muscle stem cells. Mech. Dev. 34, 155-164.

Schaller, D., Wittmann, C., Spicher, A., Müller, F. and Tobler, H. (1990). Cloning and analysis of three new homeobox genes from the nematode *Caenorhabditis elegans*. Nucl. Acids Res. 18, 2033-2036.

Schena, M. and Davis, R.W. (1992). HD-Zip proteins: members of an *Arabidopsis* homeodomain protein superfamily. Proc. Natl. Acad. Sci. U.S.A. 89, 3894-3898.

Schenk, M.A., Bode, H.R. and Steele, R.E. (1993). Expression of Cnox-2, a HOM/HOX homeobox gene in hydra, is correlated with axial pattern formation. Development 117, 657–667.

Schier, A.F. and Gehring, W.J. (1992). Direct homeodomain-DNA interaction in the autoregulation of the *fushi tarazu* gene. Nature 356, 804-807.

Schierwater, B., Murtha, M., Dick, M., Ruddle, F.H. and Buss, L.W. (1991). Homeoboxes in Cnidarians. J. of Exp. Zool. 260, 413-416.

Schneitz, K., Spielmann, P. and Noll, M. (1992). Molecular genetics of *aristaless*, a prd-type homeobox gene involved in the morphogenesis of proximal and distal pattern elements in a subset of appendages in *Drosophila*. Genes Dev. 7, 114–129.

Schöler, H.R. (1991). Octamania: The POU factors in murine development. TIG 7, 323-329.

Schubert, F.R., Nieselt-Struwe, K. and Gruss, P. (1993). The *Antennapedia*-type homeobox genes have evolved from three precursors separated early in metazoan evolution. Proc. Natl. Acad. Sci. U.S.A. 90, 143–147.

Schulz, B., Banuett, F., Dahl, M., Schlesinger, R., Schäfer, W., Martin, T., Herskowitz, I. and Kahmann, R. (1990). The b Alleles of *U. maydis*, whose combinations program pathogenic development, code for polypeptides containing a homeodomain-related motif. Cell 60, 295-306.

Schummer, M., Scheurlen, I., Schaller, C. and Galliot, B. (1992). HOM/HOX homeobox genes are present in hydra (*Chlorohydra viridissima*) and are differentially expressed during regeneration. EMBO J. 11, 1815-1823.

Scott, M.P. (1992) Vertebrate homeobox gene nomenclature. Cell 71, 551-553.

Scott, M.P., Tamkun, J.W. and Hartzell III, G.W. (1989). The structure and function of the homeodomain. Biochim. Biophys. Acta review on cancer 989, 25-48.

Scott, M.P. and Weiner, A.J. (1984). Structural relationships among genes that control development: Sequence homology between the *Antennapedia*, *Ultrabithorax*, and *fushi tarazu* loci in *Drosophila*. Proc. Natl. Acad. Sci. U.S.A. 81, 4115-4119.

Shankland, M., Martindale, M.Q., Nardelli-Haefliger, D., Baxter, E. and Price, D.J. (1991). Origin of segmental identity in the development of the leech nervous system. Development Supplement 2, 29-38.

Shepherd, J.C.W., McGinnis, W., Carrasco, A.E., De Robertis, E.M. and Gehring, W.J. (1984). Fly and frog homoeo domains show homologies with yeast mating type regulatory proteins. Nature 310, 70-71.

Simeone, A., Acampora, D., Gulisano, M., Stornaiuolo, A. and Boncinelli, E. (1992). Nested expression domains of four homeobox genes in developing rostral brain. Nature 358, 687-690.

Singh, G., Kaur, S., Stock, J.L., Jenkins, N.A., Gilbert, D.J., Copeland, N.G. and Potter, S.S. (1991). Identification of 10 murine homeobox genes. Proc. Natl. Acad. Sci. U.S.A. 88, 10706-10710.

Stankis, M.M., Specht, C.A., Yang, H., Giasson, L., Ullrich, R.C. and

Novotny, C.P. (1992). The Aα mating locus *Schizophyllum commune* encodes two dissimilar multiallelic homeodomain proteins. Proc. Natl. Acad. Sci. U.S.A. 89, 7169-7173.

Stickland, J.E., Sharpe, C.R., Turner, P.C. and Hames, B.D. (1992). A *Xenopus borealis* homeobox gene expressed preferentially in posterior ectoderm. Gene 116, 269-273.

Struhl, K. (1989). Molecular mechanisms of transcriptional regulation in yeast. Annu. Rev. Biochem. 58, 1051-1077.

Sturm, R.A. and Herr, W. (1988). The POU domain is a bipartite DNA-binding structure. Nature 336, 601-604.

Sulston, J., Du, Z., Thomas, K., Wilson, R., Hillier, L., Staden, R., Halloran, N., Green, P., Thierry-Mieg, J., Qiu, L., Dear, S., Coulson, A., Craxton, M., Durbin, R., Berks, M., Metzstein, M., Hawkins, T., Ainscough, R. and Waterston, R. (1992). The C. *elegans* genome sequencing project: a beginning. Nature 356, 37-41.

Tabin, C.J. (1992). Why we have (only) five fingers per hand: Hox genes and the evolution of paired limbs. Development 116, 289-296.

Thomas, P.Q. and Rathjen, P.D. (1992). HES-1, a novel homeobox gene expressed by murine embryonic stem cells, identifies a new class of homeobox genes. Nucl. Acids Res. 20, 5840.

Treacy, M.N., Neilson, L.I., Turner, E.E., He, X. and Rosenfeld, M.G. (1992). Twin of I-POU: a two amino acid difference in the I-POU homeodomain distinguishes activator from an inhibitor of transcription. Cell 68, 491-505.

Treisman, J., Harris, E. and Desplan, C. (1991). The paired box encodes a second DNA-binding domain in the Paired homeo domain protein. Genes Dev. 5, 594-604.

Treisman, J., Harris, E., Wilson, D. and Desplan, C. (1992). The Homeodomain: a new face for the helix-turn-helix? BioEssays 14, 145-150.

Tymon, A.M., Kües, U., Richardson, W.V.J. and Casselton, L.A. (1992). A fungal mating type protein that regulates sexual and asexual development contains a POU-related domain. EMBO J. 11, 1805-1813.

Vaessin, H., Grell, E., Wolff, E., Bier, E., Jan, L.Y. and Jan, Y.N. (1991). *prospero* is expressed in neuronal precursors and encodes a nuclear protein that is involved in the control of axonal outgrowth in *Drosophila*. Cell 67, 941-953.

Vollbrecht, E., Veit, B., Sinha, N. and Hake, S. (1991). The developmental gene Knotted-1 is a member of a maize homeobox gene family. Nature 350, 241-243.

Walther, C. and Gruss, P. (1991). *Pax-6*, a murine paired box gene, is expressed in the developing CNS. Development 113, 1435-1449.

Walther, C., Guenet, J.-L., Simon, D., Deutsch, U., Jostes, B., Goulding, M.D., Plachov, D., Balling, R. and Gruss, P. (1991). Pax: a murine multigene family of paired box-containing genes. Genomics 11, 424-434.

Waterston, R., Martin, C., Craxton, M., Huynh, C., Coulson, A., Hillier, L., Durbin, R., Green, P., Shownkeen, R., Halloran, N., Metzstein, M., Hawkins, T., Wilson, R., Berks, M., Du, Z., Thomas, K., Thierry-Mieg, J. and Sulston, J. (1992). A survey of expressed genes in *Canorhabditis elegans*. Nature Genet. 1, 114-123.

Webster, P.J. and Mansour, T.E. (1992). Conserved classes of homeodomains in *Schistosoma mansoni*, an early bilateral metazoan. Mech. Dev. 38, 25-32.

Wharton, K.A., Yedvobnick, B., Finnerty, V.G. and Artavanis-Tsakonas, S. (1985). opa: a novel family of transcribed repeats shared by the notch locus and other developmentally regulated loci in *D. melanogaster*. Cell 40, 55-62.

Wolberger, C., Vershon, A.K., Liu, B., Johnson, A.D. and Pabo, C.O. (1991). Crystal structure of MATα2 Homeodomain-operator complex suggests a general model for homeodomain-DNA interactions. Cell 67, 517-528.

Wright, C.V.E., Cho, K.W.Y., Fritz, A., Bürglin, T.R. and De Robertis, E.M. (1987). A Xenopus laevis gene encodes both homeobox-containing and homeobox-less transcripts. EMBO J. 6, 4083-4094.

Wysocka-Diller, J.W., Aisemberg, G.O., Baumgarten, M., Levine, M. and Macagno, E.R. (1989). Characterization of a homologue of bithorax-complex genes in the leech *Hirudo medicinalis*. Nature 341, 760-763.

Xu, Y., Baldassare, M., Fisher, P., Rathbun, G., Oltz, E., Yancopoulos, G.D., Jessell, T.M. and Alt, F. (1992). LH-2: a LIM/homeodomain gene expressed in developing lymphocytes and neural cells. Proc. Natl. Acad. Sci. U.S.A. 90, 227–231.

Yuan, Y.-L.O. and Fields, S. (1991). Properties of the DNA-binding domain of the *Saccharomyces cerevisiae* STE12 protein. Mol. Cell. Biol. 11, 5910-5918.

■ *Thomas R. Bürglin:*
Dept. of Molecular Biology,
Massachusetts General Hospital, and
Dept. of Genetics,
Harvard Medical School,
Wellman 8,
Fruit Street,
Boston, MA 02114,
USA

Descriptions of the Homeobox Genes

Functional analysis of the homeodomain-related proteins of the $A\alpha$ locus of *Schizophyllum commune*. Proc. Natl. Acad. Sci. USA 89, 7174-7178.

Stankis, M.M., Specht, C.A., Yang, H., Giasson, L., Ullrich, R.C. and Novotny, C.P. (1992). The $A\alpha$ mating locus of *Schizophyllum commune* encodes two dissimilar, multiallelic, homeodomain proteins. Proc. Natl. Acad. Sci. USA 89, 7169-7173.

$A\alpha1$ [Sc]

Species: *Schizophyllum commune*
Chromosomal location: -
Other names: *Z*
Cognate genes: -
Type of homeobox: atypical
Accession number: M80789 ($A\alpha1$), M80824 ($A\alpha3$) and M80822 ($A\alpha4$)

■ Origin and description

The *Z* gene is one of two genes at the $A\alpha$ mating type locus of S. *commune*. It is present in two of the alleles sequenced ($A\alpha3$ and $A\alpha4$); it is absent in $A\alpha1$ (Stankis et al., 1992). The *Z* gene contains an ORF of 891 amino acids in *Z3* and 930 amino acids in *Z4*. The *Z* polypeptide contains two highly acidic regions of about 30 amino acids each rich in glutamate and aspartate. The *Z3* polypeptide contains an atypical homeodomain at position 111 to 189, with 34 extra amino acids between helix 2 and 3, and 3 extra amino acids between helix 1 and 2.

■ Genetics, function

Genetic analysis indicates that introducing the *Y3* allele into *Y4 Z4* ($A\alpha4$) strains triggers the *A* developmental pathway. *Y4* likewise induces the *A* pathway when introduced into *Y3 Z3* ($A\alpha3$) strains. Introduction of the *Z3* allele or the *Z4* allele into the $A\alpha1$ strain, which is naturally deleted for *Z*, activates the *A* pathway. These results have led to the hypothesis that activation of the *A* pathway is mediated by a heteromultimeric protein that results from the interaction of *Z* and *Y* polypeptides encoded by different $A\alpha$ alleles, for example, *Y1-Z3* or *Y4-Z3*, etc. The interaction of two dissimilar homeodomain proteins to form a functional heteromultimer in S. *commune* is remarkably similar to the interaction proposed for the *bE* and *bW* polypeptides encoded by the multiallelic *b* locus of *U. maydis* and to the interaction of a1 and a2 encoded by *MAT* in S. *cerevisiae* (reviewed by Herskowitz, 1992).

■ References

Giasson, L., Specht, C.A., Milgrim, C., Novotny, C.P. and Ullrich, R.C. (1989). Cloning and comparison of $A\alpha$ mating-type alleles of the Basidiomycete Schizophyllum commune. Mol. Gen. Genet. 218, 72-77.

Herskowitz, I. (1992). Yeast branches out. Nature 357, 190-191.

Novotny, C.P., Stankis, M.M., Specht, C.A., Yang, H., Ullrich, R.C. and Giasson, L. (1991). The $A\alpha$ mating type locus of *Schizophyllum commune*. In: More Gene Manipulations in Fungi; eds. Bennett, J.W. and Lasure, L.L. (Academic Press, New York) pp. 234-257.

Specht, C.A., Stankis, M.M., Giasson, L. and Novotny, C.P. (1992).

$A\alpha2$ [Cc]

Species: *Coprinus cinereus*
Chromosomal location: -
Other names: -
Cognate genes: -
Type of homeobox: -
Accession number: -

■ Origin and description

Molecular analysis of the A42 mating factor of *Coprinus cinereus* revealed the presence of seven genes, two of which are in the $A\alpha$ locus and five in the $A\beta$ locus (Mutasa et al., 1989; Kües et al., 1992). The $A\alpha$ genes are designated $A\alpha1$ and $A\alpha2$; the $A\beta$ genes are designated $A\beta1$, $A\beta2$, $A\beta3$, $A\beta4$ and $A\beta5$. (Alleles at each of these genes are indicated by a dash followed by a number, for example $A\beta1$-1). The polypeptide encoded by $A\alpha2$ contains a typical homeodomain, though the complete sequence of this polypeptide has not been published (Kües et al., 1992; Kües and Casselton, 1992).

■ Genetics, function

Coprinus cinereus is a filamentous Basidiomycete fungus, closely related to *Schizophyllum commune*. They share many morphological features. C. *cinereus* has two types of filaments: haploid and dikaryotic. The latter can be distinguished from the former by a characteristic structure, the clamp connection. Four multiallelic mating type loci ($A\alpha$, $A\beta$, $B\alpha$ and $B\beta$) regulate maintenance of the dikaryotic mycelium and differentiation of fruiting bodies (the typical mushroom) from this mycelium (Casselton, 1978). Formation of the dikaryon results from cell fusion between two haploid filaments that must differ at $A\alpha$ or $A\beta$ and at $B\alpha$ or $B\beta$. As in S. *commune* the A and B factors control independent but complementary pathways in development of the dikaryon. *A* controls clamp cell formation and synchronized division of the nuclei in the dikaryon; *B* controls nuclear migration and clamp cell fusion. The multiallelic $A\alpha$ and $A\beta$ loci are very closely linked; the multiallelic $B\alpha$ and $B\beta$ loci are linked to each other but unlinked to the A loci.

Genetic analysis indicates that four of the seven genes identified ($A\alpha2$, $A\beta1$, $A\beta2$ and $A\beta4$) determine specificity of *A*. Specifically, the presence in a cell of any two different alleles for any of these specificity genes is sufficient to trigger the *A* developmental pathway (Kües et al., 1992).

■ References

Casselton, L. (1978). Dikaryon formation in higher basidiomycetes. In the filamentous fungi (ed. J.E. Smith and D.R. Berry), vol. 3, pp. 275-297.

Kües, U. and Casselton, L. (1992). Homeodomains and regulation of sexual development in basidiomycetes. Trends Genet. 8, 154-155.

Kües, U., Richardson, W.V.J., Tymon, A.M., Mutasa, E.S., Gottgens, B., Gaubatz, S., Gregoriades, A. and Casselton, L. (1992). The combination of dissimilar alleles of the *Aα* and *Aβ* gene complexes, whose proteins contain homeodomain motifs, determines sexual development in the mushroom *Coprinus cinereus*. Genes Dev. 6, 568-577.

Mutasa, E.S., Tymon, A.M., Gottgens, B., Mellon, F.M., Little, P.F.R. and Casselton, L. (1990). Molecular organization of an *A* mating type factor of the basidiomycete fugus *Coprinus cinereus*. Curr. Genet. 18, 223-229.

AαY [Sc]

Species: *Schizophyllum commune*

Chromosomal location: -

Other names: *Y*

Cognate genes: -

Type of homeobox: -

Accession number: M80789 (Aα1), M80824 (Aα3) and M80822 (Aα4)

■ Origin and description

Three alleles of the mating type locus *Aα* (*Aα1*, *Aα3*, and *Aα4*) were cloned and sequenced (Giasson et al., 1988; Stankis et al., 1992). The *Aα3* and *Aα4* alleles contain two divergently transcribed genes designated *Y* and *Z*. The *Aα1* allele contains only the *Y* gene. The *Y* gene contains an open reading frame of 891 amino acids in the *Y1* allele, 926 in *Y3* and 928 in *Y4*. *Y4* is 54% identical to *Y1* and 49% identical to *Y3*. The C terminal region of the *Y* polypeptide contains a basic region rich in lysine and arginine and an adjacent region rich in serine. A typical homeodomain is found at position 147-206 for *Y1* and *Y3* and at position 152-216 for *Y4*.

■ Genetics, function

Schizophyllum commune is a wood-rotting filamentous Basidiomycete fungus (Novotny et al., 1991). There are two types of filaments: haploid and dikaryotic. The latter can be distinguished from the former by a characteristic structure, the clamp connection. Four multiallelic mating type loci (*Aα*, *Aβ*, *Bα* and *Bβ*) regulate maintenance of the dikaryotic mycelium and differentiation of fruiting bodies

(the typical mushroom) from this mycelium (Novotny et al., 1991). Formation of the dikaryon results from cell fusion between two haploid filaments that must differ at *Aα* or *Aβ* and at *Bα* or *Bβ*. *A* and *B* control independent but complementary pathways in development of the dikaryon: *A* controls clamp cell formation and synchronized division of the nuclei in the dikaryon; *B* controls nuclear migration and clamp cell fusion. The *A* loci (*Aα* and *Aβ*) are linked; the *B* loci (*Bα* and *Bβ*) are linked to each other but unlinked to the *A* loci.

Fusion between *S. commune* hyphae is promiscuous and can occur between strains of any mating type. After fusion, if the mating partners differ at both *A* and *B*, the two mating cells exchange their nuclei, a process controlled by *B*. The incoming nucleus then migrates along the length of the recipient filament (a process controlled by *B*) until it reaches the tip cell, thus establishing dikaryosis. Maintenance of this dikaryotic state is assured by a very specialized type of cell division which occurs at the tip cell and which involves formation of the clamp cell. This structure ensures that both the tip cell and the penultimate cell inherit the two different nuclei, which eventually will fuse and undergo meiosis in the mushroom. Synchronized division of the nuclei and clamp cell formation are controlled by *A*; fusion of the clamp cell and movement of the nuclei are controlled by *B*.

■ References

Giasson, L., Specht, C.A., Milgrim, C., Novotny, C.P. and Ullrich, R.C. (1989). Cloning and comparison of *Aα* mating-type alleles of the Basidiomycete Schizophyllum commune. Mol. Gen. Genet. 218, 72-77.

Novotny, C.P., Stankis, M.M., Specht, C.A., Yang, H., Ullrich, R.C. and Giasson, L. (1991). The *Aα* mating type locus of *Schizophyllum commune*. In: More Gene Manipulations in Fungi; eds. Bennett, J.W. and Lasure, L.L. (Academic Press, New York) pp. 234-257.

Specht, C.A., Stankis, M.M., Giasson, L. and Novotny, C.P. (1992). Functional analysis of the homeodomain-related proteins of the *Aα* locus of *Schizophyllum commune*. Proc. Natl. Acad. Sci. USA 89, 7174-7178.

Stankis, M.M., Specht, C.A., Yang, H., Giasson, L., Ullrich, R.C. and Novotny, C.P. (1992). The *Aα* mating locus of *Schizophyllum commune* encodes two dissimilar, multiallelic, homeodomain proteins. Proc. Natl. Acad. Sci. USA 89, 7169-7173.

Aβ1, Aβ4 [Cc]

Species: *Coprinus cinereus*

Chromosomal location: -

Other names: -

Cognate genes: -

Type of homeobox: -

Accession number: X62336 (*Aβ1-1*)

Origin and description

$A\beta1$ and $A\beta4$ are two of the five genes present in the $A\beta$ locus of *C. cinereus* (Kües et al., 1992). They are specificity genes, that is, the presence of different alleles for any of the specificity genes is sufficient to activate the *A* pathway.

$A\beta1$-1 contains an open reading frame of 681 amino acids. An atypical homeodomain is found at position 165 to 227, with 3 extra amino acids between helix 1 and 2. The amino terminal portion of the protein contains 30% serine and threonine, which could be an activation domain. Other putative activation domains can be found in the C terminal portion of the polypeptide (Tymon et al., 1992).

The polypeptide encoded by $A\beta4$ contains an atypical homeodomain with 3 extra amino acids between helix 1 and 2. The complete sequence of $A\beta4$ is not published.

Genetics and function

Coprinus cinereus is a filamentous Basidiomycete fungus, closely related to *Schizophyllum commune*. They share many morphological features. C. cinereus has two types of filaments: haploid and dikaryotic. The latter can be distinguished from the former by a characteristic structure, the clamp connection. Four multiallelic mating type loci ($A\alpha$, $A\beta$, $B\alpha$ and $B\beta$) regulate maintenance of the dikaryotic mycelium and differentiation of fruiting bodies (the typical mushroom) from this mycelium (Casselton, 1978). Formation of the dikaryon results from cell fusion between two haploid filaments that must differ at $A\alpha$ or $A\beta$ and at $B\alpha$ or $B\beta$. As in *S. commune* the *A* and *B* factors control independent but complementary pathways in development of the dikaryon. *A* controls clamp cell formation and synchronized division of the nuclei in the dikaryon; *B* controls nuclear migration and clamp cell fusion. The multiallelic $A\alpha$ and $A\beta$ loci are very closely linked; the multiallelic $B\alpha$ and $B\beta$ loci are linked to each other but unlinked to the A loci.

Genetic analysis indicates that four of the seven genes identified ($A\alpha2$, $A\beta1$, $A\beta2$ and $A\beta4$) determine specificity of *A*. Specifically, the presence in a cell of any two different alleles for any of these specificity genes is sufficient to trigger the *A* developmental pathway (Kües et al., 1992).

Four of the seven genes identified at the $A\alpha$ and $A\beta$ loci of *C. cinereus* ($A\alpha2$, $A\beta1$, $A\beta2$ and $A\beta4$) are the determinants of specificity of the *A* developmental pathway. In particular, the presence of different alleles at any of the specificity genes is sufficient to activate A developmental steps. This observation has led to the hypothesis that heteroallelic polypeptides interact to form a functional heteromultimeric regulatory protein. The demonstrated interaction of $\alpha1$ and $\alpha2$ in *S. cerevisiae*, and the genetic evidence for interaction of bW and bE in *U. maydis* and $A\alpha Y$ and $A\alpha Z$ in *S. commune* (reviewed in Banuett, 1992; and also in Herskowitz, 1992) provide precedent that the interaction in *C. cinereus* is between a polypeptide containing a typical homeodomain and one containing an atypical homeodomain.

References

Casselton, L. (1978). Dikaryon formation in higher basidiomycetes. In the filamentous fungi (ed. J.E. Smith and D.R. Berry), vol. 3, pp. 275-297.

Kües, U. and Casselton, L. (1992). Homeodomains and regulation of sexual development in basidiomycetes. Trends Genet. 8, 154-155.

Kües, U., Richardson, W.V.J., Tymon, A.M., Mutasa, E.S., Gottgens, B., Gaubatz, S., Gregoriades, A. and Casselton, L. (1992). The combination of dissimilar alleles of the $A\alpha$ and $A\beta$ gene complexes, whose proteins contain homeodomain motifs, determines sexual development in the mushroom *Coprinus cinereus*. Genes Dev. 6, 568-577.

Mutasa, E.S., Tymon, A.M., Gottgens, B., Mellon, F.M., Little, P.F.R. and Casselton, L. (1990). Molecular organization of an *A* mating type factor of the basidiomycete fugus *Coprinus cinereus*. Curr. Genet. 18, 223-229.

Tymon, A.M., Kües, U., Richardson, W.V.J. and Casselton, L. (1992). A fungal mating type protein that regulates sexual and asexual development contains a POU-related domain. EMBO J. 11, 1805-1813.

$A\beta2$ [Cc]

Species: *Coprinus cinereus*
Chromosomal location: -
Other names: -
Cognate genes: -
Type of homeobox: -
Accession number: -

Origin and description

$A\beta2$ is one of the five genes at the $A\beta$ locus of *C. cinereus*. It is also one of the specificity genes as are $A\beta1$, $A\beta4$ and $A\alpha2$. The polypeptide encoded by $A\alpha2$ contains a typical homeodomain (Kües et al., 1992), though the complete sequence of the gene is not yet published.

Genetics and function

Four of the seven genes identified at the $A\alpha$ and $A\beta$ loci of *C. cinereus* ($A\alpha2$, $A\beta1$, $A\beta2$ and $A\beta4$) are the determinants of specificity of the *A* developmental pathway. In particular, the presence of different alleles at any of the specificity genes is sufficient to activate *A* developmental steps. This observation has led to the hypothesis that heteroallelic polypeptides interact to form a functional heteromultimeric regulatory protein. The demonstrated interaction of $\alpha1$ and $\alpha2$ in *S. cerevisiae*, and the genetic evidence for interaction of bW and bE in *U. maydis* and $A\alpha Y$ and $A\alpha Z$ in *S. commune* (reviewed in Banuett, 1992; and also in Herskowitz, 1992) provide precedent that the interaction in *C. cinereus* is between a polypeptide containing a typical homeodomain and one containing an atypical homeodomain.

References

Casselton, L. (1978). Dikaryon formation in higher basidiomycetes. In the filamentous fungi (ed. J.E. Smith and D.R. Berry), vol. 3, pp. 275-297.

Herskowitz, I. (1992). Yeast branches out. Nature 357, 190-191.

Kües, U. and Casselton, L. (1992). Homeodomains and regulation of sexual development in basidiomycetes. Trends Genet. 8, 154-155.

Kües, U., Richardson, W.V.J., Tymon, A.M., Mutasa, E.S., Gottgens, B., Gaubatz, S., Gregoriades, A. and Casselton, L. (1992). The combination of dissimilar alleles of the $A\alpha$ and $A\beta$ gene complexes, whose proteins contain homeodomain motifs, determines sexual development in the mushroom *Coprinus cinereus*. Genes Dev. 6, 568-577.

Mutasa, E.S., Tymon, A.M., Gottgens, B., Mellon, F.M., Little, P.F.R. and Casselton, L. (1990). Molecular organization of an *A* mating type factor of the basidiomycete fugus *Coprinus cinereus*. Curr. Genet. 18, 223-229.

Tymon, A.M., Kües, U., Richardson, W.V.J. and Casselton, L. (1992). A fungal mating type protein that regulates sexual and asexual development contains a POU-related domain. EMBO J. 11, 1805-1813.

AbdAAf [ar]

Species: *Artemia franciscana*
Chromosomal location: -
Other names: *abdominal-A*
Cognate genes: *abd-A*
Type of homeobox: *Antp* class
Accession number: X70076

■ Origin and description

Cloned by PCR with degenerate homeobox primers, followed by inverse PCR of 1 kb fragment from genomic DNA. Amino acid conservation in the homeodomain and immediately flanking regions uniquely idntifies this gene as a homologue of *Drosophila abd-A*.

■ References

Averof, M. and Akam, M. (1993). Hom/Hox genes in a crustacean; implication for the origin of insect and crustacean body plans. Current biology 3, 73–78.

AbdASg [Sg]

Species: *Schistocerca gregaria*
Chromosomal location: -
Other names: *abdominal-A*
Cognate genes: *abd-A*
Type of homeobox: *Antp*-type
Accession number: X54674

■ Origin and description

Cloned by low stringency screening of a *Schistocerca gregaria* genomic library using *Drosophila* and *Schistocerca* homeobox probes. Amino acid conservation extends both upstream and downstream of the homeodomain, and uniquely identifies this gene as a homologue of *Drosophila abd-A*.

■ Expression

Expressed in the abdomen from 22% embryogenesis onwards. The anterior limit lies within abdominal segment 1, an identical anterior boundary to that of *abd-A* in *Drosophila*. Initially expression extends posteriorly to A10, but later (40% embryogenesis) protein disappears from A9 and A10.

■ References

Akam, M., Dawson, I. and Tear, G. (1988). Homeotic genes and the control of segment identity. Development 104, 123-133.

Tear, G., Akam, M. and Martinez-Arias, A. (1990). Isolation of an abdominal-A gene from the locust *Schistocerca gregaria* and its expression during embryogenesis. Development 110, 915–925.

abd-A [d]

Species: *Drosophila melanogaster*
Chromosomal location: Right arm of chromosome 3, bands 89E1-2
Other names: *abdominal-A*
Cognate genes: -
Type of homeobox: *Antp*-type
Accession number: X54453

■ Origin and function

The first mutation that affect the *abdominal-A* gene (*abd-A*) was described in 1978 by Ed. Lewis. *abd-A* is one of the three homeotic genes of the bithorax complex (BX-C;

Sanchez-Herrero et al., 1985; Tiong et al., 1985). *abd-A* mutations are lethal. In the dead embryos, PS7 to PS9 are transformed into copies of PS6 (and to a lesser extend, PS10 to 13 are partly transformed into PS9). Cuticular analysis of mosaic clones reveal similar transformations in adult. Thus the *abd-A* product is required for proper parasegmental identity of parasegments 7 to 9 that form the second through 4th abdominal segments of an adult fly. (Lewis, 1978; Morata et al., 1983; Sanchez-Herrero et al., 1985; Tiong et al., 1985). Complementation studies with mutations that affect the regulation of *abd-A* and other observations suggest that *abd-A* is also involved to assign proper identity of PS10 (Karch et al., 1985; Celniker et al., 1990).

The *abd-A* gene was cloned by chromosomal walking (Karch et al., 1985). The transcription unit is spread over a 20 kilobases region of DNA which is defined by the DNA lesions associated with *abd-A* lethal mutations (Karch et al., 1985; 1990). The mature *abd-A* transcript is composed of at least 8 exons (Karch et al., 1990). Like in *Ubx*, the *abd-A* transcription unit contains a micro-exon but so far alternative splicing generating multiple forms of the *abd-A* product was not detected (F. Karch and S. Sakondju, unpublished). The homeobox homology is found near the middle of the 330 amino acids sequence protein. Unlike *Ubx* or *Antp*, the *abd-A* homeobox is separated from the 3' coding region by a short intron.

As for all the homeotic mutations in flies, the change in parasegmental identity found in *abd-A* mutations suggest that *abd-A* product control developmental pathways by regulating subordinate target genes responsible for parasegment-specific morphogenesis (see Garcia-Bellido, 1977). This is best demonstrated for *Ubx*. The *abdA* gene resembles the *Antp* and *Ubx* genes in several respects. The three homeoboxes are quite similar; there are only four substitutions between *abdA* and *Ubx* (two are conservative), and five substitutions between *abdA* and *Antp* (three are conservative). The homology with *Ubx*, extends 9 residues beyond the carboxyl end of the 60 amino acids homeobox. There is another short peptide, Tyr-Pro-Trp-Met (YPWM) encoded within the 5th exon of *abdA* that is conserved in *Ubx, Antp, Dfd, Scr, lab* and other vertebrate homeobox containing genes (Mavilio et al., 1986; Krumlauf et al., 1987; Wilde and Akam, 1987). As in *Ubx* and *Antp*, the *abdA* YPWM peptide is separated from the homeobox exon by a small "micro" exon. Wilde and Akam (1987) reported a sequence Met-(Asn)-Ser-Tyr-Phe (MXSYF) lying at the amino terminus of the *Ubx* protein that is also found close to the amino terminus of the *Antp* and is the initiation sequence in the mouse gene *Hox 2.1* (Krumlauf et al., 1987). In *abdA*, a related sequence, Met-Asn-Ser-Tyr-Gln, starting at amino acid 37 is found. The *abdA opa* $(CAA/CAG)_n$ encodes a stretch of poly-Gln downstream of the homeobox. Similar repeats are found in *AbdB, Ubx, Antp, Dfd, en, ftz, lab* where they encode strings of either Gln, Ala or Ser. Aside from these features, the *abdA* protein is highly diverged from those of *Ubx* and *Antp*.

The *abdA* and *Ubx* products appear to have similar functions in the developing fly. The second abdominal segment (PS7) is similar to the first abdominal segment (PS6) where *Ubx* is expressed in both larvae and adults. In some specific cell types, the *Ubx* and *abdA* products have identical effects. The *Ubx* product is needed to connect the dorsal tracheal trunks on the first instar larva in PS5 and PS6, and *abdA* does the same in PS7 through PS12 (Lewis, 1978). Almost all cells of the peripheral nervous system are morphologically unchanged between PS6 and PS7 despite the expression of *abdA* in many of these cells in PS7. The similarity of *abdA* and *Ubx* is also demonstrated by the functioning of the hybrid protein in the *Ubx^C1* mutation. This deletion apparently makes a fusion protein whose amino terminal half is derived from *abdA* and carboxy terminal half from *Ubx*. The *Ubx^C1* hybrid protein can carry out wild type function in PS5 and partially rescue the *abx/bx* mutations (Casanova et al., 1988; Rowe and Akam, 1988). Thus the homeobox portion of the protein appears to be important, but not exclusive, for the character of the cellular transformations that the homeotic protein can specify.

The *abdA* protein does function differently than *Ubx* in some cell types. The five lateral chordotonal cells of the peripheral nervous system express *abdA* from parasegment 7 to 12. Analogous groups of three chordotonal cells are found more dorsally in PS5 and 6; these cells are expressing *Ubx* protein. In *abdA* mutants, the 5 abdominal chordotonal cells are transformed into the thoracic type group of three. Thus the *abdA* product can direct the precursor cells of the chordotonal cells in a different pathway of development in PS7 to 12 (Karch et al., 1990).

Where it is expressed (from PS7 to 12, see below), the *abd-A* product represses expression of *Ubx* (Struhl and White, 1985; Harding et al., 1985). The regulatory role of *abd-A* has also been reproduced in yeast where it behaves as activator of transcription (Samson et al., 1989).

■ Expression

The *abd-A* gene by itself is not sufficient to achieve proper identity of PS7 to PS9. A region of DNA that spans from 10 kb downstream of the transcription unit to 30 kb upstream of the promoter is required in *cis*. This 60 kb domain of DNA includes the regions defined by the *iab-2, iab-3* and *iab-4* mutations that affect respectively PS7, PS8 and PS9 (the 2nd, 3rd and 4th abdominal segments; Lewis, 1978, Karch et al., 1985, 1990).

The pattern of expression of *abd-A* during embryonic development has been studied using antibodies directed against the *abd-A* protein (Karch et al., 1990; Macias et al., 1990). *abd-A* is expressed from PS7 to 12 which are the parasegments affected by *abd-A* mutations. The pattern is much more uniform from parasegment to parasegment than are the patterns of *Ubx* or *Antp*. This is expected since PS7 to PS12 are morphologically quite similar in young larvae. However, there are discernable pattern differences in PS7, 8 and 9. The *iab-2,3*, and *4* mutant classes define positive *cis*-regulatory elements that induce expression of *abd-A* in parasegments 7, 8 and

9 (abdominal segments 2, 3 and 4), respectively (Karch et al., 1990; Sanchez-Herrero, 1991). Once a pattern of *abd-A* expression is turned on in a given parasegment, it remains on in the more posterior parasegments, so that the complex pattern of expression is built up in the successive parasegments (Peifer et al., 1987; Karch et al., 1990).

■ References

Busturia, A., Casanova, J., Sanchez-Herrero, E., González, R. and Morata, G. (1989). Genetic structure of the *abd-A* gene of *Drosophila*. Development 107, 575-583.

Casanova, J., Sanchez-Herrero, E., and Morata, G. (1988). Developmental analysis of a hybrid gene composed of parts of the *Ubx* and *abd-A* genes of *Drosophila*. EMBO J. 7, 1097-1105.

Celniker, S.R. Sharma S., Keelan, D.J. and Lewis, E.B. (1990). The molecular genetics of the bithorax complex of *Drosophila*: *cis*-regulation in the *Abdominal-B* domain. EMBO J. 9, 4277-4286.

Harding, K., Wedeen, C., McGinnis, W. and Levine, M. (1985). Spatially regulated expression of homeotic genes in *Drosophila*. Science 229, 1236-1242.

Karch, F., Weiffenbach, B., Peifer, M., Bender, W., Duncan, I., Celniker, S., Crosby, M. and Lewis, E.B. (1985). The abdominal region of the bithorax complex. Cell 43, 81-96.

Karch, F., Bender, W., and Weiffenbach, B. (1990). *abdA* expression in *Drosophila* embryos. Genes Dev 4., 1573-1587.

Krumlauf, R., Holland, P.W.H., McVey, J.H. and Hogan, B.L.M. (1987). Developmental and spatial patterns of expression of the mouse homeobox gene, *Hox2.1*. Development 99, 603-617.

Lewis, E.B. (1978). A gene complex controlling segmentation in *Drosophila*. Nature 276, 565-570.

Macias, A., Casanova, J. and Morata, G. (1990). Expression and regulation of the *abd-A* gene of *Drosophila*. Development 110, 1197-1207.

Mavilio, F., Simeone, A., Giampaolo, A., Faiella, A., Zappavigna, V., Acampora, D., Poiana, G., Russo, G., Peschle, C. and Boncinelli, E. (1986). Differential and stage-related expression in embryonic tissues of a new human homeobox gene. Nature 324, 664-668.

Morata, G., Botas, J., Kerridge, S. and Struhl, G. (1983). Homeotic transformations of the abdominal segments of *Drosophila* caused by breaking or deleting a central portion of the bithorax complex. J. Embryol. Exp. Morph. 78, 319-341.

Peifer, M., Karch, F. and Bender, W. (1987). The bithorax complex: control of segmental identity. Genes Dev. 1, 891-898.

Rowe, A. and Akam, M. (1988). The structure and expression of a hybrid homeotic gene. EMBO J. 7, 1107-1114.

Samson, M.L., Jackson-Grusby, L. and Brent, R. (1989). Gene activation and DNA binding by *Drosophila* Ubx and *abd-A* proteins. Cell 57, 1045-1052.

Sanchez-Herrero, E., Vernos, I., Marco, R. and Morata, G. (1985). Genetic organization of the *Drosophila* bithorax complex. Nature 313, 108-113.

Sanchez-Herrero, E. (1991). Control of the expression of the bithorax complex *abdominal-A* and *Abdominal-B* by *cis*-regulatory regions in *Drosophila* embryos. Development 111, 437-449.

Struhl, G. and White, A.H. (1985). Regulation of the *Ultrabithorax* gene of *Drosophila* by other bithorax complex genes. Cell 43, 507-519.

Tiong, S.Y.K., Bone, L.M. and Whittle, J.R.S. (1985). Recessive lethal mutations within the bithorax complex in *Drosophila*. Mol. Gen. Genet. 200, 335-342.

Wilde, C.D. and Akam, M. (1987). Conserved sequence elements in the 5′ region of the *Ultrabithorax* transcription unit. EMBO J. 6, 1393-1401.

Abd-B [d]

Species: *Drosophila melanogaster*
Chromosomal location: Right arm of chromosome 3, bands 89E1-2
Other names: *Abdominal-B*
Cognate genes: -
Type of homeobox: *Abd-B*, *Antp*-class
Accession number: X51663, X51669, X51670, X16134

■ Origin and function

The *Abdominal-B* gene (*Abd-B*) is one of the three homeotic genes of the Bithorax complex (BX-C) in *Drosophila*. It was first described and named by Sanchez-Herrero et al., 1985 (see also Tiong et al., 1985). *Abd-B* mutations are haplo-insufficient and thus harbor a visible phenotype as heterozygous flies (*Abd-B/+*; they are sterile and males show a small 7th tergite. Homozygous *Abd-B* mutations are lethal. Dead embryos of the strongest alleles have parasegments 10 to 14 (PS10 to 14) transformed into copies of PS9 (PS10 to 14 will form the 5th through 8th abdominal segments). Thus the function of the *Abd-B* product is to assign proper identity of PS10 to 14 (Sanchez-Herrero et al., 1985; Tiong et al., 1985; 1988; Casanova et al., 1986).

The *Abd-B* gene was cloned by chromosomal walking and by homeobox homology (Karch et al., 1985; Regulski et al., 1985). The transcription unit is complex with multiple transcripts initiating at alternate promoters (Celniker and Lewis, 1987; DeLorenzi et al., 1988; Sanchez-Herrero and Crosby, 1988; Kuziora and McGinnis, 1988; Celniker et al., 1989; Zavortink and Sakonju, 1989). These transcripts give rise to two types of proteins, which differ in length at their N-terminus. The homeobox homology is found at the C-terminus. A short *Abd-B* transcript (class A or m element) encodes for the long *Abd-B* protein (55×10^3D). This product is required for morphological diversity of PS10 to 13 (Casanova et al., 1986; DeLorenzi et al., 1988; 1990; Sanchez-Herrero and Crosby, 1988; Kuziora and McGinnis, 1988; Celniker et al., 1989; 1990; Zavortink and Sakonju, 1989; Boulet et al., 1991). In PS14, another class of *Abd-B* transcripts (class B, C and gamma) initiating from different upstream promoters are activated (DeLorenzi et al., 1988; Sanchez-Herrero and Crosby, 1988; Kuziora and McGinnis, 1988; Celniker et al., 1989; 1990; Zavortink and Sakonju, 1989; Boulet et al., 1991). These classes of transcripts code for the r element (or regulatory element) described by Casanova et al., 1986. They encode the short form of the

Abd-B product (truncated at the N-terminus) of 30×10^3D.

As for all the homeotic mutations in flies, the change in parasegmental identity found in *Abd-B* mutations suggest that *Abd-B* products control developmental pathways by regulating subordinate target genes responsible for parasegment-specific morphogenesis (see Garcia-Bellido, 1977). Indeed, the finding of the homeobox sequence strongly supports the idea that the *Abd-B* product is a regulator of transcription. In embryos, *Abd-B* has been shown to repress expression of *Ubx* and *abd-A* in PS13 and 14 (Harding et al., 1985; Struhl and White, 1985; Karch et al., 1990; Macias et al., 1990). Activation of transcription by *Abd-B* protein has been demonstrated in transfection experiments in mammalian tissue culture cells (Thali et al., 1988; Ali and Bienz, 1991).

■ Expression

The short *Abd-B* transcription unit (A class) which spans about 10 kb is not sufficient to provide proper identity of PS10 to 13 and a large 3′ end regulatory region of 60 kb is required in *cis*. This region is defined by the *iab-5*, *iab-6* and *iab-7* mutations which affect the identity of respectively PS10, PS11 and PS12 (the 5th, 6th and 7th abdominal segments; Lewis, 1978; Kuhn et al., 1981; Awad et al., 1981; Karch et al., 1985; Duncan, 1987; Gyurkovics et al., 1991; Celniker et al., 1989; 1991). The subdivision in discrete independent parasegment-specific *cis*-regulatory regions is demonstrated by staining of wild type and mutant embryos with antibody directed against *Abd-B* (Celniker et al., 1989; 1990; DeLorenzi et al., 1990; Boulet et al., 1991; Sanchez-Herrero, 1991). The *iab-5*, 6 and 7 parasegment-specific elements regulate expression of the class A *Abd-B* product in PS10, 11 and 12 respectively. The existence of a PS13-specific regulatory element (*iab-8*) remains to be demonstrated. In PS14, *Abd-B* transcripts (class B, C and gamma) initiating from at least three other promoters are activated; it is also not clear whether PS14-specific regulatory sequences regulate expression of these transcripts.

■ References

Ali, N. and Bienz, M. (1991). Functional dissection of *Drosophila Abdominal-B* protein. Mech. Dev. 35, 55-64.

Awad, A.A.M., Gausz, J., Gyurkovics, H. and Parducz, A. (1981). A new homeotic mutation, SGA-62, in *Drosophila melanogaster*. Acta Biol. Acad. Sci. Hung. 32, 219-228.

Boulet, A., Lloyd, A. and Sakonju, S. (1991). Molecular definition of the morphogenetic and regulatory functions and the *cis*-regulatory elements of the *Drosophila Abd-B* homeotic gene. Development 111, 393-405.

Casanova, J., Sanchez-Herrero, E. and Morata, G. (1986). Identification and characterization of a parasegment specific regulatory element of the *Abdominal-B* gene of D*rosophila*. Cell 47, 627-636.

Celniker, S.E. and Lewis, E.B. (1987). *Transabdominal*, a dominant mutant of the bithorax complex, produces a sexually dimorphic segmental transformation in *Drosophila*. Genes Dev. 1, 111-123.

Celniker, S.E., Keelan, D.J. and Lewis, E.B. (1989). The molecular genetics of the Bithorax Complex of *Drosophila*: Characterization of the products of the *Abdominal-B* domain. Genes Dev. 3, 1424-1435.

Celniker, S.R., Sharma, S., Keelan, D.J. and Lewis, E.B. (1990). The molecular genetics of the bithorax complex of *Drosophila*: *cis*-regulation in the *Abdominal-B* domain. EMBO J. 9, 4277-4286.

DeLorenzi, M., Ali, N., Saari, G., Henry, C., Wilcox, M. and Bienz, M. (1988). Evidence that the *Abdominal-B* r element function is conferred by a *trans*-regulatory homeoprotein. EMBO J. 7, 3223-3231.

DeLorenzi, M. and Bienz, M. (1990). Expression of *Abdominal-B* homeoproteins in *Drosophila* embryos. Development, 108, 323-329.

Duncan, I. (1987). The bithorax complex. Annu. Rev. Genet. 21, 285-319.

Garcia-Bellido, A. (1977). Homeotic and atavic mutations in insects. Am. Zool. 17, 613-629.

Gyurkovics, H., Gausz, J., Kummer, J. and Karch, F. (1990). A new homeotic mutation in the *Drosophila* bithorax complex removes a boundary separating two domains of regulation. EMBO J. 9, 2579-2585.

Harding, K., Wedeen, C., McGinnis, W. and Levine, M. (1985). Spatially regulated expression of homeotic genes in *Drosophila*. Science 229, 1236-1242.

Karch, F., Weiffenbach, B., Peifer, M., Bender, W., Duncan, I., Celniker, S., Crosby, M. and Lewis, E.B. (1985). The abdominal region of the bithorax complex. Cell 43, 81-96.

Karch, F., Bender, W. and Weiffenbach, B. (1990). *abdA* expression in *Drosophila* embryos. Genes Dev. 4, 1573-1587.

Kuhn, D.T., Woods, D.F. and Andrew, D. (1981). Linkage analysis of the tumorous-head (*tuh-3*) gene in *Drosophila melanogaster*. Genetics 99, 99-107.

Kuziora, M.A. and McGinnis, W. (1988). Different transcripts of the *Drosophila AbdB* gene correlate with distinct genetic subfunctions. EMBO J. 7, 3233-3244.

Lewis, E.B. (1978). A gene complex controlling segmentation in *Drosophila*. Nature 276, 565-570.

Macias, A., Casanova, J. and Morata, G. (1990). Expression and regulation of the *abd-A* gene of *Drosophila*. Development 110, 1197-1207.

Regulski, M., Harding, K., Kostriken, R., Karch, F., Levine, M. and McGinnis, W. (1985). Homeobox genes of the *Antennapedia* and bithorax complexes of *Drosophila*. Cell 43, 71-80.

Sanchez-Herrero, E., Vernos, I., Marco, R. and Morata, G. (1985). Genetic organization of the *Drosophila* bithorax complex. Nature 313, 108-113.

Sanchez-Herrero, E. and Crosby, M.A. (1988). The *Abdominal-B* gene of *Drosophila melanogaster*: overlapping transcripts exhibit two different spatial distributions. EMBO J. 7, 2163-2173.

Sanchez-Herrero, E. (1991). Control of the expression of the bithorax complex *abdominal-A* and *Abdominal-B* by *cis*-regulatory regions in *Drosophila* embryos. Development 111, 437-449.

Struhl, G. and White, A.H. (1985). Regulation of the *Ultrabithorax* gene of *Drosophila* by other bithorax complex genes. Cell 43, 507-519.

Thali, M., Müller, M.M., DeLorenzi, M., Matthias, P. and Bienz, M. (1988). *Drosophila* homeotic genes encode transcriptional activators similar to mammalian OTF-2. Nature 336, 598-601.

Tiong, S.Y.K., Bone, L.M. and Whittle, J.R.S. (1985). Recessive lethal mutations within the bithorax complex in *Drosophila*. Mol. Gen. Genet. 200, 335-342.

Tiong, S.Y.K., Gribbin, M.C. and Whittle, J.R.S. (1988). Mutational dissection of gene expression in the abdominal region of the bithorax complex of *Drosophila* in imaginal tissue. Roux's Arch. Dev. Biol., 197, 131-140.

Zavortink, M. and Sakonju, S. (1989). The morphogenetic and regulatory functions of the *Drosophila Abdominal-B* gene are encoded in overlapping RNAs transcribed from separate promoters. Genes Dev. 3, 1969-1981.

AbdBSg [Sg]

Species: *Schistocerca gregaria* (African Desert Locust)

Chromosomal location: -

Other names: *Abdominal-B*

Cognate genes: *Abd-B*

Type of homeobox: *Abd-B*-type

Accession number: X69161

■ Origin and description:

Cloned by low stringency screening of a *Schistocera gregaria* genomic library using *Drosophila* and *Schistocerca* homeobox probes. Amino acid conservation extends both upstream and downstream of the homeodomain, and uniquely identifies this gene as a homologue of *Drosophila Abd-B*.

■ Expression

Expressed from c. 28% of embryogenesis onwards in the most posterior part of the abdomen, before this is visibly segmented. Initially expression is limited to segment A11, but later expression extends anteriorly, appearing in the ectoderm of A9 and posterior A8 by 40% of embryogenesis. This domain remains unchanged until at least 55% development; later stages have not been examined. Mesodermal expression is seen in the somatic mesoderm of A10 at least, and in the hindgut visceral mesoderm. In the CNS ganglion VIII shows high levels of expression.

■ References

Kelsh, R.N. (1991). Ph.D. Thesis, University of Cambridge.

Kelsh, R.N., Dawson, I. and Akam, M.E. (1993). An analysis of Abdominal-B expression in the Locust *Schistocerca gregaria*. Development 117, 293–302.

*AHox*1 [Hal]

Species: *Halocynthia roretzi* (ascidian)

Chromosomal location: -

Other names: *Hlx [Hal]*

Cognate genes: -

Type of homeobox: Related to the *Drosophila H2.0*

Accession number: -

■ Origin and description

This gene was isolated from a *Halocynthia* genomic library (Saiga et al. 1991) by cross-homology with the homeobox of an *Antennapedia* type homeobox gene of silk moth (Hara and Suzuki, unpublished). Conceptual translation of the gene shows that it encodes a protein consisting of 741 amino acids. The homeobox of *AHox*1 shows 70% similarity to that of the *Drosophila H2.0* (Barad et al. 1988) but only 47% similarity to that of the *Antp* at a deduced amino acid sequence level. It may also be related to the murine homeobox gene *Hlx* (Allen et al. 1991).The *AHox*1 homeobox is interrupted by two introns; one is at the position 9 and the other after position 44.

■ Expression

The size of the *AHox*1 transcript is about 2.8 kB. In the early development of *Halocynthia*, the expression of *AHox*1 is quite limited. When embryos develop to the larval stage, the *AHoxa-7* transcripts can evidently be detected. After metamorphosis, the expression level increases as development proceeds up to day 7 and remains high thereafter. Upon *in situ* hybridization with sections of day 7 juvenile, positive signals are seen at the epithelium of future digestive tract.

In adult tissues, the digestive tract, digestive gland (liver) and coelomic cells are the major sites of the expression. In gonad, body wall muscle and pharyngeal epithelium, the expression of *AHox*1 is relatively weak.

■ References

Allen, J.D., Lints, T., Jenkins, N.A., Copeland, N.G., Strasser, A., Harvey, R.P. and Adams, J.M. (1991). Novel murine homeobox gene on chromosome 1 expressed in specific hematopoietic lineages and during embryogenesis. Genes Dev. 5, 509-520.

Barad, M., Jack, T., Chadwick, R. and McGinnis, W. (1988). A novel, tissue-specific, *Drosophila* homeobox gene. EMBO J. 7, 2151-2161.

Saiga, H., Mizokami, A., Makabe, K.W., Satoh, N. and Mita, T. (1991). Molecular cloning and expression of a novel homeobox gene of *AHox*1 of the ascidian, *Halocynthia roretzi*. Development 111, 821-828.

Ahox1 [ax]

Species: *Ambystoma mexicanum* (axolotl)
Chromosomal location: -
Other names: -
Cognate genes: *Hox*a-1, *Ghox-lab*
Type of homeobox: *Antp*-type, related to the *Drosophila* gene *labial* (*lab*)
Accession number: X15575

■ Origin and description

A genomic fragment containing the *Ahox*1 homeobox was isolated from a partial axolotl genomic library enriched for homeobox sequences, using the *Antp* homeobox from *Drosophila* as a probe, and low stringency hybridization conditions (Whiteley and Armstrong, 1990).

■ Expression

Transcripts of 1.6 and 2.2 kb that hybridize to *Ahox*1 are detected during the neurula and tail bud stages of development. These have been further localized to the posterior two thirds of the neural plate in *neurulae* embryos, and in tail bud embryos from the anterior margin of the gill bulge to the pronephric duct (Whiteley and Armstrong, 1991a).

■ Function

The possible function of this gene was explored by ectopic expression of a 560 bp *Kpn*I fragment containing the *Ahox*1 homeobox under the control of the mouse hsp68 promoter (Whiteley and Armstrong, 1991a). Expression from this promoter had previously been shown to be constitutive in the axolotl (Whiteley and Armstrong, 1991b), and injected DNA sequences was retained in high copy number (100 or more per genome) until late in development. Injection of such a construct resulted in 20% of the injected embryos with an abnormal anterior neural plate. Later in development, these embryos had small heads, no eyes, and appeared to lack the normal regionalization of the brain. We argue that even if the fusion protein containing the *Ahox*1 homeobox was not sufficiently complete to retain regulatory activity, it might perturb development by competing with the normal *Ahox*1 protein for the latter's target sequences, or the protein might be sufficiently complete to retain normal regulatory activity, in which case, its ectopic expression could likewise perturb development. We infer from the phenotype observed in injected embryos that the *Ahox*1 gene functions in the regional specification of neural structures, and that this is intimately tied in with the forces that shape the neural plate, since the neural plate of affected embryos is distinctively narrow along its entire length.

■ References

Whiteley, M. and Armstrong, J.B. (1990). Isolation and characterization of a developmentally regulated homeobox sequence in the Mexican axolotl. Biochem. Cell. Biol. 68, 622-629.

Whiteley, M. and Armstrong, J.B. (1991a). Ectopic expression of a genomic fragment containing a homeobox causes neural defects in the axolotl. Biochem. Cell. Biol. 69, 366-374.

Whiteley, M. and Armstrong, J.B. (1991b). On the origin of the mesoderm in the Mexican axolotl, *Ambystoma mexicanum*. Can. J. Zool. 69, 1221-1225.

al [d]

Species: *Drosophila melanogaster*
Chromosomal location: Chromosome 2, 21C1,2
Other names: *aristaless*
Cognate genes: -
Type of homeobox: *prd*-type
Accession number: -

■ Origin and description

Viable mutations in the *aristaless* (*al*) gene lead to an aberrant morphogenesis of both the most proximal and most distal regions in a subset of the adult appendages (Stern and Bridges, 1926; Lewis, 1945). These phenotypic effects have been attributed to an altered growth pattern of the corresponding imaginal discs (Tokunaga and Stern, 1968; Schneitz et al., 1993). In addition, *al* belongs to a set of genes which also includes *expanded* and *dachsous*. All three genes are closely linked and interact with the nearby located but distinct *R* locus (Golubovsky and Kulakov, 1978; Korochkina and Golubovsky, 1978). Molecular identification of *al* was achieved by mapping an inversion breakpoint (*In(2L)al^{130}*; E. Lewis, unpublished) to the *al* transcription unit and by genomic rescue experiments (Schneitz et al., 1993). The putative Al protein contains an extended *prd*-type homeodomain but no paired-domain. Based on a comparison of the *al* homeodomain with other homeodomain sequences, its closest relatives (75% identity) are the murine *Pax7* gene (Jostes et al., 1991) and the *smox-3* gene of the flatworm *Schistosoma mansoni* (Webster and Mansour, 1992).

■ Expression

An *al* transcript pattern is first detectable as segmentally repeated epidermal spots at mid-embryogenesis (Schneitz et al., 1993). Expression of *al* in the head evolves into a more complex pattern persisting throughout the remaining embryogenesis, suggesting a role in the development of specific larval head organs. Trunk expression is modified during subsequent stages as well. Most conspicuously, *al* is later expressed in thoracic segments in a horseshoe-type of pattern surrounding the distal portion of the developing leg imaginal discs. This finding suggests that *al* plays a role in early leg disc development. Besides its epidermal pattern, *al* transcripts are found in the endoderm of the anterior midgut during late embryogenesis. The maintenance of this endoderm expression is under the control of an induction process mediated by the mesodermally expressed homeotic genes *Sex combs reduced* and *Antennapedia* (Schneitz and Noll, in preparation). A striking bimodal *al* pattern observed in third instar imaginal antennal, leg, and wing discs indicates that *al* is under the control of a prepattern shared at least among these three types of discs.

■ References

Golubovsky, M.D. and Kulakov, L.A. (1978). Regulator gene in *Drosophila*? Dros. Inf. Serv. 53, 133-134.

Jostes, B., Walther, C. and Gruss, P. (1991). The murine paired box gene, *Pax7*, is expressed specifically during the development of the nervous and muscular system. Mech. Dev. 33, 27-37.

Korochkina, L.S. and Golubovsky, M.D. (1978). Cytogenetic analysis of induced mutations on the left end of the second chromosome of *D. melanogaster*. Dros. Inf. Serv. 53, 197-200.

Lewis, E.B. (1945). The relation of repeats to position effect in *Drosophila melanogaster*. Genetics 30, 137-166.

Schneitz, K., Spielmann, P. and Noll, M. (1993). Molecular genetics of *aristaless*, a *prd*-type homeobox gene involved in the morphogenesis of proximal and distal pattern elements in a subset of appendages in *Drosophila*. Genes Dev. 7, 114-129.

Stern, C. and Bridges, C.B. (1926). The mutants of the extreme left end of the second chromosome of *Drosophila melanogaster*. Genetics 11, 503-530.

Tokunaga, C. and Stern, C. (1969). Determination of bristle direction in *Drosophila*. Dev. Biol. 20, 411-425.

Webster, P.J. and Mansour, T.E. (1992). Conserved classes of homeodomains in *Schistosoma mansoni*, an early bilateral metazoan. Mech. Dev. 38, 25-32.

Alhb-1 [Al]

Species: *Ascaris lumbricoides*
Chromosomal location: -
Other names: *AHB-1*
Cognate genes: -
Type of homeobox: *Antp*-type
Accession number: -

■ Origin and description

Isolated using *Antp* and *ftz* as probes under low stringency hybridization conditions (Spicher et al., 1988). This gene was used to isolate *C. elegans* homeobox genes (Schaller et al., 1990). The homeodomain is of the *Antp*-type. The sequence was inadvertently presented in Boncinelli et al. (1989).

■ References

Spicher, A., Schaller, D., Müller, F. and Tobler, H. (1988). Homeobox containing genes in nematodes. Experientia, 44, A27, abstract CMB 81.

Schaller, D., Wittmann, C., Spicher, A., Müller, F. and Tobler, H. (1990). Cloning and analysis of three new homeobox genes from the nematode *Caenorhabditis elegans*. Nucl. Acids Res. 18, 2033-2036.

Boncinelli, E., Acampora, D., Pannese, M., D'Esposito, M., Somma, R., Gaudino, G., Stornaiuolo, A., Cafiero, M., Faiella, A. and Simeone, A. (1989). Organization of human class I homeobox genes. Genome 31, 745-756.

Antp [d]

Species: *Drosophila melanogaster*
Chromosomal location: Chromosome 3, map position 3-47.5
Other names: *Antennapedia, Nasobemia*
Cognate genes: -
Type of homeobox: Prototype of the *Antp*-class
Accession number: -

■ Origin and description

The *Antennapedia* (*Antp*) gene was first identified as a dominant mutation which partially transforms the antennae on the head of the fly into legs (Le Calvez, 1948). In the earliest reports, *Antp* was erroneously considered to be allelic to *spineless-aristapedia* (*ss*[a]) (Balkaschine, 1929), which maps more distally on the third chromosome and transforms only the most distal part of the antenna, the arista, into tarsal segments with claws. In contrast to *ss*[a], *Antp* also transforms the more proximal parts of the antenna into leg structures, in extreme cases to complete middle legs. Genetically *Antp* maps to position 47.5 on the third chromosome, which cytologically corresponds to band 84B1-2 in the giant polytene chromosomes of the salivary glands. Most dominant *Antp* mutations are chromosome rearrangements with one breakpoint in the *Antp* locus at 84B1-2. They are homozygous lethal with the exception of *Antp*[Ns] (Nasobemia) (Gehring, 1966) and *Antp*[72j] (Baker, 1987), which are homozygous viable. The recessive loss-of-function mutants (null mutants) are lethal at late embryonic stages and show homeotic transformations of the mesothoracic segment towards the prothoracic, and the prothoracic towards the labial segment (see below).

The *Antp* gene was cloned independently in two laboratories (Garber et al., 1983; Scott et al., 1983) by chromosomal walking and the exons and parts of the introns were sequenced (Schneuwly et al., 1986; Laughon et al., 1986; Stroeher et al., 1986). The gene spans more than 100 kb and contains two independently regulated promoters (P1, P2) and two polyadenylation sites (A1, A2). The two transcription units measure 103 kb and 36 kb respectively, and encode four major transcripts which share the same protein coding regions. The *Antp* gene consists of eight exons; P1 transcripts initiate in exon 1, P2 transcripts in exon 3. The P1 leader contains eight AUG codons followed by short open reading frames, the P2 leader 15 preceding the AUG that is used for initiation of translation. The mapping of the exons by hybridization of the cDNA clones to the chromosomal DNA led to the discovery of the homeobox (see Introduction) which is located in exon 8 very close to the intron/exon boundary. Additional transcripts are generated by differential splicing (Bermingham and Scott, 1988; Stroeher et al., 1988); exon 6 encoding 13 amino acids is optional and exon 7 contains at its 3' end two different splicing donor sites which are 12 nucleotides apart. When the longer form of exon 7 is used, an additional four amino acids are encoded, located immediately upstream of the homeobox. Therefore, a total of 16 transcripts can be generated, which encode four proteins differing by 4, 13 or 17 amino acids in length. The protein contains an evolutionarily conserved MXSYF sequence close to the amino terminus, a glutamine-rich region (M- or Opa-repeats), a YPWM sequence which is also found in vertebrate proteins of the *Antp*-class, and, of course, a homeodomain which serves as the prototype for the *Antp*-class of homeodomains. The three-dimensional structure of the *Antp* homeodomain and the complex between the homeodomain and the DNA binding site have been determined by NMR spectroscopy (Qian et al., 1989; Otting et al., 1990).

Antp is the most distal member of the homeobox genes which form the *Antennapedia*-Complex (ANT-C) in *Drosophila* (see Kaufman et al., 1990). It is preceded in proximal direction by *fushi tarazu* (*ftz*), *Sex combs reduced* (*Scr*), *Deformed* (*Dfd*), *bicoid* (*bcd*), *zerknüllt* (*zen* 1 and 2), *proboscipedia* (*pb*) and *labial* (*lab*). The homeotic selector genes *Antp*, *Scr*, *Dfd*, *pb* and *lab* are arranged in the same order along the chromosome, as they are expressed along the antero-posterior body axis, with the exception of *pb*.

■ Expression

The expression of *Antp* transcripts was studied by *in situ* hybridization (Hafen et al., 1983; Levine et al., 1983; Hafen et al., 1984; Martinez-Arias, 1986; Harding et al., 1985; Wedeen et al., 1986; Ingham and MartinezArias, 1986; Bermingham et al., 1990) and analyzed by reporter gene (β-galactosidase) expression (Engström et al., 1992), whereas protein expression was studied by using antibody probes (Carroll et al., 1986a; Wirz et al., 1986). During embryogenesis the transcripts originating from the two promoters were analyzed separately by Bermingham et al. (1990) and by Engström et al. (1992) for P1. P1 transcripts are first detected at the blastoderm stage in a belt encircling the embryo in an area which later gives rise to the thorax, i.e. parasegments (PS) 4 to 6. P2 transcripts are expressed in a sharply defined belt in PS4 in the ventral part of PS6, which later gives rise to the corresponding mesodermal cells in PS6, and weakly in a belt in PS14. During gastrulation and early germband elongation the P1 promoter is expressed in the ectoderm of PS4, the ectoderm and mesoderm of PS5 and anterior PS6, and much more weakly from posterior PS6 through PS12. At this stage PS2 transcripts continue to be present at high levels in PS4. In PS6 only the mesodermal cells invaginating through the ventral furrow expressed P2 transcripts. At the extended germband stage P1 transcripts are found in the epidermis of PS4, PS5 and anterior PS6, in the mesoderm of PS5 and anterior PS6, and the ventral nervous system in PS4 and 5. P2 transcripts are restricted to a subset of epidermal cells ventrally in PS3 to 5 and to a small cluster of lateral cells in each PS from 4 to 14, probably in the primordial cells of sensory organs. In the ventral nervous system a subset of cells expresses P2 transcript in each neuromere from the first thoracic (T1) to the last abdominal (A9) segment. After germband retraction at late embryonic stages, P1 transcripts are detected around the anterior spiracles which form at the T1/T2 boundary. In the ventral nervous system P1 transcripts accumulate mainly in PS4, at lower levels in PS5 and at very low levels in PS6 to 12. In contrast to PS2 no PS1 transcription is detected in PS13 and 14. All somatic muscles of T3 express P1 transcripts and a group of cells of the visceral mesoderm surrounding the anterior midgut are also labelled by P1 probes. These cells are forming the anterior constriction of the midgut.

In the imaginal discs of the third instar larvae *Antp* transcription has been analyzed by *in situ* hybridization using probes specific for either P1 or P2 (Jorgensen and Garber, 1987) and by studying reporter gene expression under the control of P1 (Engström et al., 1992). *Antp* is transcribed in all thoracic imaginal discs, but no expression is detected in eye-antennal discs. The strongest expression is found in the humeral (dorsal prothoracic) disc. *Antp* transcripts are detected in all three pairs of leg discs; P1 transcripts accumulate mainly in the posterior compartment of the prothoracic leg disc, mostly in the anterior compartment of the mesothoracic leg disc and throughout the metathoracic leg disc, but the expression pattern does not coincide with compartment boundaries. In the wing disc mostly the primordial cells for prescutum, mesopleura and pteropleura show P1 expression. In the haltere disc the presumptive metanotum and scabellum primordia express P1 transcripts. The P2 expression pattern differs from that of P1 only in the wing and mesothoracic leg disc, where P2 transcripts are more uniformly distributed than those of P1. In addition to the imaginal discs, all three thoracic ganglia of the ventral nervous system express *Antp* P1 transcripts, T1 and T2 more strongly than T3, and weak expression is found in specific cells of the neuromeres from A1 to A7.

Antp transcripts are also found in the adult fly, most prominently in the T2 ganglion, but also in the ganglia of T1 and T3. Neurons innervating the indirect flight muscles and the muscles of the second legs also show expression. Prominent expression is also found in cells anterior to the thoracic spiracle (as in embryos and larvae) and in the metathoracic leg muscles.

The expression pattern of the ANTP protein is consistent with the pattern of transcription (Carroll et al., 1986a; Wirz et al., 1986). In young embryos it is limited to the thoracic segments in the epidermis, whereas it is found in all neuromeres of the head, thorax, and abdomen. Towards the end of embryogenesis, ANTP accumulates mainly in the ventral nervous system in certain cells of neuromeres T1 to anterior T3 with a gap in posterior T2. The protein expression pattern is essentially the same as P1 directed expression of the β-galactosidase reporter gene (Engström et al., 1992). ANTP was also detected in the adult, at least in the thoracic ganglion.

Therefore, the *Antp* gene is expressed from early embryonic to adult stages and in the same location, e.g. in PS4 of the embryo, around the anterior spiracle in the late embryo, the humeral disc (around the spiracle) of the larvae and in the same spiracle of the adult. Similar observations are made for the T3 segment in the embryo, the metathoracic leg imaginal disc and finally the adult muscles of the third leg. This indicates that the same gene is used in the same region at different developmental stages.

■ Function

Recessive loss-of-function mutations ("null mutations") are lethal at late embryonic stages. The homozygous mutant embryos show homeotic transformations mainly of the mesothoracic segment (T2) towards T1 with a "beard" of denticles which is characteristic for the prothorax (T1). Furthermore, T1 is transformed towards the labial segment. Both of these transformations occur in the anterior direction. Partial loss-of-function mutants, for example *Antp*^PW/*Antp*^PW, which lacks promoter P1 but retains an intact P2 transcription unit, survive until late larval stages and show the same segmental transformations and have defective anterior spiracles (Schneuwly and Gehring, 1985). The cells which are necrotic in the homozygous mutant, are those which strongly express *Antp* in the wildtype.

X-ray induced somatic clones of *Antp*⁻/*Antp*⁻ cells indicate that *Antp*⁺ is required for the proper development of the legs and the dorsal pro- and mesothorax (Struhl, 19081; Abbott and Kaufman, 1986); Wagner-Bernholz et al., 1992). Only clones induced in the mesothoracic leg disc show homeotic transformation and form antennal structures. In transplantation experiments of *Antp*^PW/*Antp*^PW imaginal discs into wildtype hosts no homeotic transformations of the mesothoracic leg discs were observed, but the wing discs showed transformation into head structures, again in the anterior direction (Schneuwly and Gehring, 1985). The partial loss-of-function mutants, which have a defective P1 and retain an intact P2 transcription unit, partially or completely complement null mutants (Abbott and Kaufman, 1986). However, it is not clear to what extent the P2 transcription unit can substitute for P1, since the reciprocal mutants which leave P1 intact and inactivate P2 have not been studied.

Dominant gain-of-function mutants were shown to result in the ectopic expression of *Antp* (Jorgensen and Garber, 1987; Schneuwly et al., 1987b) and lead to a transformation of the antennae into second (mesothoracic) legs and the dorsal part of the head capsule into dorsal mesothoracic structures (Duncan and Lewis, 1982). These transformations are in the opposite direction from those observed in recessive loss-of-function mutants as was first described for the genes of the Bithorax-Complex (Lewis, 1978). Since loss-of-function mutants lack a mesothoracic segment and gain-of-function mutants generate additional mesothoracic structures in the head region, it can be concluded that *Antp* primarily specifies the mesothoracic segment. This was confirmed by constructing an artificial dominant gain-of-function mutant by inserting a wildtype *Antp* cDNA into a heatshock vector and P-element mediated germline transformation (Scheuwly et al., 1987a). Upon heatshock of early third instar larvae the same homeotic transformations, i.e. antennae to mesothoracic legs and dorsal head capsule to dorsal mesothorax were obtained as in strong dominant gain-of-function mutants. Heat induction at early embryonic stages leads to the transformation of head and T1 segments towards T2, and as a consequence to a failure of head involution (Gibson and Gehring, 1988). This clearly shows that *Antp* specifies the second thoracic segment. However, the effects are restricted to certain developmental stages, when the determinative events normally occur, and they are also confined to anterior segments. The anterior confinement is probably due to the fact that the genes of the Bithorax-Complex (BX-C) are epistatic in the posterior segments. Deletion of the BX-C genes alone results in the transformation of all the posterior segments towards T2 and the concurrent expression of *Antp* at high levels in all the posterior segments (Hafen et al., 1984; Struhl and White, 1985). In contrast to the posterior boundary, the anterior boundary of *Antp* is not affected by the corresponding homeotic gene *Scr* (Wirz et al., 1986).

Most of these effects have been studied in the epidermis, where convenient cuticular markers are available. However, *Antp* also has important functional roles in interior tissues such as the nervous system and the mesoderm. Homeotic transformation of antennal imaginal discs also affects the cell fate of the sensory neurons which fail to establish the proper connection to the central nervous system. Furthermore, *Antp* expression is required in the visceral mesoderm for the formation of the formation of a constriction in the anterior midgut.

■ References

Abbott, M.K. and Kaufman, T.C. (1986). The relationship between the functional complexity and the molecular organization of the *Antennapedia* locus of *Drosophila melanogaster*. Genetics 114, 919-942.

Baker, B. (1987). Dros. Inf. Serv. this reference will be communicated in the week of September 6-10th.

Balkaschina, E.I. (1929). Ein Fall der Erbhomöosis (die Genoraviation "aristopedia") der *Drosophila melanogaster*. Arch. Entw. Mech. 115, 448-463.

Bermingham, Jr., J.R. and Scott, M.P. (1988). Developmentally regulated alternative splicing of transcripts from the *Drosophila* homeotic gene *Antennapedia* can produce four different proteins, EMBO J. 7, 3211-3222.

Bermingham, Jr., J.R., Martinez-Arias, A., Petitt, M.G. and Scott, M.P. (1990). Different patterns of transcription from the two *Antennapedia* promoters during *Drosophila* embryogenesis. Development 109, 553-566.

Carroll, S.B., Laymon, R., McCuthcheon, M., Riley, P. and Scott, M.P. (1986). The localization and regulation of *Antennapedia* protein expression in *Drosophila* embryos. Cell 47, 113-122.

Duncan, I. and Lewis, E.B. (1982). Genetic control of body segment differentiation in *Drosophila* (eds. S. Subtelney and P.B. Green). Developmental order, its origin and regulation: 40th Symp. Dev. Biol. A. Liss, New York, pp. 533-554.

Engström, Y., Schneuwly, S. and Gehring, W.J. (1992). Spatial and temporal expression of an *Antennapedia/lacZ* gene construct integrated into the endogenous *Antennapedia* gene of *Drosophila melanogaster*. Roux's Arch. Dev. Biol. 201, 65-80.

Garber, R.L., Kuroiwa, A. and Gehring, W.J. (1983). Genomic and cDNA clones of the homeotic locus *Antennapedia* in *Drosophila*. EMBO J. 2, 2027-2036.

Gehring, W. (1966). Bildung eines vollständigen Mittelbeines mit Sternopleura in der Antennenregion bei der Mutante *Nasobemia* (*Ns*) von *Drosophila melanogaster*. Jul. Klaus. Arch. 41, 44-54.

Gibson, G. and Gehring, W.J. (1988). Head and thoracic transformations caused by ectopic expression of *Antennapedia* during *Drosophila* development. Development 102, 657-675.

Hafen, E., Levine, M., Garber, R.L. and Gehring, W.J. (1983). An improved *in situ* hybridization method for the detection of cellular RNAs in *Drosophila* tissue sections and its application for localizing transcript of the homeotic *Antennapedia* gene complex. EMBO J. 2, 617-623.

Hafen, E., Levine, M. and Gehring, W.J. (1984). Regulation of *Antennapedia* transcript distribution by the bithorax complex in *Drosophila*. Nature 307, 287-289.

Harding, K., Wedeen, C., McGinnis, W. and Levine, M. (1985). Spatially regulated expression of homeotic genes in *Drosophila*. Science 229, 1236-1242.

Ingham, P. and Martinez-Arias, A. (1986). The correct activation of *Antennapedia* and bithorax complex genes requires the *fushi tarazu* gene. Nature 324, 592-597.

Jorgensen, E.M. and Garber, R.L. (1987). Function and misfunction of the two promoters of the *Drosophila Antennapedia* gene. Genes Dev. 1, 544-555.

Kaufman, T.C., Seeger, M.A. and Olsen, G. (1990). Molecular and genetic organization of the *Antennapedia* gene complex of *Drosophila melanogaster*. Adv. Genet. 27, 309-362.

Laughon, A., Boulet, A.M., Bermingham, Jr. J.R., Laymon, R.A. and Scott, M.P. (1986). Structure of transcripts from the homeotic *Antennapedia* gene of *Drosophila melanogaster*: two promoters control the major protein-coding region. Mol. Cell. Biol. 6, 4676-4689.

Le Calvez, J. (1948). In (3R) SSAr: Mutation *Aristapedia*, hétérozygote dominante, homozygote lethal chez *Drosophila melanogaster*. Bull. Biol. France Belg. 82, 97-113.

Levine, M., Hafen, E., Garber, R.L. and Gehring, W.J. (1983). Spatial distribution of *Antennapedia* transcripts during *Drosophila* development. EMBO J. 2, 2037-2046.

Lewis, E.B. (1978). A gene complex controlling segmentation in *Drosophila*. Nature 276, 565-570.

Martinez-Arias, A. (1986). The *Antennapedia* gene is required and expressed in parasegments 4 and 5 of the *Drosophila* embryo. EMBO J. 5, 135-141.

Otting, G., Qian, Y.Q., Billeter, M., Müller, M., Affolter, M., Gehring, W.J. and Wüthrich, K. (1990). Protein-DNA contacts in the structure of a homeodomain-DNA complex determined by nuclear magnetic resonance spectroscopy in solution. EMBO J. 9, 3085-3092.

Qian, Y.-Q., Billeter, M., Otting, G., Müller, M., Gehring, W.J. and Wüthrich, K. (1989). The structure of the *Antennapedia* homeodomain determined by NMR spectroscopy in solution: Comparison with prokaryotic repressors. Cell 59, 573-580.

Schneuwly, S. and Gehring, W.J. (1985). Homeotic transformation of thorax into head: Developmental analysis of a new *Antennapedia* allele in *Drosophila melanogaster*. Dev. Biol. 108, 377-386.

Schneuwly, S., Kuroiwa, A., Baumgartner, P. and Gehring, W.J. (1986). Structural organization and sequence of the homeotic gene *Antennapedia* of *Drosophila melanogaster*. EMBO J. 5, 733-739.

Schneuwly, S., Klemenz, R. and Gehring, W.J. (1987a). Redesigning of the body plan of *Drosophila* by ectopic expression of the homeotic gene *Antennapedia*. Nature 325, 816-818.

Schneuwly, S., Kuroiwa, A. and Gehring, W.J. (1987b). Molecular analysis of the dominant homeotic *Antennapedia* phenotype. EMBO J. 6, 201-206.

Scott, M.P., Weiner, A.J., Polisky, B.A., Hazelrigg, T.I., Pirrotta, V., Scalenghe, F. and Kaufman, T.C. (1983). The molecular organization of the *Antennapedia* complex of *Drosophila*. Cell 35, 763-776.

Stroeher, V.L., Jorgensen, E.M. and Garber, R.L. (1986). Multiple transcripts from the *Antennapedia* gene of *Drosophila melanogaster*. Mol. Cell. Biol. 6, 4667-4675.

Stroeher, V.L., Gaiser, J.C. and Garber, R.L. (1988). Alternative RNA splicing that is spatially regulated: Generation of transcripts from the *Antennapedia* gene of *Drosophila melanogaster* with different protein-coding regions. Mol. Cell. Biol. 8, 4143-4154.

Struhl, G. (1981). A homeotic mutation transforming leg to antenna in *Drosophila*. Nature 292, 635-638.

Struhl, G. and White, R. (1985). Regulation of the *Ultrabithorax* gene of *Drosophila* by other bithorax complex genes. Cell 43, 507-519.

Wagner-Bernholz, J.T., Wilson, C., Gibson, G., Schuh, R. and Gehring, W.J. (1991). Identification of target genes of the homeotic gene *Antennapedia* by enhancer detection. Genes Dev. 5, 2467-2480.

Wedeen, G., Harding, K. and Levine, M. (1986). Spatial regulation of *Antennapedia* and bithorax gene expression by the *Polycomb* locus in *Drosophila*. Cell 44, 739-748.

Wirz, J., Fessler, L.I. and Gehring, W.J. (1986). Localization of the *Antennapedia* protein in *Drosophila* embryos and imaginal discs. EMBO J. 5, 3327-3334.

Species: *Artemia franciscana*
Chromosomal location: -
Other names: *Antennapedia*
Cognate genes: -
Type of homeobox: *Antp*-class
Accession number: -

■ Origin and description

Cloned by PCR with degenerate homeobox primers, followed by inverse PCR of 0.8 kb fragment from genomic DNA. Amino acid conservation in the homeodomain uniquely identifies this gene as a homologue of *Drosophila Antp*.

■ References

Averof, M. and Akam, M. (1993). Hom/Hox genes in a crustacean; Implication for the origin of insect and crustacean body plans. Current biology 3, 73–78.

Species: *Drosophila subobscura*
Chromosomal location: -
Other names: *Antennapedia*
Cognate genes: -
Type of homeobox: *Antp*-class
Accession number: -

■ Origin and description

The gene was isolated on the basis of its sequence homology to the *Antp* gene of *Drosophila melanogaster* and sequenced from exon 4 to 8 (Hooper et al., 1992). A similarly high degree of sequence conservation was found between *Drosophila subobscura* and *Drosophila melanogaster* as for *Drosophila virilis*. The homeobox carries 21 silent mutational changes as compared to *Drosophila melanogaster* and 20 relative to *Drosophila virilis*, whereas the amino acid sequences of the homeodomain of all three species are identical.

■ References

Hooper, J.E., Perez-Alonso, M., Bermingham, J.R., Prout, M., Rocklein, B.A., Wagenbach, M., Edström, J.E., de Frutos, R. and Scott, M.P. (1992). Evolutionary conservation of *Antennapedia* homeotic gene sequences. Genetics 132, 453–469.

Species: *Drosophila virilis*
Chromosomal location: -
Other names: *Antennapedia*
Cognate genes: -
Type of homeobox: *Antp*-class
Accession number: -

■ Origin and description

The *Antp* gene of *Drosophila virilis* was cloned on the basis of its sequence homology to the *Drosophila melanogaster* gene (Hooper et al., 1992). The overall sequence organization of the *Antp* genes from the two species which are thought to have been evolutionarily separated for about 60 million years is very similar. As in *Drosophila melanogaster*, the *Drosophila virilis Antp* gene contains two alternative promoters (P1 and P2) and two alternative polyadenylation sites (A1 and A2) which yield four different major transcripts. The two transcription units measure 121 kb and 41 kb, as compared to 103 kb and 36 kb in *Drosophila melanogaster*. In the promoter regions there are blocks of conserved sequences that range in size from 8 to 134 b interspersed with diverged sequences. These blocks are arranged colinearly. As in *Drosophila melanogaster*, there is a large array of AUG codons upstream of the protein coding region: 3 for P1 with open reading frames (ORF) ranging from 21 to 51 b long, and 14 AUG codons for P2 with ORFs from 18 to 336. In *Drosophila melanogaster* there are eight for P1 and 15 for P2, but neither the positions, sizes, nor sequences of the ORFs are conserved between the two species. This indicates that ORFs per se are not important, but the large array of AUG codons may serve a regulatory function in translation initiation as in other systems (Thireos et al., 1984; Hinnebusch, 1984). A comparison of the exon sequences between the two species shows that the highest degree of conservation is in the protein coding region, but there is also a remarkable degree of conservation in the 3′ trailer and numerous blocks of conserved sequences in the 5′ leader. Also, the region around the alternative splice site in exon 7 is highly conserved, which suggests that alternative splicing may also occur in *Drosophila virilis*. There are also striking co-linear homologies through the 100 kb of introns.

The deduced protein sequences are extremely well conserved in the N-terminal 104 amino acids including the MXSYF motif, and the C-terminal 126 amino acids including the homeodomain, which is identical in the two species, even though the DNA sequence of the homeobox contains 29 silent changes.

References

Hinnebusch, A.G. (1984). Evidence for translational regulation of the activator of general amino acid control in yeast. Proc. Natl. Acad. Sci. USA 81, 6442-6446.

Hooper, J.E., Perez-Alonso, M., Bermingham, J.R., Prout, M., Rocklein, B.A., Wagenbach, M., Edström, J.E., de Frutos, R. and Scott, M.P. (1992). Evolutionary conservation of *Antennapedia* homeotic gene sequences. Genetics 132, 953–969.

Thireos, G., Penn, M.D. and Greer, H. (1984). 5′ untranslated sequences are required for the translational control of a yeast regulatory gene. Proc. Natl. Acad. Sci. USA 81, 5096-5100.

ATBF1 [h]

Species: *Homo sapiens*
Chromosomal location: -
Other names: -
Cognate genes: -
Type of homeobox: -
Accession number: D90395

■ Origin and description

ATBF1 cDNA was isolated by screening λgt11 cDNA expression libraries prepared from HuH7 (human hepatoma) mRNA with a DNA fragment containing AT-rich sequence (AT-motif) found in the human α-fetoprotein gene enhancer. Amino acid sequence of *ATBF1* deduced from the cDNA sequence revealed that this is the largest DNA-binding protein known to date, containing four homeodomains (HD1 to HD4), 17 zinc finger motifs, and a number of segments potentially involved in transcriptional regulation. HD1 has some homology to *mec-3* homeodomain (37%). HD2, HD3, and HD4 have homologies to lin-11 homeodomain (40%, 43%, and 42%, respectively).

HD1 is unique in that it is preceded by a POU-specific domain A-box.

■ Expression

The expression pattern of *ATBF1* has not been studied. *ATBF1* mRNA was detected, by using reverse transcriptase-linked PCR, in all cell lines so far studied which include HuH7, HepG2, HeLa, T24, J82, HTB9, and MCF7. Although *ATBF1* mRNA is detected in HeLa cells, nuclear extract prepared from HeLa cells cannot protect the AT-motif in foot print analysis. This suggests that functional *ATBF1* is not produced in HeLa cells.

■ Genetics, function

Based on the observation that *ATBF1* binds to human α-fetoprotein gene enhancer, it is speculated that this protein is involved in transcriptional regulation of α-fetoprotein gene.

■ References

Morinaga, T., Yasuda, H., Hashimoto, T., Higashio, K. and Tamaoki, T. (1991). A human α-fetoprotein enhancer-binding protein, ATBF1, contains four homeodomains and seventeen zinc fingers. Mol. Cell. Biol. 11, 6041-6049.

Athb-1A; Athb-1B [At]

Species: *Arabidopsis thaliana*
Chromosomal location: -
Other names: -
Cognate genes: -
Type of homeobox: -
Accession number: -

■ Origin and description

Athb-1A and *Athb-1B* are two cDNA clones which overlap the *Athb-1C* cDNA.

■ References

Ruberti, I., Sessa, G., Lucchetti, S. and Morelli, G. (1991). A novel class of plant proteins containing a homeodomain with a closely linked leucine zipper motif. EMBO J. 7, 1787-1791.

bap [d]

Species: *Drosophila melanogaster*
Chromosomal location: Chromosome 3R, 93E1, 2
Other names: *bagpipe, NK-3*
Cognate genes: -
Type of homeobox: *NK*-type
Accession number: M2791

■ Origin and description

The gene was isolated by hybridizing a genomic library with degenerate oligonucleotides corresponding to homeobox regions encoding parts of the presumed recognition helix (Kim and Nirenberg, 1989). A sequence motif 14-25 aa to the C-terminus of the homeodomain is also found in several other homeobox genes of the NK type (Price et al., 1992). The *bap* transcription unit starts ~6.5 kb downstream of *tinman* (*tin*), towards the distal side of the chromosome (Azpiazu and Frasch, in preparation).

■ Expression

bap starts to be expressed at stage 10 of embryogenesis. The expression occurs in cells of the dorsal mesoderm and is observed in 11 segmental patches, extending from parasegments 2-12 with respect to the epidermis. These cells appear to correspond to primordial cells of the midgut musculature. *bap* mRNA expression ceases before the dorsal mesoderm splits into the visceral and somatic mesoderm. At stage 11 *bap* starts to be also expressed in the prospective foregut and hindgut mesoderm (Azpiazu and Frasch, in preparation). The expression of *bap* is regulated by *tin* and segmentation genes (Azpiazu and Frasch, unpublished).

■ Function

In embryos mutant for an EMS-induced allele, *bap*[208], the visceral mesoderm is segmentally interrupted. Thus, in wild type *bap* appears to specify the fate of dorsal mesodermal cells to develop into visceral mesoderm. In late mutant embryos, the midgut musculature is strongly reduced and midgut morphology is abnormal (Azpiazu and Frasch, in preparation).

■ References

Azpiazu, N. and Frasch, M. in preparation.
Kim, Y. and Nirenberg, M. (1989). *Drosophila* NK-homeobox genes. Proc. Natl. Acad. Sci. USA 86, 7716-7720.
Price, M., Lazzaro, D., Pohl, T., Mattei, M.-G., Rüther, U., Olivo, J.-C., Duboule, D. and DiLauro, R. (1992). Regional expression of the homeobox gene Nkx2.2 in the developing mammalian forebrain. Neuron 8, 241-255.

Bar H1 [d]

Species: *Drosophila melanogaster/ Drosophila ananassae*

Chromosomal location: 16A on the X chromosome in the case of the *Dm BarH1* gene. The *Om* (1D) locus of the X chromosome in the case of the *Da BarH1* gene.

Other names: -

Cognate genes: *BarH2* of *D. melanogaster*

Type of homeobox: WY type

Accession numbers: M59965 for the *Dm BarH1*. M59962-3 for the *Da BarH1*.

■ Origin and description

The *Da BarH1* gene was initially identified in the cloned *Om* (1D) region using the M-repeat oligonucleotide as a probe. The *Dm* counterpart was isolated by cross-homology with the Da *BarH1*. cDNA structure of the *Dm* gene along with the genomic structures of *Da* and *Dm BarH1* genes were examined. A unique feature of the homeodomains encoded by *BarH1* genes is that the phenylalanine residue in helix 3, conserved in all metazoan homeodomains so far examined, is replaced by a tyrosine residue. *BarH1* and its cognate *BarH2* (Higashijima et al., 1992) from a small complex at the *Bar* locus of *D. melanogaster*.

■ Expression

Two distinct *BarH1* transcripts, 3.1 kb and 0.8 kb long were found. The 3.1 kb RNA was expressed mainly in embryos and pupae, while the 0.8 kb RNA was expressed throughout development. In the normal eye disc, cell-type-specific expression of *BarH1* was observed; the *BarH1* protein was expressed at the morphogenetic furrow, two photoreceptors R1/R6 and a pair of primary pigment cells and appeared to be functionally required for the normal eye development (Higashijima et al., 1992). The *BarH1* overexpression caused ommatidia to decrease in number as in the case of the *Bar* mutation.

■ Genetics, function

The *BarH1* is required at least for the normal eye development. It is one of the genes causing the *B* mutation.

■ References

Tetsuya Kojima., Satoshi Ishimaru, Shin-ichi Hiogashijima, Eiji Takayama, Hiroshi Akimaru, Masaki Sone, Yasufumi Emori and Kaoru Saigo. (1991). Identification of a different-type homeobox gene, BarH1, possibly causing Bar (B) and Om (Om) mutations in *Drosophila*. Proc. Natl. Acad. Sci. USA 88, 4343-4347.
Shin-ichi Higashijima, Tatsuo Michiue, Yasufumi Emori, and Kaoru Saigo. (1992). Subtype determination of *Drosophila* embryonic external sensory organs by redundant homeobox genes BarH1 and BarH2. Gene Dev. 6, 1005–1018.

bE [Um]

Species: *Ustilago maydis* (corn smut fungus)

Chromosomal location: -

Other names: *b*

Cognate genes: -

Type of homeobox: -

Accession number: M30648 (*b*1), M30649 (*b*2) M30650 (*b*3) M30651 (*b*4) X54069 (*b*5) X54070 (*b*6), X54071 (*b*6)

■ Origin and description

The multiallelic *b* locus of *Ustilago maydis* contains two divergently transcribed genes, *bE* and *bW*, separated by a short intergenic region of 240 bp. Thus, the *b*1 allele consists of *bE*1 and *bW*1; *b*2 consists of *bE*2 and *bW*2, etc. There is no recombination between *bE* and *bW*. There are estimated to be 25 *b* alleles. The *bE*2 gene was cloned by functional complementation: a DNA segment carrying *bE*2 is sufficient to confer filamentous growth and tumor-inducing ability on a diploid strain homozygous for *b*1 (*bW*1 *bE*1/*bW*1 *bE*1), which normally is capable only of yeast-like growth and is non-pathogenic (Schulz et al., 1990). Seven *bE* alleles have been cloned and sequenced: *bE*1-*bE*4 by Schulz et al. (1990); *bE*5-*bE*7 by Kronstad and Leong (1990). All these alleles contain an open reading frame (ORF) of 410 amino acids. Comparison of the ORFs from different alleles shows that the amino terminal region of 110 amino acids is variable (63% identity) and the remainder of the ORF (the constant region) is highly conserved (93% identity). The constant region contains an atypical homeodomain at position 107-180 (Schulz et al., 1990), with 15 extra amino acids between helix 2 and 3, and 3 extra amino acids between helix 1 and 2.

■ Genetics, function

Ustilago maydis is a Basidiomycete fungus that infects corn and induces the formation of tumors. Growth in the plant and tumor formation are essential in order for the fungus to acquire competence to undergo meiosis. The fungus is dimorphic, with a haploid, unicellular, yeast-like form that is non-pathogenic, and a filamentous, dikaryotic form that is pathogenic and needs the plant for its growth. The multiallelic *b* locus (25 naturally occurring alleles) controls the dimorphic transition and pathogenicity: the presence of different *b* alleles is necessary for filamentous growth and for tumor induction. It is hypothesized that self-nonself monitoring occurs by interaction of *bE* and *bW* polypeptides contributed by different b alleles. This interaction results in formation of a heteromultimeric protein that governs expression of genes for filamentous growth and tumor induction (see review by Banuett, 1992). Precedent for interaction of homeodomain proteins is provided by the α1-α2 heteromultimer of Saccharomyces cerevisiae (reviewed by Herskowitz, 1992).

■ References

Banuett, F. (1992). *Ustilago maydis*, the delightful blight. Trends in Genet. 8, 174-180.

Gillissen, B., Bergemann, J., Sandmann, C., Schroeer, B., Bolker, M. and Kahmann, R. (1992). A two-component regulatory system for self-nonself recognition in *Ustilago maydis*. Cell 66, 647-657.

Herskowitz, I. (1992). Yeast branches out. Nature 357, 190-191.

Kronstad, J.W. and Leong, S.A. (1990). The *b* mating type locus of *Ustilago maydis* contains variable and constant regions. Genes Dev. 4, 1384-1395.

Kües, U., Richardson, W.V.J., Tymon, A.M., Mutasa, E.S., Göttgens, B., Gaubatz, S., Gregoriades, A. and Casselton, L.A. (1992). The combination of dissimilar alleles of the Aα and Aβ gene complexes, whose proteins contain homeodomain motifs, determines sexual development in the mushroom *Coprinus cinereus*. Genes Dev. 6, 568-577.

Schulz, B., Banuett, F., Dahl, M., Schlesinger, R., Schafer, W., Martin, T., Herskowitz, I. and Kahmann, R. (1990). The *b* alleles of *U. maydis*, whose combinations program pathogenic development, code for polypeptides containing a homeodomain-related motif. Cell 60, 295-306.

bcd [d]

Species: *Drosophila melanogaster*

Chromosomal location: 84 A

Other names: *bicoid*

Cognate genes: -

Type of homeobox: *bicoid*-type

Accession number: -

■ Origin and description

The *bicoid* gene was genetically characterized during screens for maternal effect mutations in *Drosophila* (Frohnhöfer and Nüsslein-Volhard, 1986). Embryos from females mutant for *bicoid* (*bcd*) lack head and thorax, and instead develop a duplication of posteriormost structures, the telson, at the anterior. Transplantation experiments indicate that *bcd* is responsible for the organization of the anterior half of the *Drosophila* embryo (Frohnhöfer et al., 1986). *bcd* was cloned during homology screens for paired repeats (Frigerio et al., 1986; Berleth et al., 1988).

■ Expression

bcd is maternally transcribed during oogenesis, and the transcript is transported into the oocyte, where it becomes strictly localized at the anterior pole (Berleth et al., 1988; St. Johnston et al., 1989). Sequences in the 3 prime untranslated region of the *bcd* mRNA appear to be involved in attaching the mRNA to elements of the cytosceleton (Macdonald and Struhl, 1988; Macdonald, 1990; Pokrywka and Stephenson, 1991; Seeger and Kaufman, 1990). The *bcd* protein product (*Bicoid*) is translated from this localized messenger RNA and forms a concentration gradient along the anterior-posterior axis of the early embryo, with its highpoint at the anterior tip of the embryo (Driever and Nüsslein-Volhard, 1988a).

■ Function

Bicoid has been demonstrated to determine pattern in the anterior half of the *Drosophila* embryo in a concentration dependent manner, e.g. determining both the "quality" head or thorax as well as the position of these structures along the axis (Frohnhöfer and Nüsslein-Volhard, 1986; Driever and Nüsslein-Volhard, 1988b); Driever et al., 1990). Thus *Bicoid* fits the definition of a morphogen. The *Bicoid* homeodomain binds to the promoter of the zygotic target gene hunchback (Driever and Nüsslein-Volhard, 1989) and activates hunchback transcription in the anterior half of the embryo (Driever et al., 1989a and 1989b; Struhl et al., 1989). The affinity of binding sites within the target promoter appears to determine the extent of the target gene's expression domain along the anterior posterior axis within the graded field of *Bicoid* distribution (Driever et al., 1989a). Several genes other than hunchback have been implied to be regulated by *Bicoid*, too (Finkelstein and Perimon, 1990).

The binding specificity of *Bicoid* for DNA (consensus: TCTAATCCC) is at variant when compared to those of other homeodomain proteins. A single amino acid at position 9 of the binding helix appears to be responsible for the class of sequences recognized by homeodomains (Hanes and Brent, 1989; Treisman et al.,1989).

The *Bicoid* morphogen provides the link between the maternal genome and the onset and regulation of region specific gene expression in the anterior of the *Drosophila* embryo.

■ References

Berleth, T., Burri, M., Thoma, G., Bopp, D., Richstein, S., Frigerio, G., Noll, M. and Nüsslein-Volhard, C. (1988). The role of localization of *bicoid* RNA in organizing the anterior pattern of the *Drosophila* embryo. EMBO J. 7, 1749-1756.

Driever, W. and Nüsslein-Volhard, C. (1988a). A gradient of *bicoid* protein in *Drosophila* embryos. Cell 54, 83-93.

Driever, W. and Nüsslein-Volhard, C. (1988b). The *bicoid* protein determines position in the *Drosophila* embryo in a concentration dependent manner. Cell 54, 95-104.

Driever, W. and Nüsslein-Volhard, C. (1989). *bicoid* protein, a positive regulator of hunchback transcription in the early *Drosophila* embryo. Nature 337, 138-143.

Driever, W., Thoma, G. and Nüsslein-Volhard, C. (1989a). Determination of spatial domains of zygotic gene expression in the *Drosophila* embryo by the affinity of binding site for the bicoid morphogen. Nature 340, 363-367.

Driever, W., Ma, J., Nüsslein-Volhard, C. and Ptashne, M. (1989b). Rescue of bicoid-mutant *Drosophila* embryos by *bicoid* fusion proteins containing heterologous activating sequences. Nature 342, 149-154.

Driever, W., Siegel, V. and Nüsslein-Volhard, C. (1990). Autonomous determination of anterior structures in the early *Drosophila* embryo by the *bicoid* morphogen. Development 109, 811-820.

Finkelstein, R. and Perrimon, N. (1990). The orthodenticle gene is regulated by bicoid and torso and specifies *Drosophila* head development. Nature 346, 485-488.

Frigerio, G., Burri, M., Bopp, D., Baumgartner, S. and Noll, M. (1986). Structure of the segmentation gene paired and the *Drosophila* PRD gene set as part of a gene network. Cell 47, 735-746.

Frohnhöfer, G. and Nüsslein-Volhard, C. (1986). Organization of anterior pattern in the *Drosophila* embryo by the maternal gene *bicoid*. Nature 324, 120-125.

Frohnhöfer, H.G., Lehmann, R. and Nüsslein-Volhard, C. (1986). Manipulating the anterioposterior pattern of the *Drosophila* embryo. J. Embryol. Exp. Morphol. 97, 169-179.

Frohnhöfer, H.G. and Nüsslein-Volhard, C. (1987). Maternal genes required for the anterior localization of *bicoid* activity in the embryo of *Drosophila*. Genes Dev. 1, 880-890.

Hanes, S.D. and Brent, R. (1989). DNA specificity of the *bicoid* activator protein is determined by homeo domain recognition helix residue 9. Cell 57, 1275-1283.

Macdonald, P.M. and Struhl, G. (1988). *Cis*-acting sequences responsible for anterior localization of *bicoid* mRNA in *Drosophila* embryos. Nature, 336, 595-598.

Macdonald, P.M. (1990). *bicoid* mRNA localization signal: phylogenetic conservation of function and RNA secondary structure. Development 110, 161-171.

Pokrywka, N.J. and Stephenson, E.C. (1991). Microtubules mediate the localization of *bicoid* RNA during *Drosophila* oogenesis. Development, 113, 55-66.

Seeger, M.A. and Kaufman, T.C. (1990). Molecular analysis of the *bicoid* gene from *Drosophila* pseudoobscura: identification of conserved domains within coding and noncoding regions of the *bicoid* mRNA. EMBO J. 9, 2977-2987.

St. Johnston, D., Driever, W., Berleth, T., Richstein, S. and Nüsslein-Volhard, C. (1989). Mutiple steps in the localization of *bicoid* mRNA to the anterior pole of the *Drosophila* oocyte. Development Supplement, 13-19.

Struhl, G., Struhl, K. and Macdonald, P. (1989). The gradient morphogen *bicoid* is a concentration-dependent transcriptional activator. Cell 57, 1259-1273.

Treisman, J., Gönczy, P., Vashishtha, M., Harris, E. and Desplan, C. (1989). A single amino acid can determine the DNA binding specificity of homeo domain proteins. Cell 59, 553-562.

bW [Um]

Species: *Ustilago maydis* (corn smut fungus)
Chromosomal location: -
Other names: *b*
Cognate genes: -
Type of homeobox: -
Accession number: M84179 (*bW*1), M84180 (*bW*3), M84181 (*bW*4), M84182 (*bW*2)

■ Origin and description

The *bW* gene is one of two genes present in the multiallelic *b* locus of *Ustilago maydis*. Deletion analysis of the *bE* gene led to identification of the adjacent *bW* gene. *bW* is multiallelic: four different *b* alleles have been cloned and sequenced (Gillissen et al., 1990). Each of these alleles contains an open reading frame of 626 amino acids. Comparison of the ORFs of these four alleles shows that the *bW* polypeptide is organized in similar fashion to the *bE* polypeptide: the 130 amino acids at the amino terminal end are variable (46% identity) and the remainder of the polypeptide (the constant region) is highly conserved (96% identity). The constant region contains a typical homeodomain at position 136-195. Despite the similarity in organization of the *bW* and *bE* polypeptides, they do not share any similarity at the amino acid level except for some amino acids in the homeodomain region.

■ Genetics, function

A strain containing two different *b* alleles, for example *b*1 and *b*2 (*bW*1, *bE*1 and *bW*2, *bE*2), is filamentous and induces tumors in corn. A strain containing only 2 different *bW* alleles and deleted for *bE* (*bW*1 *bE*1 and *bW*2 *bE*2) is neither filamentous nor pathogenic. A strain deleted for *bW* and containing different *bE* alleles (*bW*1, *bE*1 and *bW*2-*bE*2) is also non-filamentous and non-pathogenic. Thus, different *bE* alleles alone or different *bW* alleles alone are not sufficient for filamentous growth and tumor induction. However, a strain containing only one *bE* allele and one *bW* allele (*bW*1, *bE*1 and *bW*2, *bE*2) exhibits filamentous growth and is pathogenic, as long as *bE* and *bW* are contributed by different *b* strains. This indicates that the functional *b* regulatory protein is a heteromultimer consisting of *bW* and *bE* polypeptides contributed by different *b* alleles (for example, *bW*2-*bE*1 or *bW*1-*bE*3 heteromultimers). It is proposed that the variable regions of the *bW* and *bE* polypeptides are the determinants for generation of a functional protein. Specifically, it is proposed that the variable regions from different *b* alleles determine the proper positioning of

helilx 3 domains of the heteromultimer on the DNA (see Banuett, 1992).

■ References

Banuett, F. (1992). *Ustilago maydis*, the delightful blight. Trends in Genet. 8, 174-180.

Gillissen, B., Bergemann, J., Sandmann, C., Schroeer, B., Bolker, M. and Kahmann, R. (1992). A two-component regulatory system for self-nonself recognition in *Ustilago maydis*. Cell 66, 647-657.

Herskowitz, I. (1992). Yeast branches out. Nature 357, 190-191.

Kronstad, J.W. and Leong, S.A. (1990). The *b* mating type locus of *Ustilago maydis* contains variable and constant regions. Genes Dev. 4, 1384-1395.

Schulz, B., Banuett, F., Dahl, M., Schlesinger, R., Schafer, W., Martin, T., Herskowitz, I. and Kahmann, R. (1990). The *b* alleles of *U. maydis*, whose combinations program pathogenic development, code for polypeptides containing a homeodomain-related motif. Cell 60, 295-306.

cad [d]

Species: *Drosophila melanogaster*
Chromosomal location: 38E
Other names: *caudal*
Cognate genes: *Cdx-1, Chox-cad, Xcad*
Type of homeobox: *cad*-type
Accession number: -

■ Origin and description

The *caudal* (*cad*) gene was isolated by cross hybridization to the *Ubx* homeo box (Mlodzik et al., 1985) and independently also by cross hybridization by Levine et al. (1985) and Macdonald and Struhl (1986). The *cad* homeodomain shares 58% homology to the *ftz* homeo domain, and 53% or 52% with *Antp* and *Ubx* homeo domains, respectively.

■ Expression:

caudal is expressed both maternally and zygotically as a 2.4 Kb and 2.6 Kb transcript, respectively. The maternal expression is detected in the nurse cells during oogenesis and the mRNA accumulates in the developing oocyte. At the end of oogenesis *cad* mRNA is evenly distributed throughout the oocyte. There is no Cad protein detected during oogenesis. Cad protein is detected beginninig with the nuclear division cycles 6 or 7 just prior to the formation of the syncitial blastoderm. From this stage on to the 13 nuclear cycle Cad protein distribution forms a concentration gradient throughout the embryo along the antero-posterior axis with its maximum at the posterior end. The RNA remains uniformly distributed throughout the cytoplasm to the 12 nuclear cleavage, later it follows the protein gradient and is then distributed as a concentration gradient itself. Nevertheless, the protein gradient precedes the RNA gradient by several nuclear cycles. Following the final nuclear cleavage during cellularization of the blastoderm the maternally derived *cad* RNA disappears and with it the protein translated from it. Cad also accumulates in the pole cells during their formation and is present there for several hours of embryogenesis. However there is no de novo expression in the pole cells and all protein present goes back to the maternally provided RNA. The formation of the Cad protein gradient does not depend on zygotic gene expression. In older unfertilized eggs, an antero-posterior gradient is detected similar to that normally present in embryos of the same stage (Mlodzik et al., 1985; Macdonald and Struhl, 1986; Mlodzik and Gehring, 1987).

Zygotic expression begins during cellularization of the blastoderm as a stripe in the posterior region of the embryo corresponding to the anlagen of parts of the telson and the hindgut. At the extended germ band stage and throughout the rest of embryogenesis *cad* is expressed in the posterior midgut, the Malpighian tubules, the anal plate and at weaker levels during germ band extension in pair-rule like stripes (Mlodzik et al., 1985; Macdonald and Struhl, 1986; Mlodzik and Gehring, 1987). In 3rd instar larvae *cad* is expressed in the Malpighian tubules, the posterior midgut, the gonads of both sexes and in the anlagen of the anal plates and the hindgut within the genital disc (Mlodzik and Gehring, 1987).

■ Function:

caudal is required for segment specification in the abdominal and posterior thoracic regions. The posterior part of embryos lacking both maternal and zygotic gene product are severely shortened with variable deletions of many of the posterior segments, the odd-numbered abdominal segments are more resistant to deletion. Zygotic function can largely rescue these defects. Embryos from wild-type mothers that lack zygotic function display an absence of cuticular structures of the posterior terminalia, in particular the anal tuft, parts of the anal pads and the terminal sense organs, all structures that are derived from a cryptic 10th abdominal segment (Macdonald and Struhl, 1986). Ectopic expression of *cad* throughout the embryo at the blastoderm stage leads to phenotypes that are somewhat opposite to the phenotypes observed in *cad*- embryos (Mlodzik et al., 1990). Analysis of the *ftz* cis-acting elements has suggested that *cad* is directly activating *ftz* transcription (Dearolf et al., 1989).

■ References

Dearolf, C., Topol, J. and Parker, C. (1989). The *caudal* gene product is a direct activator of *fushi tarazu* transcription during *Drosophila* embryogenesis. Nature 341, 3430-343.

Levine, M., Harding, K., Wedeen, C., Doyle, H., Hoey, T. and Radomska, H. (1985). Expression of the homeo box gene family in Drosophila. Cold Spring Harb. Symp Quant. Biol. 50, 209-222.

Macdonald, P.M. and Struhl, G. (1986). A molecular gradient in early Drosophila embryos and its role in specifying the body pattern. Nature 324, 537-545.

Mlodzik, M., Fjose, A. and Gehring, W.J. (1985). Isolation of *caudal*, a *Drosophila* homeo box-containing gene with maternal expression, whose transcripts form a concentration gradient at the pre-blastoderm stage. EMBO J. 4, 2961-2969.

Mlodzik, M. and Gehring, W.J. (1987). Expression of the *caudal* gene in the germ line of Drosophila: formation of an RNA and protein gradient during early embryogenesis. Cell 48, 465-478.

Mlodzik, M., Gibson, G. and Gehring, W.J. (1990). Effects of ectopic expression of *caudal* during *Drosophila* development. Development 109, 271-277.

Cbmec-3 [cb]

Species: *Caenorhabditis briggsae*
Chromosomal location: -
Other names: -
Cognate genes: -
Type of homeobox: LIM class
Accession number: L02877

■ Origin and description

Isolated using the *C. briggsae* gene *mec-3* (Xue et al., 1992) to identify conserved sequence elements. See *mec-3* for further description.

■ References

Xue, D., Finney, M., Ruvkun, G. and M. Chalfie, M. (1992). Regulation of the *mec-3* gene by the *C. elegans* homeoproteins UNC-86 and MEC-3. EMBO J. 11, 4969-4979.

CDP [h]

Species: *Homo sapiens*
Chromosomal location: 7q22
Other names: -
Cognate genes: *Drosophila cut, Clox-1*
Type of homeobox: -
Accession number: M74099

■ Origin and description

CDP (CCAAT displacement protein) is a ubiquitous eukaryotic repressor protein. It was first described as a putative repressor of sperm-specific histone H2B expression in sea urchin by Busslinger et al. (1977). *CDP* was purified from

HeLa cell nuclear extract by DNA affinity chromatography using a binding-site oligonucleotide from the promoter of the X-linked cytochrome gene, gp91-phox (Neufeld et al., 1992). Purified *CDP* is a dimer of 180-190 kD. Antibody against the purified protein was used to obtain cDNA clones from a human cDNA expression library in λ gt11. Sequencing revealed a cDNA of 5.3 kb, encoding a predicted protein of 1505 amino acids, containing a homeobox and three "*Cut* repeat" domains closely homologous to the *Drosophila* homeoprotein *cut* (Blochlinger et al., 1988). The two proteins are 50% identical and 68% highly conserved over 380 amino acids in these domains (Neufeld et al., 1992). In the remainder of the two molecules, there is no obvious homology. A related partial cDNA from canine heart, designated *Clox-1*, was subsequently reported. *Clox* and *CDP* differ at the C-terminus, possibly due to alternative splicing (Andres et al., 1992). *CDP* is implicated as a repressor of terminally-differentiated gene expression in hematoietic cells of myeloid lineage (Skalnik et al., 1991). Its homology to *cut*, which acts to specify cell fate in several *Drosophila* tissues during embryonic development, suggests that *CDP* may play a role in control of gene expression in other mammalian tissues.

■ Expression

Northern blot analysis with portions of *CDP* cDNA as probes reveals mRNA species of 12, 8, (possibly 6), 3.4 and 1.8 kb. Full length CDP protein is presumably encoded by one or more of the larger mRNA species. The 3.4 kb species is alternatively spliced, and shares its N-terminal 400 amino acids with *CDP*, but diverges entirely thereafter, so that it is lacking the Cut repeats and homeodomain. *CDP* mRNA and DNA binding activity is ubiquitous in mammalian cell lines tested to date (Superti-Furga et al, 1989; Strauss, E; Dkalnik, D.G., Orkin, S.H. unpublished). It is widespread in embryonic tissues as well (Andres et al., 1992).

■ References

Andres, V., Nadal-Ginard, B. and Mahdavi, V. (1992). Clox, a mammalian homeobox gene related to *Drosophila* cut, encodes DNA binding regulatory proteins differentially expressed during development. Development 116, 321-334.

Barberis, A., Superti-Furga, G. and Busslinger, M. (1987). Mutually exclusive interaction of the CCAAT-binding factor and of a displacement protein with overlapping sequences of a histone gene promoter. Cell 50, 347-359.

Blochlinger, K., Bodmer, R., Jack, J., Jan, L.Y. and Jan, Y.N. (1988). Primary structure and expression of a produt from cut, a locus involved in specifying sensory organ identity in *Drosophila*. Nature 333, 629-635.

Neufeld, E.J., Skalnik, D.G., Lievens, P.M.-J. and Orkin, S.H. (1992). Human CCAAT displacement protein is homologous to the *Drosophila* homeodomain protein, cut. Nature Genet. 1, 50-55.

Scherer, S.W., Neufeld, E.J., Lievens, P.M.-J., Orkin, S.H. and Tsui, L.-C. (1993). Regional localization of the CCAAT displacement protein (CDP) gene to band q22 of chromosome 7. Genomics, 15, 695-696.

Skalnik, D.G., Strauss, E.C. and Orkin, S.H. (1991). CCAAT displacement protein as a repressor of the myelomonocytic-specific gp91-phox gene promoter. J. Biol. Chem. 266, 16736-16741.

Superti-Furga, G., Barberis, A., Schreiber, E. and Busslinger, M. (1989). The protein CDP, but not CP1, footprints on the CCAAT region of the gamma-globin gene in unfractionated B-cell extracts. Biochim. Biophys. Acta 1007, 237-242.

cdxA [c]

Species: *Gallus gallus* (chicken)
Chromosomal location: -
Other names: *Chox-cad; Chox-4*
Cognate genes: *Xcad 2*
Type of homeobox: *cad*-type
Accession numbers: X57760 and X57761

■ Origin and description

This gene was cloned during a screen of a chicken genomic library with a mixed homeobox probe from which it preferentially hybridized to the *scr* probe (Rangini et al., 1989), its characterization is described in Frumkin et al. (1991). The homeobox sequence of the *Chox-cad* gene is interrupted by an intron between amino acids 44 and 45 of the homeodomain. The cDNA for this gene was isolated from a stage 12-13 embryonic cDNA library. The cDNA clone is 2486 bp long and it corresponds to the 2.6 kb transcript seen in northern analysis. This cDNA clone codes for a 248 amino acid long homeodomain protein and it contains 250 and 1492 bp of 5' and 3' untranslated sequences respectively. The caudal family includes at present the genes *caudal* from *Drosophila* (Mlodzik et al., 1985), *ceh-3* from *C. elegans* (Bürglin et al., 1989), *Cdx1* and *Cdx2* from mice (Duprey et al., 1988; James and Kazenwadel, 1991) and *Xcad1* and *Xcad2* from *Xenopus* (Blumberg et al., 1991). The *Chox-cad* protein has been expressed in *E. coli* as a fusion protein with the glutathione-S-transferase protein and this fusion protein has been shown to specifically bind to DNA, target sequences for the *CHox-cad* protein have been isolated (Fainsod et al., 1991).

■ Expression

The temporal pattern of expression has been determined by northern analysis during the first ten days of embryonic development. The onset of *CHox-cad* transcript accumulation in the embryo correlates with the onset of gastrulation and by day 5 of incubation no *CHox-cad* transcripts can be detected (Frumkin et al., 1991). This gene encodes only a 2.6 kb transcript. Northern analysis of embryos at the end of primitive streak stages and early head fold stages subdivided into anterior and posterior regions at the level of Hensen's node revealed that the *CHox-cad* transcripts are localized posterior to Hensen's node. In late primitive streak/early head fold embryos *CHox-cad* transcripts exhibit a predominantly posterior localization and lower transcript levels extending anteriorly. The tissue distribution in the embryo shows *CHox-cad* transcripts in the epiblast, primitive streak and the definitive endodermal cells. At 3 days of incubation *CHox-cad* transcripts are only detected in the epithelial lining of the digestive tract and the yolk sac which is of endodermal origin (Frumkin et al., 1991).

■ References

Blumberg, B., Wright, C.V.E., De Robertis, E.M. and Cho, K.W.Y. (1991). Organizer-specific homeobox genes in *Xenopus laevis* embryos. Science 253, 194-196.

Bürglin, T.R., Finney, M., Coulson, A. and Ruvkun, G. (1989). *Caenorhabditis elegans* has scores of homeobox-containing genes. Nature 341, 239-243.

Duprey, P., Chowdhury, K., Dressler, G.R., Balling, R., Simon, L.D., Guenet, J. and Gruss, P. (1988). A mouse gene homologous to the *Drosophila* gene *caudal* is expressed in epithelial cells from the embryonic intestine. Genes Dev. 2, 1647-1654.

Fainsod, A., Margalit, Y., Haffner, R. and Gruenbaum, Y. (1991). Non-immunological precipitation of protein-DNA complexes using glutathione-S-transferase fusion proteins. Nucl. Acids Res. 19, 4005.

Frumkin, A., Rangini, Z., Ben-Yehuda, A., Gruenbaum, Y. and Faisod, A. (1991). A chicken caudal homologue, *CHox-cad*, is expressed in the epiblast with posterior localization and in the early endodermal lineage. Development 112, 207-219.

James, R. and Kazenwadel, J. (1991). Homeobox gene expression in the intestinal epithelium of adult mice. J. Biol. Chem. 266, 3246-3251.

Mlodzik, M., Fjose, A. and Gehring, W.J. (1985). Isolation of *caudal*, a *Drosophila* homeo box-containing gene with maternal expression, whose transcripts form a concentration gradient at the pre-blastoderm stage. EMBO J. 4, 2961-2969.

Rangini, Z., Frumkin, A., Shani, G., Guttmann, M., Eyal-Giladi, H., Gruenbaum, Y. and Fainsod, A. (1989). The chicken homeobox genes *Chox 1* and *3*: Cloning, sequencing and expression during embryogenesis. Gene 76, 61-74.

Cdx-1 [m]

Species: *Mus musculus*
Chromosomal location: Chromosome 18
Other names: -
Cognate genes: *cad, Chox-cad, Xcad-2*
Type of homeobox: *cad*-type
Accession number: M37163

■ Origin and description

This gene was isolated by cross-homology with the *Drosophila* caudal gene. *Cdx-1* is linked closely to *fim2* on chromosome 18 (Duprey et al., 1988). Another caudal related gene could be isolated in the mouse, *Cdx-2* (James and Kazenwadel, 1991). Based on sequence homology in- and outside the homeodomain, it is more or less likely that *Chox-cad* (Frumkin et al., 1991) and *Xcad-2* (Blumberg et al., 1991) represents the cognate genes of *Cdx-1*.

■ Expression

In situ and Northern blot analyses revealed two phases of *Cdx-1* expression. The first started at day 7.5 p.c. of embryonic development in the posterior portion of the embryo. It is expressed in the primitive ectoderm and mesoderm and later in the neuroectoderm, the somites and the developing limb. At day 13 p.c. no *Cdx-1* transcript could be detected (Meyer and Gruss, in preparation). The second phase of *Cdx-1* expression started on day 14 p.c. in the intestine and can be there detected also in adult tissue (Duprey et al., 1988).

■ References

Blumberg, B., Wright, C.V.E., DeRobertis, E.M. and CHo, K.W.Y. (1991). Organizer-specific homeobox genes in *Xenopus laevis* embryos. Science 253, 194-196.

Duprey, P., Chowdhury, K., Dressler, G.R., Balling, R., Simon, D., Guenet, J.-L., and Gruss, P. (1988). A mouse gene homologous to the *Drosophila* gene *caudal* is expressed in epithelial cells from the embryonic intestine. Genes Dev. 2, 1647-1654.

Frumkin, A., Rangini, Z., Ben-Yehuda, A., Gruenbaum, Y. and Fainsod, A. (1991). A chicken *caudal* homologue, *CHox-cad*, is expressed in the epiblast with posterior localization and in the early endodermal lineage. Development 112, 207-219.

James, R. and Kazenwadel, J. (1991). Homeobox gene expression in the intestinal epithelium of adult mice. J. Biol. Chem. 266, 3246-3251.

Cdx-4 [m]

Species: *Mus musculus*
Chromosomal location: -
Other names: -
Cognate genes: -
Type of homeobox: *cad*-type
Accession number: L08061

■ Origin and description

This gene was cloned from an 8.5 d.p.c. murine embryonic cDNA library at very low stringency with a homeobox-containing probe from the endoderm specific homeobox gene *XlHbox 8* (Wright et al., 1988). Sequence analysis revealed that the *Cdx4* homeodomain belongs to the *caudal* family together with a number of homeodomains from other organisms such as the murine *Cdx1* and *Cdx2* (Duprey et al., 1988; James and Kazenwadel, 1991), hamster *Cdx3* (German et al., 1992), chicken *CHox-cad* (Frumkin et al., 1991), *Xenopus Xcad1* and *Xcad2* (Blumberg et al, 1991), zebrafish *Zfcad1* (Joly et al., 1992), and *C. elegans ceh-3* (Waring and Kenyon, 1991) genes. The genomic locus for *Cdx-4* has been isolated from libraries derived from 129 and C3Hf strains. It is comprised of three exons, with introns in almost identical positions to the related *CHox-cad* gene (Frumkin et al., 1991). Among the vertebrate caudal-related homeoprotein sequences, in addition to the homeodomains, there are 3 blocks of conservation: one at the N-terminus, another halfway towards the homeodomain, and a diverged version of the hexapeptide motif lying just upstream of the homeodomain.

■ Expression

Expression has been studied by northern blot, *in situ* RNA localization, and by immunohistochemistry (Gamer and Wright, in preparation). *Cdx-4* mRNA and protein are first seen at 7.0-7.5 d.p.c. in the entire allantois and posterior primitive streak. Between this time and ~8.5 d.p.c., expression in the allantois diminishes, and *Cdx-4* transcripts and protein spread anteriorly into the posterior third (or so) of the embryo. At 8.0 d.p.c., *Cdx-4* is expressed in the neural tube, presomitic and lateral plate mesoderm, and endoderm of the hindgut. In early embryos, the *Cdx-4* protein is in a concentration gradient with maximum levels at the posterior of the embryo-reminiscent of the posterior to anterior gradient of *caudal* protein seen in *Drosophila* embryos. In the paraxial mesoderm, *Cdx-4* expression begins to disappear in a rostral-caudal direction, as the somites bud off. Expression in the neural tube, lateral plate, and endoderm, also retracts in a rostral-caudal direction, but somewhat delayed from that in the paraxial mesoderm. At 9.5 d.p.c., *Cdx-4* is still expressed in the tip of the growing tail. To the level of analysis so far, *Cdx-4* is not expressed after ~10 d.p.c., or in the embryonic gut at later stages, and in this latter respect may contrast *Cdx-1* and *Cdx-2* genes (Duprey et al., 1988; James and Kazenwadel, 1991).

■ References

Blumberg, B., Wright, C.V.E., DeRobertis, E.M. and CHo, K.W.Y. (1991). Organizer-specific homeobox genes in *Xenopus laevis* embryos. Science 253, 194-196.

Duprey, P., Chowdhury, K., Dressler, G.R., Balling, R., Simon, L.D., Guenet, J. and Gruss, P. (1988). A mouse gene homologous to the *Drosophila* gene *caudal* is expressed in epithelial cells from the embryonic intestine, Genes Dev. 2, 1647-1654.

Frumkin, A., Rangini, Z., Ben-Yehuda, A., Gruenbaum, Y. and Fainsod, A. (1991). A chicken *caudal* homologue, CHox-cad, is expressed in the epiblast with posterior localization and in the early endodermal lineage. Development 112, 207-219.

Gamer, L. and Wright, C.V.E. (in preparation). Murine *cdx-4* bears striking similarities to the *Drosophila caudal* gene in its homeodomain sequence and early expression pattern.

German, M.S., Wang, J., Chadwick, R.B. and Rutter, W.J. (1992). Synergistic activation of the insulin gene by a LIM-homeodomain protein and a basic helix-loop-helix protein: building a functional insulin minienhancer complex. Genes Dev. 6, 2165-2176.

James, R. and Kazenwadel, J. (1991). Homeobox gene expression in the intestinal epithelium of adult mice. J. Biol. Chem. 266, 3246-3251.

Joly, J.-S., Maury, M., Joly, C., Duprey, P., Boulekbache, H. and Condamine, H. (1992). Expression of a zebrafish *caudal* homeobox gene correlates with the establishment of posterior cell lineages at gastrulation. Differentiation 50, 75-87.

Waring, D.A. and Kenyon, C. (1991). Regulation of cellular responsiveness to inductive signals in the developing *C. elegans* nervous system. Nature 350, 712-715.

Wright, C.V.E., Schnegelsberg, P. and DeRobertis, E.M. (1988). XlHbox8: a novel *Xenopus* homeo protein restricted to a narrow band of endoderm. Development 104, 787-794.

ceh-1 [ce]

Species: *Caenorhabditis elegans* (N2)
Chromosomal location: X
Other names: *JM L1001*
Cognate genes: -
Type of homeobox: related to *NK-1/S59*
Accession number: X52810

■ Origin and description

Isolated using an *Antennapedia*-type derived oligonucleotide with *C. elegans* codon usage bias (Hawkins and McGhee, 1990).

■ References

Hawkins, N.C. and McGhee, J.D. (1990). Homeobox containing genes in the nematode *Caenorhabditis elegans*. Nucl. Acids Res. 18, 6101-6106.

ceh-2 [ce]

Species: *Caenorhabditis elegans* (N2)
Chromosomal location: I
Other names: -
Cognate genes: -
Type of homeobox: -
Accession number: PIR: S05703

Origin and description

Isolated using highly degenerate oligonucleotides for the most conserved region of helix 3 (Bürglin et al., 1989). Homeobox sequence incomplete.

References

Bürglin, T.R., Finney, M., Coulson, A. and Ruvkun, G. (1989). *Caenorhabditis elegans* has scores of homeobox-containing genes. Nature 341, 239-243.

ceh-5 [ce]

Species: *Caenorhabditis elegans* (N2)
Chromosomal location: I
Other names: -
Cognate genes: -
Type of homeobox: -
Accession number: PIR: SO5706

Origin and description

Isolated using highly degenerate oligonucleotides for the most conserved region of helix 3 (Bürglin et al., 1989). Novel class of homeodomain.

References

Bürglin, T.R., Finney, M., Coulson, A. and Ruvkun, G. (1989). *Caenorhabditis elegans* has scores of homeobox-containing genes. Nature 341, 239-243.

ceh-6 [ce]

Species: *Caenorhabditis elegans* (N2)
Chromosomal location: I
Other names: -
Cognate genes: -
Type of homeobox: POU-III subclass
Accession number: PIR: SO5707

Origin and description

Isolated using highly degenerate oligonucleotides for the most conserved region of helix 3 (Bürglin et al., 1989). *ceh*-6 is a member of the POU-III subclass and thus is most closely related to genes such as *Brn*-1, *Brn*-2, *Tst*-1.

References

Bürglin, T.R., Finney, M., Coulson, A. and Ruvkun, G. (1989). *Caenorhabditis elegans* has scores of homeobox-containing genes. Nature 341, 239-243.

ceh-7 [ce]

Species: *Caenorhabditis elegans* (N2)
Chromosomal location: II
Other names: -
Cognate genes: -
Type of homeobox: -
Accession number: PIR: SO5708

Origin and description

Isolated using highly degenerate oligonucleotides for the most conserved region of helix 3 (Bürglin et al., 1989). *ceh*-7 appears to represent a novel class of homeobox.

References

Bürglin, T.R., Finney, M., Coulson, A. and Ruvkun, G. (1989). *Caenorhabditis elegans* has scores of homeobox-containing genes. Nature 341, 239-243.

ceh-8 [ce]

Species: *Caenorhabditis elegans* (N2)
Chromosomal location: I
Other names: -
Cognate genes: -
Type of homeobox: -
Accession number: PIR: SO5709

Origin and description

Isolated using highly degenerate oligonucleotides for the most conserved region of helix 3 (Bürglin et al., 1989). The homeobox sequence is incomplete, possibly it is related to the human gene *S8*.

References

Bürglin, T.R., Finney, M., Coulson, A. and Ruvkun, G. (1989). *Caenorhabditis elegans* has scores of homeobox-containing genes. Nature 341, 239-243.

ceh-9 [ce]

Species: *Caenorhabditis elegans* (N2)
Chromosomal location: -
Other names: -
Cognate genes: -
Type of homeobox: NEC class
Accession number: X52811

■ Origin and description

Isolated using an *Antennapedia*-type derived oligonu-cleotide with *C. elegans* codon usage bias (Hawkins and McGhee, 1990). Most closely related to the sea urchin homeobox gene *Tghbox 5*, NEC class.

■ References

Hawkins, N.C. and McGhee, J.D. (1990). Homeobox containing genes in the nematode *Caenorhabditis elegans*. Nucl. Acids Res. 18, 6101-6106.

ceh-10 [ce]

Species: *Caenorhabditis elegans* (N2)
Chromosomal location: III
Other names: -
Cognate genes: -
Type of homeobox: *prd*-type
Accession number: X52812

■ Origin and description

Isolated using an *Antennapedia*-type derived oligonu-cleotide with *C. elegans* codon usage bias (Hawkins and McGhee, 1990). The homeodomain is distantly related to the *prd*-class, thus grouped into the *prd*-like class.

■ Expression

A major 1.85 kb transcript, and a minor 1.25 kb transcript are present in embryos, and decrease during the larval stages (Hawkins and McGhee, 1990).

■ References

Hawkins, N.C. and McGhee, J.D. (1990). Homeobox containing genes in the nematode *Caenorhabditis elegans*. Nucl. Acids Res. 18, 6101-6106.

ceh-11 [ce]

Species: *Caenorhabditis elegans* (N2)
Chromosomal location: III
Other names: -
Cognate genes: -
Type of homeobox: divergent *Abd-B*
Accession number: X17075, X52813

■ Origin and description

Isolated using an *Antennapedia*-type derived oligonu-cleotide with *C. elegans* codon usage bias (Hawkins and McGhee, 1990).

Isolated using an *Ascaris* homeobox as probe (Schaller et al., 1990). This homeobox is part of the *C. elegans* cluster and appears to correspond to the *Abd-B* type genes (Bürglin et al., 1991, Kenyon and Wang, 1991).

■ Expression

A weak 1.6 kb transcript can be detected in embryos (Hawkins and McGhee, 1990).

■ Genetics, function

ceh-11 might correspond to the gene *egl-5* (Hawkins and McGhee, 1990; Schaller et al., 1990; Chisholm, 1991).

■ References

Bürglin, T.R., Ruvkun, G., Coulson, A., Hawkins, N.C., McGhee, J.D., Schaller, D., Wittmann, C., Müller, F. and Waterston, R.H. (1991). Nematode homeobox cluster. Nature 351, 703-703.
Chisholm, A. (1991). Control of cell fate in the tail region of *C. elegans* by the gene *eg*1-5. Development 111, 921-932.
Hawkins, N.C. and McGhee, J.D. (1990). Homeobox containing genes in the nematode *Caenorhabditis elegans*. Nucl. Acids Res. 18, 6101-6106.
Kenyon, C. and Wang, B. (1991). A cluster of *Antennapedia*-class homeobox genes in a nonsegmented animal. Science 253, 516-517.
Schaller, D., Wittmann, C., Spicher, A., Müller, F. and Tobler, H. (1990). Cloning and analysis of three new homeobox genes from the nematode *Caenorhabditis elegans*. Nucl. Acids Res. 18, 2033-2036.

ceh-12 [ce]

Species: *Caenorhabditis elegans* (N2)
Chromosomal location: I
Other names: -
Cognate genes: -
Type of homeobox: -
Accession number: X17076

■ Origin and description

Isolated using an *Ascaris* homeobox as probe (Schaller et al., 1990). Homeobox sequence incomplete.

■ References

Schaller, D., Wittmann, C., Spicher, A., Müller, F. and Tobler, H. (1990). Cloning and analysis of three new homeobox genes from the nematode *Caenorhabditis elegans.* Nucl. Acids Res. 18, 2033-2036.

ceh-13 [ce]

Species: *Caenorhabditis elegans* (N2)
Chromosomal location: III
Other names: -
Cognate genes: -
Type of homeobox: *labial*-type
Accession number: X17077

■ Origin and description

Isolated using an *Ascaris* homeobox as probe (Schaller et al., 1990). This homeobox is part of the *C. elegans* cluster and appears to correspond to the *labial* type genes (Bürglin et al., 1991, Kenyon and Wang, 1991).

■ References

Bürglin, T.R., Ruvkun, G., Coulson, A., Hawkins, N.C., McGhee, J.D., Schaller, D., Wittmann, C., Müller, F. and Waterston, R.H. (1991). Nematode homeobox cluster. Nature 351, 703.
Kenyon, C. and Wang, B. (1991). A cluster of *Antennapedia*-class homeobox genes in a nonsegmented animal. Science 253, 516-517.
Schaller, D., Wittmann, C., Spicher, A., Müller, F. and Tobler, H. (1990). Cloning and analysis of three new homeobox genes from the nematode *Caenorhabditis elegans.* Nucl. Acids Res. 18, 2033-2036.

ceh-14 [ce]

Species: *Caenorhabditis elegans* (N2)
Chromosomal location: X
Other names: -
Cognate genes: -
Type of homeobox: LIM class
Accession number: PIR: SO5710

■ Origin and description

Isolated using highly degenerate oligonucleotides for the most conserved region of helix 3 (Bürglin et al., 1989). The homeodomain resembles the ones of *mec*-3 and *lin*-11, and two copies of the LIM-motif are found upstream of the homeodomain (T. Bürglin, personal communication).

■ References

Bürglin, T.R., Finney, M., Coulson, A. and Ruvkun, G. (1989). *Caenorhabditis elegans* has scores of homeobox-containing genes. Nature 341, 239-243.

ceh-15 [ce]

Species: *Caenorhabditis elegans* (N2)
Chromosomal location: III
Other names: *hom-2/hom-3*
Cognate genes: -
Type of homeobox: most similar to *Dfd*
Accession number: -

■ Origin and description

Isolated using degenerate primers for PCR (Kamb et al., 1989). This homeobox is part of the *C. elegans* cluster and probably corresponds to the *Dfd* type genes (Bürglin et al., 1991, Kenyon and Wang, 1991).

■ References

Bürglin, T.R., Ruvkun, G., Coulson, A., Hawkins, N.C., McGhee, J.D., Schaller, D., Wittmann, C., Müller, F. and Waterston, R.H. (1991). Nematode homeobox cluster. Nature 351, 703.
Kamb, A., Weir, M., Rudy, B., Varmus, H. and Kenyon, C. (1989). Identification of genes from pattern formation, tyrosine kinase and pottassium channel families by DNA amplification. Proc. Natl. Acad. Sci. USA 86, 4372-4376.
Kenyon, C. and Wang, B. (1991). A cluster of *Antennapedia*-class homeobox genes in a nonsegmented animal. Science 253, 516-517.

Species: *Caenorhabditis elegans* (N2)
Chromosomal location: III
Other names: *en-3/en-5*
Cognate genes: -
Type of homeobox: related to *engrailed*
Accession number: -

■ Origin and description

Isolated using degenerate primers for PCR. The sequence of the PCR product corresponds most closely to the *engrailed*-class genes (Kamb et al., 1989). Incomplete homeobox sequence.

■ References

Kamb, A., Weir, M., Rudy, B., Varmus, H. and Kenyon, C. (1989). Identification of genes from pattern formation, tyrosine kinase and pottassium channel families by DNA amplification. Proc. Natl. Acad. Sci. USA 86, 4372-4376.

ceh-17 [ce]

Species: *Caenorhabditis elegans* (N2)
Chromosomal location: -
Other names: *prd-1/prd-2*
Cognate genes: -
Type of homeobox: related to *paired*
Accession number: -

■ Origin and description

Isolated using degenerate primers for PCR. The sequence of the PCR product corresponds most closely to the *prd*-class genes (Kamb et al., 1989). Incomplete homeobox sequence.

■ References

Kamb, A., Weir, M., Rudy, B., Varmus, H. and Kenyon, C. (1989). Identification of genes from pattern formation, tyrosine kinase and potassium channel families by DNA amplification. Proc. Natl. Acad. Sci. USA 86, 4372-4376.

ceh-18 [ce]

Species: *Caenorhabditis elegans* (N2)
Chromosomal location: X
Other names: -
Cognate genes: -
Type of homeobox: POU
Accession number: -

■ Origin and description

Isolated using highly degenerate oligonucleotides for the most conserved region of helix 3 as described in Bürglin et al., 1989. The homeodomain and POU-specific domain are different from other POU-subclasses described thus far. In addition, the linker region is longer than in other POU-genes (S. Hird, M. Finney, D. Greenstein and G. Ruvkun, personal communication).

■ Expression

Antibodies raised against *ceh*-18 stain many different cell types, e.g. epidermal and intestinal cells, starting in embryogenesis and continuing throughout larval development into adult animals. However, very few neuronal cells stain. To generalize, it appears that all staining cells have some kind of epithelial character (D. Greenstein, personal communication).

■ References

Bürglin, T.R., Finney, M., Coulson, A. and Ruvkun, G. (1989). *Caenorhabditis elegans* has scores of homeobox-containing genes. Nature 341, 239-243.

ceh-19 [ce]

Species: *Caenorhabditis elegans* (N2)
Chromosomal location: IV
Other names: -
Cognate genes: -
Type of homeobox: -
Accession number: Z11794, Z11795

■ Origin and description

Isolated using PCR with degenerate oligonucleotides (Naito et al., 1992). Currently the homeodomain is not closely related to any other known homeobox genes and appears to constitute a new class.

■ References

Naito, M., Kohara, Y. and Kurosawa, Y. (1992). Identification of a homeobox-containing gene located between *lin-45* and *unc-24* on chromosome IV in the nematode *Caenorhabditis elegans*. Nucleic Acids Res. 2967-2969.

ceh-20 [ce]

Species: *Caenorhabditis elegans* (N2)
Chromosomal location: III
Other names: -
Cognate genes: -
Type of homeobox: atypical, PBC class
Accession number:

■ Origin and description

Identified in the random cDNA sequencing project (Waterston et al., 1992) based on sequence similarity to *PBX2*. The complete sequence revealed an atypical homeodomain similar to the ones of *PBX1*, *PBX2* and *PBX3* (Bürglin and Ruvkun, 1992), although *PBX1*, *PBX2* and *PBX3* are more similar to each other than to *ceh-20*. Upstream of the homeodomain a conserved motif was found, termed the PBC domain, which can be further sub-divided into an A and B region (Bürglin and Ruvkun, 1992).

■ References

Bürglin, T.R. and Ruvkun, G. (1992). A new motif in *PBX* genes. Nature Genet. 1, 319–320.
Waterston, R., Martin, C., Craxton, M., Huynh, C., Coulson, A., Hillier, L., Durbin, R., Green, P., Shownkeen, R., Halloran, N., Metzstein, M., Hawkins, T., Wilson, R., Berks, M., Du, Z., Thomas, K., Thierry-Mieg, J. and Sulston, J. (1992). A survey of expressed genes in *Caenorhabditis elegans*. Nature Genet. 1, 114-123.

Cfla [d]

Species: *Drosophila melanogaster*
Chromosomal location: 65D1-3
Other names: -
Cognate genes: -
Type of homeobox: POU
Accession number: X58435

Origin and description

cfla was initially isolated by its binding to a dopamine neuron specific regulatory element of the *Drosophila* dopa decarboxylase gene and was also isolated by cross-hybridization with a POU hybridization probe (Treacy and al., 1991). The POU-specific and POU-homeodomains of *cfla* show strong similarity to several vertebrate POU proteins that are expressed in the developing CNS. *cfla* apparently binds DNA as a dimer. *cfla* DNA binding and transcriptional activation can be disrupted by the formation of inactive heterodimers with the I-POU protein *in vitro* and in transfected cells (Treacy et al., 1991; 1992).

■ References

Johnson, W.A. and Hirsh, J. (1990). A *Drosophila* "POU Protein" binds to a sequence element regulating gene expression in specific dopaminergic neurons. Nature, 343, 467-470.
Treacy, M.N., He, X. and Rosenfeld, M.G. (1991). I-POU: a POU-domain protein that inhibits neuron-specific gene. Nature 350, 577-584.
Treacy, M.N., Nelson, L.I. et al. (1992). Twin of I-POU: A two amino acid difference in the I-POU homeodomain distinguishes an activator from an inhibitor of transcription. Cell 68, 491-505.

Chox-E [c]

Species: *Gallus gallus* (chicken)
Chromosomal location: -
Other names: *Chlx-A*
Cognate genes: -
Type of homeobox: *H2.0*-type
Accession numbers: X59548 and X59549

■ Origin and description

This gene was cloned during a screen of a chicken genomic library with a mixed homeobox probe from which it preferentially hybridized to the *Hoxc-8* probe (Rangini et al., 1989), its characterization is described in Rangini et al. (1991). The homeobox sequence of the *Chox E* gene is interrupted by an intron at position 44-45 of the homeodomain. Two overlapping phages were isolated that contain the *Chox E* homeobox, and together they encompass about 24 kb of genomic DNA. The *Chox E* homeodomain belongs to a small family of homeobox genes together with the *Drosophila H2.0* (Barad et al., 1988) the murine *Hlx* (Allen et al., 1991) and the *Halocynthia roretzi AHox1* (Saiga et al., 1991) genes. The *CHox E* gene codes for three specific transcripts 1.5, 1.65 and 2.5 kb in size. A small cDNA clone of 328 bp was isolated and this clone maps one of the *CHox E* poly A addition signals. About 2 kb upstream from the *CHox E* homeobox there is a CR1 repeat which was studied in some detail as these repetitive elements may play a regulatory role in the chicken genome (Shapira et al., 1991).

■ Expression

The temporal pattern of expression has been determined by northern analysis during the first ten days of embryonic development (Rangini et al., 1991). The onset of *Chox E* expression is between the first and second days of incubation but high transcript levels are only reached at about day 3. After day 5 of incubation the transcript levels decrease. Analysis of the spatial pattern of expression between days 3 to 12 was performed at all developmental stages studied. *Chox E* expression is restricted to the central nervous system (Rangini et al., 1991). At day 3 of development *Chox E* expression extends from the rhombencephalic isthmus through all the rhombomeres to the caudal end of the spinal cord. The expression at these stages is restricted in a dorso-ventral plane to the dorsal half of the basal plate, the intermediate plate, the source of the mesoneurons. At days 5 and 8 the antero-posterior and dorso-ventral restrictions remain the same and the only change takes place in the medio-lateral plane. Medio-laterally *Chox E* expression becomes restricted towards the central canal always being expressed only in the ventricular zone and later ependymal cells before their differentiation. At day 12 of incubation no *Chox E* transcripts could be detected as the expressing region is only an embryonic temporary region.

■ References

Allen, J.D., Lints, T., Jenkins, N.A., Copeland, N.G., Strasser, A., Harvey, R.P. and Adams, J.M. (1991). Novel murine homeobox gene on chromosome 1 expressed in specific hematopoietic lineages and during embryogenesis. Genes Dev. 5, 509-520.

Barad, M., Jack, T., Chadwick, R. and McGinnis, W. (1988). A novel, tissue-specific, *Drosophila* homeobox gene. EMBO J. 7, 2151-2161.

Rangini, Z., Frumkin, A., Shani, G., Guttmann, M., Eyal-Giladi, H., Gruenbaum, Y. and Fainsod, A. (1989). The chicken homeobox genes *Chox 1* and *3*: Cloning, sequencing and expression during embryogenesis. Gene 76, 61-74.

Rangini, Z., Ben-Yehuda, A., Shapira, E., Gruenbaum, Y. and Fainsod, A. (1991). *CHox E* a chicken homeogene of the *H2.0* type exhibits dorso-ventral restriction in the proliferating region of the spinal cord. Mech. Dev. 35, 13-24.

Saiga, H., Mizokami, A., Makabe, K.W., Satoh, N. and Mita, T. (1991). Molecular cloning and expression of a novel homeobox gene AHox1 of the ascidian, *Halocynthia roretzi*. Development 111, 821-828.

Shapira, E., Yarus, S. and Fainsod, A. (1991). Genomic organization and expression during embryogenesis of the chicken repeat. Genomics 10, 931-939.

Chox-M [c]

Species: *Gallus gallus* (chicken)
Chromosomal location: -
Other names: -
Cognate genes: -
Type of homeobox: related to *msh*
Accession number: -

■ Origin and description

This gene is expressed in BM-2 cells, a chicken myeloblast cell line transformed by the avian myeloblastosis virus. *CHox-M* was isolated by screening a λgt11 cDNA expression library derived from BM-2 cells with two binding site probes from the chicken myelomonocytic growth factor (*cMGF*) promotor.

The homeobox of *CHox-M* is similar to *K8*, *msh-2*, *GHox-8* and *Hox-7.1*. It is most closely related to *CHox-7* (Fainsod & Gruenbaum, 1989) which shares 58 out of the 60 homeodomain amino acids. No similarity between the protein sequence outside of the homeodomain with other proteins has yet been found.

■ Function

A function is presently not known. However, binding to the *cMGF* promotor implicates a role of *CHox-M* in transcriptional regulation of *cMGF* expression.

■ References

Fainsod, A. and Gruenbaum, Y. (1989). A chicken homeobox gene with developmentally regulated expression. FEBS Lett. 250, 381-385.

Leidert , P. and Leutz, A. (manuscript in preparation).

Chox-3 [c]

Species: *Gallus gallus* (chicken)
Chromosomal location: -
Other names: *Sax-1*
Cognate genes: -
Type of homeobox: *NK1*-type
Accession number: M23065

■ Origin and description

This gene was cloned during a screen of a chicken genomic library with a mixed homeobox probe from which it preferentially hybridized to the *Hoxa-3* probe, its characterization is described in Rangini et al. (1989). In order to characterize the homeobox and flanking sequences 967 bp of genomic sequence were determined. Sequence analysis revealed a new type of homeodomain which together with the *Drosophila NK1* homeodomain (Kim and Nirenberg, 1989) forms a novel family. Assuming the absence of introns downstream from the *Chox-3* homeobox sequence, termination of the protein product occurs 107 amino acids downstream from the homeodomain.

■ Expression

The temporal pattern of expression of the *Chox-3* gene was determined by northern analysis using three different probes hybridized to RNA extracted from embryos during the first five days of development. The homeobox containing probe recognized five transcripts 1.3, 1.9, 2.6, 5.6 and 7.9 kb respectively. A probe upstream to the homeobox recognizes only three of the transcripts 1.3, 1.9 and 5.6 kb in size. The probe downstream to the homeobox hybridizes only to the 1.9 kb transcript. With all three probes there is a marked decrease in transcript abundance between the first and second days of incubation. Preliminary *in situ* hybridization studies suggest that *Chox-3* transcripts are localized to the developing spinal cord from the onset of its differentiation.

■ References

Kim, Y. and Nirenberg, M. (1989). *Drosophila* NK-homeobox genes. Proc. Natl. Acad. Sci. USA 86, 7716-7720.

Rangini, Z., Frumkin, A., Shani, G., Guttmann, M., Eyal-Giladi, H., Gruenbaum, Y. and Fainsod, A. (1989). The chicken homeobox genes Chox 1 and 3: Cloning, sequencing and expression during embryogenesis. Gene 76, 61-74.

Chox-7   [c]

Species: *Gallus gallus* (chicken)
Chromosomal location: -
Other names: *Ovx-1*
Cognate genes: -
Type of homeobox: -
Accession number: X16159

■ Origin and description

This gene was cloned during a screen of a chicken genomic library with a mixed homeobox probe from which it preferentially hybridized to the *Hoxa-3* and *scr* probes (Rangini et al., 1989), its characterization is described in Fainsod and Gruenbaum (1989). In order to characterize the homeobox and flanking sequences 374

bp of genomic sequence were determined. Sequence analysis revealed a new type of homeodomain which represents a new sub-family for which no other members have been reported, with homologies in the range of 50-60% to other homeodomains. Putative translation of the sequenced genomic region revealed that the region codes for one contiguous open reading frame that includes the homeodomain.

■ Expression

Northern analysis of embryonic RNA from the first five days of incubation with the *Chox-7* homeobox probe revealed hybridization to four transcripts (Fainsod and Gruenbaum, 1989). The four transcripts are 1.4, 1.9, 2.4 and 3.5 kb respectively and they exhibit changes in abundance during development. At day 1 of incubation the most abundant is the 1.4 kb transcript and by day 3 the most abundant is the 1.9 kb transcript. Preliminary studies of the spatial pattern of expression suggest that the *Chox-7* gene is expressed in the hindbrain and in the otic vesicle.

■ References

Fainsod, A. and Gruenbaum, Y. (1989). A chicken homeobox gene with developmentally regulated expression. FEBS Lett. 250, 381-385.

Rangini, Z., Frumkin, A., Shani, G., Guttmann, M., Eyal-Giladi, H., Gruenbaum, Y. and Fainsod, A. (1989). The chicken homeobox genes *Chox 1* and *3*: Cloning, sequencing and expression during embryogenesis. Gene 76, 61-74.

Clox-1   [can]

Species: *Canis* (dog)
Chromosomal location: -
Other name: -
Cognate genes: *cut*, *CDP*
Type of homeobox: *cut*-type
Accession number: -

■ Origin and description

Clox-1 was isolated as a cDNA clone. In addition to the *cut*-type homeobox, it contains the three cut-repeats. It encodes a nuclear protein.

■ Expression

Western blot analyses revealed that many *Clox*-like nuclear proteins exist. Cotransfection experiments showed that Clox proteins can function as repressors of tissue-specific gene transcription.

■ References

Andres, V., Nadal-Ginard, B. and Mahdavi, V. (1992). *Clox*, a mammalian homeobox protein related to the *Drosophila cut*, encodes DNA-binding regulatory proteins differentially expressed during development. Development, 116, 321-334.

cnox1　[Hc]

Species:　*Chlorohydra viridissima* (Cniderian)
Chromosomal location:　-
Other names:　-
Cognate genes:　-
Type of homeobox:　*Antp*-type, related to the *Drosophila labial* gene
Accession number:　X64625

■ Origin and description

This gene was isolated from a cDNA library by cross hybridization with a 50 guessmer encoding the third helix of the *Antp* homeodomain according to the hydra codon usage.

■ Expression

By quantitative PCR, its maximal expression is detected within the first hours during head regeneration of a *C.v.* multiheaded mutant.

■ References

Schummer, M., Scheurlen, I., Schaller, C. and Galliot, B. (1992). HOM/HOX homeobox genes are present in hydra (*Chlorohydra viridissima*) and are differentially expressed during regeneration. EMBO J. 11, 1815-1823.

Cnox-1　[Ed]

Species:　*Eleutheria dichotoma* (Cnidarian)
Chromosomal location:　-
Other names:　-
Cognate genes:　-
Type of homeobox:　*Antp*-type; also showing similarity to *SU-HB-en*
Accession number:　-

■ Origin and description

The cnidarian homeobox gene *Cnox*-1 was isolated in two steps: (1) A central 77bp fragment of the homeobox was amplified using PCR with degenerate primers specific for conserved regions of the *Antp*-type homeobox, and the amplification products subcloned and sequenced (Schierwater et al., 1991); (2) A genomic *Hind*III library in γ NM1149 was prepared, screened with the *Cnox*-1 homeobox fragment and the gene cloned and sequenced (Schierwater et al., in preparation). PAUP analysis shows *Cnox*-1 to cluster on a sister branch to other *Antp*-class homeoboxes.

■ References

Schierwater, B., Murtha, M., Dick, M., Ruddle, F.H., Buss, L.W. (1991). Homeoboxes in Cnidarians. J. Exp. Zool. 260, 413-416.

cnox2　[Hc]

Species:　*Chlorohydra viridissima* (Cnidarian)
Chromosomal location:　-
Other names:　-
Cognate genes:　-
Type of homeobox:　*Antp*-type, related to the *Drosophila Deformed* gene
Accession number:　X64626

■ Origin and description

This gene was isolated from a cDNA library by cross hybridization with a 50 guessmer encoding the third helix of the *Antp* homeodomain according to the hydra codon usage.

■ Expression

By quantitative PCR, its maximal expression is detected the first day after cutting during head regeneration of a *C.v.* multiheaded mutant.

■ References

Schummer, M., Scheurlen, I., Schaller, C. and Galliot, B. (1992). HOM/HOX homeobox genes are present in hydra (*Chlorohydra viridissima*) and are differentially expressed during regeneration. EMBO J. 11, 1815-1823.

Cnox-2-Ed [Ed]

Species: *Eleutheria dichotoma* (Cnidarian)
Chromosomal location: -
Other names: -
Cognate genes: *Cnox-2-Hs*
Type of homeobox: *Antp*-type
Accession number: -

■ Origin and description

The cnidarian homeobox gene *Cnox-2-Ed* was isolated in two steps: (1) A central 77bp fragment of the homeobox was amplified using PCR with degenerate primers specific for conserved regions of the *Antp*-type homeobox, and the amplification products subcloned and sequenced (Schierwater et al., 1991); (2) A genomic *Hind*III library in γ NM1149 was prepared, screened with the *Cnox-2-Ed* homeobox fragment and the gene cloned (Schierwater et al., in preparation). PAUP analysis shows *Cnox-2* to be embedded within the *Antp*-class lineage and to be related to the *Hoxb-4* and *Hoxb-1* cognate groups.

■ References

Schierwater, B., Murtha, M., Dick, M., Ruddle, F.H., Buss, L.W. (1991). Homeoboxes in Cnidarians. J. Exp. Zool. 260, 413-416.

Cnox-2-Hs [Hs]

Species: *Hydractinia symbiolongicarpus* (Cnidarian)
Chromosomal location: -
Other names: -
Cognate genes: *Cnox-2-Ed*
Type of homeobox: *Antp*-type
Accession number: -

■ Origin and description

The cnidarian homeobox gene *Cnox-2-Hs* was isolated in two steps: (1) A central 77bp fragment of the homeobox was amplified using PCR with degenerate primers specific for conserved regions of the *Antp*-type homeobox, and the amplification products subcloned and sequenced (Schierwater et al., 1991); (2) A genomic *Hind*III library in γ NM1149 was prepared, screened with the *Cnox-2-Hs* homeobox fragment and the gene cloned (Schierwater et al., in preparation). PAUP analysis shows *Cnox-2* to e

embedded within the *Antp*-class lineage and to be related to the *Hoxb-4* and *Hoxb-1* cognate groups.

■ References

Schierwater, B., Murtha, M., Dick, M., Ruddle, F.H., Buss, L.W. (1991). Homeoboxes in Cnidarians. J. Exp. Zool. 260, 413-416.

cnox3 [Hc]

Species: *Chlorohydra viridissima* (Cnidarian)
Chromosomal location: -
Other names: -
Cognate genes: -
Type of homeobox: related to the *Drosophila BarH1* and *Distal less* genes
Accession number: X64627

■ Origin and description

This gene was isolated from a cDNA library by cross hybridization with a 50 guessmer encoding the third helix of the *Antp* homeodomain according to the hydra codon usage. The cnox3 homeodomain displays a particular feature: the Phe50 in the third α-helix is replaced by a Tyr.

■ Expression

By quantitative PCR, its maximal expression is detected the second day after cutting during head regeneration of a *C.v.* multiheaded mutant.

■ References

Schummer, M., Scheurlen, I., Schaller, C. and Galliot, B. (1992). HOM/HOX homeobox genes are present in hydra (*Chlorohydra viridissima*) and are differentially expressed during regeneration. EMBO J. 11, 1815-1823.

Species: *Chlorohydra viridissima* (Cnidarian)
Chromosomal location: -
Other names: -
Cognate genes: -
Type of homeobox: -
Accession number: X64628

■ Origin and description

This gene was isolated from a cDNA library by cross hybridization with a 50 guessmer encoding the third helix of the Antp homeodomain according to the hydra codon usage. As in case of *cnox3*, an "atypical" Tyr50 residue is also found in *cnox4* homeodomain.

■ References

Schummer, M., Scheurlen, I., Schaller, C. and Galliot, B. (1992). HOM/HOX homeobox genes are present in hydra *(Chlorohydra viridissima)* and are differentially expressed during regeneration. EMBO J. 11, 1815-1823.

CS8A [c]

Species: *Gallus gallus* (chicken)
Chromosomal location: -
Other names: -
Cognate genes: *S8* (mouse)
Type of homeobox: *paired*-type
Accession number: -

■ Origin and description

Using mouse *S8* cDNA as a probe, cDNAs were cloned from a chick embryo library, encoding a homeodomain protein very similar to mouse *S8*. The homeodomain is identical to that of mouse *S8*, homology outside the homeodomain being extensive as well.

■ References

F.Meijlink, unpublished.

CS8B [c]

Species: *Gallus gallus* (chicken)
Chromosomal location: -
Other names: -
Cognate genes: -
Type of homeobox: *paired*-type
Accession number: -

■ Origin and description

Using mouse *S8* cDNA as a probe, a cDNA was cloned from a chick embryo fibroblast library, encoding a homeodomain protein very similar to mouse *S8*. The homeodomain is >90% identical to that of mouse *S8*. It is not a true homologue of *S8*, since CS8A is more similar.

■ References

F. Meijlink, unpublished.

cut [d]

Species: *Drosophila melanogaster*
Chromosomal location: 7B on X chromosome
Other names: -
Cognate genes: *CDP, Clox-1*
Type of homeobox: *cut*
Accession number: -

■ Origin and description

Originally isolated as a mutation affecting the wings.

■ Expression

In PNS, es organs, and md neurons. It is also expressed in a subset of cells in CNS, Malpighian tubule and tracheal histoblasts.

■ Genetics, function

Required for es organ to acquire correct identity, loss of function mutations also cause defects in Malpighian tubules and trachea.

■ References

Blochlinger, K., Bodmer, R., Jack, J., Jan, L.Y. and Jan, Y.N. (1988). Primary structure and expression of a product from *cut*, a *locus* involved in specifying sensory organ identity in *Drosophila*. Nature 333, 629-635.

Blochlinger, K., Bodmer, R., Jan, L.Y. and Jan, Y.N. (1990). Patterns of expression of *cut*, a protein required for external sensory organ development, in wild-type and *cut* mutant *Drosophila* embryos. Genes Dev.4, 1322-1331.

Blochlinger, K., Jan, L.Y. and Jan, Y.N. (1991). Transformation of sensory organ identity by ectopic expression of cut in Drosophila. Genes Dev. 5, 1124-1131.

Bodmer, R., Barbel, S., Shepherd, S., Jack, J.W., Jan, L.Y. and Jan, Y.N. (1987). Transformation of sensory organs by mutations of the *cut locus* of D. melanogaster. Cell 51, 293-307.

Jack, J.W. (1985). Molecular organization of the *cut locus* of Drosophila melanogaster. Cell 42, 869-876.

Cymec-3 [Cv]

Species: *Caenorhabditis vulgarensis*
Chromosomal location: -
Other names: -
Cognate genes: -
Type of homeobox: LIM-class
Accession number: X63956

■ Origin and description

Isolated using the *C. elegans* gene *mec-3* (Way et al., 1991) to identify conserved sequence elements. See *mec-3* for further description.

■ References

Way, J.C., Wang, L., Run, J.-Q. and Wang, A. (1991). The *mec-3* gene contains *cis*-acting elements mediating positive and negative regulation in cells produced by asymmetric cell division in *Caenorhabditis elegans*. Genes Dev. 5, 2199-2211.

Dfd [d]

Species: *Drosophila melanogaster*
Chromosomal location: 84A4,5 (*Antennapedia* complex)
Other names: *Deformed*, r11 complementation group, 99
Cognate genes: Vertebrate group 4 *Hox* genes, *XHox1A* (Xenopus), *ceh-15*
Type of homeobox: *Deformed* sub-class, *Antp*-class
Accession number: -

■ Origin and description

The *Dfd* gene was originally identified by Eleth Cattell in 1913 in Cambridge, England, as a dominant mutation affecting head development (Bridges and Morgan, 1923). The dominant phenotype, which is variable, involves the loss of ventral regions of the adult eye. Work from T. Kaufman's group later placed *Dfd* in the middle of the *Antennapedia* complex (Wakimoto and Kaufman, 1981; Hazelrigg and Kaufman, 1983). The DNA encoding the *Dfd* locus was first roughly identified by molecular break-points that resulted in *Dfd* mutations (Scott et al., 1983), and identification of *Dfd* coding sequences was facilitated by using homeobox probes from the *Antp* and *Ubx* loci (McGinnis et al., 1984; Regulski et al., 1985 and 1987). The *Dfd* transcription unit spans approximately 11 kb and contains 5 exons. No alternatively spliced mRNAs have been found. The *Dfd* protein contains 586 amino acids and has distinctive protein motifs outside the homeodomain including His-Pro repeats, and Gln-rich, and Asn-rich regions.

■ Expression

In embryos, *Dfd* transcripts and protein are initially expressed in a blastoderm stripe that includes all or part of parasegment 0, and all of parasegment 1 (posterior mandibular and anterior maxillary segment) (Chadwick and McGinnis, 1987; Martinez-Arias et al., 1987; Jack et al., 1988). During gastrulation, the expression pattern expands to include the posterior maxillary segment, as well as cells on the ventral aspect of the embryo in the anterior mandibular/hypopharyngeal region (which are probably all incorporated into the CNS). Expression persists in ventral and ventral posterior maxillary epidermis and in the anterior subesophageal ganglion cells until the end of embryogenesis.

During larval stages, *Dfd* is expressed in the subesophageal ganglion, and in discrete regions of the eye-antennal imaginal disc. Fate maps of the eye-antennal disc that are correlated with *Dfd* expression patterns indicate that the *Dfd* expressing cells will develop into ventral head (rostral membrane) and maxillary palpi. *Dfd* expression persists in localized fashion in the adult CNS.

■ Function

During embryonic development, *Dfd* mutants fail to develop maxillary cirri, the cirri-associated ventral organs, and maxillary mouth hooks (Merrill et al., 1987; Regulski, et al., 1987). Mutants may also be missing the dorsal lateral papillae of the antennal sense organ, which are believed to develop from mandbular neuroectoderm. Ectopic expression of *Dfd* protein in embryos induces ectopic autoregulated expression of the endogenous *Dfd* locus in other segments. In labial and thoracic segment, the persistent ectopic expression of *Dfd* induces the development of ectopic maxillary cirri and mouth hooks (Kuziora and McGinnis, 1988).

In adults, somatic recombination experiments have shown that *Dfd* function is required in ventral and posterior regions of the adult head capsule, and in the maxillary palps (Merrill et al., 1987). Posterior head cells that are mutant for *Dfd* are transformed to thoracic identities, and ventral head cells that mutant for *Dfd* are apparently deleted.

Transient ectopic expression of *Dfd* protein in imaginal cells induces a phenotype much like the original *Dfd*^{dominant} allele, in which the ventral region of the eye is deleted (Kuziora and McGinnis, 1988; Chadwick et al., 1990).

■ References

Bridges, C.B. and Morgan, T.H. (1923). The third chromosome group of mutant characters of *Drosophila melanogaster*. Carnegie Inst. Wash. Pub. No. 327, 93.

Chadwick, R. and McGinnis, W. (1987). Temporal and spatial distribution of transcripts from the *Deformed* gene of *Drosophila*. EMBO J. 6, 779-789.

Chadwick, R., Jones, B., Jack, T. and McGinnis, W. (1990). Ectopic expression from the *Deformed* gene triggers a dominant defect in *Drosophila* adult head development. Devel. Biol. 141, 130-140.

Hazelrigg, T. and Kaufman, T.C. (1983). Revertants of dominant mutations associated with the *Antennapedia* gene complex of *Drosophila melanogaster*. Cytology and genetics. Genetics 105, 581-600.

Jack, T., Regulski, M. and McGinnis, W. (1988). Pair-rule segmentation genes regulate the expression of the homeotic selector gene, *Deformed*. Genes Dev. 2, 635-651.

Kuziora, M.A. and McGinnis, W. (1988). Autoregulation of a *Drosophila* homeotic selector gene. Cell 55, 477-485.

Martinez-Arias, A., Ingham, P.W., Scott, M.P. and Akam, M.E. (1987). The spatial and temporal deployment of *Dfd* and *Scr* transcripts during development of *Drosophila*. Devel. 100, 673-686.

McGinnis, W., Jack, T., Chadwick, R., Regulski, M., Bergson, C., McGinnis, N. and Kuziora, M.A. (1990). Establishment and maintenance of position-specific expression of the *Drosophila* homeotic selector gene *Deformed*. Adv. Genet. 27, 363-402.

Merrill, V.K.L., Turner, F.R. and Kaufman, T.C. (1987). A genetic and developmental analysis of mutations in the *Deformed* locus in *Drosophila melanogaster*. Dev. Biol. 122, 379-395.

Regulski, M., Harding, K., Kostriken, R., Karch, F., Levine, M. and McGinnis, W. (1985). Homeobox genes of the *antennapedia* and bithorax complexes of *Drosophila*. Cell 43, 71-80.

Regulski, M., McGinnis, N., Chadwick, R. and McGinnis, W. (1987). Developmental and molecular analysis of *Deformed*: a homeotic gene controlling *Drosophila* head development. EMBO J. 6, 767-777.

Scott, M.P., Weiner, A.J., Hazelrigg, T.I., Polisky, B.A., Pirotta, V., Scalenghe, F. and Kaufman, T.C. (1983). The molecular organization of the *antennapedia* locus of *Drosophila*. Cell 35, 763-776.

Wakimoto, B.T. and Kaufman, T.C. (1981). Analysis of larval segmentation in lethal genotypes associated with the *antennapedia* gene complex in *Drosophila melanogaster*. Dev. Biol. 81, 51-64.

DfdAf [ar]

Species: *Artemia franciscana*
Chromosomal location: -
Other names: *Deformed*
Cognate genes: *Dfd*
Type of homeobox: *Dfd*-class
Accession number: X70078

■ Origin and description

Cloned by PCR with degenerate homeobox primers, followed by inverse PCR of 1.3 kb fragment from genomic DNA. Amino acid conservation in the homeodomain and immediately flanking regions uniquely identifies this gene as a homologue of *Drosophila Dfd*.

■ References

Averof, M. and Akam, M. (1993). Hom/Hox genes in a crustacean; implication for the origin of insect and crustacean body parts. Current biology 3, 73–78.

Dll [d]

Species: *Drosophila melanogaster*
Chromosomal location: chromosome 2R, 60E5,6
Other names: *Distal-less, Brista*
Cognate genes: *Dlx-1, Dlx-2, Tes1*
Type of homeobox: *Dll*-type
Accession number: -

■ Origin and description

The gene was isolated by chromosomal walking (Cohen et al. 1989). The transcription unit was mapped with respect to several Distal-less mutant alleles which involve chromosomal rearrangements. The transcript consists of 7 exons spanning 20 kb. Essential *cis*-regulatory control elements are located in an additional 15 kb of upstream and 5 kb of downstream flanking sequences, as well as in the large intron that divides the homeobox. Two transcript size classes are produced (Cohen et al. 1989). Both encode the same protein, and differ only in 3' untranslated sequences.

■ Expression

Distal-less transcript is expressed in the limb primordia of the *Drosophila* embryo. *Distal-less* provides the earliest molecular marker that labels the presumptive limbs, and is expressed well before any other molecular or morphological sign of limb development (Cohen, 1990). The earliest expression occurs in two incomplete dorsal stripes in the head region at blastoderm stage. One stripe corresponds to the future antennal limb primordia. The second is a common stripe of expression for the presumptive maxillary and *labial* primordia. Early in gastrulation a pair of spots corresponding to the presumptive labrum appears anterior to the antennal primordium. During germ band extension the fused maxillary and *labial* stripe resolves into discrete patches of expression in these primordia. By the end of germ band extension (stage) the leg primordia begin to express *Distal-less*. This pattern of expression in 7 patches of cells (per side) persists throughout embryogenesis. The patches are the primordia of the larval limbs. *Distal-less* expression also marks the nascent leg imaginal discs in the thoracic segments (Cohen et al. 1991). *Distal-less* is also expressed in the maxillary cirri in the embryo (O'Hara et al. in preparation) as well as in optic formation centre of the larval brain, and at the base of the posterior spiracles.

Distal-less is also expressed in the imaginal discs. In the leg *Distal-less* is expressed in a central cluster of cells until the third instar. During 3rd instar the transcript forms a step gradient along the proximal distal axis of the leg (Cohen & Cohen, submitted). *Distal-less* is also expressed in the larval antenna, brain and along the margin of the wing disc.

■ Function

Distal-less is required for the development of all limb structures in the larva and the adult (Sunkel and Whittle, 1987; Cohen and Jürgens, 1989 a,b). The gene functions as a developmental switch to promote the development of limb structures above a default state of body wall. The fate of presumptive limb cells can be switched to body wall by eliminating *Distal-less* in genetically mosaic limb primorida. Homozygous mutants die as embryos or early larvae. The limbs of the gnathal (head) segments are associated with the larval head skeleton. In mutants the head skeleton is often collapsed or twisted and may compress the esophagus (Cohen and Jürgens, 1989 a). The organization of the proximal distal axis of the limbs depends on *Distal-less* in a concentration dependent manner (Cohen and Jürgens, 1989 a). The specification of positional values along the proximal-distal axis of the leg appears to be driven by regional differences in *Distal-less* concentration (Cohen & Cohen, submitted).

■ References

Cohen, S.M. (1990). Specification of limb development in the *Drosophila* embryo by positional cues from segmentation genes. Nature 343, 173-177.

Cohen, S.M. and Jürgens, G. (1989a). Proximal-distal pattern formation in *Drosophila*: cell autonomous requirement for Distal-less gene activity in limb development. EMBO J. 8, 2045-2055.

Cohen, S.M. and Jürgens, G. (1989b). Proximal-distal pattern formation in *Drosophila*: graded requirement for Distal-less gene activity during limb development in *Drosophila*. Roux's Archiv. Dev. Biol. 198, 157-169.

Cohen, B., Wimmer, E. and Cohen, S.M. (1991). Early development of the leg and wing primorida in the *Drosophila* embryo. Mech. Devel. 33, 229-240.

Cohen, S.M., Brönner, G., Küttner, F., Jürgens, G. and Jäckle, H. (1989). Distal-less encodes a homeodomain required for limb development in *Drosophila*. Nature 338, 432-434.

Sunkel, C.E. and Whittle, J.R.S. (1987). Brista: a gene involved in the specification and differentiation of distal cephalic and thoracic structures in *Drosophila melanogaster*. Roux's Arch. Dev. Biol. 196, 124-132.

Dlx-1 [m]

Species: *Mus musculus*
Chromosomal location: Chromosome 2
Other names: *Dlx*
Cognate genes: *Distal-less* (*Dll*)
Type of homeobox: *Dll*-type
Accession number: -

■ Origin and description

This gene was cloned from a genomic cosmid library by cross-homology with the mouse *Nkx-2.1* (*TTF-1*) gene (Price et al., 1991). A cDNA clone was obtained from a neonatal brain library. Its homeobox sequence is more related to the *Drosophila Distal-less* gene than to the *NK-2* family of genes. It is one of probably 4 members of the mammalian *Dlx* gene family (*Dlx-1* to *Dlx-4*, M. Price, R. DiLauro and E. Boncinelli, personal communication) which may have a clustered organization (two tandems; Boncinelli, personal communication).

■ Expression

Dlx-1 is expressed during fetal development in a variety of places. It is strongly expressed at mid-gestation in the forebrain, especially in the diencephalon and the adjacent telencephalic areas (the future striatum). In the diencephalon, the expression is restricted to the ventral thalamus as well as to a restricted part of the hypothalamus (around the optic chiasma). The *Dlx-1* expression domain suggests that this gene may be involved in the definition of a broad domain during forebrain morphogenesis (Price et al., 1991; 1992).

Dlx-1 is also expressed in facial (arches) mesenchyme and, later on, in the teeth buds. It is strongly expressed during ocular development, in the retina and in several derivatives of the neural crest such as nerve trunks from the peripheric nervous system. In addition, a weak expression was detected in the limb apical ectodermal ridge (Dollé et al., 1992).

■ References

Dollé, P., Price, M. and Duboule, D. (1992). Expression of the mouse *Dlx-1* homeobox gene during facial, ocular and limb development. Differentiation 49, 93-100.

Price, M., Lemaistre, M., Pischetola, M., DiLauro, R. and Duboule D. (1991). A mouse Distal-less related homeobox gene shows a restricted expression in the developing forebrain. Nature 351, 748-751.

Price, M., Lazzaro, D., Pohl, T., Mattei, M-G., Ruether, U., Olivo, J-C., Duboule, D. and DiLauro R. (1992). Regional expression of the homeobox gene *Nkx-2.2* in the developing mammalian forebrain. Neuron 8, 241-255.

Dlx-2 [m]

Species: *Mus musculus*

Chromosomal location: -

Other names: *Tes-1*

Cognate genes: *Distal-less* (*Dll*) *Drosophila*

Type of homeobox: *Dll*-type

Accession number: GenBank No. M80540

■ Origin and description

The *Dlx-2* gene was cloned using a subtractive hybridization protocol which was designed to isolate cDNAs that are encoded by genes which are preferentially expressed during development of the telencephalon (Porteus et al., 1992). The subtractive cDNA library was screened at low-stringency with oligonucleotide probes homologous to several classes of homeodomains. *Dlx-2* was identified using these approaches (Porteus et al., 1991). The homeobox is most closely related to that of the murine *Dlx-1*

(Price et al., 1991), *Dlx-2* and *Dlx-3* (Robinson et al., 1991) and the *Drosophila Distal-less* (*Dll*) (Cohen et al., 1989) genes. *Dlx-1* and *Dlx-2* were also identified by Robinson et al., 1991).

■ Expression

The mouse *Dlx-2* gene is expressed greater the twenty-fold more in the mid-gestational (E15) then the adult telencephalon. *In situ* RNA hybridization shows that the expression of this gene is under strong temporal and spatial control. It is expressed only in the head of the mouse, and the level of expression is at least 20-fold greater in the embryo than in the adult. At E11.5, it is expressed in the forebrain in a spatially restricted pattern: expression is detected in the ventral telencephalon (in the ganglionic eminence) and in two zones in the ventral diencephalon. This overall pattern is maintained through mid to late gestation, until E17.5 when expression in the dorsal neuroepithelium is first detectable. Postnatally, expression is found at low levels in several ventral forebrain nuclei, the olfactory bulb, and in the hippocampus. In the adult brain, low levels of expression are detectable using northerns, but not by *in situ* hybridization. Expression is also found in the ameloblast precursor cells (Porteus et al., 1991; Robinson et al. 1991).

■ References

Cohen, S.M., Bronner, G., Kuttner, F., Jurgens, G. and Jackle, H. (1989). Distal-less encodes a homeodomain protein required for limb development in *Drosophila*. Nature 338, 432-434.

Porteus, M.H., Bulfone, A., Ciaranello, R.D. and Rubenstein, J.L.R. (1991). Isolation and characterization of a novel cDNA encoding a homeodomain that is developmentally regulated in the ventral forebrain. Neuron 7, 221-229.

Porteus, M.H., Brice, A.E.J, Bulfone, A., Usdin, T.B., Ciaranello, R.D. and Rubenstein, J.L.R. (1992). Isolation and characterization of a library of cDNA clones that are preferentially expressed in the embryonic telencephalon. Molecular Brain Research 12, 7-22.

Price, M., Lemaistre, M., Pischetola, M., Di Lauro, R. and Duboule, D. (1991). A mouse gene related to Distal-less shows a restricted expression in the developing forebrain. Nature 351, 748-751.

Robinson, G.W., Wray, S. and Mahon, K.A. (1991). Spatially restricted expression of a member of a new family of murine Distal-less homeobox genes in the developing forebrain. The New Biologist 3, 1183-1194.

Dth-1   [Dt]

Species: *Dugesia [Girardia] tigrina*
(Planarian)
Chromosomal location: -
Other names: -
Cognate genes: -
Type of homeobox: *NK-2*-type
Accession number: X56499

■ Origin and description

This gene was isolated (Garcia-Fernandez et al., 1991) by cross-homology with degenerated oligonucleotides HB-1 (Bürglin et al., 1989). It contains a homeobox related to the Planarian *Dth-2* gene, *Drosophila NK-2* homeobox type and *Nkx-2* class of vertebrate genes.

■ Expression

The expression of *Dth-1* gene by Northern-blot analysis in intact and regenerating adult organisms shows a continuous expression with a slight increase at 5 days of regeneration. *In situ* hybridization analysis shows a specific expression in the gastrodermis, intestinal cells (endodermal origin) of intact adults.

■ References

Bürglin, T.R., Finney, M., Coulson, A. and Ruvkun, G. (1989). *Caenorhabdiris elegans* has scores of homeobox-containing genes. Nature 341, 239-243.
Garcia-Fernandez, J., Baguñà, J. and Saló, E. (1991). Planarian homeobox genes: Cloning analysis and expression. Proc. Natl. Acad. Sci. USA 88, 7338-7342.

Dth-2   [Dt]

Species: *Dugesia [Girardia] tigrina*
(Planarian)
Chromosomal location: -
Other names: -
Cognate genes: -
Type of homeobox: *NK-2*-type
Accession number: X56500

■ Origin and description

This gene was isolated (Garcia-Fernandez et al., 1991) by cross-homology with degenerated oligonucleotides HB-1 (Bürglin et al., 1989). It contains a homeobox related to the Planarian *Dth-1* gene, *Drosophila NK-2* homeobox type and *Nkx-2* class of vertebrate genes.

■ Expression

The expression of *Dth-2* by Northern-blot analysis in intact and regenerating adult organisms shows a continuous expression with a slight increase at 5 days of regeneration. *In situ* hybridization analysis shows a specific expression in the dorsal peripheral parenchyma of intact adults.

■ References

Bürglin, T.R., Finney, M., Coulson, A. and Ruvkun, G. (1989). *Caenorhabdiris elegans* has scores of homeobox-containing genes. Nature 341, 239-243.
Garcia-Fernandez, J., Baguñà, J. and Saló, E. (1991). Planarian homeobox genes: Cloning analysis and expression. Proc. Natl. Acad. Sci. USA 88, 7338-7342.

E5   [d]

Species: *Drosophila melanogaster*
Chromosomal location: 3rd chromosome, 88A1,2
Other names: -
Cognate genes: mouse and human *Emx*
Type of homeobox: *empty spiracles*-type
Accession number: -

■ Origin and description

The *E5* gene was originally cloned by homeobox homology to the *even-skipped* homeobox sequence (Dalton et al., 1989). The *E5* gene maps near (in a cytogenetic sense) the *empty-spiracles* gene, and is very closely related to *empty-spiracles* in its homeodomain sequence (52 matches in 60 residues). The molecular distance between the two genes is not yet known. A 2.2 *E5* cDNA has been sequenced and used to detect transcript distribution in *Drosophila* embryos.

The *E5* homeodomain sequence is highly diverged from *Antp* (less than 50% identity), but *E5* and *empty spiracles* homeodomains are both closely related to recently isolated vertebrate homeodomain sequences found in the mouse and human *Emx* genes (Simeone et al., 1992).

■ Expression

The *E5* homeobox gene is expressed in *Drosophila* embryos in a pattern that includes metamerically reiterated groups of cells in every segment that probably overlaps in some cells with the late metameric pattern of *empty spiracles* expression (Dalton et al., 1989; Walldorf and Gehring, 1992). No expression is detected at stages earlier than extended germ band.

■ References

Dalton, D., Chadwick, R. and McGinnis, W. (1989). Expression and embryonic function of *empty spiracles*: A *Drosophila* homeobox gene with dual patterning functions on the anterior-posterior axis. Genes Dev. 3, 1940-1956.

Simeone, A., Gulisano, M., Acampora, D., Stornaiuolo, A., Rambaldi, M.and Boncinelli, E. (1992a). 2 vertebrate homeobox genes related to the *Drosophila*-empty spiracles gene are expressed in the embryonic cerebral-cortex. EMBO J. 11, 2541-2550.

Walldorf, U. and Gehring, W. (1992). *empty spiracles*, a gap gene containing a homeobox involved in *Drosophila* head development. EMBO J. 11, 2247-2259.

EgHbx1 [eg]

Species: *Echinococcus granulosus* (tapeworm)
Chromosomal location: -
Other names: -
Cognate genes: -
Type of homeobox: *NK-1*-type
Accession number: EMBL X66817

■ Origin and description

This gene was isolated from a genomic library using a degenerate oligonucleotide (HB-1) (Oliver et al., 1992 submitted) against the conserved third helix. It is a member of the *NK-1* type of genes (Kim and Nirenberg, 1989).

■ References

Kim, Y. and Nirenberg, M. (1989). *Drosophila* NK-homeobox genes. Proc. Natl. Acad. Sci. USA 86, 7716-7720.

Oliver, G., Vispo, M., Mailhos, A., Martínez, C., Sosa-Pineda, B., Fielitz, W. and Ehrlich, R. (1992). Homeobox containing genes in flatworms. Gene 121, 337–342.

EgHbx2 [eg]

Species: *Echinococcus granulosus* (tapeworm)
Chromosomal location: -
Other names: -
Cognate genes: -
Type of homeobox: *NK-3*-type
Accession number: EMBL X66818

■ Origin and description

This gene was isolated from a genomic library using a degenerate oligonucleotide (HB-1) (Oliver et al., 1992 submitted) against the conserved third helix. It is a member of the *NK* type of genes (Kim and Nirenberg, 1989).

■ References

Kim, Y. and Nirenberg, M. (1989). *Drosophila* NK-homeobox genes. Proc. Natl. Acad. Sci. USA 86, 7716-7720.

Oliver, G., Vispo, M., Mailhos, A., Martínez, C., Sosa-Pineda, B., Fielitz, W. and Ehrlich, R. (1992). Homeobox containing genes in flatworms. Gene 121, 337-342.

EgHbx3 [eg]

Species: *Echinococcus granulosus* (tapeworm)
Chromosomal location: -
Other names: -
Cognate genes: -
Type of homeobox: *NK-2*-type
Accession number: EMBL X66819

■ Origin and description

This gene was isolated from a genomic library using a degenerate oligonucleotide (HB-1) (Oliver et al., 1992 submitted) against the conserved third helix. It is a member of the *NK-2* type of genes (Kim and Nirenberg, 1989).

■ References

Kim, Y. and Nirenberg, M. (1989). *Drosophila* NK-homeobox genes. Proc. Natl. Acad. Sci. USA 86, 7716-7720.

Oliver, G., Vispo, M., Mailhos, A., Martínez, C., Sosa-Pineda, B., Fielitz, W. and Ehrlich, R. (1992). Homeobox containing genes in flatworms. Gene 121, 337–342.

EgHbx4 [eg]

Species: *Echinococcus granulosus* (tapeworm)
Chromosomal location: -
Other names: -
Cognate genes: -
Type of homeobox: *goosecoid*-type
Accession number: EMBL X66820

■ Origin and description

This gene was isolated from a genomic library using a degenerate oligonucleotide (HB-1) (Oliver et al., 1992 submitted) against the conserved third helix. It is similar to the dorsal lip specific *goosecoid* gene isolated from *Xenopus* (Blumberg et al., 1991).

■ References

Blumberg, B., Wright, C.V.E., De Robertis, E.M. and Cho, K.W.Y. (1991). Organizer-specific homeobox genes in *Xenopus laevis* embryos. Science 253, 194-196.

Oliver, G., Vispo, M., Mailhos, A., Martínez, C., Sosa-Pineda, B., Fielitz, W. and Ehrlich, R. (1992). Homeobox containing genes in flatworms. Gene 121, 337–342.

ems   [d]

Species: *Drosophila melanogaster*
Chromosomal location: 88A
Other names: *empty spiracles, E4, W13*
Cognate genes: -
Type of homeobox: related to the *msh/NK*-type
Accession number: -

■ Origin and description

This gene was isolated by cross-homology with the *even-skipped* (*eve*) homeobox (Dalton et al., 1989) and independently with the muscle segment homeobox (*msh*). It contains a homeobox most closely related to the gene *E5* (Dalton et al., 1989).

■ Expression

First expression of *ems* is seen at the syncytial blastoderm stage in a single anterior band which is at the cellular blastoderm stage repressed on the ventral side. Later during embryogenesis it is expressed in the clypeolabrum and in lateral regions of each segment. This metameric pattern consists of bilateral patches of neuroblasts and ectodermal cells around the tracheal pits. Additional expression is seen in the optic lobe and the posterior spiracles. Expression in the nervous system and the posterior spiracles stays on until the end of embryogenesis.

■ Genetics, function

The mutant *ems* was isolated in a genetic screen for embryonic lethals (Jürgens et al., 1984). From the five different alleles isolated are now four analyzed molecularly (Dalton et al., 1989; Walldorf and Gehring, 1992). The *ems* mutant shows head defects, the antennal sense organ and part of the cephalopharyngeal skeleton like the dorsal arms, vertical plates and lateralgräten are missing. All these deleted structures derive from the preantennal, antennal and intercalary segments. In A8 the posterior spiracles are devoid of the Filzkörper. Regarding the phenotype in the head, a deletion of structures from three adjacent segments, *ems* could be classified as a gap gene of the head region (Cohen and Jürgens, 1990; Walldorf and Gehring, 1992). It is most likely one of the genes X which are postulated target genes of the maternal effect gene *bicoid* (Driever et al., 1989).

■ References

Cohen, S.M. and Jürgens, G. (1990). Mediation of *Drosophila* head development by gap-like segmentation genes. Nature 346, 482-485.

Dalton, D., Chadwick, R. and McGinnis, W. (1989. Expression and embryonic function of empty spiracles: a *Drosophila* homeobox gene with two patterning functions on the anterior-posterior axis of the embryo. Genes Dev. 3, 1940-1956.

Driever, W., Thoma, G. and Nüsslein-Volhard, C. (1989). Determination of spatial domains of zygotic gene expression in the Drosophila embryo by the affinity of binding sites for the bicoid morphogen. Nature 340, 363-367.

Jürgens, G., Wieschaus, E., Nüsslein-Volhard, C. and Kluding, H. (1984). Mutations affecting the pattern of the larval cuticle in Drosophila melanogaster. 2. Zygotic loci on the third chromosome. Wilhelm Roux's Arch. Dev. Biol. 193, 283-295.

Walldorf, U. and Gehring, W.J. (1992). Empty spiracles, a gap gene containing a homeobox involved in *Drosophila* head development. EMBO J., 11, 2247-2259.

EMX1   [h]

Species: *Homo sapiens*
Chromosomal location: -
Other names: -
Cognate genes: *Emx-1* (mouse)
Type of homeobox: *ems*-type
Accession number: -

■ Origin and description

This gene was isolated by cross-homology with the *Drosophila ems* homeobox. (Simeone et al., 1992a). A cDNA library prepared from 8-week human embryos was screened at low stringency conditions with a short *ems* genomic sequence including the homeobox (Dalton et al., 1989). Two classes of homologous cDNA clones, termed *EMX1* and *EMX2*, were found. Using these cDNA clones a human genomic library constructed in cosmids was screened.

■ Expression

Emx-1 is expressed during mouse development in the dorsal telencephalon, almost exclusively in the presumptive cerebral cortex from E9.0 (Simeone et al., 1992a , 1992b).

■ References

Dalton, D., Chadwick, R. and McGinnis, W. (1989). Expression and embryonic function of *empty spiracles*: A *Drosophila* homeobox gene with dual patterning functions on the anterior-posterior axis. Genes Dev. 3, 1940-1956.

Simeone, A., Gulisano, M., Acampora, D., Stornaiuolo, A., Rambaldi, M.and Boncinelli, E. (1992a). 2 vertebrate homeobox genes related to the *Drosophila*-empty spiracles gene are expressed in the embryonic cerebral-cortex. EMBO J. 11, 2541-2550.

Simeone, A., Acampora, D., Gulisano, M., Stornaiuolo, A. and Boncinelli, E. (1992b). Nested expression domains of four homeobox genes in developing rostral brain. Nature 358, 687-690.

EMX2 [h]

Species: *Homo sapiens*
Chromosomal location: -
Other names: -
Cognate genes: *Emx-2* (mouse)
Type of homeobox: *ems*-type
Accession number: -

■ Origin and description

This gene was isolated by cross-homology with the *Drosophila ems* homeobox. (Simeone et al., 1992a). A cDNA library prepared from 8-week human embryos was screened at low stringency conditions with a short *ems* genomic sequence including the homeobox (Dalton et al., 1989). Two classes of homologous cDNA clones, termed *EMX1* and *EMX2*, were found. Using these cDNA clones as probes a human genomic library constructed in cosmids was screened.

■ Expression

By *in situ* hybridization *Emx-2* was found to be specifically expressed in the cerebral cortex and in some embryonic neuroectodermal areas including olfactory placodes and olfactory epithelia. Furthermore, a peculiar expression of this gene has been observed in the ectodermal regions in the snout and limbs as well as in metanephric tubular epithelia and the genital ridge (Simeone et al. 1992a, 1992b)

■ References

Dalton, D., Chadwick, R. and McGinnis, W. (1989). Expression and embryonic function of *empty spiracles*: A *Drosophila* homeobox gene with dual patterning functions on the anterior-posterior axis. Genes Dev. 3, 1940-1956.

Simeone, A., Gulisano, M., Acampora, D., Stornaiuolo, A., Rambaldi, M.and Boncinelli, E. (1992a). 2 vertebrate homeobox genes related to the *Drosophila*-empty spiracles gene are expressed in the embryonic cerebral-cortex. EMBO J. 11, 2541-2550.

Simeone, A., Acampora, D., Gulisano, M., Stornaiuolo, A. and Boncinelli, E. (1992b). Nested expression domains of four homeobox genes in developing rostral brain. Nature 358, 687-690.

en [d]

Species: *Drosophila melanogaster*
Chromosomal location: chromosome 2R, band 48A
Other names: *engrailed*
Cognate genes: *En* (mouse, chicken, human, *Xenopus*, zebrafish, sea urchin, Bombyx, grasshopper, leech, Brachiopod, brook lamprey)
Type of homeobox: *engrailed*
Accession number: -

■ Origin and description

The *engrailed* and *invected* are neighbouring genes which were cloned by walking from the met-tRNA gene in 48B (Kuner et al., 1985) and by cross-hybridization with a homeobox probe (Fjose et al., 1985). *Engrailed* was localized with numerous inversion and translocation breakpoint mutations (Kuner et al., 1985). The homeobox is rather divergent from the *Antp* prototype and is interrupted by an intron (Poole et al., 1985). A model of the *Drosophila* engrailed homeodomain has been proposed based on a homeodomain/DNA cocrystal (Kissinger et al., 1990).

■ Expression

Engrailed expression has been studied by Northern analysis (Drees et al., 1985), by *in situ* hybridization (Kornberg, et al., 1985; Weir and Kornberg, 1985), with antibodies (DiNardo et al., 1985; Karr et al., 1989; Patel, 1989), and with promoter-LacZ fusions integrated into the germline (Hama et al., 1990; Kassis, 1990). *Engrailed* is expressed in the posterior developmental compartments of the embryo, larva, imaginal discs, and histoblasts, in the clypeolabrum of the embryo, in the hindgut of the embryo and larva, and in the central nervous system and brain of the embryo and larva.

■ References

DiNardo, S., Kuner, J., Theis, J. and O'Farrell, P. (1985). Development of embryonic pattern in *D. melanogaster* as revealed by accumulation of the nuclear *engrailed* protein. Cell 43, 59-69.

Drees, B., Ali, Z., Soeller, W., Coleman, K., Poole, S. and Kornberg, T. (1987). The transcription unit of the *Drosophila* engrailed locus: an unusually small portion of a 70,000 bp gene. EMBO J. 6, 2803-2809.

Fjose, A., McGinnis, W. and Gehring, W. (1985). Isolation of a homeobox-containing gene from the *engrailed* region of *Drosophila* and the spatial distribution of its transcript. Nature 313, 284-289.

Hama, C., Ali, Z. and Kornberg, T.B. (1990). Region-specific recombination and expression are directed by portions of the *Drosophila engrailed* promoter. Genes and Dev. 4, 1079-1093.

Karr, T.L., Weir, M.J., Ali, Z. and Kornberg, T. (1989). Patterns of *engrailed* protein in early *Drosophila* embryos. Development 105, 605-612.

Kassis, J.A. (1990). Spatial and temporal control elements of the *Drosophila engrailed* gene. Genes and Dev. 4, 433-443.

Kissinger, C.R., Liu, B., Martin-Blanco, E., Kornberg, T.B. and Pabo, C.O. (1990). Structure of an *engrailed* homeodomain/DNA complex at 2.6Å resolution: a framework for understanding homeodomain/DNA interactions. Cell 63, 579-590.

Kornberg, T., Siden, I., O'Farrell, P. and Simon, M. (1985). The *engrailed* locus of *Drosophila*: *in situ* localization of transcripts reveals compartment-specific expression. Cell 40, 45-53.

Kuner, J., Nakanishi, M., Ali, Z., Drees, B., Gustavson, E., Theis, J., Kauvar, L., Kornberg, T. and O'Farrell, P. (1985). Molecular cloning of *engrailed*, a gene involved in the development of pattern in *Drosophila* melanogaster. Cell 42, 309-316.

Patel, N.H., Martin-Blanco, E., Coleman, K., Poole, S., Ellis, M.C., Kornberg, T.B. and Goodman, C.S. (1989). Expression of *engrailed* proteins in arthropods, annelids, and chordates. Cell 58, 955-968.

Poole, S.J., Kauvar, L., Drees, B. and Kornberg, T. (1985). The *engrailed* locus of *Drosophila*: structural analysis of an embryonic transcript. Cell 40, 37-43.

Weir, M.P. and Kornberg, T. (1985). Patterns of *engrailed* and fushi tarazu transcripts reveal novel intermediate stages in *Drosophila* segmentation. Nature 318, 433-439.

en [Lp]

Species: *Lampetra planeri* (brook lamprey)
Chromosomal location: -
Other names: *engrailed*
Cognate genes: *en*
Type of homeobox: *en*-type
Accession number: X59122

■ Origin and description

Homeobox region only, cloned from genomic DNA via PCR using degenerate primers for en-like genes (Holland and Williams, 1990).

■ References

Holland, P.W.H. and Williams, N.A. (1990). Conservation of engrailed-like homeobox sequences during vertebrate evolution. FEBS Lett. 277, 250-252.

en [Sa]

Species: *Schistocerca americana* (grasshopper)
Chromosomal location: -
Other names: *engrailed, GH-en*
Cognate genes: -
Type of homeobox: *en*-type
Accession number: -

■ Origin and description

This gene was cloned by screening a grasshopper embryo expression library (λgt11) with the monoclonal antibody MAb 4D9 which recognizes engrailed-related proteins in a wide variety of organisms (Patel et al., 1989a). The homeobox region was sequenced and revealed an engrailed class homeobox. Comparisons made with the available sequence suggest that the grasshopper engrailed gene is equally related to both *Drosophila engrailed* and *invected*. Genomic Southern analysis suggests that this is the only engrailed-class gene in the grasshopper genome.

■ Expression

The expression pattern of grasshopper *engrailed* was determined by both *in situ* hybridization and antibody staining. As in *Drosophila*, grasshopper *engrailed* is expressed in a striped pattern during segmentation and in a subset of neurons during neurogenesis (Patel et al., 1989a). Unlike *Drosophila*, however, the *engrailed* stripes in grasshopper appear long after the blastoderm stage and arise sequentially along the length of the germband (Patel et al., 1989b).

■ References

Patel, N.H., Martin-Blanco, E., Coleman, K.G., Poole, S.J., Ellis, M.C., Kornberg, T.B., Goodman, C.S. (1989a). Expression of engrailed proteins in arthropods, annelids, and chordates. Cell 58, 955-968.

Patel, N.H., Kornberg, T.B. and Goodman, C.S. (1989b). Expression of engrailed during segmentation in grasshopper and crayfish. Development 107, 201-212.

en [Tg]

Species: *Tripneustes gratilla* (sea urchin)
Chromosomal location: -
Other names: *SU-HB-en*
Cognate genes: *en* (*Drosophila*)
Type of homeobox: related to *engrailed*
Accession number: M19709

■ Origin and description

TgHbox-en was isolated from a *Tripneustes gratilla* genomic library in λ charon 4, using the mouse en-1 gene as a probe. The sea urchin genomic library was screened extensively for other *engrailed* clones without success. It appears that the sea urchin contains a single engrailed gene in contrast to both vertebrates and insects and represents the primitive condition. The duplication of the engrailed gene in vertebrates and insects may have occurred independently after the divergence of the two groups from each other.

■ Expression

Northern transfers were prepared from RNA extracted from the various embryonic stages and from most adult tissues. The expression of *TgHbox-en* was only detected in the adult tissue around the Aristotle's lantern which contains the central nerve ring. The messenger RNAs were 4.7 and 5.3 kb. Rare transcripts were present in ovary, testis and coelomocytes.

■ Reference

Dolecki, G.J. and Humphreys, T. (1988). An engrailed class homeobox gene in sea urchins. Gene 64, 21-31.

en [Tr]

Species: *Terebratulina retusa* (Brachiopod)
Chromosomal location: -
Other names: -
Cognate genes: *en*
Type of homeobox: *en*-type
Accession number: -

■ Origin and description

Homeobox region only, cloned from genomic DNA via PCR using degenerate primers for en-like genes (Holland et al. 1991).

■ References

Holland, P.W.H., Williams, N.A. and Lanfear, J. (1991). Cloning of segment polarity gene homologues from the unsegmented brachiopod Terebratulina retusa (Linnaeus). FEBS Lett. 2, 211–213.

En-A [Mg]

Species: *Myxine glutinosa* (hagfish)
Chromosomal location: -
Other names: -
Cognate genes: *en*
Type of homeobox: *en*-type
Accession number: X59120

■ Origin and description

Homeobox region only, cloned from genomic DNA via PCR using degenerate primers for en-like genes (Holland and Williams, 1990).

■ References

Holland, P.W.H. and Williams, N.A. (1990). Conservation of engrailed-like homeobox sequences during vertebrate evolution. FEBS Lett. 277, 250-252.

En-B [Mg]

Species: *Myxine glutinosa* (hagfish)
Chromosomal location: -
Other names: *engrailed*
Cognate genes: *en*
Type of homeobox: *en*-type
Accession number: X59121

■ Origin and description

Homeobox region only, cloned from genomic DNA via PCR using degenerate primers for en-like genes (Holland and Williams, 1990).

■ References

Holland, P.W.H. and Williams, N.A. (1990). Conservation of engrailed-like homeobox sequences during vertebrate evolution. FEBS Lett. 277, 250-252.

Species: *Mus musculus*

Chromosomal location: Chromosome 1, 0.28 cM distal to *Dh*

Other names: *Mo-en.1*

Cognate genes: *EN1* (human; chromosome 2q13-q21.2), *En-1* (chicken and *Xenopus*), *Eng1* (zebrafish)

Type of homeobox: *en*-type

Accession number: -

■ Origin and description

The *En-1* gene was cloned by screening a mouse genomic DNA library at low stringency with a *Drosophila en* homeobox region DNA probe (Joyner et al., 1985). The protein encoded by *En-1* contains five regions of homology that are conserved in all *en*-like genes cloned from organisms as diverse as Platyhelminthes, Annelids, Arthropods, and Chordates (reviewed in Joyner and Hanks, 1991; Logan et al., 1992). The largest conserved region is the homeodomain. Most organisms contain one or two *en*-like genes, with the exception of zebrafish which contains 3. In all vertebrates the two genes are referred to as *En-1* and *En-2*, except in human (*EN1, 2*) and zebrafish (*Eng1,2,3*). Based on amino acid sequence comparisons, the third zebrafish gene, *Eng3*, appears to have arisen form a recent duplication of *Eng2* (Ekker et al., 1992).

■ Expression

The expression of *En-1* (and its cognate genes) has been studied by RNA *in situ* and antibody staining in mouse, chicken, *Xenopus* and zebrafish. Two antibodies have been used: αEnhb-1, which was raised against a C-terminal homeodomain fragment of the mouse *En-2* protein and recognizes *En-1* and *En-2* proteins in mouse, chicken and *Xenopus* and *Eng2* and *Eng3* in zebrafish, and 4D9 which was raised against *Drosophila inv* protein and recognizes the homeodomains of chicken and *Xenopus En-2* and zebrafish, *Eng1, 2, 3*. The expression patterns of the vertebrate cognate genes are all similar (reviewed in Joyner and Hanks, 1991). In mouse, *En-1* is first expressed at the one somite stage in two dorsal patches in the neural folds (McMahon et al., 1992). Expression soon expands to a band of cells that later marks the mid-hindbrain junction region. *En-1* continues to be expressed in this region throughout embryonic and early postnatal development. However, there is a gradual transition from a spatially restricted band to cell type-specific expression in groups of neurons involved in motor control (Davis and Joyner, 1988; Davidson et al., 1988; Davis et al., 1991). In the cerebellum, *En-1* is primarily expressed in the midline region during embryogenesis and early postnatal development (Millen and Joyner, unpublished). In the adult, *En-1* is not expressed in the cerebellum (Davis and Joyner, 1988). At the 18-20 somite stage *En-1* begins also to be expressed in 2 lateral stripes along the hindbrain and spinal cord, in a stripe of cells in the somites (midway in the dorsal/ventral axis of the dermamyotome) and the ventral ectoderm of the limbs. *En-1* is also expressed later in the schlerotome in the cells that will form the vertebrae and in body and head muscle (Davis and Joyner, 1988; Davidson et al., 1988; Davis et al., 1991; Sassoon, unpublished).

■ Function

A mouse mutant (*En-1*[hd]) lacking the homeobox exon of *En-1* has recently been made by gene targeting in ES cells (Wurst and Joyner, unpublished). The preliminary analysis of the phenotype of homozygous mutants demonstrates that *En-1* is required for development of the mid-hindbrain junction region, normal development of the vertebral column, ribs and sternum, and patterning of the limbs. Most *En-1*[hd] homozygous mice die within a day of birth and do not feed. The mutants are distinguishable at birth by a distinct limb abnormality that includes digit fusions and an extra digit. Brain sections of mutant embryos or newborns shows that in the midline the coliculi are reduced and the cerebellum is absent, although in lateral regions, some cerebellar tissue is present.

■ References

Davidson, D., Graham, E., Sime, C. and Hill, R. (1988). A gene with sequence similarity to *Drosophila engrailed* is expressed during the development of the neural tube and vertebrae in the mouse. Development 104, 305-316.

Davis, C.A., Holmyard, D.P., Millen, K.J. and Joyner, A.L. (1991). Examining pattern formation in mouse, chicken and frog embryos with an *En*-specific antiserum. Development 111, 287-298.

Davis, C.A. and Joyner, A.L. (1988). Expression patterns of the homeo box-containing genes *En-1* and *En-2* and the proto-oncogene *int-1* diverge during mouse development. Genes and Development 2, 1736-1744.

Ekker, M., Wegner, J., Akimenko, M.A. and Westerfield, M. (1992). Coordinate embryonic expression of three zebrafish *engrailed* genes. Development 116, 1001-1010.

Hill, R.E., Hall, A.E., Simte, C.M. and Hastie, N.D. (1987). A mouse homeo box-containing gene maps near a developmental mutation. Cytogenet. Cell Genet. 44, 171-174.

Joyner, A.L. and Hanks, M. (1991). The *engrailed* genes: Evolution of funcion. Seminars in Developmental Biology 2, 435-445.

Joyner, A.L., Kornberg, T., Coleman, K.G., Cox, D.R., Martin, G.R. (1985). Expression during embryogenesis of a mouse gene with sequence homology to the *Drosophila engrailed* gene. Cell 43, 29-37.

Joyner, A.L., Martin, G.R. (1987). *En-1* and *En-2*, two mouse genes with sequence homology to the *Drosophila engrailed* gene: expression during embryogenesis. Genes and Development 1, 29-38.

Kohler, A., Logan, C., Joyner, A.L. and Muenki, M. (1993). Regional assignment of the human homeobox-containing gene *EN1* to chromosome 2q13-q21.2. Genomics, 15, 233–235.

Logan, C., Willard, H.F., Rommens, J.M., Joyner, A.L. (1989).

Chromosomal localization of the human homeo box-containing genes, *EN1* and *EN2*. Genomics 4, 206-209.

Logan, C., Hanks, M.C., Noble-Topham, S., Nallainathan, D., Provart, N.J. and Joyner, A.L. (1992). Cloning and sequence comparison of the mouse, human and chicken *engrailed* genes reveal potential functional domains and regulatory regions. Developmental Genetics 13, 345-358.

Martin, G.R., Richman, M., Reinsch, S., Nadeau, J. and Joyner, A.L. (1990). Mapping of the two mouse *engrailed*-like genes: close linkage of *En-1* to *dominant hemimelia* (*Dh*) on Chromosome 1 and of *En-2* to *hemimelic extra-toes* (*Hx*) on Chromosome 5. Genomics 6, 302-308.

McMahon, A.P., Joyner, A.L., Bradley, A. and McMahon, J.A. (1992). The mid-hindbrain phenotype of *wnt-1⁻/wnt-1⁻* mice results from stepwise deletion of *engrailed* expressing cells by 9.5 days *post-coitum*. Cell 69, 581-595.

En-2 [m]

Species: *Mus musculus*

Chromosomal location: chromosome 5, 1.1 cM proximal to *Hx*

Other names: *Mo-en.2*

Cognate genes: *EN2* (human, chromosome 7q36) *En-2* (chicken, *Xenopus*), *Eng2*, *Eng3*, (zebrafish)

Type of homeobox: *en*-type

Accession number: -

■ Origin and description

The *En-2* gene was cloned by screening a mouse genomic DNA library with a mouse *En-1* homeobox probe (Joyner and Martin, 1987). The protein encoded by *En-2* contains five regions of homology that are conserved in all *en*-like genes cloned from organisms as diverse as Platyhelminthes, Annelids, Arthropods, and Chordates (reviewed in Joyner and Hanks, 1991; Logan et al., 1992). The largest conserved region is the homeodomain. Most organisms contain one or two *en*-like genes, with the exception of zebrafish which contains three. In all vertebrates the two genes are referred to as *En-1* and *En-2*, except in human (EN1 and 2) and zebrafish (*Eng1,2,3*). Based on amino acid sequence comparisons, the third zebrafish gene, *Eng3*, appears to have arisen from a recent duplication of *Eng2* (Ekker et al., 1992).

■ Expression

The expression of *En-2* (and its vertebrate cognate genes) has been studied by RNA *in situ* and antibody staining in mouse, chicken, *Xenopus* and zebrafish. Two antibodies have been used: αEnhb-1, which was raised against a C-terminal homeodomain fragment of the mouse *En-2* protein and recognizes *En-1* and *En-2* proteins in mouse, chicken and *Xenopus* and *Eng2* and *Eng3* in zebrafish and 4D9 which was raised against the *Drosophila inv* protein and recognizes the homeodomains of chicken and *Xenopus En-2* and zebrafish, *Eng1, 2, 3*. The expression patterns of the cognate genes are all similar (reviewed in Joyner and hanks, 1991). In mouse, *En-2* is first expressed at the 4-5 somite stage in two dorsal patches in the neural folds (McMahon et al., 1992). Expression soon expands to a band of cells that later marks the mid-hindbrain junction region. *En-2* continues to be expressed in this region throughout embryonic and early postnatal development. However, there is a gradual transition from a spatially restricted band to cell type-specific expression in groups of neurons involved in motor control (Davis et al., 1988; Davis and Joyner, 1988; Davidson et al., 1988; Davis et al., 1991). In the cerebellum, *En-2* is first broadly expressed during embryogenesis and then becomes restricted to spatial domains during early postnatal stages (Millen and Joyner, unpublished). In the adult, *En-2* is expressed in the granular and molecular layers (Davis et al., 1988). *En-2* has also been detected in myoblasts in the first branchial arch (Logan et al., 1993).

■ Function

A mouse mutant (*En-2*hd) lacking the homeobox exon of *En-2* was made using gene targeting in ES cells (Joyner et al., 1991). *En-2*hd homozygous mutants are viable, can breed and show no obvious behavioral defects. Histological analysis of the brain has shown that the pattern of folds in the cerebellum is abnormal. Analysis of the development of the folds during postnatal stages has indicated that a number of fissures do not form properly causing fusions of folds (Millen and Joyner, unpublished). *En-2* is therefore required for normal patterning of the cerebellum folds. This may be due to an embryonic defect that is manifested after birth.

■ References

Davidson, D., Graham, E., Sime, C. and Hill, R. (1988). A gene with sequence similarity to *Drosophila engrailed* is expressed during the development of the neural tube and vertebrae in the mouse. Development 104, 305-316.

Davis, C.A. and Joyner, A.L. (1988). Expression patterns of the homeo box-containing genes *En-1* and *En-2* and the proto-oncogene *int-1* diverge during mouse development. Genes and Development 2, 1736-1744.

Davis, C.A., Holmyard, D.P., Millen, K.J. and Joyner, A.L. (1991). Examining pattern formation in mouse, chicken and frog embryos with an *En*-specific antiserum. Development 111, 287-298.

Davis, C.A., Noble-Topham, S.E., Rossant, J., Joyner, A.L. (1988). Expression of the homeo box containing gene *En-2* delineates a specific region of the developing mouse brain. Genes and Development 2, 361-371.

Ekker, M., Wegner, J., Akimenko, M.A. and Westerfield, M. (1992). Coordinate embryonic expression of three zebrafish *engrailed* genes. Development 116, 1001-1010.

Fjose, A., Njolstad, P.R., Nornes, S., Molven, A. and Krauss, S. (1992). Structure and early embryonic expression of the zebrafish *engrailed*-2 gene. Mech. Dev. 39, 51-62.

Hemmati-Brivanlou, A., de la Torre, J.R., Holt, C., Stewart, R.M. and Harland, R.M. (1991). Cephalic expression and molecular characterization of *Xenopus En*-2. Development 111, 715-724.

Joyner, A.L., Herrup, K., Auerbach, B.A., Davis, C.A. and Rossant, J. (1991). Subtle cerebellar phenotype in mice homozygous for a targeted deletion of the *En-2* homeobox- Science 251, 1239-1243.

Joyner, A.L. and Hanks, M. (1991). The *engrailed* genes: Evolution of function. Seminars in Developmental Biology 2, 435-445.

Joyner, A.L., Kornberg, T., Coleman, K.G., Cox, D.R., Martin, G.R. (1985). Expression during embryogenesis of a mouse gene with sequence homology to the *Drosophila engrailed* gene. Cell 43, 29-37.

Joyner, A.L., Martin, G.R. (1987). *En-1* and *En-2*, two mouse genes with sequence homology to the *Drosophila engrailed* gene: expression during embryogenesis. Genes and Development 1, 29-38.

Logan, C., Hanks, M.C, Noble-Topham, S., Nallainathan, D., Provart, N.J. and Joyner, A.L. (1992). Cloning and sequence comparison of the mouse, human and chicken *engrailed* genes reveal potential functional domains and regulatory regions. Developmental Genetics 13, 345-358.

Logan, C., Khoo, W., Cado, D. and Joyner, A.L. (1993). Two enhancer regions in the mouse *En-2* locus direct expression to the mid/hindbrain region and mandibular myoblasts. Development, 117, 905–918.

Logan, C., Willard, H.F., Rommens, J.M. and Joyner, A.L. (1989). Chromosomal localization of the human homeo box-containing genes, EN1 and EN2. Genomics 4, 206-209.

Martin, G.R., Richman, M., Reinsch, S., Nadeau, J. and Joyner, A.L. (1990). Mapping of the two mouse *engrailed*-like genes: close linkage of *En-1* to *dominant hemimelia* (*Dh*) on Chromosome 1 and of *En-2* to *hemimelic extra-toes* (*Hx*) on Chromosome 5. Genomics 6, 302-308.

McMahon, A.P., Joyner, A.L., Bradley, A. and McMahon, J.A. (1992). The mid-hindbrain phenotype of *wnt-1*⁻/*wnt-1*⁻ mice results from stepwise deletion of *engrailed* expressing cells by 9.5 days *post-coitum*. Cell 69, 581-595.

Poole, S.J., Law, ML., Kao, F.G. and Lau, Y.F. (1989). Isolation and chromosomal localization of the human *En-2* gene. Genomics 4, 225-231.

En-1a [x]

Species: *Xenopus laevis*
Chromosomal location: -
Other names: *engrailed*
Cognate genes: *En-1* (mouse, human)
Type of homeobox: *en*-type
Accession number: X59123

■ Origin and description

Homeobox region only, cloned from genomic DNA via PCR using degenerate primers for en-like genes (Holland and Williams, 1990).

■ References

Holland, P.W.H. and Williams, N.A. (1990). Conservation of engrailed-like homeobox sequences during vertebrate evolution. FEBS Lett. 277, 250-252.

En-1b [x]

Species: *Xenopus laevis*
Chromosomal location: -
Other names: *engrailed*
Cognate genes: *En-1* (mouse, human)
Type of homeobox: *en*-type
Accession number: X59124

■ Origin and description

Homeobox region only, cloned from genomic DNA via PCR using degenerate primers for en-like genes (Holland and Williams, 1990).

■ References

Holland, P.W.H. and Williams, N.A. (1990). Conservation of engrailed-like homeobox sequences during vertebrate evolution. FEBS Lett. 277, 250-252.

En-1 [zf]

Species: *Brachydanio rerio* (zebrafish)
Chromosomal location: -
Other names: *engrailed*
Cognate genes: *En-1* (mouse, human)
Type of homeobox: *en*-type
Accession number: X59125

■ Origin and description

Homeobox region only, cloned from genomic DNA via PCR using degenerate primers for en-like genes (Holland and Williams, 1990).

■ References

Holland, P.W.H. and Williams, N.A. (1990). Conservation of engrailed-like homeobox sequences during vertebrate evolution. FEBS Lett. 277, 250-252.

Species: *Brachydanio rerio* (zebrafish)
Chromosomal location: -
Other names: -
Cognate genes: *En-2* (mouse, human)
Type of homeobox: *en*-type
Accession number: X59126

■ Origin and description

Homeobox region only, cloned from genomic DNA via PCR using degenerate primers for en-like genes (Holland and Williams, 1990).

■ References

Holland, P.W.H. and Williams, N.A. (1990). Conservation of engrailed-like homeobox sequences during vertebrate evolution. FEBS Lett. 277, 250-252.

■ Expression

The developmental expression of *eng*-2 has been studied by Northern blotting (Fjose et al., 1988) and *in situ* hybridization (Njølstad and Fjose, 1988; Fjose et al., 1992). Transcripts are detected within a transverse stripe at the midbrain-hindbrain junction and in a small group of cells within each myotome.

■ References

Fjose, A., Eiken, H.G., Njølstad, P.R., Molven, I. and Hordvik, I. (1988). A zebrafish *engrailed*-like sequence expressed during embryogenesis. FEBS Lett. 231, 355-360.

Fjose, A., Njølstad, P.R., Nornes, S., Molven, A. and Krauss, S. (1992). Structure and early embryonic expression of the zebrafish *eng*-2 gene. Mech. Dev. 39, 51–62.

Hemmati-Brivanlou, A., de la Torre, J.R., Holt, C. and Harland, R.M. (1991). Cephalic expression and molecular characterization of *Xenopus En*-2. Development 111, 715-724.

Njølstad, P.R. and Fjose, A. (1988). *In situ* hybridization patterns of zebrafish homeobox gene homologous to *Hox*-2.1 and *En*-2 of mouse. Biochem. Biophys. Res. Commun. 157, 426-432.

Patel, N.H., Martin-Blanco, E., Coleman, K.G., Poole, S.J., Ellis, M.C., Kornberg, T. and Goodman, C.S. (1989). Expression of engrailed proteins in Arthropods, Annelids and Chordates. Cell 58, 955-968.

Species: *Brachydanio rerio* (zebrafish)
Chromosomal location: -
Other names: *zf-en*
Cognate genes: *En-2* (mouse), *XEn-2* (*Xenopus*)
Type of homeobox: *en*-type`
Accession number: -

■ Origin and description

The *eng*-2 gene was cloned by cross-hybridization with the *engrailed* homeobox sequence (Fjose et al., 1988). A genomic lambda clone (C28) was isolated which contains a homeobox closely related to the mouse *En*-2 gene. Among several cDNA clones which were identified by immunohistochemical screening of a zebrafish embryonic expression library with the 4D9 antibody (Patel et al., 1989), one clone (En2-c1) was found to derive from the *eng*-2 gene (Fjose et al., 1992). This clone contains the entire protein coding region and the corresponding coding sequences were identified in the genomic (C28) clone. The predicted zebrafish *Eng*-2 protein shares 65.3% identity with the XEn-2 homologue in *Xenopus* (Hemmati-Brivanlou et al., 1991; Fjose et al., 1992).

Species: *Drosophila melanogaster*
Chromosomal location: 63F (chromosome 3L)
Other names: -
Cognate genes: -
Type of homeobox: -
Accession number: -

■ Origin and description

This gene had been isolated by PCR using degenerate primers derived from the homeobox of the *cut* gene (R. Bodmer, unpublished). The homeodomain is unrelated to any other, the closest relative is the *gsbBSH4* homeodomain (46% amino acid identity).

■ Expression

esh-1 is first expressed at the germband extended stage in ectodermal stripes in the dorsal half of each body segment. Later expression expands to the ventral ectoderm and becomes weaker.

EST01828 [h]

Species: *Homo sapiens*
Chromosomal location: -
Other names: -
Cognate genes: -
Type of homeobox: *otd*-related
Accession number: M85320

■ Origin and description

Randomly cloned from a cDNA library (Adams et al., 1992). The homeobox sequence is incomplete, since the clone contains an intron in helix 3, in the same position as *unc-4* and *ceh-10* (Bürglin and Barnes, 1992). The best match of *EST01828* is to *otd*.

■ References

Adams, M.D., Dubnick, M., Kerlavage, A.R., Moreno, R., Kelley, J.M., Utterback, T.R., Nagle, J.W., Fields, C. and Venter, J.C. (1992). Sequence identification of 2375 human brain genes. Nature 355, 632-634.
Bürglin, T.R. and Barnes, T.M. (1992). Introns in ESTs. Nature, 357, 367.

eve [d]

Species: *Drosophila melanogaster*
Chromosomal location: Chromosome 2R, 46C
Other names: *even-skipped, S72*
Cognate genes: *Xhox3* (*Xenopus*) *Evx1, 2* (mouse), *EVX1,2* (human), *eveC* (cnidaria), *even-skipped* (grasshopper)
Type of homeobox: *eve*-type
Accession number: X05138, M14767

■ Origin and description

Mutations in *even-skipped* were identified through their embryonic lethal segmentation phenotype (Nüsslein-Volhard and Wieschaus, 1980). The gene was isolated in low stringency screens with homeobox sequences from several homeotic genes (Macdonald et al., 1986; Frasch et al., 1987). The gene contains a 71bp intron and encodes a protein with 376aa. The homeodomain is located in the N-terminal half. The mutations eveID19 (ts) and eveIIR59 caused single amino acid substitutions in the homeo-domain (Frasch et al., 1988).

■ Expression

In the blastoderm stage, *eve* is expressed in seven transverse stripes between 15% and 70% egg length. The pattern is complementary to that of *fushi tarazu*. During gastrulation, the *eve* stripes form sharp anterior limits which define the parasegmental borders. After gastrulation, each stripe splits into two, resulting in a 14-striped pattern with seven strong and seven weak stripes. The striped expression disappears during the germ band elongation stage (Macdonald et al., 1986; Frasch et al., 1987). In subsequent stages of embryogenesis, *eve* is expressed in the anal pad, in a subset of neurons of the CNS, in the pericardial cells of the heart, and in a subset of dorsal muscles (Frasch et al., 1987). The striped expression of *eve* is controlled by distinct regulatory elements in the promoter region (Harding et al., 1989; Goto et al., 1989). Its expression is regulated by gap genes and the pair rule genes *hairy, runt,* and *eve* itself (Frasch and Levine, 1987). Binding sites for gap gene products, *eve* and *bicoid* have been characterized in 5′ regulatory elements of the *eve* promoter (Harding et al., 1989; Stanojevic et al., 1991).

■ Function

eve functions during segmentation and neurogenesis. In weak *eve* mutants, the odd-numbered parasegments are deleted. Strong mutants lack both odd-numbered and even-numbered parasegments and are therefore unsegmented (Nüsslein-Volhard et al., 1985). *eve* represses the segment polarity gene *wingless* and is required to activate expression of the segment polarity gene *engrailed* (Ingham et al., 1988; Frasch et al., 1988). The expression patterns of pair rule genes are also dependent on *eve* function. Experiments involving ectopic expression of *eve* in early embryos suggested that *wingless, fushi tarazu, odd-skipped, runt* and *paired* are direct targets of *eve*, while *engrailed* and *hairy* may be indirect targets (Manoukian and Krause, 1992). In the CNS, *eve* is required for the normal development of the aaC and RP2 neurons. The pCC neurons that also express *eve* appear to develop normally in the absence of *eve* function (Doe et al., 1988). *In vitro*, the eve protein binds with equal preference to A+T rich and G+C rich consensus sequences (Hoey et al., 1988). *In vitro* transcription and co-transfection experiments in tissue culture cells showed that *eve* can act as a transcriptional repressor (Biggin and Tjian, 1989).

■ References

Biggin, M. and Tjian, R. (1989). A purified *Drosophila* homeodomain protein represses transcription *in vitro*. Cell 58, 433-440.
Doe, C.Q., Smouse, D. and Goodman, C.S. (1988). Control of neuronal fate by the *Drosophila* segmentation gene even-skipped. Nature 333, 376-378.
Frasch, M., Hoey, T., Rushlow, C., Doyle, H.J. and Levine, M. (1987). Characterization and localization of the even-skipped protein of *Drosophila*. EMBO J. 6, 749-759.
Frasch, M. and Levine, M. (1987). Complementary patterns of even-skipped and fushi tarazu expression involve their

differential regulation by a common set of segmentation genes in *Drosophila*. Genes Dev. 1, 981-995.

Frasch, M.R.W., Tugwood, J. and Levine, M. (1988). Molecular analysis of even-skipped mutants in *Drosophila* development. Gen. Dev. 2, 1824-1838.

Goto, T., Macdonald, P. and Maniatis, T. (1989). Early and late periodic patterns of even-skipped expression are controlled by distinct regulatory elements that respond to different spatial cues. Cell 57, 413-422.

Harding, K., Hoey, T., Warrior, R. and Levine, M. (1989). Autoregulatory and gap response elements of the even-skipped promoter of *Drosophila*. EMBO J. 8, 1205-1212.

Hoey, T., Warrior, R., Manak, J. and Levine, M. (1988). DNA-binding activities of the *Drosophila melanogaster* even-skipped protein are mediated by its homeodomain and influenced by protein context. Mol. Cell. Biol. 8, 4598-4607.

Ingham, P.W., Baker, N.E. and Martinez-Arias, A. (1988). Regulation of segment polarity genes in the *Drosophila* blastoderm by fushi tarazu and even-skipped. Nature 331, 73-75.

Macdonald, P.M., Ingham, P. and Struhl, G. (1986). Isolation, structure and expression of even-skipped: a second pair-rule gene of *Drosophila* containing a homeobox. Cell 47, 721-734.

Manoukian, A.S. and Krause, H.M. (1992). Concentration-dependent activities of the even-skipped protein in *Drosophila* embryos. Genes Dev. 6, 1740-1751.

Nüsslein-Volhard, C., Kluding, H. and Jürgens, G. (1985). Genes affecting the segmental subdivision of the *Drosophila* embryo. Cold Spring Harb. Symp. Quant. Biol. 50, 145-154.

Nüsslein-Volhard, C. and Wieschaus, E. (1980). Mutations affecting segment number and polarity in *Drosophila*. Nature 287, 795-801.

Stanojevic, D., Small, S. and Levine, M. (1991). Regulation of a segmentation stripe by overlapping activators and repressors in the *Drosophila* embryo. Science 254, 1385-1387.

eve [Sa]

Species: *Schistocerca americana* (grasshopper)
Chromosomal location: -
Other names: *even-skipped, GH-eve*
Cognate genes: -
Type of homeobox: *eve*-type
Accession number: Z11845

■ Origin and description

This gene was cloned by first isolating a portion of the homeobox by PCR amplification using grasshopper embryo first strand synthesis cDNA and oligos designed to amplify *even-skipped* class homeoboxes. This PCR product was then used to screen a grasshopper cDNA library. The longest cDNA, 2140 bp, was completely sequenced. Translation of the sequence reveals an *even-skipped* type homeodomain as well as conserved regions located C-terminal to the homeodomain (Patel et al., 1992).

■ Expression

Antibodies made against the grasshopper *even-skipped* protein were used to examine its expression during development. As in *Drosophila*, grasshopper *even-skipped* is expressed in a subset of neurons, the dorsal mesoderm, and the anal pads during mid-embryogenesis. Early in development, however, grasshopper *even-skipped* does not show any obvious pair-rule pattern of expression like its *Drosophila* counterpart. Instead, grasshopper *even-skipped* is expressed in the posterior portions of the germband well ahead of the developing *engrailed* stripes (Patel et al., 1992; Patel, 1992). This posterior expression pattern is reminiscent of the expression of the mouse *Evx-1* gene (Bastian and Gruss, 1990; Dush and Martin, 1992).

■ References

Bastian, H. and Gruss, P. (1990). A murine even-skipped homologue, Evx-1, is expressed during early embryogenesis and neurogenesis in a biphasic manner. EMBO J. 9, 1839-1852.

Dush, M.K. and Martin, G.R. (1992). Analysis of mouse Evx genes: Evx-1 displays a graded expression in the primitive streak. Dev. Biol. 151, 273-287.

Patel, N.H., Ball, E.E. and Goodman, C.S. (1992). Changing role of even-skipped during the evolution of insect pattern formation. Nature 357, 339-342.

Patel, N.H. (1992). Evolution of insect pattern formation: a molecular analysis of short germband segmentation. In Spradling, A.C. (ed.): Evolutionary conservation of developmental mechanisms. New York: John-Wiley, pp. 85-110.

Evx-1 [h]

Species: *Homo sapiens*
Chromosomal location: 7p14-p21
Other names: -
Cognate genes: *Evx-1* (mouse), *Xhox 3* (*Xenopus*)
Type of homeobox: *eve*-related
Accession number: X60655

■ Origin and description

A human genomic library in EMBL3 was screened with *EVX2* sequence at low stringency. Restriction analysis of positive clones revealed that one of these, *EVX1*, did not contain *EVX2* sequences. Subcloning and sequencing showed that it actually contained sequences different from *EVX2* and closely related to the murine *Evx-1* reported sequences.

■ Expression

The expression of *EVX1* was analyzed in human 7-week embryos by Northern blot hybridization of polyadenylated RNA. A 3 Kb transcript is detectable in the CNS. *EVX1* expression was further investigated in NT2/D1 cells using the RNase protection technique. *EVX1* is weakly expressed and properly spliced in differentiated NT2/D1 cells after eight days of retinoic acid.

■ References

Bastian, H. and Gruss, P. (1990). A murine even-skipped homologue, Evx-1, is expressed during early embryogenesis and neurogenesis in a biphasic manner. EMBO J. 9, 1839-1852.

D'Esposito, M., Morelli, F., Acampora, D., Migliaccio, E., Simeone, A. and Boncinelli, E. (1991). EVX2, a human homeobox gene homologous to the even-skipped segmentation gene, is localized at the 5′ end of HOX4 locus on chromosome 2. Genomics 10, 43-50.

Faiella, A., D'Esposito, M., Rambaldi, M., Acampora, D., Balsofiore, S., Stornaiuolo, A., Mallamaci, A., Migliaccio, E., Gulisano, M., Simeone, A. and Boncinelli, E. (1991). Isolation and mapping of EVX1, a human homeobox gene homologous to *even skipped*, localized at the 5′ end of HOX 1 locus on chromosome 7. Nucl. Acids Res. 19, 6541-6545.

Evx-1 [m]

Species: *Mus musculus*
Chromosomal location: Chromosome 6
Other names: -
Cognate genes: *eve, Xhox-3*
Type of homeobox: *eve*-type
Accession number: -

■ Origin and description

This gene was isolated by cross-homology with the *Drosophila even-skipped* (*eve*) homeobox (Frasch et al., 1988). It is linked to the *HOXA* complex on chromosome 6 (Bastian et al., 1992). A second gene, *Evx-2*, with nearly 100% homology inside the homeodomain and extended aminoacid conservation outside of the homeodomain were isolated (Bastian et al., 1991). The cognate *Xenopus* gene, *Xhox-3*, shows 100% aminoacid identity inside of the homeodomain (Ruiz i Altaba and Melton, 1989a).

■ Expression

The expression of the *Evx-1* gene was studied by *in situ* and Northern blot hybridization. During embryogenesis, *Evx-1* shows a biphasic expression pattern. From days 7 to 9 p.c. *Evx-1* expression emerges at the posterior end of the embryo within the primitive ectoderm, and later in the mesoderm and neuroectoderm. From days 10 to 12.5 p.c. *Evx-1* transcript is restricted to specific cells within the neural tube and the hindbrain along their entire lengths. No signal could be obtained in adult tissues (Bastian et al., 1990).

■ References

Bastian, H. and Gruss, P. (1990). A murine even-skipped homologue, *Evx-1*, is expressed during early embryogenesis and neurogenesis in a biphasic manner. EMBO J. 9, 1839-1852.

Bastian, H., Gruss, P., Duboule, D., Izpisúa-Belmonte, J.-C. (1992). The murine even-skipped-like gene *Evx-2* is closely linked to the *Hox-4* complex, but is transcribed in the opposite direction. Mammalian Genome 3, 241–243.

Frasch, M., Warrior, R., Tugwood, J. and Levine, M. (1988). Molecular analysis of *even-skipped* mutants in *Drosophila* development. Genes. Dev. 2, 1824-1838.

Ruiz i Altaba, A. and Melton, D.A. (1989a). Bimodal and graded expression of the *Xenopus* homeobox gene *Xhox-3* during embryonic development. Development 106, 173-183.

evx-1 [x]

Species: *Xenopus laevis*
Chromosomal location: -
Other names: *Xhox3*
Cognate genes: Possibly *Evx-1*
Type of homeobox: *eve*-type
Accession number: D10455

■ Origin and description

The *Xhox3* gene was isolated in a genomic screen with a probe derived from the homeobox of the *Drosophila* gene *even skipped*. The homeobox region is split by an intron in helix 3. *Xhox3* is the first vertebrate gene isolated with a homeobox of the *eve* type (Ruiz i Altaba and Melton, 1989a). *Xhox3* encodes a protein with the homeodomain located at the center. NH2-terminal to the homeodomain there are a series of histidines and cysteines that resemble but do not fit the LIM domain consensus. In this region there is also a long stretch of acidic residues that can be predicted to for an alpha helix (Ruiz i Altaba et al., 1991). A similar gene has been detected in mice (*Evx-1*; Bastian and Gruss, 1990).

■ Expression

Xhox3 shows a bimodal expression pattern in embryonic development. *Xhox3* RNA is first expressed zygotically at the mid-blastula transition and is then detected in a graded manner along the A-P axis of the mesoderm with highest levels posteriorly. At the tailbud stage, new expression is detected in the developing CNS with highest expression in the hindbrain (Ruiz i Altaba and Melton, 1989a; Ruiz i Altaba, 1990). *Xhox3* protein is first detected with polyclonal Abs in the mesoderm of late gastrula-early neurula embryos where it shows a graded expression along the A-P and D-V axes with highest levels in the pos-terior-dorsal (circumblastoporal) region (Ruiz i Altaba et al., 1991). There is very low expression, if any, in the prechordal plate. As in the mesoderm, *Xhox3* protein expression correlates that of the RNA as seen both by RNase protections following dissections and whole-mount *in situ* hybridizations (not shown). In the CNS there is a very low level of expression in a subset of spinal cord neurons but higher expression in the hindbrain and later in the midbrain. In the hindbrain of tadpoles and metamorphosing larvae, *Xhox3* expression exhibits different rhombomeric and graded patterns (Ruiz i Altaba et al., 1991). *Xhox3* is also expressed in a gradient in the proctodeum and developing gut and in the tail bud (Ruiz i Altaba and Melton, 1989a; Ruiz i Altaba et al., 1991).

■ Function

Xhox3 expression is a rapid response to mesoderm-inducing factors and the level of induced *Xhox3* expression was used to discover an interaction of peptide growth factors and homeobox genes in mesodermal patterning (Ruiz i Altaba and Melton, 1989c). Overexpression of *Xhox3* by injection of synthetic RNA into developing embryos results in a series of axial defects that derive primarily from actions on the patterning of the mesoderm (Ruiz i Altaba and Melton, 1989b). Injected tadpoles exhibit reduced anterior structures and sometimes lack a head. The axial mesoderm is abnormal resulting in defects in the overlying CNS. Tail bud growth is also affected. Interference with the function of *Xhox3* by injection of specific antibodies into developing embryos results in a series of posterior defects (Ruiz i Altaba et al., 1991). These defects appear to derive from an interference with the development of the posterior mesoderm. Thus *Xhox3* functions in patterning of the A-P mesoderm and appears to be required for the normal development of the posterior region. This function is independent of the effects of retinoic acid on axial pattern since the early expression of *Xhox3* is not affected by early RA-treatments (Ruiz i Altaba and Jessell, 1991). The function of *Xhox3* in other regions of the embryo has not been studied.

■ References

Ruiz i Altaba, A. (1990). Neural expression of the *Xenopus* homeobox gene Xhox3: evidence for a patterning neural signal that spreads through the ectoderm. Development 108, 595-604.

Ruiz i Altaba, A., Choi, T. and Melton, D.A. (1991). Expression of the Xhox3 homeobox protein in *Xenopus* embryos: blocking its early function suggests the requirement of Xhox3 for normal posterior development. Dev. Growth Diff. 33, 651-669.

Ruiz i Altaba, A. and Jessell, T.M. (1991a). Retinoic acid modifies mesodermal patterning in early *Xenopus* embryos. Genes Dev. 5, 175-187.

Ruiz i Altaba, A. and Melton, D.A. (1989a). Bimodal and graded expression of the *Xenopus* homeobox gene Xhox3 during embryonic development. Development 106, 173-183.

Ruiz i Altaba, A. and Melton, D.A. (1989b). Involvement of the *Xenopus* homeobox gene Xhox3 in pattern formation along the anterior-posterior axis. Cell 57, 317-326.

Ruiz i Altaba, A. and Melton, D.A. (1989c). Interaction between peptide growth factors and homeobox genes in the establishment of anterior-posterior polarity in frog embryos. Nature 341, 33-38.

Evx-2 [m]

Species: *Mus musculus*
Chromosomal location: Chromosome 2
Other names: -
Cognate genes: *Evx-2*
Type of homeobox: *eve*-type
Accession number: -

■ Origin and description

This gene was isolated by cross-homology with the *Drosophila even-skipped* gene (Frasch et al., 1988). It constitutes together with *Evx-1* (Bastian et al., 1990), *Xhox-3* (Ruiz i Altaba et al., 1989a), *EVX2* (D'Esposito, M. et al, 1991) and *eve* a family of related genes based on similar homeodomain sequences. The chromosomal location could be mapped in linkage to the *HOXD* complex on chromosome 2 (Bastian et al., 1992).

■ References

Bastian, H. and Gruss, P. (1990). A murine even-skipped homologue, *Evx-1*, is expressed during early embryogenesis and neurogenesis in a biphasic manner. EMBO J. 9, 1839-1852.

Bastian, H., Gruss, P., Duboule, D., Izpisúa-Belmonte, J.-C. (1992). The murine even-skipped-like gene Evx-2 is closely linked to the *Hox-4* complex, but is transcribed in the opposite direction. Mammalian Genome 3, 241–243.

D'Esposito, M., Morelli, F., Acampora, D., Migliaccio, E., Simone, A. and Boncinelli, E. (1991). EVX2, a human homeobox gene homologous to the even-skipped homeotic gene, is localized at the 5′ end of the *Hox-4* locus on chromosome 2. Genomics 10, 43-50.

Frasch, M., Warrior, R., Tugwood, J. and Levine, M. (1988). Molecular analysis of *even-skipped* mutants in *Drosophila* development. Genes. Dev. 2, 1824-1838.

Ruiz i Altaba, A. and Melton, D.A. (1989a). Bimodal and graded expression of the *Xenopus* homeobox gene Xhox-3 during embryonic development. Development 106, 173-183.

ftz [d]

Species: *Drosophila melanogaster*
Chromosomal location: Chromosome 3
map position 3-47.5 Salivary gland
chromosome band 84B1-2
Other names: *fushi tarazu*
Cognate genes: -
Type of homeobox: *Antp*-class
Accession number: -

■ Origin and description

Fushi tarazu (ftz) was first isolated as an embryonic lethal mutation (Wakimoto et al., 1984). Animals survive until the end of embryogenesis and show a pair-rule mutant phenotype; they lack all the even-numbered parasegments from PS2 to 14, i.e. the posterior compartment (P) of T1, for example, and the adjacent anterior compartment (a) of T2, pT3/aA1/etc., resulting in an embryo with only half the number of segments. On the basis of this phenotype, B. Wakimoto named the mutation *fushi tarazu*, which in Japanese means segment deficient or not enough segments. The transcription unit was identified on the chromosomal walks between *Antp* and *Scr* (Garber et al., 1983; Scott et al., 1983) and then cloned and sequenced (Kuroiwa et al., 1984; Laughon and Scott, 1984). Its identity was confirmed by P-element mediated germline transformation (Hiromi et al., 1985). In addition to the coding region which is just over 2 kb in length with two exons and a single 150 bp intron, at least 6 kb of 5' flanking sequences are required for rescue of the *ftz*- phenotype. This indicates that the essential *cis*-regulatory control regions are considerably larger than the coding region. The protein (398 amino acids) has a predicted molecular weight of 43,000 and contains a homeobox which is closely related to the one of *Antp*. Both the mRNA and the protein have short half-lives. The PEST sequences found in the protein might account for its instability (Duncan, 1986). FTZ is heavily phosphorylated on serine and threonine residues (Krause et al., 1988; Krause and Gehring, 1989).

■ Expression

A breakthrough in the understanding of segmentation genes came from *in situ* hybridization studies using *ftz* as a probe, revealing for the first time a pattern of seven stripes on the blastoderm reflecting the embryonic fate map (Hafen et al., 1984). The position and width of the stripes coincide with the even-numbered parasegmental anlagen, which are missing in *ftz*- animals. The stripes cover both the presumptive ectoderm and the mesoderm. The formation of the stripes is preceded at syncytial blastoderm stages by weak labelling of the entire area which later will be resolved into the seven stripes. FTZ protein synthesis lags slightly behind, but is clearly detectable at the cellular blastoderm stage by immunostaining in the nuclei of the cells within the stripes (Carroll and Scott, 1985). At later germband extension stages, expression in specific cells of the CNS becomes detectable (Carroll and Scott, 1985; Hiromi and Gehring, 1987). In contrast to the early embryonic pair-rule pattern, the corresponding neuroblasts and ganglion mother cells are labelled in every segment (Doe et al., 1988). At late embryonic stages, FTZ is expressed in a narrow ring around the hindgut (Krause and Gehring, 1989). No expression has been detected during postembryonic development.

■ Function

Loss-of-function mutations result in embryonic lethality and exhibit a pair-rule phenotype (Jürgens et al., 1984).

All even-numbered parasegments from PS2 to PS14 are missing. The mutant phenotype coincides with the expression pattern. These observations indicate that *ftz* has an essential role in the segmentation process. X-ray induced clones of homozygous *ftz*- cells develop normally with respect to cuticular structures when induced after the blastoderm stage, indicating that *ftz* is required only at early embryonic stages for the segmentation process and becomes dispensable at later stages in the epidermis. Dominant gain-of-function mutants exhibit variable homeotic transformations, e.g. from posterior haltere to posterior wing (*tz*[Rpl]) or from the first to the third abdominal segment (*ftz*[Ual]) (Duncan, 1986). Artificial dominant gain-of-function mutants, in which *ftz* can be expressed ectopically in all cells of the embryo by heat-shock induction, have an "anti-*ftz*" phenotype, i.e. the uneven-numbered parasegments are missing (Struhl, 1985). However, the nature of this phenotypic effect is not understood since the same phenotype can be induced with a *ftz* gene lacking the homeobox (Fitzpatrick et al., 1992).

The function of the *ftz* gene has been studied by germline transformation using *ftz-lacZ* fusions and a large number of *ftz* mutant variants (Hiromi and Gehring, 1987; Furukubo-Tokunaga et al., 1992). In the 5' flanking sequences three major *cis*-regulatory elements have been identified; the zebra element which is closest to the site of transcription initiation, the neurogenic region, which is further upstream, and the upstream enhancer element. The zebra element is capable of directing reporter gene (*lacZ*) expression in a pattern of seven stripes. The neurogenic region is required for the expression in the CNS, and the enhancer enhances the expression in the stripes and has no effect on the expression in the CNS. The enhancer by itself can direct reporter gene expression in a pattern of seven stripes and has an autoregulatory function (see below).

By constructing an artificial *ftz* gene lacking the neurogenic region and introducing this construct into the germline of *ftz*-flies, it could be shown that *ftz* has an essential role in neurogenesis (Doe et al., 1988). In such mutants, which undergo nearly normal segmentation, but do not express *ftz* in the CNS, neurons that normally express *ftz* do not establish the correct axonal connections and undergo a change in cell fate. This experiment proves that *ftz* is involved in determining cell fate in the nervous system. Besides the function of homeobox genes in the specification of the body plan, these genes appear to serve a crucial function in neurogenesis.

The DNA binding specificity of the *ftz* homeodomain has been studied extensively, both *in vitro* and *in vivo*. *In vitro* DNA binding studies, methylation and ethylation interference experiments and NMR spectroscopy (Percival-Smith et al., 1990; Otting et al., 1990) have shown that position 9 in the recognition helix of the homeodomain is important for binding specificity. In *ftz* the amino acid at this position is glutamine (Q50) and contacts the two bases CC preceding the ATTA motif found in the consensus binding site. In *bicoid (bcd)* the corresponding amino acid is lysine (K50) and the consensus binding site is

GGATTA. The affinity of the *ftz* homeodomain is almost two orders of magnitude lower for GGATTA than for CCATTA. However, high affinity can be restored by introducing lysine instead of glutamine at position 9 of the recognition helix (Percival-Smith et al., 1990). These *in vitro* data provided the basis for examining the homeodomain-DNA interaction *in vivo* by studying the expression of *ftz* in the embryo. For this purpose, the enhancer of *ftz* was used (Schier and Gehring, 1992). As mentioned above, the *ftz* enhancer when fused to a *lacZ* reporter gene can generate a β-galactosidase expression pattern of seven stripes in the early embryo. However, in *ftz⁻* embryos no stripes are formed indicating that FTZ protein is required for the expression in stripes. The simplest interpretation of this result is that *ftz* interacts directly with its enhancer and enhances the expression of its own gene via an autocatalytic positive-feedback loop. In order to find out whether the interaction is direct or indirect, a second site-suppression experiment was carried out. The *ftz* enhancer contains multiple *ftz in vitro* binding sites. Mutating three of these binding sites to GGATTA strongly reduces enhancer activity *in vivo*. The effect of this downmutation can be suppressed by introducing the corresponding mutation in *ftz*, leading to the substitution of glutamine by lysine. The *ftz* Gln50->Lys mutant recognizes the GGATTA binding sites and the expression in the seven stripes is restored (Schier and Gehring, 1992). This experiment provides definitive evidence for a direct positive autoregulatory interaction *in vivo*. The FTZ protein is a transcriptional activator that binds directly to its target sites in the enhancer, and enhances transcription of its own gene.

■ References

Carroll, S. and Scott, M. (1985). Localization of the *fushi tarazu* protein during *Drosophila* embryogenesis. Cell 43, 47-57.

Doe, C.Q., Hiromi, Y., Gehring, W.J. and Goodman, C.S. (1988). Expression and function of the segmentation gene *fushi tarazu* during *Drosophila* neurogenesis. Science 239, 170-175.

Duncan, I. (1986). Control of bithorax complex functions by the segmentation gene *fushi tarazu* of *D. melanogaster*. Cell 47, 297-309.

Fitzpatrick V.D., Percival-Smith, A., Ingles, C.J. and Krause, H.M. (1992). Homeodomain-independent activity of the *fushi tarazu* polypeptide in *Drosophila* embryos. Nature 356, 610-612.

Furukubo-Tokunaga, K., Müller, M., Affolter, M., Pick, L., Kloter, U. and Gehring, W.J. (1992). *In vivo* analysis of the helix-turn-helix motif of the *fushi tarazu* homeodomain of *Drosophila melanogaster*. Genes and Dev. 6, 1082-1096.

Garber, R.I., Kuroiwa, A. and Gehring, W.J. (1983). Genomic and cDNA clones of the homeotic locus *Antennapedia* in *Drosophila*. EMBO J. 2, 2027-2036.

Hafen, E., Kuroiwa, A. and Gehring, W.J. (1984). Spatial distribution of transcripts from the segmentation gene *fushi tarazu* during *Drosophila* embryonic development. Cell 37, 833-841.

Hiromi, Y. and Gehring, W.J. (1987). Regulation and function of the *Drosophila* segmentation gene *fushi tarazu*. Cell 50, 963-974.

Hiromi, Y., Kuroiwa, A. and Gehring, W.J. (1985). Control elements of the *Drosophila* segmentation gene *fushi tarazu*. Cell 43, 603-613.

Jürgens, G., Wieschaus, E., Nüsslein-Volhard, C. and Kluding, H. (1984). Mutations affecting the pattern of the larval cuticle in *Drosophila melanogaster*. II. Zygotic loci on the third chromosome. Roux's Arch. Dev. Biol. 193, 283-295.

Krause, H.M., Klemenz, R. and Gehring W.J. (1988). Expression, modification and localization of the *fushi tarazu* protein in *Drosophila* embryos. Genes and Dev. 2, 1021-1036.

Krause, H.M. and Gehring, W.J. (1989). Stage-specific phosphorylation of the *fushi tarazu* protein during *Drosophila* development. EMBO J. 8, 1197-1204.

Kuroiwa, A., Hafen, E. and Gehring, W.J. (1984). Cloning and transcriptional analysis of the segmentation gene *fushi tarazu* in *Drosophila*. Cell 37, 825-831.

Laughon, A. and Scott, M.P. (1984). Sequence of a *Drosophila* segmentation gene: Protein structure homology with DNA-binding proteins. Nature 310, 23-31.

Otting, G., Qian, Y.Q., Billeter, M., Müller, M., Affolter, M., Gehring, W.J. and Wüthrich, K. (1990). Protein-DNA contacts in the structure of a homeodomain-DNA complex determined by nuclear magnetic resonance spectroscopy in solution. EMBO J. 9, 3085-3092.

Percival-Smith, A., Müller, M., Affolter, M. and Gehring, W.J. (1990). The interaction with DNA of wild-type and mutant *fushi tarazu* homeodomains. EMBO J. 9, 3967-3974.

Schier, A.F. and Gehring, W.J. (1992). Direct homeodomain-DNA interaction in the autoregulation of the *fushi tarazu* gene. Nature 356, 804-807.

Scott, M.P., Weiner, A.J., Polisky, B.A., Hazelrigg, T.I., Pirrotta, V., Scalenghe, F. and Kaufman, T.C. (1983). The molecular organization of the *Antennapedia* complex of *Drosophila*. Cell 35, 763-776.

Struhl, G. (1985). Near-reciprocal phenotypes caused by inactivation or indiscriminate expression of the *Drosophila* segmentation gene *ftz*. Nature 318, 677.

Wakimoto, B.T., Turner, F.R. and Kaufman, T.C. (1984). Defects in embryogenesis in mutants associated with the *Antennapedia* gene complex of *Drosophila melanogaster*. Devel. Biol. 102, 147-172.

ftz [dh]

Species: *Drosophila hydei*

Chromosomal location: -

Other names: *fushi tarazu*

Cognate genes: -

Type of homeobox: *Antp*-class

Accession number: X56038

■ Origin and description

The *fushi tarazu (ftz)* gene of *Drosophila hydei* was cloned on the basis of its sequence homology to sequences from the first exon of the *Drosophila melanogaster* gene (Maier et al., 1990). The overall sequence organization of the *ftz* genes from these two species is very similar but the *Drosophila melanogaster* gene is inverted relative to the one in *Drosophila hydei*. The breakpoints are close to the boundaries delimiting the functional *ftz* gene as defined by transformation experiments in *Drosophila melanogaster*. Sequencing of the *cis*-regulatory region revealed strong homologies over the entire 6 kb of the *ftz* upstream control region with the best match in the enhancer element which contains blocks of highly conserved sequences. The more proximal neurogenic and the zebra element are also conserved, whereas the region that was found to be dispensable in *Drosophila melanogaster* (Hiromi and Gehring, 1987) is absent in *Drosophila hydei*.

■ Function

Transformation experiments show that *Drosophila hydei ftz* gene products can restore *ftz* function in *Drosophila melanogaster* and that the transacting factors from *Drosophila melanogaster* can recognize and control the *cis*-regulatory elements of *Drosophila hydei* (Maier et al., 1990). Therefore, the regulatory network has been conserved.

■ References

Hiromi, Y. and Gehring, W.J. (1987). Regulation and function of the *Drosophila* segmentation gene *fushi tarazu*. Cell 50, 963-974.

Maier, D., Preiss, A. and Powell, J.R. (1990). Regulation of the segmentation gene *fushi tarazu* has been functionally conserved in *Drosophila*. EMBO J. 9, 3957-3966.

gsb, *gsbn*   [d]

Species: *Drosophila melanogaster*

Chromosomal location: Chromosome 2, 60E9, F1

Other names: *gooseberry*, *gsb-neuro*, *BSH*9, *gsb-d* and *gsb-BSH*4, *gsb-p*

Cognate genes: -

Type of homeobox: *prd*-type

Accession number: -

■ Origin and description

A search for genes sharing homologous domains with the *paired* (*prd*) gene discovered two genes rather than one at the *gooseberry* (*gsb*) locus (Bopp et al., 1986), genetically characterized to belong to the segment-polarity class of segmentation genes (Nüsslein-Volhard and Wieschaus, 1980). Both genes, originally called *gsb-BSH*9 and *gsb-BSH*4, share a paired domain and an extended homeodomain with *prd*, which are both of the *prd*-type. They have been identified by two overlapping *gsb* deficiencies as two genes that exhibit a segmentally repeated expression pattern (Bopp et al., 1986; Baumgartner et al., 1987; Côté et al., 1987) regulated by pair-rule genes

(Baumgartner, 1988; Bopp et al., 1989). The two transcription units of *gsb* are divergently transcribed and separated by about 10 kb (Li et al., pers. com.). All known *gsb* mutants are deficiencies (Nüsslein-Volhard et al., 1984; Côté et al., 1987). Their cuticular phenotype exhibits a typical segment-polarity pattern (Nüsslein-Volhard and Wieschaus, 1980) while in the central nervous system, most strikingly, posterior commissures are nearly nonexistent (Patel et al., 1989).

■ Expression

The two *gsb* genes are expressed at single segment periodicity in different though partly overlapping subsets of cells in the posterior half of each segment during embryonic development (Bopp et al., 1986; Côté et al., 1987). Their expression patterns evolve in a rather complex way during cellularization, gastrulation, and germ band elongation (Baumgartner et al., 1987; Côté et al., 1987; Gutjahr et al., 1993). One gene, *gsb* (previously called *gsb-BSH*9 or *gsb-distal*), is predominantly expressed in the epidermis, but also in cells of the developing mesoderm and central nervous system. The other gene, *gsb neuro* (*gsbn*, previously named *gsb-BSH*4 or *gsb-proximal*), is expressed in specific neuroblasts and probably their progeny. At late developmental stages (after germ band retraction), a low expression of *gsbn* is observed in epidermal cells (Baumgartner et al., 1987; Gutjahr et al., 1993).

■ References

Baumgartner, S. (1988). Patterns of *paired* and *gooseberry* transcripts in wild-type and segmentation mutant embryos imply a combinatorial regulation of segmentation genes in *Drosophila*. Thesis, University of Basel.

Baumgartner, S., Bopp, D., Burri, M. and Noll, M. (1987). Structure of two genes at the gooseberry locus related to the *paired* gene and their spatial expression during *Drosophila* embryogenesis. Genes Dev. 1, 1247-1267.

Bopp, D., Burri, M., Baumgartner, S., Frigerio, G. and Noll, M. (1986). Conservation of a large protein domain in the segmentation gene *paired* and in functionally related genes of *Drosophila*. Cell 47, 1033-1040.

Bopp, D., Jamet, E., Baumgartner, S., Burri, M. and Noll, M. (1989). Isolation of two tissue-specific *Drosophila* paired box genes, *pox meso* and *pox neuro*. EMBO J. 8, 3447-3457.

Côté, S., Preiss, A., Haller, J., Schuh, R., Kienlin, A., Seifert, E. and Jäckle, H. (1987). The *gooseberry-zipper* region of *Drosophila*: five genes encode different spatially restricted transcripts in the embryo. EMBO J. 6, 2793-2801.

Gutjahr, T., Frei, E. and Noll, M. (1993). Complex regulation of early *paired* expression: Initial activation by gap genes and pattern modulation by pair-rule genes. Development, 117, 609-623.

Nüsslein-Volhard, C. and Wieschaus, E. (1980). Mutations affecting segment number and polarity in *Drosophila*. Nature 287, 795-801.

Nüsslein-Volhard, C., Wieschaus, E. and Kluding, H. (1984). Mutations affecting the pattern of the larval cuticle in *Drosophila melanogaster*. I. Zygotic loci on the second chromosome. Roux's Arch. Dev. Biol. 193, 267-282.

Patel, N.H., Schafer, B., Goodman, C.S. and Holmgren, R. (1989). The role of segment polarity genes during *Drosophila* neurogenesis. Genes Dev. 3, 890-904.

gsc [c]

Species: *Gallus gallus*
Chromosomal location: -
Other names: *goosecoid*
Cognate genes: see *gsc*
Type of homeobox: *prd*-type; related to Drosophila *bicoid* and *gooseberry*
Accession number: X70471

■ Origin and description

Chick *goosecoid* was isolated by cross-homology with the *Xenopus goosecoid* cDNA (Blumberg et al., 1991). Its homeodomain is most closely related to that of the two Drosophila genes *bicoid* (Frigerio et al. 1986) and *gooseberry* (Bopp et al., 1986). Most notably it contains a lysine residue in position 50 of the homeodomain, as does the homologous genes in *Xenopus* (Blumberg et al., 1991), mouse (Blum et al. 1992), zebrafish (S. Schulte-Merker, E. M. De Robertis and C. Nüsslein-Volhard, in preparation) and human (M. Blum and E. M. De Robertis, unpublished). This predicts a DNA binding specificity similar to that of Drosophila *bicoid* (Hanes and Brent, 1989; Treisman et al., 1989; Blumberg et al., 1991).

■ Expression

Chicken *goosecoid* expression is first detected in the unincubated egg associated with Koller's sickle, a crescent-shaped thickening located at the edge of the posterior marginal zone. This early pattern of expression is confined to a small group of cells, previously unnoticed, located betwen the epiblast and the forming hypoblast. Later, expression is found in Hensen's node, traditionally considered the chick organizer, and finally expression leaves the node to occupy the anteriormost region of the head process, cells that contribute to the prechordal plate at the midline of the pharyngeal mesendoderm (Izpisùa-Belmonte et al., 1993).

■ Function

Like the frog (Cho et al., 1991) and mouse (Blum et al., 1992) genes chick *goosecoid* is a marker for the organizer tissue of the gastrula. Chick *goosecoid* mRNA is able, when injected into xenopus embryos, of generating a secondary axis. Transplantation of Koller's sickle or anterior primitive streak cells expressing *goosecoid* between quail donors and chick hosts shows that these cells have the ability to form an ectopic primitive streak containing host cells that also expresses *goosecoid*. Moreover, fate mapping experiments indicates that these regions expressing *goosecoid* in the early chick are related by cell lineage (Izpisùa-Belmonte et al., 1993).

■ References

Blum, M., Gaunt, S., Cho, K. W. Y., Steinbeisser, H., Bittner, D., Blumberg, B., and De Robertis, E. M. (1992) Gastrulation in the mouse: the role of the homeobox gene goosecoid. Cell 69, 1097-1106.

Blumberg, B., Wright, C.V.E., De Robertis, E.M. and Cho, K.W.Y. (1991). Organizer-specific homeobox genes in *Xenopus laevis* embryos. Science 253, 194-196.

Bopp, D. Burri, M., Baumgartner, S., Frigerio, G. and Noll, M. (1986). Conservation of a large protein domain in the segmentation gene *paired* and in functionally related genes of Drosophila. Cell 47, 1033-1040.

Cho, K.W.Y., Blumberg, B., Steinbeisser, H. and De Robertis, E.M. (1991). Molecular nature of Spemann's organizer: the role of the *Xenopus* homeobox gene *goosecoid*. Cell 67, 1111-1120.

Frigerio, G. Burri, M., Bopp, D., Baumgartner, S. and Noll, M. (1986). Structure of the segmentation gene *paired* and the Drosophila PRD gene set as part of a gene network. Cell 47, 735-746.

Hanes, S. D. and Brent, R. (1989). DNA specificity of the *bicoid* activator protein is determined by homeodomain recognition helix residue 9. Cell 57, 1275-1283.

Izpisùa-Belmonte, J.-C., De Robertis, E. M., Storey, K. G. and Stern, C. D. (1993). The homeobox gene *goosecoid* and the origin of organizer cells in the early chick blastoderm. Cell 74, 645–666.

Treisman, J., Gönczy, P., Vashishtha, M., Harris, E. and Desplan, C. (1989). A single amino acid can determine the DNA binding specificity of homeodomain proteins. Cell 59, 553-562.

gsc [m]

Species: *Mus musculus*
Chromosomal location: Chromosome 12
Other names: *goosecoid*
Cognate genes: see *gsc*
Type of homeobox: *prd*-type; related to Drosophila *bicoid* and *gooseberry*
Accession number: M85271

■ Origin and description

Mouse *goosecoid* was isolated by cross-homology with the *Xenopus goosecoid* cDNA (Blum et al., 1992, Blumberg et al., 1991). Its homeodomain is most closely related to that of the two *Drosophila* genes *bicoid* (Frigerio et al. 1986) and *gooseberry* (Bopp et al., 1986). Most notably it contains a lysine residue in position 50 of the homeodomain, as do the homologous genes in *Xenopus* (Blumberg et al., 1991), chick (Izpisùa-Belmonte et al., 1993), zebrafish (S. Schulte-Merker, E. M. De Robertis and C. Nüsslein-Volhard, in preparation), and human (M. Blum and E. M. De Robertis, unpublished). This predicts a DNA binding specificity similar to that of Drosophila *bicoid* (Hanes and Brent, 1989; Treisman et al., 1989; Blumberg et al., 1991). The genomic organization is conserved between *Xenopus* and mouse: both genes consist of three exons and two introns, with an intron in the home-odomain (Blum et al., 1992). The gene was mapped to mouse chromosome 12 (D. Simon-Chazottes, J.-L. Guénet, E. M. De Robertis and M. Blum, in preparation).

■ Expression

goosecoid is expressed in a biphasic manner during development. Its mRNA is transiently detected over a period of ten hours during early gastrulation. The signal is seen from E6.4 to E6.8, first in the just-forming primitive streak mesoderm, then always in its anteriormost aspect (Blum et al., 1992). *goosecoid* expression thus may mark a resident population of cells at the anterior end of the streak, which presumably have stem cell characteristics and may play a crucial role in generating and maintaining the activity of the primitive streak (Lawson et al., 1991, Lawson, 1992). *goosecoid* is reexpressed at mid-embryogenesis, starting at E10.5, in cranial neural crest-derived head mesenchyme and limb buds (Gaunt et al., 1993). *goosecoid* stains the medial and lateral nasal processes, branchial arches I and II, the proximal limb buds and adjacent lateral body wall. Within branchial arch II there is a sharp boundary such that only the anterior third of this arch is stained (Gaunt et al., 1993). Expression of *goosecoid* along the main body axis is therefore anterior to that of the Hox genes. Expression persists through E14.5 in derivatives of these structures, namely the nasal chamber, mandible, palate, tongue, lower jaw, eustachian tube, auditory meatus, proximal limbs and lateral body wall (Gaunt et al., 1993).

■ Function

Like the frog gene, mouse *goosecoid* marks the organizer tissue of the gastrula. This is evidenced by the ability of *goosecoid*-expressing tissue to induce anterior organs in *Xenopus* hosts when analyzed by the einsteck assay (Blum et al., 1992). The peptide growth factor activin, a potent inducer of dorso-anterior mesoderm in *Xenopus* (Green and Smith, 1990), is able to induce *goosecoid* mRNA in vitro in isolated 6.4 day mouse gastrulae under serum-free conditions (Blum et al., 1992). To analyze the effect of loss-of-function of *goosecoid* during both gastrulation and organogenesis the gene is presently being evicted (G. Yamada, M. Blum, E. M. De Robertis and P. Gruss).

■ References

Blum, M., Gaunt, S., Cho, K. W. Y., Steinbeisser, H., Bittner, D., Blumberg, B. and De Robertis, E. M. (1992) Gastrulation in the mouse: the role of the homeobox gene goosecoid. Cell 69, 1097-1106.

Blumberg, B., Wright, C.V.E., De Robertis, E.M. and Cho, K.W.Y. (1991). Organizer-specific homeobox genes in *Xenopus laevis* embryos. Science 253, 194-196.

Bopp, D. Burri, M., Baumgartner, S., Frigerio, G. and Noll, M. (1986). Conservation of a large protein domain in the segmentation gene *paired* and in functionally related genes of Drosophila. Cell 47, 1033-1040.

Cho, K.W.Y., Blumberg, B., Steinbeisser, H. and De Robertis, E.M. (1991). Molecular nature of Spemann's organizer: the role of the *Xenopus* homeobox gene *goosecoid*. Cell 67, 1111-1120.

Frigerio, G. Burri, M., Bopp, D., Baumgartner, S. and Noll, M. (1986). Structure of the segmentation gene *paired* and the Drosophila PRD gene set as part of a gene network. Cell 47, 735-746.

Gaunt, S. J., Blum, M. and De Robertis, E. M.(1993) Expression of the mouse *goosecoid* gene during mid-embryogenesis may mark mesenchymal cell lineages in the developing head, limbs and ventral body wall. Development, 117, 765–778.

Green, J. B. and Smith, J. C. (1990). Graded changes in dose of a *Xenopus* activin A homologue elicit stepwise transitions in embryonic cell fate. Nature 347, 391-394.

Hanes, S. D. and Brent, R. (1989). DNA specificity of the *bicoid* activator protein is determined by homeodomain recognition helix residue 9. Cell 57, 1275-1283.

Izpisùa-Belmonte, J.-C., De Robertis, E. M., Storey, K. G. and Stern, C. D. (1993). The homeobox gene *goosecoid* and the origin of organizer cells in the early chick blastoderm. Cell, in press.

Lawson, K. A., Meneses, J.J., and Pedersen, R. A. (1991). Clonal analysis of epiblast fate during germ layer formation in the mouse. Development 113, 891-911.

Lawson, K. A. and Pedersen, R. A. (1992). Clonal analysis of cell fate during gastrulation and early neurulation in the mouse. CIBA Foundation Symposium 165, 3-27.

Treisman, J., Gönczy, P., Vashishtha, M., Harris, E. and Desplan, C. (1989). A single amino acid can determine the DNA binding specificity of homeodomain proteins. Cell 59, 553-562.

gsc [x]

Species: *Xenopus laevis*
Chromosomal location: -
Other names: *goosecoid*
Cognate genes: *gsc* (mouse, zebrafish, chicken)
Type of homeobox: *goosecoid*
Accession numbers: M63872 (type A), M81481 (type B)

■ Origin and description

goosecoid was isolated from a dorsal blastopore lip cDNA library (Blumberg et al., 1991) using a 1024-fold degenerate oligonucleotide corresponding to a highly-conserved amino acid sequence, KIWF(Q/K)NRR found in the DNA-binding helix 3 of the homeodomain (Bürglin et al., 1989). *Goosecoid* defines a novel class of homeobox-containing gene, and has an *in vitro* DNA-binding specificity identical to an important *Drosophila melanogaster* homeobox gene - the anterior morphogen, *bicoid*. To date, putative *goosecoid* homologs have been detected in genomic DNA blots from many species including: fish, turtle, snake, chicken, mouse and human. Due to the pseudo-tetraploidy of *Xenopus laevis*, two different forms of the *goosecoid* gene have been isolated, designated A and B.

■ Expression

goosecoid mRNA is expressed exclusively in the dorsal region of the late blastula and early gastrula, defining the organizer field (Blumberg et al., 1991; Cho et al., 1991). During gastrulation, *goosecoid* mRNA becomes specifically localized in the organizer. *goosecoid* expression is enhanced by LiCl, a dorsalizing agent, and inhibited by retinoic acid and ultraviolet light irradiation before first cleavage, two treatments that interfere with anterior development in *Xenopus* (Cho et al., 1991).

■ Function

Microinjection of *goosecoid* mRNA into the two ventral blastomeres of 4 cell embryos can lead to the formation of twinned embryos, including complete head structures. Microinjection of *goosecoid* mRNA into ventral blastomeres at the 32-cell stage leads to secondary embryos in which the injected cells form anterior mesoderm and recruit uninjected cells of the host to form a secondary axis, as expected from cells with Spemann's Organizer activity (Niehrs et al., 1992). *goosecoid* is believed to be part of the biochemical machinery that leads to the execution of Spemann's Organizer phenomenon (Cho et al., 1991).

■ References

Blumberg, B., Wright, C.V.E., De Robertis, E.M. and Cho, K.W.Y. (1991). Organizer-specific homeobox genes in *Xenopus laevis* embryos. Science 253, 194-196.

Bürglin, T.R., Finney, M., Coulson, A. and Ruvkun, G. (1989). *Caenorhabditis elegans* has scores of homeobox-containing genes. Nature 341, 239-243.

Cho, K.W.Y., Blumberg, B., Steinbeisser, H. and De Robertis, E.M. (1991). Molecular nature of Spemann's organizer: the role of the *Xenopus* homeobox gene *goosecoid*. Cell 67, 1111, 1120.

Niehrs, C., Cho, K.W.Y. and De Robertis, E.M. (1993). The Homeobox Genes *goosecoid* affects Gastrulation Movements and Dorsal Specification in *Xenopus* embryos. Cell 72, 491–503.

Gst-1 [m]

Species: *Mus musculus*

Chromosomal location: Chromosome 5, near (within ~3 cM) *En-2*

Other names: *MMox-A* or *MMox-B*

Cognate genes: *Chox-7*

Type of homeobox: -

Accession number: -

■ Origin and description

This gene (and a paralogous gene, *Gst-2*) was isolated by cross-homology with *XGst-A*, a *Xenopus* homeobox gene

---

induced by FGF and Activin-A during gastrulation (Frohman et al., submitted). *Gst-1* is located on chromosome 5 within ~3 cM of the *En-2* gene and contains a homeobox similar to one previously described for its cognate gene, chicken *Chox-7* (Fainsod, 1989). Gst-1 is related to (but is not a cognate of) the *Xenopus* XGst-A gene. Extensive conservation of all of the genes is observed throughout the homeobox and more downstream amino acids, although *Gst-1* and *Chox-7* are clearly more related to each other than to *Gst-2* and *XGst-A*. *Gst-1* contains a divergent antp-like homeobox (56% identical).

■ Expression

Gst-1 expression is not detected from E6.25-E10.5 (Frohman et al., submitted). At E11.5, expression is observed in a region of neuronal proliferation in the forebrain.

■ Remarks

Murtha et al. (1991) have published short amino acid sequences identical to those of *Gst-1* and *-2*. These sequences describe the region between amino acids 22 and 46 of the homeobox. In this region, the amino acid sequences are identical. Since nucleotide sequences have not been compared, it is not known which (if either) gene matches *Gst-1* and which matches *Gst-2*.

■ References

Fainsod, A. and Greunbaum, Y. (1989). A chicken homeobox gene with developmentally regulated expression. FEBS Lett. 250, 381-385.

Frohman, M.A., Northrop, J.L. and Kimelman, D. Progressive activation of homeobox gene expression during vertebrate gastrulation (submitted).

Murtha, M., Leckman, J.F. and Ruddle, F.H. (1991). Detection of homeobox genes in development and evolution. PNAS 88, 10711-10715.

Gst-2 [m]

Species: *Mus musculus*

Chromosomal location: Chromosome 1, near (within ~3 cM) *En-1*

Other names: *MMox-A* or *MMox-B*

Cognate genes: *XGst-A*

Type of homeobox: -

Accession number: -

■ Origin and description

This gene was isolated by cross-homology with *XGst-A*, a *Xenopus* homeobox gene induced by FGF and Activin-A during gastrulation (Frohman et al., submitted). *Gst-2* is located on chromosome 1 within ~3 cM of the *En*-1 gene and contains a homeobox related to one previously described for chicken *Chox*-7 (Fainsod and Greunbaum, 1989), although it is not the murine cognate gene of *Chox*-7 (the paralogous gene, *Gst-1*, is the *Chox*-7 cognate). *Gst-2* has a *Xenopus* cognate, which is *XGst-A*. Extensive conservation of *Gst-2* and *XGst-A* is observed throughout their entire coding sequence (~80% identity). *Gst-2* contains a divergent Antp-like homeobox (56% identical).

■ Expression

Gst-2 is expressed in multiple settings during development; a partial list follows: *Gst-2* is not expressed from E6.25-E6.75 (Frohman et al, submitted). It then becomes detected in primitive streak mesoderm, and then in emigrating mesoderm and the neural tube (E6.75-E8.5). It is expressed by the latter time continuously from the posterior of the embryo to the hindbrain-midbrain junction. Expression is also observed in branchial arch endoderm. By E9.5 expression along the axis dwindles (although expression remains detectable in the dorsal spinal cord) and new sites of expression are observed: in the medial portion of the otocyst adjacent to rhombomeres 5 and 6 (the future endolymphatic sac and duct) and in a stripe in the midbrain. At E11.5, expression is again observed in the otic endolymphatic sac and duct and in the midbrain. New sites include restricted expression in first arch neural crest, and expression in neuronal proliferative zones in the spinal cord hindbrain, and forebrain.

■ Remarks

Murtha et al. (1991) have published short amino acid sequences identical to those of *Gst-1* and *-2*. These sequences describe the region between amino acids 22 and 46 of the homeobox. In this region, the amino acid sequences are identical. Since nucleotide sequences have not been compared, it is not known which (if either) gene matches *Gst-1* and which matches *Gst-2*.

■ References

Fainsod, A. and Greunbaum, Y. (1989). A chicken homeobox gene with developmentally regulated expression. FEBS Lett. 250, 381-385.

Frohman, M.A., Northrop, J.L. and Kimelman, D. Progressive activation of homeobox gene expression during vertebrate gastrulation (submitted).

Murtha, M., Leckman, J.F. and Ruddle, F.H. (1991). Detection of homeobox genes in development and evolution. PNAS 88, 10711-10715.

H2.0   [d]

Species: *Drosophila melanogaster*
Chromosomal location: 2nd chromosome, 26B1
Other names: -
Cognate genes: mouse *Hlx*, ascidian *AHox1*, chicken *ChoxE*, human *HB24*
Type of homeobox: *H2.0*-type
Accession number: -

■ Origin and description

The *H2.0* gene was originally cloned by homeobox homology to the *sex combs reduced* homeobox sequence (Barad et al., 1988). The *H2.0* transcript is 1,850 bases and the H2.0 protein sequence contains 410 amino acids, including a homeodomain, a His-Gln repeat, and a Hep repeat (Allen et al., 1991). The *H2.0* homeodomain sequence is highly diverged from *Antp* (less than 40% identity), but closely related to recently isolated vertebrate homeodomain sequences found in the mouse *Hlx* gene, the human *HB24* gene and the ascidian *AHox1* gene (Allen et al., 1991; Deguchi et al., 1991; Saiga et al., 1991). The proteins encoded by these chordate and urochordate genes also share structural similarity with *H2.0* in regions outside of the homeodomain, particularly in a 30 amino acid region called the Hep motif (15/30 identities between *H2.0* and mouse *Hlx*; Allen et al., 1991). The Hep motif is also found in more diverged form in the *en* and *inv* proteins, as well as in some *prd*-class homeodomain proteins. *H2.0* is also structurally similar to the chicken *ChoxE* gene (Rangini et al., 1991), though the *H2.0/ChoxE* sequence identities are not as extensive as those between *H2.0* and mouse *Hlx* or human *HB24*. The expression pattern of the best characterized *H2.0* homolog, *Hlx*, is not obviously related to the pattern of expression in *Drosophila*. *Hlx* is expressed in cells during the hematopoietic developmental pathway, as well as in other embryonic and adult tissues. The known ascidian *H2.0*-like gene, *AHox1*, is also expressed at abundant levels in blood cells, as well as in cells of the digestive tract (Saiga et al., 1991).

■ Expression

The *H2.0* homeobox gene is expressed in *Drosophila* embryos in a pattern that includes the mesodermal cells that will give rise to the posterior midgut musculature of the *Drosophila* embryo (Barad et al., 1988). *H2.0* expression persists in these mesodermal cells until late stages of embryogenesis, until well after the cells have differentiated into the visceral musculature. In addition, *H2.0* is expressed in small patches of embryonic epidermal cells that are segmentally reiterated. During third instar stages,

H2.0 is expressed in nests of gut-associated cells (B. Jones, unpublished).

■ Function

Despite the expression pattern of the gene in the embryonic visceral mesoderm, *H2.0* is essential for its normal morphogenesis. Embryos carrying deletions for the *H2.0* gene (and a few other nearby genes) die at the end of embryogenesis, but develop posterior midguts that appear to be normal even at the level of visceral muscle actin organization (Barad et al., 1991). It is not known at present whether *H2.0* is required for adult gut muscle formation, or for other cellular phenotypes included within its expression pattern (for example, the development of the lateral patches of embryonic epidermis that express *H2.0* but whose fate is unknown).

■ References

Barad, M., Jack, T., Chadwick, R. and McGinnis, W. (1988). A novel, tissue-specific, *Drosophila* homeobox gene. EMBO J. 7, 2151-2161.

Barad, M., Erlebacher, A. and McGinnis, W. (1991). Despite expression in embryonic visceral mesoderm, *H2.0* is not essential for *Drosophila* visceral muscle morphogenesis. Dev. Genet. 12, 206-211.

H40 [Am]

Species: *Apis mellis* (honeybee)
Chromosomal location: -
Other names: -
Cognate genes: -
Type of homeobox: *msh/NK*-type
Accession number: -

■ Origin and description

This gene was isolated by cross-homology with the *empty spiracles* (*ems*) homeobox (Dalton et al., 1989) in a screen for honeybee homeobox containing genes (Walldorf et al., 1989). It is the honeybee homologue of the *Drosophila* S59/Nk1 gene (Kim and Nierenberg, 1989; Dohrman et al., 1990).

■ References

Dalton, D., Chadwick, R. and McGinnis, W. (1989). Expression and embryonic function of empty spiracles: a *Drosophila* homeobox gene with two patterning functions on the anterior-posterior axis of the embryo. Genes Dev. 3, 1940-1956.

Dohrmann, C., Azpiazu, N. and Frasch, M. (1990). A new homeobox gene is expressed in mesodermal precursor cells of distinct muscles during embryogenesis. Genes Dev. 4, 2098-2111.

Kim, Y. and Nierenberg, M. (1989). *Drosophila* NK-homeobox genes. Proc. Natl. Acad. Sci. U.S.A. 86, 7716-7720.

Walldorf, U., Fleig, R. and Gehring, W.J. (1989). Comparison of homeobox-containing genes of the honeybee and *Drosophila*. Proc. Natl. Acad. Sci. U.S.A. 86, 9971-9975.

HAT4 [At]

Species: *Arabidopsis thaliana* (plant)
Chromosomal location: -
Other names: -
Cognate genes: *HAT5, HAT22*
Type of homeobox: Homeodomain-Leucine Zipper (HD-Zip)
Accession number: M90394

■ Origin and description

This cDNA was isolated (Schena and Davis, 1992) from an *Arabidopsis* cDNA library using degenerate oligonucleotides (Bürglin et al., 1989) to homeobox sequences encoding helix 3 of the homeodomain. The cDNA encodes a 284 amino acid protein with a predicted molecular weight of 32 kilodaltons. The homeodomain contains a leucine zipper motif 8 residues 3' of the homeodomain, similar to the spacing of the basic region and leucine zipper of bZip proteins. The homeodomain-leucine zipper (HD-Zip) motif has not yet been described for homeobox genes from animals. Southern blot analysis suggests that *HAT4* is a single copy gene.

■ Expression

Message has been detected in mRNA from adult plants.

■ Remarks

Homeodomain proteins with a tightly linked leucine zipper motif (i.e. HD-Zip proteins) have not yet been described in organisms from the animal kingdom. It remains to be seen whether this configuration is indeed plant-specific, and if so what the function of such genes is in higher plants. HAT=Homeobox from Arabidopsis thaliana.

■ References

Bürglin, T.R., Finney, M., Coulson, A. and Ruvkun, G. (1989). *Caenorhabditis elegans* has scores of homeobox-containing genes. Nature 341, 239-243.

Schena, M. and Davis, R. (1992). HD-Zip Proteins: Members of an *Arabidopsis* homeodomain protein superfamily. Proc. Natl. Acad. Sci. USA 89, 3894–3898.

HAT5 [At]

Species: *Arabidopsis thaliana* (plant)
Chromosomal location: -
Other names: *Athb-1*
Cognate genes: *HAT4, HAT22*
Type of homeobox: Homeodomain-Leucine Zipper (HD-Zip)
Accession number: M90416

■ Origin and description

This cDNA was isolated (Schena and Davis, 1992) from an *Arabidopsis* cDNA library using degenerate oligonucleotides (Bürglin et al., 1989) to homeobox sequences encoding helix 3 of the homeodomain. The cDNA encodes a 272 amino acid protein with a predicted molecular weight of 30 kilodaltons. The homeodomain contains a leucine zipper motif 8 residues 3' of the homeodomain, similar to the spacing of the basic region and leucine zipper of bZip proteins. The homeodomain-leucine zipper (HD-Zip) motif has not yet been described for homeobox genes from animals. Southern blot analysis suggests that *HAT5* is a single copy gene. *HAT5* shares >99% DNA homology with *Athb-1* (Ruberti et al., 1991) and apparently corresponds to the same gene.

■ Expression

Message has been detected in mRNA from adult plants.

■ Remarks

Homeodomain proteins with a tightly linked leucine zipper motif (i.e. HD-Zip proteins) have not yet been described in organisms from the animal kingdom. It remains to be seen whether this configuration is indeed plant-specific, and if so what the function of such genes is in higher plants. HAT=Homeobox from Arabidopsis thaliana.

■ References

Bürglin, T.R., Finney, M., Coulson, A. and Ruvkun, G. (1989). *Caenorhabditis elegans* has scores of homeobox-containing genes. Nature 341, 239-243.

Roberti, I., Sessa, G., Lucchetti, S. and Morelli, G. (1991). A novel class of plant proteins containing a homeodomain with a closely linked leucine zipper motif. EMBO J. 10, 1787-1791.

Schena, M. and Davis, R. (1992). HD-Zip Proteins: Members of an *Arabidopsis* homeodomain protein superfamily. Proc. Natl. Acad. Sci. USA 89, 3894–3898.

HAT22 [At]

Species: *Arabidopsis thaliana* (plant)
Chromosomal location: -
Other names: -
Cognate genes: *HAT4, HAT5*
Type of homeobox: Homeodomain-Leucine Zipper (HD-Zip)
Accession number: M90417

■ Origin and description

This cDNA was isolated (Schena and Davis, 1992) from an *Arabidopsis* cDNA library using degenerate oligonucleotides (Bürglin et al., 1989) to homeobox sequences encoding helix 3 of the homeodomain. The partail cDNA, which lacks the 5' end, encodes 112 amino acids of the protein. The homeodomain contains a leucine zipper motif 8 residues 3' of the homeodomain, similar to the spacing of the basic region and leucine zipper of bZip proteins. The homeodomain-leucine zipper (HD-Zip) motif has not yet been described for homeobox genes from animals. Southern blot analysis suggests that *HAT22* is a single copy gene.

■ Expression

Message has been detected in mRNA prepared from adult plants.

■ Remarks

Homeodomain proteins with a tightly linked leucine zipper motif (i.e. HD-Zip proteins) have not yet been described in organisms from the animal kingdom. It remains to be seen whether this configuration is indeed plant-specific, and if so what the function of such genes is in higher plants. HAT=Homeobox from Arabidopsis thaliana.

■ References

Bürglin, T.R., Finney, M., Coulson, A. and Ruvkun, G. (1989). *Caenorhabditis elegans* has scores of homeobox-containing genes. Nature 341, 239-243.

Schena, M. and Davis, R. (1992). HD-Zip Proteins: Members of an *Arabidopsis* homeodomain protein superfamily. Proc. Natl. Acad. Sci. USA 89, 3894–3898.

Species: *Homo sapiens*
Chromosomal location: 1
Other names: -
Cognate genes: -
Type of homeobox: moderately diverged from class I homeobox genes
Accession number: M56537

■ Origin and description

Isolated from a human B lymphocyte cDNA library by the use of degenerate oligonucleotides whose sequence was from the 3′ end of the homeobox. (Deguchi et al., submitted for publication). Contains a moderately diverged homeodomain with approximately 50% identity to class I homeodomain containing proteins.

■ Expression

HB9 mRNA transcripts are present in activated lymphocytes and hematopoietic progenitors. Present in a variety of hematopoietic cell lines. *In situ* hybridization localized transcripts in spleen, tonsil (highest in the germinal centres), and thymus (highest in the medullary region). No expression in adult brain, liver, pancrease, or Hela cells. Examination of 20-week developing human tissues revealed expression in testes and brain. No transcripts were found in lung, heart, ileum, stomach, vessels, or adrenal gland. Cross hybridizing transcripts were found in developing mice (days 9.5 and 11.5) and the teratocarcinoma cell line, F9. The transcripts in F9 cells disappeared following retinoic acid treatment (Deguchi et al., submitted for publication; Deguchi et al., 1991).

■ References

Deguchi, Y., Fox, C.F. and Kehrl, J.H. Characterization of a novel human homeobox gene: Expression in activated lymphocytes and embryonic tissues. (submitted for publication).

Deguchi, Y. and Kehrl, J.H. (1991). Selective expression of two homeobox genes in CD34 positive cells from human bone marrow. Blood 78, 323.

Deguchi, Y. and Kehrl, J.H. (1991). Nucleotide sequence of a novel diverged human homeobox gene encodes a DNA binding protein. Nucleic Acid Res. 19, 3742.

Species: *Homo sapiens*
Chromosomal location: 1
Other names: -
Cognate genes: *H2.0 (Drosophila)*, *Hlx* (mouse)
Type of homeobox: *H2.0*-type
Accession number: M60721

■ Origin and description

Cloned from a human B lymphocyte cDNA library using degenerate oligonucleotides whose sequence was derived from the conserved 3′ end of class I homeobox genes. The *HB24* homeodomain is diverged from class I homeodomain but approximately 80% identical to the homeodomains of the *H2.0* gene in *Drosophila* and of the *Hlx* gene in mice.

■ Expression

HB24 mRNA transcripts are present in activated lymphocytes and hematopoietic progenitors. Transcripts were present in all the lymphoid cell lines examined. *In situ* hybridization studies revealed low level expression in tonsil, spleen, and thymus. No expression was found in Hela cells, placenta, adult lung, adult kidney, or adult brain. A limited number of human fetal tissues were examined. *HB24* transcripts were present in fetal brain and in large fetal vessels. No expression was found in 20-week lung, kidney, heart, ileum, stomach, or adrenal gland. A *HB24* cross-reactive transcript is found in retinoic treated F9 cells (Deguchi et al., 1991a; Deguchi et al., 1991b).

■ Genetics, function

Function is unknown although transfection into a human T cell line enhances the proliferative rate of the cells and increases the expression of a variety of activation genes including IL-2 and IL-2 receptor (Deguchi et al., submitted for publication).

■ References

Deguchi, Y., Moroney, J.F., Wilson, G.L., Fox, C.H., Winter, H.S. and Kehrl, J.H. (1991). Cloning of a human homeobox gene that resembles a diverged *Drosophila* homeobox gene and is expressed in activated lymphocytes. New Biol. 3, 353.

Deguchi, Y. and Kehrl, J.H. (1991). Selective expression of two homeobox genes in CD34 positive cells from human bone marrow. Blood 78, 323.

Deguchi, Y., Thevenin, C. and Kehrl, J.H. Stable expression of HB24, a diverged human homeobox gene in T lymphocytes induces genes involved in T cell activation and growth. (submitted for publication)

Species: *Mus musculus*
Chromosomal location: -
Other names: -
Cognate genes: -
Type of homeobox: Novel class, related to *XANF-1*
Accession number: L02646

■ Origin and description

This homeobox was isolated by using degenerated oligonucleotides and PCR on cDNA prepared from mouse ES cells

■ References

Thomas, P. Q and Rathjen, D. (1992). HES-1, a novel homeobox gene expressed by murine ES cells, identifies a new class of homeobox genes. Nucl. Acid. Res. 20, 5840.

HNF1   [r,m,h]

Species: *Rattus sp., Mus musc., Homo sapiens*
Chromosomal location: Mouse: 5F; Human: 12q24.3
Other names: *LF-B1, HNF1-alpha, APF, HP-1*
Cognate genes: -
Type of homeobox: Extra-large *HNF1-*type
Accession numbers: Rat: X54423 or J03170; Mouse: M57966; Human: J04771

■ Origin and description

HNF1 (Hepatic Nuclear Factor 1) was described as a sequence-specific DNA-binding protein from rat liver, that interacts with promoter or enhancer elements (consensus sequence; GTTAATNATTAAC) necessary to transcription of many genes expressed preferentially in liver, such as albumin, alpha-1-antitrypsin, alpha- and beta-fibrinogens, surface antigen of HBV virus, etc. (reviewed in Mendel and Crabtree, 1991; De Simone and Cortese, 1991; Rey-Campos and Yaniv, 1992). Protein sequence data allowed the isolation of *HNF1* cDNA clones from rat liver and rat hepatoma cell lines (Frain et al., 1989; Baumhueter et al., 1990; Chouard et al., 1990) and subsequently from other species (Bach et al., 1990; Kuo et al., 1990). The *HNF1* homeodomain is the most diverged known to date, lacking conserved residues inside and outside the third helix (Chouard et al., 1990). The best alignment between *HNF1* and other homeodomain sequences is obtained by the introduction of a 24 amino acid loop between helices 2 and 3, instead of the canonical 3 amino acids in the turn (Finney, 1990; Chouard et al., 1990; Baumhueter et al., 1990; Nicosia et al., 1990). In addition to this extra-large homeodomain, a pseudo-POU domain as well as a distal dimerization domain are required for specific DNA-binding of *HNF1*, those three structures forming the N-terminal half of the protein (Nicosia et al., 1990; Chouard et al., 1990). The C-terminal moiety of the transcription factor comprises its transactivation domain (Nicosia et al., 1990). Dimerization and transactivation by this homeodomain protein is regulated by an 11 kDa co-factor (*DCoH*: Mendel et al., 1991).

■ Expression and regulation

HNF1 was first identified in liver and differentiated hepatoma cell lines and was not found in their dedifferentiated variants which, in turn, contain another factor related to *HNF1* (the homeodomain protein called vHNF1), *HNF1* mRNA was subsequently detected in the early hepatic primordia and in the visceral endoderm of the yolk sac as well as in the adult, mainly in liver and polarized epithelia of kidney and intestine (Baumhueter et al., 1990; Blumenfeld et al., 1991; Kuo et al., 1990, Dc Simone et al., 1991). Another factor enriched in the hepatocytes, called HNF4 (or LF-A1), and related to the steroid hormone receptor superfamily was recently reported to positively control the transcription of the *HNF1* gene (Tian and Schibler, 1991; Kuo et al., 1992).

■ Function

The *HFN1* binding site was shown to be essential to the transcription of a large number of genes expressed specifically in the hepatocytes and also shown to be crucial in an intestinal context (see references in Blumenfeld et al., 1991). The factor was also shown to be a potent activator of transcription via this binding site (see Rey-Campos and Yaniv, 1991). Nevertheless, the actual function of *HNF1* itself in liver and, moreover, in intestine or kidney remains to be defined.

■ References

Bach, I., Galcheva-Gargova, Z., Mattei, M.G., Simon-Chazottes, D., Guenet, J.L., Cereghini, S. and Yaniv, M. (1990). Cloning of human Hepatic Nuclear Factor 1 (HNF1) and chromosomal localization of its gene in man and mouse. Genomics 8, 155-164.

Baumhueter, S., Mendel, D.B., Conley, P.B., Kuo, C.J., Turk, C., Graves, M.K., Edwards, C.A., Courtois, G. and Crabtree, G.R. (1990). HNF-1 shares three sequence motifs with the POU domain proteins and is identical to LFB1 and APF. Genes Dev. 4, 372-379.

Blumenfeld, M., Maury, M., Chouard, T., Yaniv, M. and Condamine, H. (1991). Hepatic Nuclear Factor 1 (HNF1) shows a wider distribution than products of its known target genes. Development 113, 589-599.

Chouard, T., Blumenfeld, M., Bach, I., Vandekerckhove, J., Cereghini, S. and Yaniv, M. (1990). A distal dimerization domain is essential for DNA-binding by the atypical HNF1 homeodomain. Nucl. Acids Res. 18, 5853-5863.

De Simone, V. and Cortese, R. (1991). Transcriptional regulation of liver-specific gene expression. Curr. Op. in Cell Biol. 3, 960-965.

De Simone, V., De Magistris, L., Lazzaro, D., Gerstner, J., Monaci, P., Nicosia, A. and Cortese, R. (1991). LFB3, a heterodimer-forming homeoprotein of the LFB1 family, is expressed in specialized epithelia. EMBO J. 10, 1435-1443.

Finney, M. (1990). The homeodomain of the transcription factor LFB1 has a 21 amino acid loop between helix 2 and helix 3. Cell 60, 5-6.

Frain, M., Swart, G., Monaci, P., Nicosia, A., Stampfli, S., Frank, R. and Cortese, R. (1989). The liver-specific transcription factor LFB1 contains a highly diverged homeobox DNA binding domain. Cell 59, 145-157.

Kuo, C.J., Conley, P.B., Hsieh, C.L., Francke, U. and Crabtree, G.R. (1990). Molecular cloning, functional expression and chromosomal localization of mouse Hepatocyte Nuclear Factor 1. Proc. Natl. Acad. Sci. USA 87, 9838-9842.

Kuo, C.J., Conley, P.B., Chen, L., Sladek, F.M., Darnell, J.E. Jr. and Crabtree, G.R. (1992). Inhibition by extinguishing loci of transcriptional hierarchy involved in cell-type specification. Nature 355, 457-461.

Mendel, D.B. and Crabtree, G.R. (1991). HNF-1, a member of a novel dimerizing homeodomain protein. J. Biol. Chem. 266, 677-680.

Mendel, D.B., Khavari, P.A., Conley, P.B., Graves, M.K., Hansen, L.P., Admon, A. and Crabtree, G.R. (1991). Characterization of a cofactor that regulates dimerization of a mammalian homeodomain protein. Science 254, 1762-1767.

Nicosia, A., Monaci, P., Tomei, L., De Francesco, R., Nuzzo, M., Stunnenberg, H. and Cortese, R. (1990). A myosin-like dimerization helix and an extra-large homeodomain are essential elements of the tripartite DNA binding structure of LFB1, Cell 61, 1225-1236.

Rey-Campos, J. and Yaniv, M. (1992). Regulation of albumin gene expression. In: Genetic intervention in diseases with unknown etiology. (T.O. Yoshida Ed.) Elsevier Science Publishers B.V. pp 1-14.

Tian, J.-M. and Schibler, U. (1991). Tissue-specific expression of the gene encoding hepatocyte Nuclear Factor 1 may involve hepatocyte Nuclear Factor 4. Genes and Dev. 5, 2225-2234.

HOXA1 [h]

Species: *Homo sapiens*

Chromosomal location: 7p14-p21

Other names: *HOX1F*

Cognate genes: *Hoxa-1, Xhoxlab-2*

Type of homeobox: *Antp*-type

Accession number: -

■ Origin and description

HOXA1 was found during a chromosome walking by the screening of a genomic library constructed in the cosmid vector pCOS2 EMBL. The cosmid, named AP2, contained the *HOXA1* homeobox. *HOXA1* lies 9 Kb downstream from *HOXA2* homeobox. *HOXA1* belongs to *HOXA* locus and forms, together with HO 2I and *HOXD1* the XIII group of paralogy (Acampora et al., 1989; Boncinelli et al., 1991).

■ Expression

The expression of *HOXA1* was studied in human terato-carcinoma cell lines (NT2/D1 and Tera-2/clone 13 cells) and in human hematopoietic cell lines (erythroleukemic: K562 and OCIM2; promyelocytic: HL60; monocytic: U937). By an RNase protection assay *HOXA1* is expressed at above 6.0 hours from induction with Retinoic Acid; the expression increases reaching the peak level at above 60 hours. In Tera-2/clone 13 cells the same activation pattern is observed (Simeone et al., 1991). By Northern blot analysis *HOXA1* shows two transcripts of 2.2 and 1.7 Kb in K562 and in HL60 cells; a single transcript of 2.2 Kb is present in OCIM2 and U937 (Magli et al., 1991).

■ References

Acampora, D., Simeone, A. and Boncinelli, E. (1991). Human HOX homeobox genes. In: Oxford Surveys on Eukaryotic Genes (MacLean, N., ed.), Oxford University Press 7, 1-28.

Boncinelli, E., Simeone, A., Acampora, D. and Mavilio, F. (1991). HOX gene activation by retinoic acid. Trends Genet. 7, 329-334.

Magli, M.C., Barba, P., Celetti, A., De Vita, G., Cillo, C. and Boncinelli, E. (1991). Coordinate regulation of HOX genes in human hematopoietic cells. Proc. Natl. Acad. Sci. USA 88, 6348-6352.

Simeone, A., Acampora, D., Nigro, V., Faiella, A., D'Esposito, M., Stornaiuolo, A., Mavilio, F. and Boncinelli, E. (1991). Differential regulation by retinoic acid of the homeobox genes of the four HOX loci in human embryonal carcinoma cells. Mech Dev. 33, 215-228.

HOXA2 [h]

Species: *Homo sapiens*

Chromosomal location: 7p14-p21

Other names: *HOX1K*

Cognate genes: *Hoxa-11*

Type of homeobox: *Antp*-type

Accession number: -

■ Origin and description

HOXA2 was found during a chromosome walking by the screening of a genomic library constructed in the cosmid vector pCOS2 EMBL. The cosmid, named AP2, contained the *HOXA2* homeobox. *HOXA2* lies 6.0 Kb downstream from *HOXA3* and 9 Kb upstream from *HOXA1* homeobox. *HOXA2* belongs to *HOXA* locus and forms, together with *HOXB2* the XII group of paralogy (Acampora et al., 1989; Boncinelli et al., 1991).

■ Expression

The expression of *HOXA2* was studied in human terato-carcinoma cell lines (NT2/D1 and Tera-2/clone 13 cells) and in human hematopoietic cell lines (erythroleukemic: K562 and OCIM2; promyelocytic: HL60; monocytic: U937). By an RNase protection assay *HOXA2* is expressed at above 25 hours from induction with retinoic acid; the expression increases reaching the peak level at 5 days. InTera-2/clone 13 cells the same activation pattern is observed (Simeone et al., 1991). By Northern blot analysis there is no expression of *HOXA2* in K562, HL60, U937 and OCIM2 cells; (Magli et al., 1991).

■ References

Acampora, D., Simeone, A. and Boncinelli, E. (1991). Human HOX homeobox genes. In: Oxford Surveys on Eukaryotic Genes (MacLean, N., ed.), Oxford University Press 7, 1-28.

Boncinelli, E., Simeone, A., Acampora, D. and Mavilio, F. (1991). HOX gene activation by retinoic acid. Trends Genet. 7, 329-334.

Magli, M.C., Barba, P., Celetti, A., De Vita, G., Cillo, C. and Boncinelli, E. (1991). Coordinate regulation of HOX genes in human hematopoietic cells. Proc. Natl. Acad. Sci. USA 88, 6348-6352.

Simeone, A., Acampora, D., Nigro, V., Faiella, A., D'Esposito, M., Stornaiuolo, A., Mavilio, F. and Boncinelli, E. (1991). Differential regulation by retinoic acid of the homeobox genes of the four HOX loci in human embryonal carcinoma cells. Mech Dev. 33, 215-228.

HOXA3 [h]

Species: *Homo sapiens*
Chromosomal location: 7p14-p21
Other names: *HOX1E*
Cognate genes: *Hoxa-3*
Type of homeobox: *Antp*-type
Accession number: -

■ Origin and description

HOXA3 was found during a chromosome walking by the screening of a genomic library constructed in the cosmid vector pCOS2 EMBL. The cosmid, named AP2, contained the *HOXA3* homeobox. *HOXA3* lies 6.0 Kb upstream from *HOXA2* and 19 Kb downstream from *HOXA4* homeobox. *HOXA3* belongs to *HOXA* locus and forms, together with *HOXB3* and *HOXD3* the XI group of paralogy (Acampora et al., 1989; Boncinelli et al., 1991).

■ Expression

The expression of *HOXA3* was studied in human terato-carcinoma cell lines (NT2/D1 and Tera-2/clone 13 cells) and in human hematopoietic cell lines (erythroleukemic: K562 and OCIM2; promyelocytic: HL60; monocytic: U937). By an RNase protection assay *HOXA3* is expressed at 50 hours from induction with retinoic acid; the expression increases reaching the peak level at 5 days. In Tera-2/clone 13 cells the same activation pattern is observed (Simeone et al., 1991). By Northern blot analysis the expression of *HOXA3* gene is barely detectable in K562, HL60, U937 and OCIM2 cells; (Magli et al., 1991).

■ References

Acampora, D., Simeone, A. and Boncinelli, E. (1991). Human HOX homeobox genes. In: Oxford Surveys on Eukaryotic Genes (MacLean, N., ed.), Oxford University Press 7, 1-28.

Boncinelli, E., Simeone, A., Acampora, D. and Mavilio, F. (1991). HOX gene activation by retinoic acid. Trends Genet. 7, 329-334.

Magli, M.C., Barba, P., Celetti, A., De Vita, G., Cillo, C. and Boncinelli, E. (1991). Coordinate regulation of HOX genes in human hematopoietic cells. Proc. Natl. Acad. Sci. USA 88, 6348-6352.

Simeone, A., Acampora, D., Nigro, V., Faiella, A., D'Esposito, M., Stornaiuolo, A., Mavilio, F. and Boncinelli, E. (1991). Differential regulation by retinoic acid of the homeobox genes of the four HOX loci in human embryonal carcinoma cells. Mech Dev. 33, 215-228.

HOXA4 [h]

Species: *Homo sapiens*
Chromosomal location: 7p14-p21
Other names: *HOX1D*
Cognate genes: *Hoxa-4, Chox1.4*
Type of homeobox: *Antp*-type
Accession number: -

■ Origin and description

HOXA4 was found during a chromosome walking by the screening of a genomic library constructed in the cosmid vector pCOS2 EMBL. The cosmid, named CD9A, contained the *HOXA4* homeobox. *HOXA4* lies 7.5 Kb downstream from *HOXA5* and 19 Kb upstream from *HOXA3* homeobox. *HOXA4* belongs to *HOXA* locus and forms, together with *HOXB4* and *HOXC4* and *HOXD4* the X group of paralogy (Acampora et al., 1989; Boncinelli et al., 1991).

■ Expression

The expression of *HOXA4* was studied in human terato-carcinoma cell lines (NT2/D1 and Tera-2/clone 13 cells) and in human hematopoietic cell lines (erythroleukemic: K562 and OCIM2; promyelocytic: HL60; monocytic: U937). A Northern blot of poly(A)+RNA from NT2/D1 cells treated with 10 µM retinoic acid for 14 days was hybridized to *HOXA4*; two transcripts of 1.9 Kb and 1.4 Kb are present (Stornaiuolo et al., 1990). By an RNase protection assay *HOXA4* is expressed at 57 hours from induction with Retinoic Acid; the expression increases reaching the peak level at 5 days. In Tera-2/clone 13 cells the same activation pattern is observed (Simeone et al., 1991). By Northern blot analysis the expression of *HOXA4* gene is barely detectable in K562, HL60 and U937 cells; no expression is present in OCIM2 cells (Magli et al., 1991).

■ References

Acampora, D., Simeone, A. and Boncinelli, E. (1991). Human HOX homeobox genes. In: Oxford Surveys on Eukaryotic Genes (MacLean, N., ed.), Oxford University Press 7, 1-28.

Boncinelli, E., Simeone, A., Acampora, D. and Mavilio, F. (1991). HOX gene activation by retinoic acid. Trends Genet. 7, 329-334.

Magli, M.C., Barba, P., Celetti, A., De Vita, G., Cillo, C. and Boncinelli, E. (1991). Coordinate regulation of HOX genes in human hematopoietic cells. Proc. Natl. Acad. Sci. USA 88, 6348-6352.

Simeone, A., Acampora, D., Nigro, V., Faiella, A., D'Esposito, M., Stornaiuolo, A., Mavilio, F. and Boncinelli, E. (1991). Differential regulation by retinoic acid of the homeobox genes of the four HOX loci in human embryonal carcinoma cells. Mech Dev. 33, 215-228.

Stornaiuolo, A., Acampora, D., Pannese, M., D'Esposito M., Morelli, F., Migliaccio, E., Rambaldi, M., Faiella, A., Nigro, V., Simeone, A. and Boncinelli, E. (1990). Human HOX genes are differentially activated by retinoic acid in embryonal carcinoma cells according to their position within the four loci. Cell Differ. Dev. 31, 119-127.

HOXA5 [h]

Species: *Homo sapiens*
Chromosomal location: 7p14-p21
Other names: *HOX1C*
Cognate genes: *Hoxa-5*
Type of homeobox: *Antp*-type
Accession number: -

■ Origin and description

HOXA5 was found during a chromosome walking by the screening of a genomic library constructed in the cosmid vector pCOS2 EMBL. The cosmid, named CD9A, contained the *HOXA5* homeobox. *HOXA5* lies 7.5 Kb upstream from *HOXA4* and 10.5 Kb downstream from *HOXA6* homeobox. *HOXA5* belongs to *HOXA* locus and forms, together with *HOXB5* and 3D the IX group of paralogy (Acampora et al., 1989; Boncinelli et al., 1991).

■ Expression

The expression of *HOXA5* was studied in human teratocarcinoma cell lines (NT2/D1 and Tera-2/clone 13 cells) and in human hematopoietic cell lines (erythroleukemic: K562 and OCIM2; promyelocytic: HL60; monocytic: U937). A Northern blot of poly(A)+RNA from NT2/D1 cells treated with 10 µM retinoic acid for 14 days was hybridized to *HOXA5*; a single transcript of 1.9 Kb is present (Stornaiuolo et al., 1990). By an RNase protection assay *HOXA5* is expressed at 65 hours from induction with retinoic acid; the expression increases reaching the peak level at 5 days. In Tera-2/clone 13 cells the same activation pattern is observed (Simeone et al., 1991). The expression of *HOXA5* gene is barely detectable in K562, HL60 and U937 cells; no expression is present in OCIM2 cells (Magli et al, 1991).

■ References

Acampora, D., Simeone, A. and Boncinelli, E. (1991). Human HOX homeobox genes. In: Oxford Surveys on Eukaryotic Genes (MacLean, N., ed.), Oxford University Press 7, 1-28.

Boncinelli, E., Simeone, A., Acampora, D. and Mavilio, F. (1991). HOX gene activation by retinoic acid. Trends Genet. 7, 329-334.

Magli, M.C., Barba, P., Celetti, A., De Vita, G., Cillo, C. and Boncinelli, E. (1991). Coordinate regulation of HOX genes in human hematopoietic cells. Proc. Natl. Acad. Sci. USA 88, 6348-6352.

Simeone, A., Acampora, D., Nigro, V., Faiella, A., D'Esposito, M., Stornaiuolo, A., Mavilio, F. and Boncinelli, E. (1991). Differential regulation by retinoic acid of the homeobox genes of the four HOX loci in human embryonal carcinoma cells. Mech Dev. 33, 215-228.

Stornaiuolo, A., Acampora, D., Pannese, M., D'Esposito M., Morelli, F., Migliaccio, E., Rambaldi, M., Faiella, A., Nigro, V., Simeone, A. and Boncinelli, E. (1990). Human HOX genes are differentially activated by retinoic acid in embryonal carcinoma cells according to their position within the four loci. Cell Differ. Dev. 31, 119-127.

HOXA6 [h]

Species: *Homo sapiens*
Chromosomal location: 7p14-p21
Other names: *HOX1B*
Cognate genes: *Hoxa-6*
Type of homeobox: *Antp*-type
Accession number: -

■ Origin and description

HOXA6 was found during a chromosome walking by the screening of a genomic library constructed in the cosmid vector pCOS2 EMBL. The cosmid, named CD9A, contained the *HOXA6* homeobox. *HOXA6* lies 9 Kb upstream from

HOXA5 and 10.5 Kb downstream from *HOXA7* homeobox. *HOXA6* belongs to *HOXA* locus and forms, together with *HOXB6* and *HOXC6* the VIII group of paralogy (Acampora et al., 1989; Boncinelli et al., 1991).

■ Expression

The expression of *HOXA6* was studied in human teratocarcinoma cell lines (NT2/D1 and Tera-2/clone 13 cells) and in human hematopoietic cell lines (erythroleukemic: K562 and OCIM2; promyelocytic: HL60; monocytic: U937). A Northern blot of poly(A)+RNA from NT2/D1 cells treated with 10 µM retinoic acid for 14 days was hybridized to HOXA6; four transcripts were present of 5.2, 3.4, 2.4 and 2.1 Kb (Stornaiuolo et al., 1990). By an RNase protection assay *HOXA6* is expressed at 85 days after induction with RA. In Tera-2/clone 13 cells the same activation pattern is present. The expression of *HOXA6* gene is barely detectable in K562, HL60 and U937 cells; no expression is present in OCIM2 cells (Magli et al., 1991).

■ References

Acampora, D., Simeone, A. and Boncinelli, E. (1991). Human HOX homeobox genes. In: Oxford Surveys on Eukaryotic Genes (MacLean, N., ed.), Oxford University Press 7, 1-28.
Boncinelli, E., Simeone, A., Acampora, D. and Mavilio, F. (1991). HOX gene activation by retinoic acid. Trends Genet. 7, 329-334.
Magli, M.C., Barba, P., Celetti, A., De Vita, G., Cillo, C. and Boncinelli, E. (1991). Coordinate regulation of HOX genes in human hematopoietic cells. Proc. Natl. Acad. Sci. USA 88, 6348-6352.
Simeone, A., Acampora, D., Nigro, V., Faiella, A., D'Esposito, M., Stornaiuolo, A., Mavilio, F. and Boncinelli, E. (1991). Differential regulation by retinoic acid of the homeobox genes of the four HOX loci in human embryonal carcinoma cells. Mech Dev. 33, 215-228.
Stornaiuolo, A., Acampora, D., Pannese, M., D'Esposito M., Morelli, F., Migliaccio, E., Rambaldi, M., Faiella, A., Nigro, V., Simeone, A. and Boncinelli, E. (1990). Human HOX genes are differentially activated by retinoic acid in embryonal carcinoma cells according to their position within the four loci. Cell Differ. Dev. 31, 119-127.

HOXA7 [h]

Species: *Homo sapiens*
Chromosomal location: 7p14-p21
Other names: *HOX1A*
Cognate genes: *Hoxa-7, r5, Xlhbox3, Xhox36, Chox1.1, Quox-1*
Type of homeobox: *Antp*-type
Accession number: EMBL X16665-X16667

■ Origin and description

HOXA7 was found during a chromosome walking by the screening of a genomic library constructed in the cosmid vector pCOS2 EMBL. The cosmid, named CD9A, contained the *HOXA7* homeobox. *HOXA7* lies 11 Kb downstream from *HOXA9* and 10.5 Kb upstream from *HOXA6* homeobox. *HOXA7* belongs to *HOXA* locus and forms, together with *HOXB7* the VII group of paralogy (Acampora et al., 1989; Boncinelli et al., 1991).

■ Expression

The expression of *HOXA7* was studied in human teratocarcinoma cell lines (NT2/D1 and Tera-2/clone 13 cells) and in human hematopoietic cell lines (erythroleukemic: K562 and OCIM2; promyelocytic: HL60; monocytic: U937). A Northern blot of poly(A)+RNA from NT2/D1 cells treated with 10 µM retinoic acid for 14 days was hybridized to *HOXA7*; two transcripts were present of 4.3 Kb and 2.4 Kb (Stornaiuolo et al., 1990). By an RNase protection assay *HOXA7* is expressed at very low, constitutive levels already in untreated cells. At 80 hours from induction with retinoic acid the expression increases reaching the peak level at 5 days. In Tera-2/clone 13 *HOXA7* remains silent (Simeone et al., 1991). The expression of *HOXA7* gene is barely detectable in K562, HL60 and U937 cells; no expression is present in OCIM2 cells (Magli et al., 1991).

■ References

Acampora, D., Simeone, A. and Boncinelli, E. (1991). Human HOX homeobox genes. In: Oxford Surveys on Eukaryotic Genes (MacLean, N., ed.), Oxford University Press 7, 1-28.
Boncinelli, E., Simeone, A., Acampora, D. and Mavilio, F. (1991). HOX gene activation by retinoic acid. Trends Genet. 7, 329-334.
Magli, M.C., Barba, P., Celetti, A., De Vita, G., Cillo, C. and Boncinelli, E. (1991). Coordinate regulation of HOX genes in human hematopoietic cells. Proc. Natl. Acad. Sci. USA 88, 6348-6352.
Simeone, A., Acampora, D., Nigro, V., Faiella, A., D'Esposito, M., Stornaiuolo, A., Mavilio, F. and Boncinelli, E. (1991). Differential regulation by retinoic acid of the homeobox genes of the four HOX loci in human embryonal carcinoma cells. Mech Dev. 33, 215-228.
Stornaiuolo, A., Acampora, D., Pannese, M., D'Esposito M., Morelli, F., Migliaccio, E., Rambaldi, M., Faiella, A., Nigro, V., Simeone, A. and Boncinelli, E. (1990). Human HOX genes are differentially activated by retinoic acid in embryonal carcinoma cells according to their position within the four loci. Cell Differ. Dev. 31, 119-127.

Species: *Homo sapiens*
Chromosomal location: 7p14-p21
Other names: *HOX1G*
Cognate genes: *Hoxa-9, XhoxB1, Chox1.7*
Type of homeobox: *Antp*-like
Accession number: X16665-X16667

■ Origin and description

HOXA9 was found during a chromosome walking by the screening of a genomic library constructed in the cosmid vector pCOS2 EMBL. The cosmid, named M128, contained the *HOXA9* homeobox. *HOXA9* lies 11 Kb upstream from *HOXA7* and 8 Kb downstream from *HOXA10* homeobox. *HOXA9* belongs to *HOXA* locus and forms, together with *HOXB9, HOXC9* and *HOXD9* the V group of paralogy (Acampora et al., 1989; Boncinelli et al., 1991).

■ Expression

The expression of *HOXA9* was studied in human teratocarcinoma cell lines (NT2/D1 and Tera-2/clone 13 cells) and in human hematopoietic cell lines (erythroleukemic: K562 and OCIM2; promyelocytic: HL60; monocytic: U937). In NT2/D1, both in stem and in 10 µM RA treated cells, no expression was found, both by Northern blot and by RNAse protection; similar results were obtained in Tera2/clone 13 (Stornaiuolo et al., 1990; Simeone et al., 1991). By Northern blot, *HOXA9* shows no transcript in K562, OCIM2 and HL60 cells; in U937 two transcripts (2.8 Kb and 1.9 Kb) are present.

■ References

Acampora, D., Simeone, A. and Boncinelli, E. (1991). Human HOX homeobox genes. In: Oxford Surveys on Eukaryotic Genes (MacLean, N., ed.), Oxford University Press 7, 1-28.

Boncinelli, E., Simeone, A., Acampora, D. and Mavilio, F. (1991). HOX gene activation by retinoic acid. Trends Genet. 7, 329-334.

Magli, M.C., Barba, P., Celetti, A., De Vita, G., Cillo, C. and Boncinelli, E. (1991). Coordinate regulation of HOX genes in human hematopoietic cells. Proc. Natl. Acad. Sci. USA 88, 6348-6352.

Simeone, A., Acampora, D., Nigro, V., Faiella, A., D'Esposito, M., Stornaiuolo, A., Mavilio, F. and Boncinelli, E. (1991). Differential regulation by retinoic acid of the homeobox genes of the four HOX loci in human embryonal carcinoma cells. Mech Dev. 33, 215-228.

Stornaiuolo, A., Acampora, D., Pannese, M., D'Esposito M., Morelli, F., Migliaccio, E., Rambaldi, M., Faiella, A., Nigro, V., Simeone, A. and Boncinelli, E. (1990). Human HOX genes are differentially activated by retinoic acid in embryonal carcinoma cells according to their position within the four loci. Cell Differ. Dev. 31, 119-127.

Species: *Homo sapiens*
Chromosomal location: 7p14-p21
Other names: *HOX1H, PL1*
Cognate genes: *Hoxa-10*
Type of homeobox: *Antp*-type
Accession number: X16665-X16667

■ Origin and description

HOXA10 was found during a chromosome walking by the screening of a genomic library constructed in the cosmid vector pCOS2 EMBL. The cosmid, named c2A, contained the *HOXA10* homeobox. *HOXA10* lies 12 Kb downstream from *HOXA11* and 8 Kb upstream from *HOXA9* homeobox. *HOXA10* belongs to *HOXA* locus and forms, together with *HOXC10* and *HOXD10* the IV group of paralogy (Acampora et al., 1989; Boncinelli et al., 1991).

■ Expression

The expression of *HOXA10* was studied in human teratocarcinoma cell lines (NT2/D1 and Tera-2/clone 13) and in human hematopoietic cell lines (erythroleukemic: K562 and OCIM2; promyelocytic: HL60; monocytic: U937). In NT2/D1, both in stem and in 10 µM RA treated cells, no expression was found, both by Northern blot and by RNAse protection; similar results were obtained in Tera2/clone 13 (Stornaiuolo et al., 1990; Simeone et al., 1991). By Northern blot, *HOXA10* shows no transcript in K562 while in OCIM2 cells a transcript of 2.0 Kb is present; *HOXA10* exhibits two transcripts (2.4 Kb and 2.0 Kb) in both HL60 and in U937 cells (Magli et al., 1991).

■ References

Acampora, D., Simeone, A. and Boncinelli, E. (1991). Human HOX homeobox genes. In: Oxford Surveys on Eukaryotic Genes (MacLean, N., ed.), Oxford University Press 7, 1-28.

Boncinelli, E., Simeone, A., Acampora, D. and Mavilio, F. (1991). HOX gene activation by retinoic acid. Trends Genet. 7, 329-334.

Magli, M.C., Barba, P., Celetti, A., De Vita, G., Cillo, C. and Boncinelli, E. (1991). Coordinate regulation of HOX genes in human hematopoietic cells. Proc. Natl. Acad. Sci. USA 88, 6348-6352.

Simeone, A., Acampora, D., Nigro, V., Faiella, A., D'Esposito, M., Stornaiuolo, A., Mavilio, F. and Boncinelli, E. (1991). Differential regulation by retinoic acid of the homeobox genes of the four HOX loci in human embryonal carcinoma cells. Mech Dev. 33, 215-228.

Stornaiuolo, A., Acampora, D., Pannese, M., D'Esposito M., Morelli, F., Migliaccio, E., Rambaldi, M., Faiella, A., Nigro, V., Simeone, A. and Boncinelli, E. (1990). Human HOX genes are differentially activated by retinoic acid in embryonal carcinoma cells according to their position within the four loci. Cell Differ. Dev. 31, 119-127.

HOXA11 [h]

Species: *Homo sapiens*
Chromosomal location: 7p14-p21
Other names: *HOX1I*
Cognate genes: *Hoxa-11, Chox1i*
Type of homeobox: *Antp*-type
Accession number: X16665-X16667

■ Origin and description

HOXA11 was found during a chromosome walking by the screening of a genomic library constructed in the cosmid vector pCOS2 EMBL. The cosmid, named c2A, contained the *HOXA11* homeobox. *HOXA11* lies 15 Kb downstream from *HOXA13* and 12 Kb upstream from *HOXA10* homeobox. *HOXA11* belongs to *HOXA* locus and forms, together with *HOXC11* and *HOXD11* the III group of paralogy (Acampora et al., 1989; Boncinelli et al., 1991).

■ Expression

The expression of *HOXA11* was studied in human teratocarcinoma cell lines (NT2/D1 and Tera-2/clone 13) and in human hematopoietic cell lines (erythroleukemic: K562 and OCIM2; promyelocytic: HL60; monocytic: U937). In NT2/D1, both in stem and in 10 µM RA treated cells, no expression was found, both by Northern blot and by RNAse protection; similar results were obtained in Tera2/clone 13 (Stornaiuolo et al., 1990; Simeone et al., 1991). By Northern blot, *HOXA11* shows no transcript neither in K562 nor in OCIM2 cells; *HOXA11* exhibits two transcripts (2.3 Kb and 2.0 Kb) in HL60 and more intensively in U937 cells (Magli et al., 1991).

■ References

Acampora, D., Simeone, A. and Boncinelli, E. (1991). Human HOX homeobox genes. In: Oxford Surveys on Eukaryotic Genes (MacLean, N., ed.), Oxford University Press 7, 1-28.

Boncinelli, E., Simeone, A., Acampora, D. and Mavilio, F. (1991). HOX gene activation by retinoic acid. Trends Genet. 7, 329-334.

Magli, M.C., Barba, P., Celetti, A., De Vita, G., Cillo, C. and Boncinelli, E. (1991). Coordinate regulation of HOX genes in human hematopoietic cells. Proc. Natl. Acad. Sci. USA 88, 6348-6352.

Simeone, A., Acampora, D., Nigro, V., Faiella, A., D'Esposito, M., Stornaiuolo, A., Mavilio, F. and Boncinelli, E. (1991). Differential regulation by retinoic acid of the homeobox genes of the four HOX loci in human embryonal carcinoma cells. Mech Dev. 33, 215-228.

Stornaiuolo, A., Acampora, D., Pannese, M., D'Esposito M., Morelli, F., Migliaccio, E., Rambaldi, M., Faiella, A., Nigro, V., Simeone, A. and Boncinelli, E. (1990). Human HOX genes are differentially activated by retinoic acid in embryonal carcinoma cells according to their position within the four loci. Cell Differ. Dev. 31, 119-127.

HOXA13 [h]

Species: *Homo sapiens*
Chromosomal location: 7p14-p21
Other names: *HOX1J*
Cognate genes: *Hoxa-13*
Type of homeobox: *Antp*-type
Accession number: -

■ Origin and description

HOXA13 was found during a chromosome walking by the screening of a genomic library constructed in the cosmid vector pCOS2 EMBL. The cosmid, named c2A, contained the *HOXA13* homeobox. *HOXA13* lies 48 Kb downstream from the *EVX1* gene and 15 Kb upstream from *HOXA11* homeobox. *HOXA13* belongs to *HOXA* locus and forms, together with *HOXC13* and *HOXD13* the I group of paralogy (Acampora et al., 1989; Boncinelli et al., 1991).

■ Expression

The expression of *HOXA13* was studied in human teratocarcinoma cell lines (NT2/D1 and Tera-2/clone 13) and in human hematopoietic cell lines (erythroleukemic: K562 and OCIM2; promyelocytic: HL60; monocytic: U937). In NT2/D1, both in stem and in 10 µM RA treated cells, no expression was found, both by Northern blot and by RNAse protection; similar results were obtained in Tera2/clone 13 (Stornaiuolo et al., 1990; Simeone et al., 1991). By Northern blot, *HOXA13* shows no transcript in K562; a faint transcript of about 5 Kb is present in OCIM2; *HOXA13* exhibits multiple sized transcripts (5.0 Kb; 3.2 Kb; 2.8 Kb; 2.3 Kb) in HL60 and more intensively in U937 cells (Magli et al., 1991).

■ References

Acampora, D., Simeone, A. and Boncinelli, E. (1991). Human HOX homeobox genes. In: Oxford Surveys on Eukaryotic Genes (MacLean, N., ed.), Oxford University Press 7, 1-28.

Boncinelli, E., Simeone, A., Acampora, D. and Mavilio, F. (1991). HOX gene activation by retinoic acid. Trends Genet. 7, 329-334.

Magli, M.C., Barba, P., Celetti, A., De Vita, G., Cillo, C. and Boncinelli, E. (1991). Coordinate regulation of HOX genes in human hematopoietic cells. Proc. Natl. Acad. Sci. USA 88, 6348-6352.

Simeone, A., Acampora, D., Nigro, V., Faiella, A., D'Esposito, M., Stornaiuolo, A., Mavilio, F. and Boncinelli, E. (1991). Differential regulation by retinoic acid of the homeobox genes of the four HOX loci in human embryonal carcinoma cells. Mech Dev. 33, 215-228.

Stornaiuolo, A., Acampora, D., Pannese, M., D'Esposito M., Morelli, F., Migliaccio, E., Rambaldi, M., Faiella, A., Nigro, V., Simeone, A. and Boncinelli, E. (1990). Human HOX genes are differentially activated by retinoic acid in embryonal carcinoma cells according to their position within the four loci. Cell Differ. Dev. 31, 119-127.

HOXB1 [h]

Species: *Homo sapiens*
Chromosomal location: 17q21-q22
Other names: *HOX2I*
Cognate genes: *Hoxb-1, GHox-lab, AHox-1*
Type of homeobox: *Antp*-type, related to *lab*
Accession number: EMBL X16667

■ Origin and description

This gene was isolated during a chromosomal walk on the *HOXB* complex around the homeobox sequences first isolated, i.e. *HOXB7, HOXB5*. The homeobox sequence was identified using an oligonucleotide representing the most conserved portion of human homeobox 5' TGGTTCCA-GAACCGGCGGATGAA 3' (Acampora et al., 1989). This gene is the most 3' located member of the *HOXB* complex. Using *HOXB1* homeobox as probe we screened a cDNA library prepared from poly(A)+ RNA of human teratocarcinoma NTERA-2 cells cultured for 14 days in 10 µM retinoic acid a cDNA clone of 1 Kb was isolated (Acampora et al., 1989). Paralogous genes on the other human *HOX* loci are *HOXA1* and *HOXD1* (Acampora et al., 1991).

■ Expression

In differentiated NTERA-2 embryonal carcinoma cells (Stornaiuolo et al., 1990), the prevalent *HOXB1* band is a 2.4 Kb transcript. The expression of the *HOXB1* gene was analyzed in embryonal carcinoma NT2/D1 cells by RNase protection assay (Simeone et al., 1990, 1991). This gene reaches 50% of maximal expression after 14 hours of exposure to RA.

■ References

Acampora, D., D'Esposito, M., Faiella, A., Pannese, M., Migliaccio, E., Morelli, F., Stornaiuolo, A., Nigro, V., Simeone, A. and Boncinelli, E. (1989). The human HOX gene family. Nucl. Acids Res. 17, 10385-10402.

Acampora, D., Simeone, A. and Boncinelli, E. (1991). Human HOX homeobox genes. In: Oxford Surveys on Eukaryotic Genes (MacLean, N. ed.) Oxford University Press, 7, 1-28.

Simeone, A., Acampora, D., Arcioni, L., Andrews, P.W., Boncinelli, E., and Mavilio, F. (1990). Sequential activation of HOX2 homeobox genes by retinoic acid in human embryonal carcinoma cells. Nature 346, 763-766.

Simeone, A., Acampora, D., Nigro, V., Faiella, A., D'Esposito, M., Stornaiuolo, A., Mavilio, F. and Boncinelli, E. (1991).Differential regulation by retinoic acid of the homeobox genes of the four HOX loci in human embryonal carcinoma cells. Mech. Dev. 33, 215-228.

Stornaiuolo, A., Acampora, D., Pannese, M., D'Esposito, M., Morelli, F., Migliaccio, E., Rambaldi, M., Faiella, A., Nigro, V.,

Simeone, A. and Boncinelli, E. (1990). Human HOX genes are differentially activated by retinoic acid in embryonal carcinoma cells according to their position within the four loci. Cell Differ. Dev. 31, 119-127.

HOXB2 [h]

Species: *Homo sapiens*
Chromosomal location: 17q21-q22
Other names: *HOX2H, K8*
Cognate genes: *Hoxb-2, NvHbox2.8*
Type of homeobox: *Antp*-type
Accession number: EMBL X16666

■ Origin and description

This gene was isolated during a chromosomal walk on the *HOXB* complex around the homeobox sequences first isolated, i.e. *HOXB7, HOXB5*. The homeobox sequence was identified using an oligonucleotide representing the most conserved portion of human homeobox 5' TGGTTCCA-GAACCGGCGGATGAA 3'. (Acampora et al., 1989). Using *HOXB2* homeobox as probe we screened a cDNA library prepared from poly(A)+ RNA of human teratocarcinoma NTERA-2 cells cultured for 14 days in 10 µM retinoic acid a cDNA clone of 1.5 Kb was isolated (Acampora et al., 1989). Paralogous genes on the other human *HOX* loci is *HOXA2* (Acampora et al., 1991).

■ Expression

In embryos (Giampaolo et al., 1989) and in differentiated NTERA-2 embryonal carcinoma cells (Stornaiuolo et al., 1990), the prevalent *HOXB2* band is a 1.7 Kb transcript. In 7-week human embryos these transcripts have been found in spinal cord, medulla oblongata and brain, a faint expression was observed in liver, kidney, lung, gut and limbs (Giampaolo et al., 1989). The expression of the *HOXB2* gene was analyzed in embryonal carcinoma NT2/D1 cells by RNase protection assay (Simeone et al., 1990, 1991). This gene reaches 50% of maximal expression after 60 hours of exposure to RA. *HOXB2* is expressed at low levels in erythroleukemic cell lines K562 and OCIM2, whereas it is silent in promyelocytic, and monocytic cell lines (Magli et al., 1991).

■ References

Acampora, D., D'Esposito, M., Faiella, A., Pannese, M., Migliaccio, E., Morelli, F., Stornaiuolo, A., Nigro, V., Simeone, A. and Boncinelli, E. (1989). The human HOX gene family. Nucl. Acids Res. 17, 10385-10402.

Acampora, D., Simeone, A. and Boncinelli, E. (1991). Human HOX homeobox genes. In: Oxford Surveys on Eukaryotic Genes (MacLean, N. ed.), Oxford University Press 7, 1-28.

Giampaolo, A., Acampora, D., Zappavigna, V., Pannese, M., D'Esposito, M., Carè, A., Faiella, A., Stornaiuolo, A., Russo, G., Simeone, A., Boncinelli, E. and Peschle, C. (1989). Differential expression of human HOX-2 genes along the anterior-posterior axis in embryonic central nervous system. Differentiation 40, 191-197.

Magli, M.C., Barba, P., Celetti, A., De Vita, G., Cillo, C. and Boncinelli, E. (1991). Coordinate regulation of HOX genes in human hematopoietic cells. Proc. Natl. Acad. Sci. USA 88, 6348-6352.

Simeone, A., Acampora, D., Arcioni, L., Andrews, P.W., Boncinelli, E. and Mavilio, F. (1990). Sequential activation of HOX2 homeobox genes by retinoic acid in human embryonal carcinoma cells. Nature 346, 763-766.

Simeone, A., Acampora, D., Nigro, V., Faiella, A., D'Esposito, M., Stornaiuolo, A., Mavilio, F. and Boncinelli, E. (1991). Differential regulation by retinoic acid of the homeobox genes of the four HOX loci in human embryonal carcinoma cells. Mech. Dev. 33, 215-228.

Stornaiuolo, A., Acampora, D., Pannese, M., D'Esposito, M., Morelli, F., Migliaccio, E., Rambaldi, M., Faiella, A., Nigro, V., Simeone, A. and Boncinelli, E. (1990). Human HOX genes are differentially activated by retinoic acid in embryonal carcinoma cells according to their position within the four loci. Cell Differ. Dev. 31, 119-127.

HOXB3 [h]

Species: *Homo sapiens*
Chromosomal location: 17q21-q22
Other names: *HOX2G*
Cognate genes: *Hoxb-3, NvHbox2.7*
Type of homeobox: *Antp*-like
Accession number: EMBL X16665

■ Origin and description

This gene was isolated during a chromosomal walk on the *HOXB* complex around the homeobox sequences first isolated, i.e. *HOXB7, HOXB5*. The homeobox sequence was identified using an oligonucleotide representing the most conserved portion of human homeobox 5' TGGTTCCA-GAACCGGCGGATGAA 3'. (Boncinelli et al., 1988). Using *HOXB3* homeobox as probe we screened a cDNA library prepared from poly(A)+ RNA of human teratocarcinoma NTERA-2 cells cultured for 14 days in 10 μM retinoic acid a cDNA clone of 1.9 Kb was isolated (Acampora et al., 1989). Paralogous genes on the other human *HOX* loci are, *HOXA3* and *HOXD3* (Acampora et al., 1991).

■ Expression

In embryos (Giampaolo et al., 1989) and in differentiated NTERA-2 embryonal carcinoma cells (Stornaiuolo et al., 1990), the prevalent *HOXB3* band is a 3.8 Kb transcript. In 7-week human embryos these transcripts have been found in spinal cord, medulla oblongata and brain, a faint expression was observed in lung, gut and limbs (Giampaolo et al., 1989). The expression of the *HOXB3* gene was analyzed in embryonal carcinoma NT2/D1 cells by RNase protection assay (Simeone et al., 1990, 1991). This gene reaches 50% of maximal expression after 60 hours of exposure to RA. *HOXB3* is expressed at low levels in erythroleukemic cell lines K562 and OCIM2, whereas it is silent in promyelocytic, and monocytic cell lines (Magli et al., 1991).

■ References

Acampora, D., D'Esposito, M., Faiella, A., Pannese, M., Migliaccio, E., Morelli, F., Stornaiuolo, A., Nigro, V., Simeone, A. and Boncinelli, E. (1989). The human HOX gene family. Nucl. Acids Res. 17, 10385-10402.

Acampora, D., Simeone, A. and Boncinelli, E. (1991). Human HOX homeobox genes. In: Oxford Surveys on Eukaryotic Genes (MacLean, N. ed.), Oxford University Press 7, 1-28.

Boncinelli, E., Somma, R., Acampora, D., Pannese, M., D'Esposito, M., Faiella, A. and Simeone, A. (1988). Organization of human homeobox genes. Hum. Reprod. 3, 880-886.

Giampaolo, A., Acampora, D., Zappavigna, V., Pannese, M., D'Esposito, M., Carè, A., Faiella, A., Stornaiuolo, A., Russo, G., Simeone, A., Boncinelli, E. and Peschle, C. (1989). Differential expression of human HOX-2 genes along the anterior-posterior axis in embryonic central nervous system. Differentiation 40, 191-197.

Magli, M.C., Barba, P., Celetti, A., De Vita, G., Cillo, C. and Boncinelli, E. (1991). Coordinate regulation of HOX genes in human hematopoietic cells. Proc. Natl. Acad. Sci. USA 88, 6348-6352.

Simeone, A., Acampora, D., Arcioni, L., Andrews, P.W., Boncinelli, E. and Mavilio, F. (1990). Sequential activation of HOX2 homeobox genes by retinoic acid in human embryonal carcinoma cells. Nature 346, 763-766.

Simeone, A., Acampora, D., Nigro, V., Faiella, A., D'Esposito, M., Stornaiuolo, A., Mavilio, F. and Boncinelli, E. (1991). Differential regulation by retinoic acid of the homeobox genes of the four HOX loci in human embryonal carcinoma cells. Mech. Dev. 33, 215-228.

Stornaiuolo, A., Acampora, D., Pannese, M., D'Esposito, M., Morelli, F., Migliaccio, E., Rambaldi, M., Faiella, A., Nigro, V., Simeone, A. and Boncinelli, E. (1990). Human HOX genes are differentially activated by retinoic acid in embryonal carcinoma cells according to their position within the four loci. Cell Differ. Dev. 31, 119-127.

HOXB4 [h]

Species: *Homo sapiens*
Chromosomal location: 17q21-q22
Other names: *HOX2F*
Cognate genes: *Hoxb-4; Xhox-1A, Chox-z*
Type of homeobox: *Antp*-type, related to *Dfd*
Accession number: -

Origin and description

This gene was isolated, using our cDNA clones as probe, during a chromosomal walk on the *HOXB* complex. The homeobox sequence was identified using an oligonucleotide representing the most conserved portion of human homeobox 5' TGGTTCCAGAACCGGCGGATGAA 3'. (Acampora et al., 1987, Boncinelli et al., 1988). Using the homeodomain of *HOXB4* as probe a 2 Kb cDNA clone was isolated from human fetal liver (Peverali et al., 1990). Paralogous genes on the other human *HOX* loci are *HOXA4, HOXC4* and *HOXD4* (Acampora et al., 1991).

Expression

We studied the expression of *HOXB4* in five neuroblastoma (NB) cell lines SK-N-BE, SK-N-SH, CHP-134, IMR-32 and LAN-1, as untreated tumour cells and during RA-induced maturation, a specific transcript of 4.4 Kb is present in these cells. In NB cells treated with retinoic acid (RA) there is an increasing expression of this gene (Peverali et al., 1990). In embryos (Giampaolo et al., 1989) and in differentiated NTERA-2 embryonal carcinoma cells (Mavilio et al., 1988; Simeone et al., 1989; Stornaiuolo et al., 1990), the prevalent *HOXB4* band is a 2.4 Kb transcript with three higher molecular weight mnor bands including one of 4.4 Kb in length. In 7-week human embryos these transcripts have been found in spinal cord, *medulla oblongata*, lung, kidney, gut and limbs. This gene appears to be expressed also in embryonic and fetal liver, where very few other *HOX* genes are expressed (Giampaolo et al., 1989). The expression of the *HOXB4* gene ws analyzed in embryonal carcinoma NT2/D1 cells by RNase protection assay (Simeone et al., 1990, 1991). This gene reaches 50% of maximal expression after 60 hours of exposure to RA. *HOXB4* is expressed in erythroleukemic cell lines K562 and OCIM2, whereas it is silent in promyelocytic, and monocytic cell lines (Magli et al., 1991).

References

Acampora, D., Pannese, M., D'Esposito, M., Simeone, A. and Boncinelli, E. (1987). Human homeobox-containing genes in development. Hum. Reprod. 2, 407-414.

Acampora, D., Simeone, A. and Boncinelli, E. (1991). Human HOX homeobox genes. In: Oxford Surveys on Eukaryotic Genes (MacLean, N., ed.), Oxford University Press 7, 1-28.

Boncinelli, E., Somma, R., Acampora, D., Pannese, M., D'Esposito, M., Faiella, A. and Simeone, A. (1988). Organization of human homeobox gnes. Hum. Reprod. 3, 880-886.

Giampaolo, A., Acampora, D., Zappavigna, V., Pannese, M., D'Esposito, M., Carè, A., Faiella, A., Stornaiuolo, A., Russo, G., Simeone, A., Boncinelli, E. and Peschle, C. (1989). Differential expression of human HOX-2 genes along the anterior-posterior axis in embryonic central nervous system. Differentiation 40, 191-197.

Magli, M.C., Barba, P., Celetti, A., De Vita, G., Cillo, C. and Boncinelli, E. (1991). Coordinate regulation of HOX genes in human hematopoietic cells. Proc. Natl. Acad. Sci. USA 88, 6348-6352.

Mavilio, F., Simeone, A., Boncinelli, E. and Andrews, P.W. (1988). Activation of four homeobox gene cluster in human embryonal carcinoma cells induced to differentiate by retinoic acid. Differentiation 37, 73-79.

Peverali, A.F., D'Esposito, M., Acampora, D., Bunone, G., Negri, M., Faiella, A., Stornaiuolo, A., Pannese, M., Migliaccio, E., Simeone, A., Della Valle, G. and Boncinelli, E. (1990). Expression of HOX homeogenes in human neuroblastoma cell culture lines. Differentiation 45, 61-69.

Simeone, A., Acampora, D., D'Esposito, M., Faiella, A., Pannese, M., Scotto, L., Montanucci, M., D'Alessandro, G., Mavilio, F. and Boncinelli, E. (1989). Posttranscriptional control of human homeobox gene expression in induced NTERA-2 embryonal carcinoma cells. Molec. Rep. Dev. 1, 107-115.

Simeone, A., Acampora, D., Arcioni, L., Andrews, P.W., Boncinelli, E. and Mavilio, F. (1990). Sequential activation of HOX2 homeobox genes by retinoic acid in human embryonal carcinoma cells. Nature 346, 763-766.

Simeone, A., Acampora, D., Nigro, V., Faiella, A., D'Esposito, M., Stornaiuolo, A., Mavilio, F. and Boncinelli, E. (1991). Differential regulation by retinoic acid of the homeobox genes of the four HOX loci in human embryonal carcinoma cells. Mech Dev. 33, 215-228.

Stornaiuolo, A., Acampora, D., Pannese, M., D'Esposito M., Morelli, F., Migliaccio, E., Rambaldi, M., Faiella, A., Nigro, V., Simeone, A. and Boncinelli, E. (1990). Human HOX genes are differentially activated by retinoic acid in embryonal carcinoma cells according to their position within the four loci. Cell Differ. Dev. 31, 119-127.

HOXB5 [h]

Species: *Homo sapiens*
Chromosomal location: 17q21-q22
Other names: *HOX2A, c10, Hu1*
Cognate genes: *Hoxb-5, Xhox-1B, Xlhbox4, zf-21*
Type of homeobox: *Antp*-type
Accession number: -

Origin and description

This gene was isolated, by cross-homology with the *Drosophila Antp* and *ftz* homeoboxes, from a cDNA library prepared from SV40 transformed human fibroblasts (Boncinelli et al., 1985; Simeone et al., 1986). Paralogous genes on the other human *HOX* loci are *HOXA5* and *HOXC5* (Acampora et al., 1991).

Expression

HOXB5 is transcribed in human embryos and fetuses at 5-10 weeks post conception. A major polyadenylated transcript of 2.1 Kb, as well as RNA species of higher relative molecular mass, are specifically expressed at a constant level in spinal cord throughout this developmental period (Simeone et al., 1986). The expression of *HOXB5* was analyzed in embryonic tissues, particularly in the central nervous system (CNS) dissected in three regions (brain, *medulla oblongata*, spinal cord), here a specific transcript of 2.1 Kb is associated with two comigrating transcripts of 5.8 Kb, expressed only in spinal cord. It is of interest that HOXB5 transcripts are more abundant in the *medulla* than in spinal cord (Giampaolo et al, 1989). The expression of the *HOXB5* gene was studied in embryonal carcinoma NT2/D1 cells by Northern blot (Mavilio et al., 1988; Simeone et al., 1989; Stornaiuolo et al., 1990) and RNase protection analysis (Simeone et al., 1990, 1991). No expression of this gene was detectable in NT2/D1 before treatment with 10 μM retinoic acid, after addition of RA to the culture medium *HOXB5* gave rise to a transcript of 2.1 Kb. This gene reaches 50% of maximal expression after 60 hours of exposure to RA. *HOXB5* is expressed at low levels in erythroleukemic cell lines K562 and OCIM2, whereas it is silent in promyelocytic, and monocytic cell lines (Magli et al., 1991).

■ References

Acampora, D., Simeone, A. and Boncinelli, E. (1991). Human HOX homeobox genes. In: Oxford Surveys on Eukaryotic Genes (MacLean, N., ed.) Oxford University Press 7, 1-28.

Boncinelli, E., Simeone, A., La Volpe, A., Faiella, A., Fidanza, V., Acampora, D. and Scotto, L. (1985). Human cDNA clones containing homeobox sequences. Cold Spring Harbor Symp. Quant. Biol. 50, 301-305.

Giampaolo, A., Acampora, D., Zappavigna, V., Pannese, M., D'Esposito, M., Carè, A., Faiella, A., Stornaiuolo, A., Russo, G., Simeone, A., Boncinelli, E. and Peschle, C. (1989). Differential expression of human HOX-2 genes along the anterior-posterior axis in embryonic central nervous system. Differentiation 40, 191-197.

Magli, M.C., Barba, P., Celetti, A., De Vita, G., Cillo, C. and Boncinelli, E. (1991). Coordinate regulation of HOX genes in human hematopoietic cells. Proc. Natl. Acad. Sci. USA 88, 6348-6352.

Mavilio, F., Simeone, A., Boncinelli, E. and Andrews, P.W. (1988). Activation of four homeobox gene cluster in human embryonal carcinoma cells induced to differentiate by retinoic acid. Differentiation 37, 73-79.

Simeone, A., Mavilio, F., Bottero, L., Giampaolo, A., Russo, G., Faiella, A., Boncinelli, E. and Peschle, C. (1986). A human homeobox gene specifically expressed in spinal cord during embryonic development. Nature 320, 763-765.

Simeone, A., Acampora, D., D'Esposito, M., Faiella, A., Pannese, M., Scotto, L., Montanucci, M., D'Alessandro, G., Mavilio, F. and Boncinelli, E. (1989). Posttranscriptional control of human homeobox gene expression in induced NTERA-2 embryonal carcinoma cells. Molec. Rep. Dev. 1, 107-115.

Simeone, A., Acampora, D., Arcioni, L., Andrews, P.W., Boncinelli, E. and Mavilio, F. (1990). Sequential activation of HOX2 homeobox genes by retinoic acid in human embryonal carcinoma cells. Nature 346, 763-766.

Simeone, A., Acampora, D., Nigro, V., Faiella, A., D'Esposito, M., Stornaiuolo, A., Mavilio, F. and Boncinelli, E. (1991). Differential regulation by retinoic acid of the homeobox genes of the four HOX loci in human embryonal carcinoma cells. Mech. Dev. 33, 215-228.

Stornaiuolo, A, Acampora, D., Pannese, M., D'Esposito, M., Morelli, F., Migliaccio, E., Rambaldi, M., Faiella, A., Nigro, V., Simeone, A. and Boncinelli, E. (1990). Human HOX genes are differentially activated by retinoic acid in embryonal carcinoma

cells according to their position within the four loci. Cell Differ. Dev. 31, 119-127.

HOXB6   [h]

Species: *Homo sapiens*
Chromosomal location: 17q21-q22
Other names: *HOX2B, Hu2*
Cognate genes: *Hoxb-6*
Type of homeobox: *Antp*-type
Accession number: -

■ Origin and description

This gene was isolated, using our cDNA clones as probe, during a chromosomal walk on the *HOXB* complex. The homeobox sequence was identified using an oligonucleotide representing the most conserved portion of human homeobox 5' TGGTTCCAGAACCGGCGGATGAA 3'. (Acampora et al., 1987, Boncinelli et al., 1988). Paralogous genes on the other human *HOX* loci are, *HOXA6* and *HOXC6* (Acampora et al., 1991).

■ Expression

The expression of *HOXB6* was analyzed in human embryonic tissues, particularly in the central nervous system (CNS) dissected in three regions (brain, *medulla oblongata*, spinal cord), here a specific transcript of 1.6 Kb is associated with two comigrating transcripts of 5.8 Kb, a minor transcript of 2.2 Kb is also detected only in spinal cord. It is of interest that *HOXB6* transcripts are more abundant in the *medulla* than in spinal cord (Giampaolo et al., 1989). Weak expression was observed in lung. The expression of the *HOXB6* gene was studied in embryonal carcinoma NT2/D1 cells by Northern blot (Mavilio et al., 1988; Stornaiuolo et al., 1990) and RNase protection (Simeone et al., 1990, 1991) analysis. No expression of this gene was detectable in NT2/D1 before treatment with 10 μM retinoic acid, after addition of RA to the culture medium *HOXB6* gave rise to a single major transcript of 1.5 Kb. This gene reaches 50% of maximal expression after 70 hours of exposure to RA. *HOXB6* is expressed in erythroleukemic cell lines K562 and OCIM2, whereas it is silent in promyelocytic, and monocytic cell lines (Magli et al., 1991).

■ References

Acampora, D., Pannese, M., D'Esposito, M., Simeone, A. and Boncinelli, E. (1987). Human homeobox-containing genes in development. Human Reproduction 2, 407-414.

Acampora, D., Simeone, A. and Boncinelli, E. (1991). Human HOX homeobox genes. In: Oxford Surveys on Eukaryotic Genes (MacLean, N., ed.) Oxford University Press 7, 1-28.

Boncinelli, E., Somma, R,. Acampora, D., Pannese, M., D'Esposito, M., Faiella, A. and Simeone, A. (1988). Organization of human homeobox genes. Human Reproduction 3, 880-886.

Giampaolo, A., Acampora, D., Zappavigna, V., Pannese, M., D'Esposito, M., Carè, A., Faiella, A., Stornaiuolo, A., Russo, G., Simeone, A., Boncinelli, E. and Peschle, C. (1989). Differential expression of human HOX-2 genes along the anterior-posterior axis in embryonic central nervous system. Differentiation 40, 191-197.

Magli, M.C., Barba, P., Celetti, A., De Vita, G., Cillo, C. and Boncinelli, E. (1991). Coordinate regulation of HOX genes in human hematopoietic cells. Proc. Natl. Acad. Sci. USA 88, 6348-6352.

Mavilio, F., Simeone, A., Boncinelli, E. and Andrews, P.W. (1988). Activation of four homeobox gene cluster in human embryonal carinoma cells induced to differentiate by retinoic acid. Differentiation 37, 73-79.

Simeone, A., Acampora, D., Arcioni, L., Andrews, P.W., Boncinelli, E. and Mavilio, F. (1990). Sequential activation of HOX2 homeobox genes by retinoic acid in human embryonal carcinoma cells. Nature 346, 763-766.

Simeone, A., Acampora, D., Nigro, V., Faiella, A., D'Esposito, M., Stornaiuolo, A., Mavilio, F. and Boncinelli, E. (1991). Differential regulation by retinoic acid of the homeobox genes of the four HOX loci in human embryonal carcinoma cells. Mech. Dev. 33, 215-228.

Stornaiuolo, A, Acampora, D., Pannese, M., D'Esposito, M., Morelli, F., Migliaccio, E., Rambaldi, M., Faiella, A., Nigro, V., Simeone, A. and Boncinelli, E. (1990). Human HOX genes are differentially activated by retinoic acid in embryonal carcinoma cells according to their position within the four loci. Cell Differ. Dev. 31, 119-127.

HOXB7 [h]

Species: *Homo sapiens*

Chromosomal location: 17q21-q22

Other names: *HOX2C, c1*

Cognate genes: *Hoxb-7, Xlhbox2*

Type of homeobox: *Antp*-type

Accession number: -

■ Origin and description

This gene was isolated, by cross-homology with the *Drosophila Antp* and *ftz* homeoboxes, from a cDNA library prepared from SV40 transformed human fibroblasts (Boncinelli et al., 1985; Simeone et al., 1987). Paralogous gene on the other human *HOX* loci is *HOXA7* (Acampora et al., 1991).

■ Expression

The expression of *HOXB7* was analyzed in human embryonic development. This gene is differentially expressed as a single major transcript of 1.6 Kb in spinal cord, backbone rudiments, limb bud, heart, and skin of human embryos and early fetuses in the 5- to 9-week postfertilization period (Simeone et al., 1987). It is expressed also in the cell line Caco-2 (cell line derived from a human colon carcinoma) and human adult intestine (Sebastio et al., 1987). In the central nervous system (CNS) dissected in three regions (brain, *medulla oblongata*, spinal cord) a single transcript of 1.6 Kb is present. It is of interest that *HOXB7* transcript is markedly more abundant in spinal cord than in *medulla* (Giampaolo et al., 1989). The expression of the *HOXB7* gene was studied in embryonal carcinoma NT2/D1 cells by Northern blot (Mavilio et al., 1988; Stornaiuolo et al., 1990) and RNase protection (Simeone et al., 1990, 1991) analysis. No expression of this gene was detectable in NT2/D1 before treatment with 10 µM retinoic acid, after addition of RA to the culture medium *HOXB7* gave rise to a single major transcript of 1.6 Kb. This gene reaches 50% of maximal expression after 90 hours of exposure to RA. The expression of *HOXB7* was studied also in erythroleukemic cell lines K562 and OCIM2, here the transcripts of 1.6 and 1.4 Kb are present whereas it is silent in promyelocytic, and monocytic cell lines (Magli et al., 1991).

■ References

Acampora, D., Simeone, A. and Boncinelli, E. (1991). Human HOX homeobox genes. In: Oxford Surveys on Eukaryotic Genes (MacLean, N., ed.) Oxford University Press 7, 1-28.

Boncinelli, E., Simeone, A., La Volpe, A., Faiella, A., Fidanza, V., Acampora, D. and Scotto, L. (1985). Human cDNA clones containing homeobox sequences. Cold Spring Harbor Symp. Quant. Biol. 50, 301-305.

Giampaolo, A., Acampora, D., Zappavigna, V., Pannese, M., D'Esposito, M., Carè, A., Faiella, A., Stornaiuolo, A., Russo, G., Simeone, A., Boncinelli, E. and Peschle, C. (1989). Differential expression of human HOX-2 genes along the anterior-posterior axis in embryonic central nervous system. Differentiation 40, 191-197.

Magli, M.C., Barba, P., Celetti, A., De Vita, G., Cillo, C. and Boncinelli, E. (1991). Coordinate regulation of HOX genes in human hematopoietic cells. Proc. Natl. Acad. Sci. USA 88, 6348-6352.

Mavilio, F., Simeone, A., Boncinelli, E. and Andrews, P.W. (1988). Activation of four homeobox gene cluster in human embryonal carinoma cells induced to differentiate by retinoic acid. Differentiation 37, 73-79.

Sebastio, G., D'Esposito, M., Montanucci, M., Simeone, A., Auricchio, S. and Boncinelli, E. (1987). Modulated expression of human homeobox genes in differentiating intestinal cells. Biochem. Biophys. Res. Commun. 146, 751-756.

Simeone, A., Mavilio, F., Acampora, D., Giampaolo, A., Faiella, A., Zappavigna, V., D'Esposito, M., Pannese, M., Russo, G., Boncinelli, E. and Peschle, C. (1987). Two human homeobox genes, c1 and c8: Structure analysis and expression in embryonic development. Proc. Natl. Acad. Sci. 84, 4914-4918.

Simeone, A., Acampora, D., Arcioni, L., Andrews, P.W., Boncinelli, E. and Mavilio, F. (1990). Sequential activation of HOX2 homeobox genes by retinoic acid in human embryonal carcinoma cells. Nature 346, 763-766.

Simeone, A., Acampora, D., Nigro, V., Faiella, A., D'Esposito, M., Stornaiuolo, A., Mavilio, F. and Boncinelli, E. (1991). Differential regulation by retinoic acid of the homeobox genes of the four HOX loci in human embryonal carcinoma cells. Mech. Dev. 33, 215-228.

Stornaiuolo, A, Acampora, D., Pannese, M., D'Esposito, M.,

Morelli, F., Migliaccio, E., Rambaldi, M., Faiella, A., Nigro, V., Simeone, A. and Boncinelli, E. (1990). Human HOX genes are differentially activated by retinoic acid in embryonal carcinoma cells according to their position within the four loci. Cell Differ. Dev. 31, 119-127.

HOXB8 [h]

Species: *Homo sapiens*
Chromosomal location: 17q21-q22
Other names: *HOX2D*
Cognate genes: *Hoxb-8*, *Xlhbox7*
Type of homeobox: *Antp*-type
Accession number: -

■ Origin and description

This gene was found during a chromosomal walk on the *HOXB* complex (Acampora et al., 1987; Boncinelli et al., 1988) The homeobox sequence (Giampaolo et al., 1989) was identified using an oligonucleotide representing the most conserved portion of human homeobox 5′ TGGTTCCAGAACCGGCGGATGAA 3′. Paralogous genes on the other human *HOX* loci are *HOXC8*, *HOXD8* (Acampora et al., 1991).

■ Expression

The expression of *HOXB8* was analyzed in embryonic tissues, particularly in the central nervous system (CNS) dissected in three regions (brain, *medulla oblongata*, spinal cord), were a single transcript of 2.3 Kb is present. It is of interest that *HOXB8* transcript is markedly more abundant in spinal cord than in *medulla* (Giampaolo et al., 1989). The expression of the *HOXB8* gene was studied in embryonal carcinoma NT2/D1 cells by Northern blot (Mavilio et al., 1988, Stornaiuolo et al., 1990) and RNase protection (Simeone et al., 1990, 1991) analysis. No expression of this gene was detectable in NT2/D1 before treatment with 10 μM retinoic acid, after addition of RA to the culture medium *HOXB8* gave rise to multiple transcripts of 3.5, 2.3, 1.7, and 1.3 Kb. This gene reaches 50% of maximal expression after 115 hours of exposure to RA. The expression of *HOXB8* was studied also in erythroleukemic cell lines K562 and OCIM2, here multiple transcripts are present whereas it is silent in promyelocytic, and monocytic cell lines (Magli et al., 1991).

■ References

Acampora, D., Pannese, M., D'Esposito, M., Simeone, A. and Boncinelli, E. (1987). Human homeobox-containing genes in development. Hum. Reprod. 2, 407-414.
Acampora, D., Simeone, A. and Boncinelli, E. (1991). Human HOX homeobox genes. In: Oxford Surveys on Eukaryotic Genes (MacLean, N., ed.), Oxford University Press 7, 1-28.
Boncinelli, E., Somma, R., Acampora, D., Pannese, M., D'Esposito, M., Faiella, A. and Simeone, A. (1988). Organization of human homeobox gnes. Hum. Reprod. 3, 880-886.
Giampaolo, A., Acampora, D., Zappavigna, V., Pannese, M., D'Esposito, M., Carè, A., Faiella, A., Stornaiuolo, A., Russo, G., Simeone, A., Boncinelli, E. and Peschle, C. (1989). Differential expression of human HOX-2 genes along the anterior-posterior axis in embryonic central nervous system. Differentiation 40, 191-197.
Magli, M.C., Barba, P., Celetti, A., De Vita, G., Cillo, C. and Boncinelli, E. (1991). Coordinate regulation of HOX genes in human hematopoietic cells. Proc. Natl. Acad. Sci. USA 88, 6348-6352.
Mavilio, F., Simeone, A., Boncinelli, E. and Andrews, P.W. (1988). Activation of four homeobox gene cluster in human embryonal carcinoma cells induced to differentiate by retinoic acid. Differentiation 37, 73-79.
Simeone, A., Acampora, D., Arcioni, L., Andrews, P.W., Boncinelli, E. and Mavilio, F. (1990). Sequential activation of HOX2 homeobox genes by retinoic acid in human embryonal carcinoma cells. Nature 346, 763-766.
Simeone, A., Acampora, D., Nigro, V., Faiella, A., D'Esposito, M., Stornaiuolo, A., Mavilio, F. and Boncinelli, E. (1991). Differential regulation by retinoic acid of the homeobox genes of the four HOX loci in human embryonal carcinoma cells. Mech Dev. 33, 215-228.
Stornaiuolo, A., Acampora, D., Pannese, M., D'Esposito M., Morelli, F., Migliaccio, E., Rambaldi, M., Faiella, A., Nigro, V., Simeone, A. and Boncinelli, E. (1990). Human HOX genes are differentially activated by retinoic acid in embryonal carcinoma cells according to their position within the four loci. Cell Differ. Dev. 31, 119-127.

HOXB9 [h]

Species: *Homo sapiens*
Chromosomal location: 17q21-q22
Other names: *HOX2E*
Cognate genes: *Hoxb-9*; *Xlhbox6*
Type of homeobox: *Antp*-type, related to the *Drosophila Abd-B* gene
Accession number: -

■ Origin and description

This gene was found during a chromosomal walk on the *HOXB* complex (Acampora et al., 1987; Boncinelli et al., 1988) The homeobox sequence (Giampaolo et al., 1989) was identified using an oligonucleotide representing the most conserved portion of human homeobox 5′ TGGTTCCAGAACCGGCGGATGAA 3′. This gene is the most 5′ located member of the *HOXB* complex. Paralogous genes on the other human *HOX* loci are *HOXA9*, *HOXC9*, *HOXD9* (Acampora et al., 1991).

■ Expression

The expression of *HOXB9* was analyzed in embryonic tissues, particularly in the central nervous system (CNS) dissected in three regions (brain, *medulla oblongata*, spinal cord), were a single transcript of 3.4 Kb is present. It is of interest that *HOXB9* transcript is markedly more abundant in spinal cord than in *medulla*. Transcription of *HOXB9* was observed in kidney and at lower level in gut (Giampaolo et al., 1989). The expression of the *HOXB9* gene was studied in embryonal carcinoma NT2/D1 cells by Northern blot (Mavilio et al., 1988; Stornaiuolo et al., 1990) and RNase protection (Simeone et al., 1990, 1991) analysis. No expression of this gene was detectable in NT2/D1 before treatment with 10 µM retinoic acid, after addition of RA to the culture medium transcript of 2.9 Kb is present. This gene reaches 50% of maximal expression after 180 hours of exposure to RA. The expression of *HOXB9* was studied also in erythroleukemic cell lines K562 and OCIM2, here a single transcript of 2.9 Kb is present whereas it is silent in promyelocytic, and monocytic cell lines (Magli et al., 1991).

■ References

Acampora, D., Pannese, M., D'Esposito, M., Simeone, A. and Boncinelli, E. (1987). Human homeobox-containing genes in development. Hum. Reprod. 2, 407-414.

Acampora, D., Simeone, A. and Boncinelli, E. (1991). Human HOX homeobox genes. In: Oxford Surveys on Eukaryotic Genes (MacLean, N., ed.), Oxford University Press 7, 1-28.

Boncinelli, E., Somma, R., Acampora, D., Pannese, M., D'Esposito, M., Faiella, A. and Simeone, A. (1988). Organization of human homeobox gnes. Hum. Reprod. 3, 880-886.

Giampaolo, A., Acampora, D., Zappavigna, V., Pannese, M., D'Esposito, M., Carè, A., Faiella, A., Stornaiuolo, A., Russo, G., Simeone, A., Boncinelli, E. and Peschle, C. (1989). Differential expression of human HOX-2 genes along the anterior-posterior axis in embryonic central nervous system. Differentiation 40, 191-197.

Magli, M.C., Barba, P., Celetti, A., De Vita, G., Cillo, C. and Boncinelli, E. (1991). Coordinate regulation of HOX genes in human hematopoietic cells. Proc. Natl. Acad. Sci. USA 88, 6348-6352.

Mavilio, F., Simeone, A., Boncinelli, E. and Andrews, P.W. (1988). Activation of four homeobox gene cluster in human embryonal carcinoma cells induced to differentiate by retinoic acid. Differentiation 37, 73-79.

Simeone, A., Acampora, D., Arcioni, L., Andrews, P.W., Boncinelli, E. and Mavilio, F. (1990). Sequential activation of HOX2 homeobox genes by retinoic acid in human embryonal carcinoma cells. Nature 346, 763-766.

Simeone, A., Acampora, D., Nigro, V., Faiella, A., D'Esposito, M., Stornaiuolo, A., Mavilio, F. and Boncinelli, E. (1991). Differential regulation by retinoic acid of the homeobox genes of the four HOX loci in human embryonal carcinoma cells. Mech Dev. 33, 215-228.

Stornaiuolo, A., Acampora, D., Pannese, M., D'Esposito M., Morelli, F., Migliaccio, E., Rambaldi, M., Faiella, A., Nigro, V., Simeone, A. and Boncinelli, E. (1990). Human HOX genes are differentially activated by retinoic acid in embryonal carcinoma cells according to their position within the four loci. Cell Differ. Dev. 31, 119-127.

HOXC4 [h]

Species: *Homo sapiens*
Chromosomal location: 12q12-12q13
Other names: *HOX3E, cp8; cp19*
Cognate genes: *Hoxc-4*
Type of homeobox: *Antp*-type
Accession number: X07495

■ Origin and description

Using a *HOXC6* cDNA as probe, a genomic region of 45 kb, represented in overlapping cosmid clones, was isolated: in this region, about 26 kb downstream of *HOXC6*, *HOXC4* was found. *HOXC4* is in the *HOXC* locus and forms the human paralogy group X with *HOXA4*, *HOXB4* and *HOXD4,* related to the *Dfd* box of *Drosophila* (Simeone et al., 1988; Boncinelli, et al., 1991).

■ Expression

The expression of *HOXC4* was studied in human embryo, human extraembryonic tissue, human teratocarcinoma, hematopoietic and neuroblastoma cell lines. By Northern blotting, at least three transcripts of this homeobox were found in human embryo of 8 weeks (Simeone et al., 1988). Three transcripts were detected also in human placenta. Two cDNAs were isolated from a placental cDNA library, cp8 and cp19, bearing the same coding region and differing from each other only because the former contains, unspliced, the second intron of the latter; interestingly, corresponding mRNAs are among the products of a large transcriptional unit comprising at least three homeoboxes of the same cluster (Simeone et al., 1988). By *in situ* hybridization, a different pattern of expression was detected for cp8 and cp19 in trophoblast and stromal cells of chorionic villi (Oudejans et al., 1990). Expression studies were carried on teratocarcinoma NT2/D1 cell line: here, *HOXC4,* switched off in stem cells, gave rise to two transcripts after treatment with RA 10 µM (Stornaiuolo et al., 1990; Simeone et al., 1991). Similar behaviour was exhibited by the homeobox in teratocarcinoma Tera2/clone 13 cell line (Simeone et al., 1991). By Northern blotting two transcripts of *HOXC4* were found in four tested human hematopoietic cell lines (erythroleukemic: K562 and OCIM2; promyelocytic: HL60; monocytic: U937) (Magli et al., 1991). The expression of *HOXC4* was studied also in five neuroblastoma cell lines (SK-N-BE; CHP-134, IMR-32, SK-N-SH, LAN-1), where only one transcript was found. *HOXC4* mRNA level is generally low in these cells, particularly in LAN-1; after RA treatment, a significant increase of *HOXC4* expression can be observed only in IMR-32, CHP-134 and SK-N-SH cells (Peverali et al., 1990).

■ Function

By cotransfection in HeLa and NIH3T3 cells, HOXC4 protein appears to be unable to transactivate the *HOXC5*-cp11T promoter: its behaviour resembles to that of HOXD4 and differs from that of HOXC5, C6, D10 and D9 proteins, which are positive transactivators of this promoter (Arcioni et al., 1991).

■ References

Arcioni, L., Simeone, A., Guazzi, S., Zappavigna, V., Boncinelli, E. and Mavilio, F. (1992). The upstream region f the human homeobox gene HOXC5 is a target for regulation by retinoic acid and HOX proteins. EMBO J. 11, 265-278.

Boncinelli, E., Simeone, A., Acampora, D. and Mavilio, F. (1991). HOX gene activation by retinoic acid. Trends Genet. 7, 329-334.

Magli, M.C., Barba, P., Celetti, A., De Vita G., Cillo, C. and Boncinelli, E. (1991). Coordinate regulation of HOX genes in human hematopoietic cells. Proc. Natl. Acad. Sci. USA 88, 6348-6352.

Oudejans, C.B.M., Pannese, M., Simeone, A., Meijer, C.I.L.M. and Boncinelli, E. (1990). The three most downstream genes of the HOX3 cluster are expressed in human extraembryonic tissues including trophoblast of androgenic origin. Development 108, 471-477.

Peverali, F.A., D'Esposito, M., Acampora, D., Bunone, G., Negri, M., Faiella, A., Stornaiuolo, A., Pannese, M., Migliaccio, E., Simeone, A., Della Valle, G. and Boncinelli, E. (1990). Expression of HOX homeogenes in human neuroblastoma cell culture lines. Differentiation 45, 61-69.

Simeone, A., Pannese, M., Acampora, D., D'Esposito, M. and Boncinelli, E. (1988). At least three human homeoboxes on chromosome 12 belong to the same transcription unit. Nucl. Acids Res. 16, 5379-5387.

Simeone, A., Acampora, D., Nigro, V., Faiella, A., D'Esposito, M., Stornaiuolo, A., Mavilio, F. and Boncinelli, E. (1991). Differential regulation by retinoic acid of the homeobox genes of the four HOX loci in human embryonal carcinoma cells. Mech. Dev. 33, 215-228.

Stornaiuolo, A., Acampora, D., Pannese, M., D'Esposito, M., Morelli, F., Migliaccio, E., Rambaldi, M., Faiella, A., Nigro, V., Simeone, A. and Boncinelli, E. (1990). Human HOX genes are differentially activated by retinoic acid in embryonal carcinoma cells according to their position within the four loci. Cell Differ. Dev. 31, 119-127.

HOXC5 [h]

Species: *Homo sapiens*

Chromosomal location: 12q12-12q13

Other names: *HOX3D, cp11*

Cognate genes: *Hoxc-5, Xlhbox5, Nvhbox3.4, ZF-25*

Type of homeobox: *Antp*-type

Accession number: -

■ Origin and description

The *HOXC5* gene was cloned from a placenta cDNA library using as probe a short DNA fragment containing the 5' exon of the cDNA clone termed c8.5111 (Simeone et al., 1988). The *HOXC* complex at least in placenta and probably in embryonic spinal cord reveals in fact a peculiar transcript organization. Three homeobox sequences, present in this chromosomal region, identify a single transcription unit. Primary transcripts from an upstream "master promoter" are alternatively processed to give mature messengers with the common 5' non coding exon (the same used in the screening) and 3' exons in term containing three different homeoboxes: *HOXC6, HOXC5* and *HOXC4*.

■ Expression

Two families of *HOXC5* cDNAs were isolated. The former (named *cp11*) correspond to that described above, the latter (named *cp11T*) is transcribed from a different promoter, proximal to the homeobox and reveals a different intron-exon organization (Arcioni et al., 1992; Acampora et al., 1991). The *HOXC5* gene is expressed in human embryonic spinal cord and placenta as a heterogenous 1.7 kb RNA band. As assayed by Northern blotting this band contains transcripts of both *cp11* and *cp11T* type and can be distinguished only by differential hybridization or RNase protection. *HOXC5* is constitutively active in a number of neuroblastoma cell lines, such as SK-N-BE (Peverali et al., 1990) and is activated by retinoic acid (RA) in the EC cell line NT2/D1. Only the *cp11T*-like type of transcript is expressed in neuroblastoma cells and induced by 10^{-5} M RA in EC cells (Simeone et al., 1991) after 80 hours of treatment.

■ References

Acampora, D., Simeone, A. and Boncinelli, E. (1991). Human HOX homeobox genes. In: Oxford Surveys on Eukaryotic Genes (MacLean, N., ed.) Oxford University Press 7, 1-28.

Arcioni, L., Simeone, A., Guazzi, S., Zappavigna, V., Boncinelli, E. and Mavilio, F. (1992). The upstream region of the human homeobox gene HOXC5 is a target for regulation by retinoic acid and HOX proteins. EMBO J. 11, 265-278.

Peverali, F.A., D'Esposito, M., Acampora, D., Bunone, G., Negri, M., Faiella, A., Stornaiuolo, A., Pannese, M., Migliaccio, E., Simeone, A., Della Valle, G. and Boncinelli, E. (1990). Expression of HOX homeogenes in human neuroblastoma cell culture lines. Differentiation 45, 61-69.

Simeone, A., Pannese, M., Acampora, D., D'Esposito, M. and Boncinelli, E. (1988). At least three human homeoboxes on chromosome 12 belong to the same transcription unit. Nucl. Acids Res. 16, 5379-5387.

Simeone, A., Acampora, D., Nigro, V., Faiella, A., D'Esposito, M., Stornaiuolo, A., Mavilio, F. and Boncinelli, E. (1991). Differential regulation by retinoic acid of the homeobox genes of the four HOX loci in human embryonal carcinoma cells. Mech. Dev. 33, 215-228.

HOXC6 [h]

Species: *Homo sapiens*
Chromosomal location: 12q12-12q13
Other names: *HOX3C, c8*
Cognate genes: *Hoxc-6; Xlhbox1, FH2, NvHbox1*
Type of homeobox: *Antp*-type
Accession number: X07495

■ Origin and description

This gene was isolated by cross-homology with the *Drosophila Antp* and *ftz* homeoboxes (Boncinelli et al., 1985). Paralogous genes on other complexes are *HOXB6* and *HOXA6*. The *HOXC6* homeobox appears to be present in at least three different mRNAs encoding three different homeoproteins (Pannese et al., 1989). Two of these are well characterized both in human (Acampora et al., 1991) and frog (Cho et al., 1988). They are named c8S (for short c8) and c8L (for long c8). c8S is fully contained in c8L but the latter differs from the former for an extradomain of 82 aa at the NH_2 terminus. The corresponding mRNAs are transcribed from two different promoters and reveal a different intron-exon organization.

■ Expression

The two mRNAs corresponding to c8S and c8L proteins (respectively 2.2 and 1.8 kb) coexist in cultured human fibroblasts and in several embryonic tissues (spinal cord, backbone, limbs and skin) although in different ratio. Analysis of *HOXC6* gene expression at least in placenta and embryonic spinal cord revealed a complex transcriptional organization of these genes (Simeone et al., 1988). The whole chromosomal region comprising at least four homeoboxes, named *HOXC8, C6, C5* and *C4* appears to be transcribed, from a "master" upstream promoter, in a single primary transcript, alternatively spliced to give mature mRNAs encoding proteins containing in their carboxyl region one or the other of the four homeodomains. The 2.2 kb mRNA correspondent to c8S is transcribed from this "master promoter". On the other hand the 1.8 mRNA correspondent to c8L is transcribed from a different promoter proximal to the *HOXC6* homeobox and respondent to retinoic acid induction in EC cell line NT2/D1 which normally does not express neither the 2.2 nor the 1.8 kb transcript (Simeone et al., 1991).

■ Function

There is some evidence that c8L works as a sequence specific transcription activator. A c8L construct under a cytomegalovirus promoter is able to activate a reporter gene present downstream from a *Drosophila* homeoprotein target sequence in cotransfection experiments into HeLa cells. The target sequence is a duplicated engrailed homeoprotein target site (TCAATTAAAT). An identical experiment with a c8S construct failed to show any activation of the same target sequence. A possible antagonistic role between the two proteins has been also suggested in frog embryos (Wright et al., 1989).

■ References

Acampora, D., Simeone, A. and Boncinelli, E. (1991). Human HOX homeobox genes. In: Oxford Surveys on Eukaryotic Genes (MacLean, N., ed.) Oxford University Press 7, 1-28.

Boncinelli, E., Simeone, A., La Volpe, A., Faiella, A., Fidanza, V., Acampora, D. and Scotto, L. (1985). Human cDNA clones containing the homeobox. Cold Spring Harbor Symp. Quant. Biol. 50, 301-305.

Cho, K.W.Y., Gotz, J., Wright, C.V.E., Fritz, A., Hardwicke, J. and De Robertis, E.M. (1988). Differential utilization of the same reading frame in a *Xenopus* homeobox gene encodes two related proteins sharing the same DNA binding specificity. EMBO J. 7, 2139-2149.

Pannese, M., Stornaiuolo, A., Acampora, D., D'Esposito, M., Cafiero, M., Somma, R., Morelli, F., Menna, A., Faiella, A. and Boncinelli, E. (1989). Human class I homeobox gene expression in teratocarcinoma cells. In: Pathology of gene expression 131-149, Frati, L. and Aaronson, S.A. eds. Raven Press.

Simeone, A., Pannese, M., Acampora, D., D'Esposito, M. and Boncinelli, E. (1988). At least three human homeoboxes on chromosome 12 belong to the same transcription unit. Nucl. Acids Res. 16, 5379-5387.

Simeone, A., Acampora, D., Nigro, V., Faiella, A., D'Esposito, M., Stornaiuolo, A., Mavilio, F. and Boncinelli, E. (1991). Differential regulation by retinoic acid of the homeobox genes of the four HOX loci in human embryonal carcinoma cells. Mech. Dev. 33, 215-228.

Wright, C.V.E., Cho, K.W.Y., Hardwicke, J., Collins, R.H. and De Robertis, E.M. (1989). Interference with function of a homeobox gene in *Xenopus* embryos produces malformations of the anterior spinal cord. Cell 59, 81-93.

HOXC8 [h]

Species: *Homo sapiens*
Chromosomal location: 12q12-12q13
Other names: *HOX3A, HOX3.1*
Cognate genes: *Hoxc-8*
Type of homeobox: *Antp*-type
Accession number: X16665-X16667

■ Origin and description

This gene was isolated through its linkage to *HOXC6* gene during "chromosome walking" experiments (Boncinelli et al., 1988). Paralogous genes on other complexes are *HOXB8* and *HOXD8* (Acampora et al., 1991).

■ Expression

HOXC8 gene expression has been studied in different hematopoietic cell lines: K562 and OCIM2 (erythroleukemic cell lines), HL60 (promyelocytic cell line), U937 (monocytic cell line). *HOXC8* gene is expressed in all four lines analyzed with a major transcript of 2.4 kb (Magli et al., 1991). *HOXC8* gene expression has been also studied in EC cell line NT2/D1 before and after a time-course induction, from 1 to 195 hours, in the continuous presence of 10^{-5} M retinoic acid (RA). RNAse protection assays reveal that *HOXC8* is undetectable in untreated cells and only weakly induced after 150 hours of exposure to RA (Simeone et al., 1991).

■ References

Acampora, D., Simeone, A. and Boncinelli, E. (1991). Human HOX homeobox genes. In: Oxford Surveys on Eukaryotic Genes (MacLean, N. ed.), Oxford University Press 7, 1-28.

Boncinelli, E., Somma, R., Acampora, D., Pannese, M., D'Esposito, M., Faiella, A. and Simeone, A. (1988). Organization of human homeobox genes. Hum. Reprod. 3, 880-886.

Magli, M.C., Barba, P., Celetti, A., De Vita, G., Cillo, C. and Boncinelli, E. (1991). Coordinate regulation of HOX genes in human hematopoietic cells. Proc. Natl. Acad. Sci. USA 88, 6348-6352.

Simeone, A., Acampora, D., Nigro, V., Faiella, A., D'Esposito, M., Stornaiuolo, A., Mavilio, F. and Boncinelli, E. (1991). Differential regulation by retinoic acid of the homeobox genes of the four HOX loci in human embryonal carcinoma cells. Mech. Dev. 33, 215-228.

HOXC9 [h]

Species: *Homo sapiens*
Chromosomal location: 12q12-12q13
Other names: *HOX3B*
Cognate genes: *Hoxc-9*
Type of homeobox: *Antp*-type, related to *Abd-B*
Accession number: -

■ Origin and description

This gene was found during a chromosomal walk on the *HOXC* complex. The homeobox sequence was identified using an oligonucleotide representing the most conserved portion of human homeobox: 5' TGGTTCCAGAACCGGCG-GATGAA 3'. This gene was identified in a cosmid clone that already had the *HOXC8* and the *HOXC10* gene.

■ Expression

The expression of the *HOXC9* gene was studied in embryonal carcinoma NT2/D1 cells by Northern blot (Stornaiuolo et al., 1990) and RNase protection analysis (Simeone et al., 1991). No expression of this gene was detectable in NT2/D1 before and after treatment with 10μM retinoic acid. The expression was studied also in erythroleukemic, promyelocytic and monocytic cell lines (Magli et al., 1991), that represents various stages of hematopoietic differentiation *HOXC9* exhibited two transcripts of 1.7 and 1.5 kb in K562 cells. It appears that during myeloid differentiation the *HOXC9* gene expression is switched off.

■ References

Magli, M.C., Barba, P., Celetti, A., De Vita, G., Cillo, C. and Boncinelli, E. (1991). Coordinate regulation of HOX genes in human hematopoietic cells. Proc. Natl. Acad. Sci. USA 88, 6348-6352.

Simeone, A., Acampora, D., Nigro, V., Faiella, A., D'Esposito, M., Stornaiuolo, A., Mavilio, F. and Boncinelli, E. (1991). Differential regulation by retinoic acid of the homeobox genes of the four HOX loci in human embryonal carcinoma cells. Mech. Dev. 33, 215-228.

Stornaiuolo, A., Acampora, D., Pannese, M., D'Esposito, M., Morelli, F., Migliaccio, E., Rambaldi, M., Faiella, A., Nigro, V., Simeone, A. and Boncinelli, E. (1990). Human HOX genes are differentially activated by retinoic acid in embryonal carcinoma cells according to their position within the four loci. Cell Differ. Dev. 31, 119-127.

HOXC10 [h]

Species: *Homo sapiens*
Chromosomal location: 12q12-12q13
Other names: *HOX3I*
Cognate genes: *Hoxc-10*
Type of homeobox: *Antp*-type, related to *Abd-B*
Accession number: -

■ Origin and description

This gene was found during a chromosomal walk on the *HOXC* complex (Stornaiuolo et al., 1990; Acampora et al., 1991). The homeobox sequence was identified using an oligonucleotide representing the most conserved portion of human homeobox: 5' TGGTTCCAGAACCGGCGGATGAA 3'. The *HOXC10* homeobox has been identified in a cosmid clone already shown to contain the *HOXC12* homeobox (Stornaiuolo et al., 1990). *HOXC10* lies 34 kb downstream from *HOXC12*.

■ Expression

The expression of the *HOXC10* gene was studied in embryonal carcinoma NT2/D1 cells by Northern blot (Stornaiuolo et al., 1990) and RNase protection (Simeone et al., 1991) analysis. No expression of this gene was detectable in NT2/D1 before and after treatment with 10 µM retinoic acid. The expression was studied also in erythroleukemic, promyelocytic and monocytic cell lines (Magli et al., 1991) representing various stages of hematopoietic differentiation. *HOXC10* exhibits two transcripts of 1.9 and 1.7 kb. It appears that during myeloid differentiation the *HOXC10* gene expression is switched off.

■ References

Acampora, D., Simeone, A. and Boncinelli, E. (1991). Human HOX hmeobox genes. In: Oxford Surveys on Eukaryotic Genes (MacLean, N., ed.), Oxford University Press 7, 1-28.

Magli, M.C., Barba, P., Celetti, A., De Vita G., Cillo, C. and Boncinelli, E. (1991). Coordinate regulation of HOX genes in human hematopoietic cells. Proc. Natl. Acad. Sci. USA 88, 6348-6352.

Simeone, A., Acampora, D., Nigro, V., Faiella, A., D'Esposito, M., Stornaiuolo, A., Mavilio, F. and Boncinelli, E. (1991). Differential regulation by retinoic acid of the homeobox genes of the four HOX loci in human embryonal carcinoma cells. Mech. Dev. 33, 215-228.

Stornaiuolo, A., Acampora, D., Pannese, M., D'Esposito, M., Morelli, F., Migliaccio, E., Rambaldi, M., Faiella, A., Nigro, V., Simeone, A. and Boncinelli, E. (1990). Human HOX genes are differentially activated by retinoic acid in embryonal carcinoma cells according to their position within the four loci. Cell Differ. Dev. 31, 119-127.

expression was found, both by Northern blot and by RNAse protection; similar results were obtained in Tera2/clone 13 (Stornaiuolo et al., 1990; Simeone et al., 1991). By Northern blot, *HOXC11* shows an abundant transcript of 1.7 kb in K562 and OCIM2; the same one is expressed at low levels in HL60; no expression is detectable in U937 (Magli et al., 1991).

■ References

Acampora, D., Simeone, A. and Boncinelli, E. (1991). Human HOX hmeobox genes. In: Oxford Surveys on Eukaryotic Genes (MacLean, N., ed.), Oxford University Press 7, 1-28.

Boncinelli, E., Simeone, A., Acampora, D. and Mavilio, F. (1991). HOX gene activation by retinoic acid. Trends Genet. 7, 329-334.

Magli, M.C., Barba, P., Celetti, A., De Vita G., Cillo, C. and Boncinelli, E. (1991). Coordinate regulation of HOX genes in human hematopoietic cells. Proc. Natl. Acad. Sci. USA 88, 6348-6352.

Simeone, A., Acampora, D., Nigro, V., Faiella, A., D'Esposito, M., Stornaiuolo, A., Mavilio, F. and Boncinelli, E. (1991). Differential regulation by retinoic acid of the homeobox genes of the four HOX loci in human embryonal carcinoma cells. Mech. Dev. 33, 215-228.

Stornaiuolo, A., Acampora, D., Pannese, M., D'Esposito, M., Morelli, F., Migliaccio, E., Rambaldi, M., Faiella, A., Nigro, V., Simeone, A. and Boncinelli, E. (1990). Human HOX genes are differentially activated by retinoic acid in embryonal carcinoma cells according to their position within the four loci. Cell Differ. Dev. 31, 119-127.

HOXC11 [h]

Species: *Homo sapiens*
Chromosomal location: 12q12-12q13
Other names: *HOX3H*
Cognate genes: *Hoxc-11*
Type of homeobox: *Antp*-type
Accession number: -

■ Origin and description

HOXC11 (Acampora et al., 1991) was found during a walk on the *HOXC* complex, by hybridization with the oligonucleotide 5' TGGTTCCAGAACCGGCGGATGAA 3', representing the most conserved part of human homeoboxes. This box is located near the 5' of *HOXC* complex and forms, together with *HOXA11* and *HOXD11*, the III *HOX* group of paralogy (Stornaiuolo et al., 1990; Boncinelli et al., 1991).

■ Expression

The expression of *HOXC11* was studied in human teratocarcinoma cell lines (NT2/D1 and Tera2/clone 13) and in human hematopoietic cell lines (erythroleukemic: K562 and OCIM2; promyelocytic: HL60; monocitic: U937). In NT2/D1, both in stem and in 10 µM RA treated cells, no

HOXC12 [h]

Species: *Homo sapiens*
Chromosomal location: 12q12-12q13
Other names: *HOX3F*
Cognate genes: -
Type of homeobox: *Antp*-type
Accession number: X16665-16667

■ Origin and description

HOXC12 was found during a walk on the *HOXC* complex, by hybridization with the oligonucleotide 5' TGGTTCCA-GAACCGGGGGATGAA 3', representing the most conserved part of human homeoboxes. This box is located near the 5' of *HOXC* complex and forms, together with *HOXD12*, the II *HOX* group of paralogy (Acampora et al., 1989; Boncinelli et al., 1991).

■ Expression

The expression of *HOXC12* was studied in human teratocarcinoma cell lines (NT2/D1 and Tera2/clone 13) and in human hematopoietic cell lines (erythroleukemic: K562 and OCIM2; promyelocytic: HL60; monocitic: U937). By both Northern blot and RNase protection, no expression was found in NT2/D1, in both stem and RA 10 μM treated cells. Similar results are obtained in Tera2/clone 13 (Stornaiuolo et al., 1990; Simeone et al., 1991). In K562 and OCIM2 *HOXC12* shows an abundant single transcript of 2.6 kb; the same one is expressed at low levels in HL60; no expression is detectable in U937 (Magli et al., 1991).

■ References

Acampora, D., D'Esposito, M., Faiella, A., Pannese, M., Migliaccio, E., Morelli, F., Stornaiuolo, A., Nigro, V., Simeone, A. and Boncinelli, E. (1989). The human HOX gene family. Nucleic Acid Res. 17, 10385-10402.
Boncinelli, E., Simeone, A., Acampora, D. and Mavilio, F. (1991). HOX gene activation by retinoic acid. Trends Genet. 7, 329-334.
Magli, M.C., Barba, P., Celetti, A., De Vita G., Cillo, C. and Boncinelli, E. (1991). Coordinate regulation of HOX genes in human hematopoietic cells. Proc. Natl. Acad. Sci. USA 88, 6348-6352.
Simeone, A., Acampora, D., Nigro, V., Faiella, A., D'Esposito, M., Stornaiuolo, A., Mavilio, F. and Boncinelli, E. (1991). Differential regulation by retinoic acid of the homeobox genes of the four HOX loci in human embryonal carcinoma cells. Mech. Dev. 33, 215-228.
Stornaiuolo, A., Acampora, D., Pannese, M., D'Esposito, M., Morelli, F., Migliaccio, E., Rambaldi, M., Faiella, A., Nigro, V., Simeone, A. and Boncinelli, E. (1990). Human HOX genes are differentially activated by retinoic acid in embryonal carcinoma cells according to their position within the four loci. Cell Differ. Dev. 31, 119-127.

HOXC13 [h]

Species: *Homo sapiens*
Chromosomal location: 12q12-12q13
Other names: *HOXC13*
Cognate genes: -
Type of homeobox: *Antp*-type
Accession number: X16665-16667

■ Origin and description

HOXC13 was found during a walk on the *HOXC* complex, by hybridization with the oligonucleotide 5' TGGTTCCAGAACCGGCGGATGAA 3', representing the most conserved part of human homeoboxes. This box is located near the 5' of *HOXC* complex and forms, together with *HOXA13* and *HOXD13*, the I *HOX* group of paralogy (Acampora et al., 1989; Boncinelli et al., 1991).

■ Expression

The expression of *HOXC13* was studied in human teratocarcinoma cell lines (NT2/D1 and Tera2/clone 13) and in human hematopoietic cell lines (erythroleukemic: K562 and OCIM2; promyelocytic: HL60; monocitic: U937). In NT2/D1, both in stem and in 10 μM RA induced cells, only a very weak expression was detected by Northern blot (Stornaiuolo et al., 1990); by RNAse protection, *HOXC13* revealed a weak expression in uninduced cells and retained the same expression level upon treatment with RA 10 μM, with a progressive decline after 150 hours; this decline does not occur in cells treated with RA 10^{-7} M (Simeone et al., 1991). Similar behaviour was found in Tera2/clone 13 (Simeone et al., 1991). In K562 and OCIM2, HOXC13 shows an abundant transcript of 2.2 kb; the same one is expressed at low level in HL60; no expression is detectable in U937 (Magli et al., 1991).

■ References

Acampora, D., D'Esposito, M., Faiella, A., Pannese, M., Migliaccio, E., Morelli, F., Stornaiuolo, A., Nigro, V., Simeone, A. and Boncinelli, E. (1989). The human HOX gene family. Nucleic Acid Res. 17, 10385-10402.
Boncinelli, E., Simeone, A., Acampora, D. and Mavilio, F. (1991). HOX gene activation by retinoic acid. Trends Genet. 7, 329-334.
Magli, M.C., Barba, P., Celetti, A., De Vita G., Cillo, C. and Boncinelli, E. (1991). Coordinate regulation of HOX genes in human hematopoietic cells. Proc. Natl. Acad. Sci. USA 88, 6348-6352.
Simeone, A., Acampora, D., Nigro, V., Faiella, A., D'Esposito, M., Stornaiuolo, A., Mavilio, F. and Boncinelli, E. (1991). Differential regulation by retinoic acid of the homeobox genes of the four HOX loci in human embryonal carcinoma cells. Mech. Dev. 33, 215-228.
Stornaiuolo, A., Acampora, D., Pannese, M., D'Esposito, M., Morelli, F., Migliaccio, E., Rambaldi, M., Faiella, A., Nigro, V., Simeone, A. and Boncinelli, E. (1990). Human HOX genes are differentially activated by retinoic acid in embryonal carcinoma cells according to their position within the four loci. Cell Differ. Dev. 31, 119-127.

HOXD1 [h]

Species: *Homo sapiens*
Chromosomal location: 2q31-q37
Other names: *HOX4G*
Cognate genes: *Hoxd-1, Xlab*
Type of homeobox: *Antp*-type, related to *lab*
Accession number: -

■ Origin and description

HOXD1 was found during a walk on the *HOXD* complex by hybridization with the oligonucleotide 5' TGGTTCCAGAACCGGCGGATGAA 3' representing the most conserved part of human homeoboxes. This homeobox is located 20 kb downstream the *HOXD3* gene and belongs, together with *HOXB1* and *HOXA1* to the first *HOX* group of paralogy (Acampora et al., 1989).

Expression

The expression of *HOXD1* was studied in human terato-carcinoma cell lines (NT2/D1 and Tera2/clone13) and in human hematopoietic cell lines (erythroleukemic K562 and OCIM2; promyelocytic HL60; monocytic U937). In the NT2/D1 human teratocarcinoma cell line the expression of the *HOXD1* gene has been studied by Northern blot (Stornaiuolo et al., 1990) and by RNase protection experiments (Simeone et al., 1991). Northern blot analysis failed to show any expression of the *HOXD1* gene both in uninduced cells, but showed a transcript of 2.4 kb in NT2/D1 cells treated with 10 µM RA for 14 days. In RNase protection assays the *HOXD1* gene is quickly and strongly activated by treatment with RA both in NT2/D1 and Tera2/clone13 cell lines. In all the hematopoietic cell lines we tested this gene is not expressed (Magli et al., 1991).

References

Acampora, D., D'Esposito, M., Faiella, A., Pannese, M., Migliaccio, E., Morelli, F., Stornaiuolo, A., Nigro, V., Simeone, A. and Boncinelli, E. (1989). The human HOX gene family. Nucleic Acids Res. 17, 10385-10402.

Magli, M.C., Barba, P., Celetti, A., De Vita G., Cillo, C. and Boncinelli, E. (1991). Coordinate regulation of HOX genes in human hematopoietic cells. Proc. Natl. Acad. Sci. USA 88, 6348-6352.

Simeone, A., Acampora, D., Nigro, V., Faiella, A., D'Esposito, M., Stornaiuolo, A., Mavilio, F. and Boncinelli, E. (1991). Differential regulation by retinoic acid of the homeobox genes of the four HOX loci in human embryonal carcinoma cells. Mech. Dev. 33, 215-228.

Stornaiuolo, A., Acampora, D., Pannese, M., D'Esposito, M., Morelli, F., Migliaccio, E., Rambaldi, M., Faiella, A., Nigro, V., Simeone, A. and Boncinelli, E. (1990). Human HOX genes are differentially activated by retinoic acid in embryonal carcinoma cells according to their position within the four loci. Cell Differ. Dev. 31, 119-127.

HOXD3 [h]

Species: *Homo sapiens*
Chromosomal location: 2q31-q37
Other names: *HOX4A, c13^{+1}*
Cognate genes: *Hoxd-3*
Type of homeobox: *Antp*-type
Accession number: -

Origin and description

HOXD3 was found during a walk on the *HOXD* complex by hybridization with the oligonucleotide 5' TGGTTCCA-GAACCGGCGGATGAA 3' representing the most conserved part of human homeoboxes. This homeobox is located 20 kb upstream from the *HOXD1* gene and belongs, together with *HOXA3* and *HOXB3* to the 11th *HOX* group of paralogy (Acampora et al., 1989).

Expression

The expression of *HOXD3* was studied in human terato-carcinoma cell lines (NT2/D1 and Tera2/clone13) and in human hematopoietic cell lines (erythroleukemic K562 and OCIM2; promielocytic HL60; monocytic U937). In the NT2/D1 human teratocarcinoma cell line the expression of the *HOXD3* gene has been studied by Northern blot (Stornaiuolo et al., 1990) and by RNase protection experiments (Simeone et al., 1991). Northern blot analysis, the expression of the *HOXD3* gene is not detectable in uninduced cells, but shows a major transcript of 2.4 kb and a minor transcript of 4.1 kb in NT2/D1 cells treated with 10 µM RA for 14 days. In RNase protection assays, *HOXD3* is gradually activated by the treatment with RA both in NT2/D1 and Tera2/clone13 cells. In all the hematopoietic cell lines we tested this gene is not expressed (Magli et al., 1991).

References

Acampora, D., D'Esposito, M., Faiella, A., Pannese, M., Migliaccio, E., Morelli, F., Stornaiuolo, A., Nigro, V., Simeone, A. and Boncinelli, E. (1989). The human HOX gene family. Nucleic Acids Res. 17, 10385-10402.

Magli, M.C., Barba, P., Celetti, A., De Vita G., Cillo, C. and Boncinelli, E. (1991). Coordinate regulation of HOX genes in human hematopoietic cells. Proc. Natl. Acad. Sci. USA 88, 6348-6352.

Simeone, A., Acampora, D., Nigro, V., Faiella, A., D'Esposito, M., Stornaiuolo, A., Mavilio, F. and Boncinelli, E. (1991). Differential regulation by retinoic acid of the homeobox genes of the four HOX loci in human embryonal carcinoma cells. Mech. Dev. 33, 215-228.

Stornaiuolo, A., Acampora, D., Pannese, M., D'Esposito, M., Morelli, F., Migliaccio, E., Rambaldi, M., Faiella, A., Nigro, V., Simeone, A. and Boncinelli, E. (1990). Human HOX genes are differentially activated by retinoic acid in embryonal carcinoma cells according to their position within the four loci. Cell Differ. Dev. 31, 119-127.

HOXD4 [h]

Species: *Homo sapiens*
Chromosomal location: 2q31-q37
Other names: *HOX4B, c13*
Cognate genes: *Hoxd-4, Chox a*
Type of homeobox: *Antp*-type
Accession number: -

Origin and description

This gene was isolated by cross-homology with the *Drosophila Antp* and *ftz* homeoboxes (Boncinelli et al.,

1985). This gene was successively mapped on the chromosome 2 (Cannizzaro et al., 1987). *HOXD4* is located in the human homeobox complex *HOXD* (Acampora et al., 1989). *HOXD4* belongs, together with *HOXA4, HOXB4* and *HOXC4* to the 10th *HOX* group of paralogy.

■ Expression

HOXD4 is expressed with multiple transcripts into a tissue and stage-specific pattern during embryogenesis (Mavilio et al., 1986). *HOXD4* shows no expression in two untreated human teratocarcinoma cell lines (NT2/D1 and Tera2/clone13) (Simeone et al., 1991), while a different basal level of *HOXD4* expression is observed in several human neuroblastoma cell lines (Peverali et al., 1990). However, in all these lines, retinoic acid treatment gradually activates *HOXD4* expression. This gene is not expressed in several hematopoietic cell lines (Magli et al., 1991). *HOXD4* promoter has been extensively studied. Two alternative promoters have been characterized (Cianetti et al., 1990), and sequences near the *HOXD4* proximal promoter act as enhancer in transgenic mice (Tuggle et al., 1990).

■ Function

In flies, *HOXD4* can specifically substitute one of the functions of its *Drosophila* orthologous gene, *Dfd* (McGinnis et al., 1990): the activation of its own promoter. A fusion gene consisting of a heat shock promoter attached to the *HOXD4* gene was introduced into the *Drosophila* genome and the effects were assayed after heat shock. In developing embryonic and larval cells, *HOXD4* specifically activates ectopic expression of the endogenous *Dfd* transcription unit and phenocopies a dominant mutant allele of *Dfd*.

■ References

Acampora, D., D'Esposito, M., Faiella, A., Pannese, M., Migliaccio, E., Morelli, F., Stornaiuolo, A., Nigro, V., Simeone, A. and Boncinelli, E. (1989). The human HOX gene family. Nucleic Acids Res. 17, 10385-10402.

Boncinelli, E., Simeone, A., La Volpe, A., Faiella, A., Fidanza, V., Acampora, D. and Scotto, L. (1985). Human cDNA clones containing the homeobox. Cold Spring Harbor Symp. Quant. Biol. 50, 301-305.

Cannizzaro, L.A., Croce, C.M., Griffin, C.A., Simeone, A., Boncinelli, E. and Huebner, K. (1987). Human homeo box-containing genes located at chromosome regions 2q31→2q37 and 12q12→ 12q13. Am J. Hum. Genet. 41, 1-15.

Cianetti, L., Di Cristofaro, A., Zappavigna, V., Bottero, L., Boccoli, G., Testa, U., Russo, G., Boncinelli, E. and Peschle, C. (1990). Molecular mechanisms underlying the expression of the human HOX-5.1 gene. Nucleic Acids Res. 18, 4361-4368.

Magli, M.C., Barba, P., Celetti, A., De Vita G., Cillo, C. and Boncinelli, E. (1991). Coordinate regulation of HOX genes in human hematopoietic cells. Proc. Natl. Acad. Sci. USA 88, 6348-6352.

Mavilio, F., Simeone, A., Giampaolo, A., Faiella, A., Zappavigna, V., Acampora, D., Poiana, G., Russo, G., Peschle, C. and Boncinelli, E. (1986). Differential and stage-related expression in embryonic tissues of a new human homeobox gene. Nature 324, 664-668.

McGinnis, N., Kuziora, M.A. and McGinnis, W. (1990). Human Hox-4.2 and *Drosophila deformed* encode similar regulatory specificities in *Drosophila* embryos and larvae. Cell 63, 969-976.

Peverali, F.A., D'Esposito, M., Acampora, D., Bunone, G., Negri, M., Faiella, A., Stornaiuolo, A., Pannese, M., Migliaccio, E., Simeone, A., Della Valle, G. and Boncinelli, E. (1990). Expression of HOX homeogenes in human neuroblastoma cell culture lines. Differentiation 45, 61-69.

Simeone, A., Acampora, D., Nigro, V., Faiella, A., D'Esposito, M., Stornaiuolo, A., Mavilio, F. and Boncinelli, E. (1991). Differential regulation by retinoic acid of the homeobox genes of the four HOX loci in human embryonal carcinoma cells. Mech. Dev. 33, 215-228.

Tuggle, C.K., Zakany, J., Cianetti, L., Peschle, C. and Nguyen-Huu, M.C. (1990). Region-specific enhancers near two mammalian homeobox genes define adjacent rostrocaudal domains in the central nervous system. Genes Dev. 4, 180-189.

HOXD8 [h]

Species: *Homo sapiens*
Chromosomal location: 2q31-q37
Other names: *HOX4E*
Cognate genes: *Hoxd-8, Chox-m*
Type of homeobox: *Antp*-type
Accession number: -

■ Origin and description

HOXD8 was found during a walk on the *HOXD* complex by hybridization with the oligonucleotide 5' TGGTTCCAGAACCGGCGGATGAA 3' representing the most conserved part of human homeoboxes. This homeobox is located 7 kb downstream from the *HOXD9* gene and belongs, together with *HOXB8* and *HOXC8* to the 6th *HOX* group of paralogy (Acampora et al., 1989).

■ Expression

The expression of *HOXD8* was studied in human teratocarcinoma cell lines (NT2/D1 and Tera2/clone13) and in human hematopoietic cell lines (erythroleukemic K562 and OCIM2; promielocytic HL60; monocytic U937). In the NT2/D1 human teratocarcinoma cell line the expression of the *HOXD8* gene has been studied by Northern blot (Stornaiuolo et al., 1990) and by RNase protection experiments (Simeone et al., 1991). Northern blot analysis failed to show any expression of the *HOXD8* gene both in uninduced cells, but showed a transcript of 2.3 kb in NT2/D1 cells treated with 10 μM RA for 14 days. In RNase protection assays the *HOXD8* gene is gradually activated by treatment with RA both in NR2/D1 and Tera2/clone13 cell lines. In all the hematopoietic cell lines we tested this gene is not expressed (Magli et al., 1991).

References

Acampora, D., D'Esposito, M., Faiella, A., Pannese, M., Migliaccio, E., Morelli, F., Stornaiuolo, A., Nigro, V., Simeone, A. and Boncinelli, E. (1989). The human HOX gene family. Nucleic Acids Res. 17, 10385-10402.

Magli, M.C., Barba, P., Celetti, A., De Vita G., Cillo, C. and Boncinelli, E. (1991). Coordinate regulation of HOX genes in human hematopoietic cells. Proc. Natl. Acad. Sci. USA 88, 6348-6352.

Simeone, A., Acampora, D., Nigro, V., Faiella, A., D'Esposito, M., Stornaiuolo, A., Mavilio, F. and Boncinelli, E. (1991). Differential regulation by retinoic acid of the homeobox genes of the four HOX loci in human embryonal carcinoma cells. Mech. Dev. 33, 215-228.

Stornaiuolo, A., Acampora, D., Pannese, M., D'Esposito, M., Morelli, F., Migliaccio, E., Rambaldi, M., Faiella, A., Nigro, V., Simeone, A. and Boncinelli, E. (1990). Human HOX genes are differentially activated by retinoic acid in embryonal carcinoma cells according to their position within the four loci. Cell Differ. Dev. 31, 119-127.

HOXD9 [h]

Species: *Homo sapiens*

Chromosomal location: 2q31-q37

Other names: *HOX4C*

Cognate genes: *Hoxd-9, Chox4.4*

Type of homeobox: *Antp*-type, related to *Abd-B*

Accession number: -

Origin and description

HOXD9 was found during a walk on the *HOXD* complex by hybridization with the oligonucleotide 5' TGGTTCCA-GAACCGGCGGATGAA 3' representing the most conserved part of human homeoboxes. This homeobox is located 5 kb downstream from the *HOXD10* gene and belongs, together with *HOXA9*, *HOXB9* and *HOXC9* to the 5th *HOX* group of paralogy (Acampora et al., 1991).

Expression

The expression of *HOXD9* was studied in human terato-carcinoma cell lines (NT2/D1 and Tera2/clone13) and in human hematopoietic cell lines (erythroleukemic K562 and OCIM2; promielocytic HL60; monocytic U937). In the NT2/D1 human teratocarcinoma cell line the expression of the *HOXD9* gene has been studied by Northern blot (Stornaiuolo et al., 1990) and by RNase protection experiments (Simeone et al., 1991). Northern blot analysis failed to show any expression of the *HOXD9* gene in uninduced cells, but showed a transcript of 2.2 kb in NT2/D1 cells treated with 10 µM RA for 14 days. In RNase protection assays the *HOXD9* gene showed a constitutive low expres-sion both in uninduced and RA induced (from 1 to 195 hours) NT2/D1 and Tera2/clone13 teratocarcinoma cell lines. In all the hematopoietic cell lines we tested this gene is not expressed (Magli et al., 1991)

References

Acampora, D., D'Esposito, M., Faiella, A., Pannese, M., Migliaccio, E., Morelli, F., Stornaiuolo, A., Nigro, V., Simeone, A. and Boncinelli, E. (1989). The human HOX gene family. Nucleic Acids Res. 17, 10385-10402.

Magli, M.C., Barba, P., Celetti, A., De Vita G., Cillo, C. and Boncinelli, E. (1991). Coordinate regulation of HOX genes in human hematopoietic cells. Proc. Natl. Acad. Sci. USA 88, 6348-6352.

Simeone, A., Acampora, D., Nigro, V., Faiella, A., D'Esposito, M., Stornaiuolo, A., Mavilio, F. and Boncinelli, E. (1991). Differential regulation by retinoic acid of the homeobox genes of the four HOX loci in human embryonal carcinoma cells. Mech. Dev. 33, 215-228.

Stornaiuolo, A., Acampora, D., Pannese, M., D'Esposito, M., Morelli, F., Migliaccio, E., Rambaldi, M., Faiella, A., Nigro, V., Simeone, A. and Boncinelli, E. (1990). Human HOX genes are differentially activated by retinoic acid in embryonal carcinoma cells according to their position within the four loci. Cell Differ. Dev. 31, 119-127.

HOXD10 [h]

Species: *Homo sapiens*

Chromosomal location: 2q31-q37

Other names: *HOX4D*

Cognate genes: *Hoxd-10, Chox4.5*

Type of homeobox: *Antp*-type, related to *Abd-B*

Accession number: -

Origin and description

HOXD10 was found during a walk on the *HOXD* complex by hybridization with the oligonucleotide 5' TGGTTCCA-GAACCGGCGGATGAA 3' representing the most conserved part of human homeoboxes. This homeobox is located 13 kb downstream from the *HOXD11* gene and belongs, together with *HOXA10* and *HOXC10* to the 4th *HOX* group of paralogy (Acampora et al., 1989).

Expression

The expression of *HOXD10* was studied in human terato-carcinoma cell lines (NT2/D1 and Tera2/clone13) and in human hematopoietic cell lines (erythroleukemic K562 and OCIM2; promielocytic HL60; monocytic U937). In the NT2/D1 human teratocarcinoma cell line the expression of the *HOXD10* gene has been studied by Northern blot (Stornaiuolo et al., 1990) and by RNase protection experiments (Simeone et al., 1991). Northern blot analysis failed to show any expression of the *HOXD10* gene both in uninduced and RA induced NT2/D1 cells. In RNase protection assays the *HOXD10* is weakly expressed in uninduced NT2/D1 cells. This level of expression progressively declines after RA treatment. In Tera2/clone13 *HOXD10* gene is expressed at constant level both in uninduced and induced cells (Simeone et al., 1991). In all the hematopoietic cell lines we tested this gene is not expressed (Magli et al., 1991).

■ References

Acampora, D., D'Esposito, M., Faiella, A., Pannese, M., Migliaccio, E., Morelli, F., Stornaiuolo, A., Nigro, V., Simeone, A. and Boncinelli, E. (1989). The human HOX gene family. Nucleic Acids Res. 17, 10385-10402.

Magli, M.C., Barba, P., Celetti, A., De Vita G., Cillo, C. and Boncinelli, E. (1991). Coordinate regulation of HOX genes in human hematopoietic cells. Proc. Natl. Acad. Sci. USA 88, 6348-6352.

Simeone, A., Acampora, D., Nigro, V., Faiella, A., D'Esposito, M., Stornaiuolo, A., Mavilio, F. and Boncinelli, E. (1991). Differential regulation by retinoic acid of the homeobox genes of the four HOX loci in human embryonal carcinoma cells. Mech. Dev. 33, 215-228.

Stornaiuolo, A., Acampora, D., Pannese, M., D'Esposito, M., Morelli, F., Migliaccio, E., Rambaldi, M., Faiella, A., Nigro, V., Simeone, A. and Boncinelli, E. (1990). Human HOX genes are differentially activated by retinoic acid in embryonal carcinoma cells according to their position within the four loci. Cell Differ. Dev. 31, 119-127.

HOXD11 [h]

Species: *Homo sapiens*
Chromosomal location: 2q31-q37
Other names: *HOX4F*
Cognate genes: *Hoxd-11, Chox-4.6*
Type of homeobox: *Antp*-type, related to *Abd-B*
Accession number: -

■ Origin and description

HOXD11 was found during a walk on the *HOXD* complex by hybridization with the oligonucleotide 5' TGGTTCCAGAACCGGCGGATGAA 3' representing the most conserved part of human homeoboxes. This homeobox is located 10 kb downstream from the *HOXD12* gene and belongs, together with *HOXA11* and *HOXC11* to the 3rd *HOX* group of paralogy (Acampora et al., 1989).

■ Expression

The expression of *HOXD11* was studied in human teratocarcinoma cell lines (NT2/D1 and Tera2/clone13) and in human hematopoietic cell lines (erythroleukemic K562 and OCIM2; promielocytic HL60; monocytic U937). In the NT2/D1 human teratocarcinoma cell line the expression of the *HOXD11* gene has been studied by Northern blot (Stornaiuolo et al., 1990) and by RNase protection experiments (Simeone et al., 1991). Northern blot analysis failed to show any expression of the *HOXD11* gene both in uninduced and RA induced NT2/D1 cells. In RNase protection assays this gene is weakly expressed in uninduced NT2/D1 cells. This level of expression progressively declines after RA treatment. In Tera2/clone13 *HOXD11* gene is expressed at constant level both in uninduced and induced cells (Simeone et al., 1991). In all the hematopoietic cell lines we tested this gene is not expressed (Magli et al., 1991).

■ References

Acampora, D., D'Esposito, M., Faiella, A., Pannese, M., Migliaccio, E., Morelli, F., Stornaiuolo, A., Nigro, V., Simeone, A. and Boncinelli, E. (1989). The human HOX gene family. Nucleic Acids Res. 17, 10385-10402.

Magli, M.C., Barba, P., Celetti, A., De Vita G., Cillo, C. and Boncinelli, E. (1991). Coordinate regulation of HOX genes in human hematopoietic cells. Proc. Natl. Acad. Sci. USA 88, 6348-6352.

Simeone, A., Acampora, D., Nigro, V., Faiella, A., D'Esposito, M., Stornaiuolo, A., Mavilio, F. and Boncinelli, E. (1991). Differential regulation by retinoic acid of the homeobox genes of the four HOX loci in human embryonal carcinoma cells. Mech. Dev. 33, 215-228.

Stornaiuolo, A., Acampora, D., Pannese, M., D'Esposito, M., Morelli, F., Migliaccio, E., Rambaldi, M., Faiella, A., Nigro, V., Simeone, A. and Boncinelli, E. (1990). Human HOX genes are differentially activated by retinoic acid in embryonal carcinoma cells according to their position within the four loci. Cell Differ. Dev. 31, 119-127.

HOXD12 [h]

Species: *Homo sapiens*
Chromosomal location: 2q31-q37
Other names: *HOX4H*
Cognate genes: *Hoxd-12, Chox4.7*
Type of homeobox: *Antp*-type, related to *Abd-B*
Accession number: -

■ Origin and description

HOXD12 was found during a walk on the *HOXD* complex by hybridization with the oligonucleotide 5' TGGTTCCAGAACCGGCGGATGAA 3' representing the most conserved part of human homeoboxes. This homeobox is located 6 kb downstream from the *HOXD13* gene and belongs, together with *HOXC12* to the 2nd *HOX* group of paralogy (D'Esposito et al., 1991).

■ Expression

The expression of *HOXD12* was studied in human terato-carcinoma cell lines (NT2/D1 and Tera2/clone13) and in human hematopoietic cell lines (erythroleukemic K562 and OCIM2; promielocytic HL60; monocytic U937). In human teratocarcinoma cell lines the expression of the *HOXD12* gene has been studied by RNase protection experiments. This gene is weakly expressed in uninduced NT2/D1 cells. This level of expression progressively declines after RA treatment. In Tera2/clone13 *HOXD12* gene is expressed at constant level both in uninduced and induced cells (Simeone et al., 1991). In all the hematopoietic cell lines we tested this gene is not expressed (Magli et al., 1991).

■ References

D'Esposito, M., Morelli, F., Acampora, D., Migliaccio, E., Simeone, A. and Boncinelli, E. (1991). EVX2, a human homeobox gene homologous to the even-skipped segmentation gene, is localized at the 5' end of HOX4 locus on chromosome 2. Genomics 10, 43-50.

Magli, M.C., Barba, P., Celetti, A., De Vita G., Cillo, C. and Boncinelli, E. (1991). Coordinate regulation of HOX genes in human hematopoietic cells. Proc. Natl. Acad. Sci. USA 88, 6348-6352.

Simeone, A., Acampora, D., Nigro, V., Faiella, A., D'Esposito, M., Stornaiuolo, A., Mavilio, F. and Boncinelli, E. (1991). Differential regulation by retinoic acid of the homeobox genes of the four HOX loci in human embryonal carcinoma cells. Mech. Dev. 33, 215-228.

HOXD13 [h]

Species: *Homo sapiens*
Chromosomal location: 2q31-q37
Other names: *HOX4I*
Cognate genes: *Hoxd-13, Chox-4.8*
Type of homeobox: *Antp*-type, related to *Abd-B*
Accession number: -

■ Origin and description

HOXD13 was found during a walk on the *HOXD* complex by hybridization with the oligonucleotide 5' TGGTTCCA-GAACCGGCGGATGAA 3' representing the most conserved part of human homeoboxes. This homeobox is located near the 5' of the *HOXD* complex and belongs, together with *HOXA13* and *HOXC13* to the 13th *HOX* group of paralogy. *EVX2*, the human gene homologous to the *Drosophila even-skipped* segmentation gene is located 13 kb upstream from *HOXD13* and in opposite direction of transcription (D'Esposito et al., 1991).

■ Expression

The expression of *HOXD13* was studied in human terato-carcinoma cell lines (NT2/D1 and Tera2/clone13) and in human hematopoietic cell lines (erythroleukemic K562 and OCIM2; promielocytic HL60; monocytic U937). In human teratocarcinoma cell lines the expression of the *HOXD13* gene has been studied by RNase protection. This gene is weakly expressed in uninduced NT/D1 cells. This level of expression progressively declines after RA treatment. In Tera2/clone13 *HOXD13* gene is expressed at constant level both in uninduced and induced cells (Simeone et al., 1991). In K562 and U937 cell lines are not expressed. *HOXD13* shows a major transcript of 2.3 kb and a minor one of 1.9 kb in OCIM2 and HL60 cell lines (Magli et al., 1991).

■ References

D'Esposito, M., Morelli, F., Acampora, D., Migliaccio, E., Simeone, A. and Boncinelli, E. (1991). EVX2, a human homeobox gene homologous to the even-skipped segmentation gene, is localized at the 5' end of HOX4 locus on chromosome 2. Genomics 10, 43-50.

Magli, M.C., Barba, P., Celetti, A., De Vita G., Cillo, C. and Boncinelli, E. (1991). Coordinate regulation of HOX genes in human hematopoietic cells. Proc. Natl. Acad. Sci. USA 88, 6348-6352.

Simeone, A., Acampora, D., Nigro, V., Faiella, A., D'Esposito, M., Stornaiuolo, A., Mavilio, F. and Boncinelli, E. (1991). Differential regulation by retinoic acid of the homeobox genes of the four HOX loci in human embryonal carcinoma cells. Mech. Dev. 33, 215-228.

Hoxa-1 [m]

Species: *Mus musculus*
Chromosomal location: Chromosome 6, bands A, B1
Other names: *Hox-1.6, Hox1-y, ERA-1*
Cognate genes: *HOXA1*
Type of homeobox: *Antp*-type, related to the *Drosophila labial* gene
Accession number: -

■ Origin and description

This gene was isolated through its linkage to the *HOXA* complex (Duboule et al., 1986; Baron et al., 1987) and, independently, by differential screening of F9 cells cDNA libraries (see below, LaRosa and Gudas, 1988). It is located at the 3' most position within the *HOXA* complex (Baron et al., 1987) and contains a homeobox related to the homeotic gene *labial* (*lab*, Mlodzick et al., 1988). Paralogous genes on other complexes are *Hoxb-1* and

Hoxd-1. *Hoxa-1* has a complex pattern of differentially spliced transcripts (Baron et al., 1987; LaRosa and Gudas, 1988) which can encode proteins with or without the homeodomain (LaRosa and Gudas, 1988).

■ Expression

In correlation with its extreme 3' position in the *HOXA* complex, *Hoxa-1* is expressed very early during development. At midday 7 p.c., it is expressed in both mesoderm and neuroectoderm from the caudal end to the level of the not yet visible rhombomere 4, in the future hindbrain (Murphy and Hill, 1991). It is later on expressed in anterior gut-associated endoderm and epithelium (Duboule and Dollé, 1989; Murphy and Hill, 1991). By day 12.5 p.c., *Hoxa-1* is no longer detected. The steady state level of transcripts for the *Hoxa-1* gene is strongly elevated in F9 cells in response to retinoic acid (Baron et al., 1987; LaRosa and Gudas, 1988). The response to RA is very rapid and allowed the cloning of this gene as an early responder to RA stimulation by differential screening of RA-treated versus non-treated F9 cells (LaRosa and Gudas, 1988, see also HOXA1).

■ Function

The function of the *Hoxa-1* gene was studied by homologous recombination in ES cells and production of null mutants (Lufkin et al., 1991; Chisaka et al., 1992). The homozygous mice die after birth, probably from anoxia and have numerous defects which are all related to structures located at the level of rhombomeres 4 to 6. They include delayed hindbrain neural tube closure, absence of certain cranial nerves and ganglia, malformed inner ears and bones of the skull (Lufkin et al., 1991; Chisaka et al., 1992). It therefore appears that *Hoxa-1* is involved either in the growth control or in the regional specification of rhombomeres 4 to 6 (M. Mark, personal communication; Dollé et al., 1993). In this area, it seems to affect principally the migration of neurogenic neural crest cells whereas mesenchymal crest cells from the same level appear to be controlled by another *Hox* gene (Lufkin et al., 1991; see *Hoxa-3*). The absence of the Hoxa-1 product results in the absence, or partial development, of the motor nucleus of the facial nerves (cranial nerve VI, VII), the malformation of the ganglia of the facial-acoustic nerve (VII-VIII) and glossopharyngeal-vagus (IX-X), the spiral and vestibular ganglia (ga8; gv8) and the roots of the acoustic nerve (M. Mark and P. Chambon, personal communication).

■ References

Baron, A., Featherstone, M. F., Hill, B. Hall, A., Galliot, B. and Duboule, D. (1987). Hox1-6, a new mouse homeobox containing gene member of the Hox-1 complex. EMBO J. 6, 2977-2086.

Chisaka, O., Musci, T. S. and Capecchi, M. R. (1992). Developmental defects of the ear, cranial nerves and hindbrain resulting from targeted disruption of the mouse homeobox gene Hox-1.6. Nature 355, 516-521.

Dollé, P., Lufkin, T., Krumlauf, R., Mark, M., Duboule, D. and Chambon, P. (1993). Local alterations of Krox-20 and Hox gene expression in the hindbrain of Hox-1.6 null embryos. Proc. Natl. Acad. Sci. U.S.A. 90, 7666–7670.

Duboule, D., Baron, A., Mähl, P. and Galliot, B. (1986). A new homeo-box is present in overlapping cosmid clones which define the mouse Hox-1 locus. EMBO J. 5, 1973-1980.

Duboule, D. and Dollé P. (1989). The structural and functional organization of the murine HOX gene family resembles that of *Drosophila* homeotic genes. EMBO J. 8, 1497-1505.

LaRosa, G. J. and Gudas, L. J. (1988a). Early retinoic acid-induced F9 teratocarcinoma stem cell gene ERA-1: alternate splicing creates transcripts for a homeobox-containing protein and one lacking the homeobox. Mol. Cell. Biol. 8, 3906-3917.

Lufkin, T., Dierich, A., LeMeur, M., Mark, M. and Chambon, P. (1991). Disruption of the Hox-1.6 homeobox gene results in defects in a region corresponding to its rostral domain of expression. Cell 66, 1105-1120.

Mlodzik, M., Fjose, A. and Gehring, W. J. (1988). Molecular structure and spatial expression of a homeobox gene from the labial region of the *Antennapedia*-complex. EMBO J. 7, 2569-2578.

Murphy, P. and Hill, R. E. (1991). Expression of the mouse labial-like homeobox-containing genes, Hox-2.9 and Hox-1.6 during segmentation of the hindbrain. Development 111, 61-74.

Hoxa-1   [x]

Species: *Xenopus laevis*
Chromosomal location: -
Other names: *Xhox.lab2*
Cognate genes: *Hoxa-1*
Type of homeobox: *Antp*-type, related to *Drosophila labial*
Accession number: -

■ Origin and description

Isolated by homology to the chicken *Ghox.lab* homeodomain (Sundin et al., 1990; Sive and Cheng, 1991). It is a member of the *labial* gene family that corresponds to *Hoxa-1* (LaRosa and Gudas, 1988) in the homeodomain and immediately upstream "labial domain" (97% identity), but is highly divergent in the remainder of the protein, another potential block of identity, as the extreme amino terminal has not been examined for lack of a full length cDNA clone.

■ Expression

Expressed from early neurula (stage 14) through hatching (stage 35), with RNA levels peaking through neurula and tailbud stages. By dissection of embryos, RNA is localized in both dorsal and ventral regions. These regions include dorsal mesoderm and ectoderm, particularly in the middle of the embryo (trunk) in early neurula. Strongly induced by exogenous retinoic acid (RA) during gastrula and neurula stages (up to 50-fold in dorsal ectoderm and mesoderm).

■ Function

Unknown. May be partly responsible for mediating the axial truncation caused by RA, since it becomes induced in anterior and posterior regions concomitant with anterior and posterior truncation, after RA application (Durston et al., 1989; Sive et al., 1990).

■ References

Blumberg, B., Wright, C.V.E., de Robertis, E.M. and Cho, K.W.Y. (1991). Organizer-specific homeobox genes in *Xenopus laevis*. Science 253, 194-196.

Durston, A.J., Timmermans, J.P.M., Hage, W.J., Hendriks, H.F.J., de Vries, N.J., Heideveld, M. and Nieuwkoop, P.D. (1989). Retinoic acid causes an anteroposterior transformation in the developing central nervous system. Nature 340, 140-144.

LaRosa, G.J. and Gudas, L.J. (1987). Early retinoic acid-induced F9 teratocarcinoma stem cell gene ERA-1: alternate splicing creates transcripts for a homeobox-containing protein and one lacking the homeobox. Mol. Cell Biol. 8, 3906-3917.

Sive, H.L., Draper, B.W., Harland, R.M. and Weintraub, H. (1990). Identification of a retinoic acid-sensitive period during primary axis formation in *Xenopus laevis*. Genes Dev. 4, 932-942.

Sive, H.L. and Cheng, P.F. (1991). Retinoic acid perturbs the expression of *Xhox.lab* genes and alters mesodermal determination in *Xenopus laevis*. Genes Dev. 5, 1321-1332.

Sundin, O.H., Busse, H.G., Rogers, M.B., Gudas, L.J. and Eichele, G. (1990). Region-specific expression in early chick and mouse embryos of *Ghox.lab* and Hox1.6, vertebrate homeobox-containing genes related to *Drosophila labial*. Development 108, 47-58.

Hoxa-2 [m]

Species: *Mus musculus*

Chromosomal location: Chromosome 6, bands A, B1

Other names: *Hox-1.11*

Cognate genes: *HOXA2*

Type of homeobox: *Antp*-type, related to *Drosophila proboscipedia* gene

Accession number: -

■ Origin and description

A human homologue of the gene, *HOXA2*, (Acampora et al., 1989; Simeone et al., 1990; Boncinelli et al., 1991) was used to probe a cosmid clone from the *HOXA* complex originally characterised to contain the *Hoxa-1* and *Hoxa-3* genes (Duboule et al., 1986; Baron et al., 1987). A region between the two previously identified genes hybridized to the probe and it was verified that this corresponded to

the mouse *Hoxa-2* gene (Hunt et al., 1991). The sequence is nearly identical to the human *HOXA2* homeobox and demonstrates that it is another member of the mouse *pb* related paralogous group.

■ Expression

In situ hybridisation with the mouse probe showed that at 9.0 dpc the gene was expressed up to a rhombomere boundary at r2/3 like the *Hoxb-2* paralogue (Hunt et al., 1991). In addition it showed high levels of expression in r3 and r5. At earlier stages there are more anterior domains of expression in the mouse which are reported to correspond to expression in r2. This domain is transient and disappears around 9.0 dpc.

■ References

Acampora, D., D'Esposito, M., Faiella, A., Pannese, M., Migliaccio, E., Morelli, F., Stornaiuolo, A., Nigro, V., Simeone, A. and Boncinelli, E. (1989). The human HOX gene family. N.A.R. 17, 10385-10402.

Baron, A., Featherstone, M. S., Hill, R. E., Hall, A., Galliot, B. and Duboule, D. (1987). Hox-1.6: a mouse homeo-box-containing gene member of the Hox-1 complex. EMBO J. 6, 2977-2986.

Boncinelli, E., Simeone, A., Acampora, D. and Mavilio, F. (1991). HOX gene activation by retinoic acid. TIG 7, 329-334.

Duboule, D., Baron, A., Mahl, P. and Galliot, B. (1986). A new homeo-box is present in overlapping cosmid clones which define the mouse Hox-1 locus. EMBO J. 5, 1973-1980.

Hunt, P., Gulisano, M., Cook, M., Sham, M., Faiella, A., Wilkinson, D., Boncinelli, E. and Krumlauf, R. (1991). A distinct *Hox* code for the branchial region of the head. Nature 353, 861-864.

Simeone, A., Acampora, D., Arcioni, L., Andrews, P. W., Boncinelli, E. and Mavilio, F. (1990). Sequential activation of HOX2 homeobox genes by retinoic acid in human embryonal carcinoma cells. Nature 346, 763-766.

Hoxa-4 [m]

Species: *Mus musculus*

Chromosomal location: Chromosome 6, band A, B1

Other names: *Hox-1.4, Hox-1.3, HBT-1, MH-3*

Cognate genes: *HOXA4*

Type of homeobox: *Antp*-type, related to the *Drosophila Deformed* gene

Accession number: X66861

■ Origin and description

This gene has been independently isolated by cross hybridization with *Drosophila* homeobox probes from a mouse cosmid genomic library (Duboule et al., 1986), from adult mouse testis cDNA librabries (Wolgemuth et al., 1986; Rubin et al., 1986; Galliot et al., 1989) and from a fetal mouse cDNA library (Galliot et al., 1989). *Hoxa-4* has been located on overlapping cosmid clones containing several homeoboxes defining the *HOXA* complex (Duboule et al., 1986). The *Hoxa-4* homeobox is related to that of the *Deformed* gene; the paralogous genes on other complexes are *Hoxb-4* and *Hoxd-4*. A minimal promoter extending 360 bp upstream to the transcriptional site is able to promote transcription in transfected cells (Galliot et al., 1989). This GC-rich promoter region contains binding sites for the transcription factor Sp1 and the segment specific gene *Krox 20* (Galliot et al., 1989; Chavrier et al., 1990).

■ Expression

In adult, *Hoxa-4* is mostly expressed in male germ cells during late meiosis (Rubin et al., 1986; Wolgemuth et al., 1986; 1987) but exhibits no detectable expression, by Northern blot analysis, in other tissues tested (Rubin et al., 1986; Wolgemuth et al., 1986; Duboule et al., 1986). During prenatal development, the expression pattern of *Hoxa-4* was studied by *in situ* hybridization (Toth et al., 1987; Gaunt et al., 1988; Galliot et al., 1989; Gaunt et al., 1989). A strong expression was observed at all developmental stages in the central nervous system with a sharp anterior boundary within the middle portion of the floor of the myelencephalon. *Hoxa-4* transcripts are also detected in all the sclerotome-derived prevertebrae with the exception of the first vertebral anlage, thus displaying an anterior boundary between metameres C1 and C2 at the level of the 6th somite. *Hoxa-4* expression in various, non-somitic mesodermal derivatives, includes the whole mesonephric column, the metanephric parenchyma, the mesenchymal components of the digestive and respiratory tracts and the mesenchyme of the limb buds. The anterior limit of *Hoxa-4* expression in non-segmented mesoderm (trachea, base of larynx) is consistent with that seen between metameres C1 and C2 since no expression is detected in structures derived from the cephalic somitomeres. Thus *Hoxa-4* belongs to a sub-family of homeogenes expressed from a rather anterior level. Comparison of the respective expression of *Hoxa-4*, *Hoxb-4* and *Hoxd-4* revealed that, in spite of close similarities in the positions of their transcripts domains, these three genes display striking stage- and tissue-dependent differences in the relative abundance of their transcripts suggesting that their expression might be coordinately regulated (Gaunt et al., 1989).

■ Genetics, Function

Using a transgenic approach, a high level of *Hoxa-4* expression in gut has been linked to the appearance of megacolons (Wolgemuth et al., 1989).

■ References

Bucan, M., Yang-Feng, T., Colberg-Poley, A.M., Wolgemuth, D.J., Guenet, J.L., Francke, U. and Lehrach, H. (1986). Genetic and cytogenetic localization of the homeo box containing genes on mouse chromosome 6 and human chromosome 7. EMBO J. 5, 2899-2905.

Chavrier, P., Vesque, C., Galliot, B., Vigneron, M., Dollé, P., Duboule, D. and Charnay, P. (1990). The segment-specific gene Krox-20 encodes a transcription factor with binding sites in the promoter region of the Hox-1.4 gene. EMBO J. 9, 1209-1218.

Duboule, D., Baron, A., Mähl, P. and Galliot, B. (1986). A new homeo-box is present in overlapping cosmid clones which define the mouse Hox-1 locus. EMBO J. 5, 1973-1980.

Galliot, B., Dolle, P., Vigneron, M., Featherstone, M. S., Baron, A. and Duboule, D. (1989). The mouse Hox-1.4 gene: primary structure, evidence for promoter activity and expression during development. Development 107, 343-359.

Gaunt, S.J., Sharpe, P.T., and Duboule, D. (1988). Spatially restricted domains of homeo-gene transcripts in mouse embryos: relation to a segmented body plan. Development, 104 (Supplement) 169-181.

Gaunt, S. J., Krumlauf, R. and Duboule, D. (1989). Mouse homeogenes within a subfamily Hox-1.4, -2.6 and -5.1 dispaly similar anteroposterior domains of expression in the embryo, but show stage- and time-dependent differences in their regulation. Development 107, 131-141.

Rubin, M.R., Toth, L.E., Patel, M.D., d'Eustachio, P. and Nguyen-Huu, M.C. (1986). A mouse homeobox gene is expressed in spermatocytes and embryos. Science 233, 663-667.

Toth, L.E., Slawin, K.L., Pintar, J.E. and Nguyen-Huu, C.M. (1987). Region-specific expression of mouse homeobox genes in the embryonic mesoderm and central nervous system. Proc. Natl. Acad. Sci. U.S.A. 84, 6790-6794.

Wolgemuth, D.J., Engelmyer, E., Duggel, R.N., Gizang-Ginsberg, E., Mutter, G.L., Ponzetto, C., Viviano, C. and Zakeri, Z.F. (1986). Isolation of a mouse cDNA coding for a devlopmentally regulated, testis-specific transcript containing homeo box homolgy. EMBO J. 5, 1229-1235.

Wolgemuth, D. J., Viviano, C.M., Gizang-Ginsberg, E., Frohman, M.A., Joyner, A. and Martin, G. R. (1987). Differential expression of the mouse homeobox-containing the Hox-1.4 during male germ cell differentiation and embryonic development. Proc. Natl. Acad. Sci. U.S.A. 84, 5813-5817.

Wolgemuth, D.J., Behringer, R.R:, Mostoller, M.P., Brinster, R.L. and Palmiter, R.D. (1989). Transgenic mice overexpressing the mouse homoeobox-containing gene Hox-1.4 exhibit abnormal gut development. Nature 337, 464-467.

Hoxa-4 [c]

Species: *Gallus gallus* (chicken)

Chromosomal location: -

Other names: *Chox1-4*

Cognate genes: *HOXA4*, *Hoxa-4*

Type of homeobox: *Antp*-type, related to the *Drosophila Deformed* gene

Accession number: X17245

■ Origin and description

The chicken *hoxa-4* gene was isolated by means of cross homology to *Drosophila Antennapedia* homeobox sequence. The membership of this gene to the *HOXA* complex was identified by a chromosome walk on the whole chicken *HOXA* complex (Sasaki et al., 1990, Sasaki et al., 1992 and Kuroiwa et al., unpublished). The *Hoxa-4* mRNA is 2 kB large and encodes a homeodomain containing protein whose MW is 33.222 D.

■ Expression

The expression of the chicken *Hoxa-4* gene was already observed in the 2 day old embryo and continued at the earliest to 4-week old chick. The *Hoxa-4* transcripts show a graded distribution along rostro-caudal axis in spinal cord and has the highest concentration at the rostral side. The expression was observed in lung and testis but not in liver and kidney (Sasaki et al., 1992).

■ Function

An *E. coli* produced full length Hoxa-4 protein binds to 5′ upstream region of *Hoxa-4* and *Hoxd-4* genes and artificial NP sequence *in vitro* (Sasaki et al., 1990 and 1992). In transient transfection experiment, the Hoxa-4 protein showed little activation to NP6 dependent transcription despite binding the sequence *in vivo* (Sasaki et al., 1992).

■ References

Sasaki, H., Yokoyama, E. and Kuroiwa, A. (1990). Specific DNA binding of the two chicken Deformed family homeodomain proteins, *Chox*-1.4 and *Chox*-a. Nucl. Acids Res. 18, 1739-1747.
Sasaki, H., Yamamoto, M. and Kuroiwa, A. (1992). Cell type dependent transcription regulation by chick homeodomain proteins. Mech. of Dev. 37, 25–36.

Hoxa-5 [m]

Species: *Mus musculus*
Chromosomal location: Chromosome 6, bands B3-C
Other names: *Hox-1.3*, m2, *Hox 1-x*
Cognate genes: *HOXA5*
Type of homeobox: *Antp*-class, related to *Scr*
Accession number: Y00208, M28021, M36604

■ Origin and description

Hoxa-5 was identified by low stringency hybridization screening of genomic (Colberg-Poley et al., 1985; Duboule et al., 1986) and cDNA libraries (Odenwald et al., 1987) with *Drosophila Antp* class homeobox probes. Analysis of both the *Hoxa-5* cDNA and genomic sequences indicates that the major transcript (~1.85 Kb) encodes a 270 amino acid *Antp* class homeodomain protein (Odenwald et al., 1987; Fibi et al., 1988 and Zakany et al., 1988). *Hoxa-5* contains at least one intron (960 bp) located 21 bp upstream of its homeobox. The protein is rich in serine (14%), glycine (10%), and proline (7.8%) amino acid residues and like many other *Antp* class homeobox genes, the homeodomain is located close to the carboxyl terminus of the protein (16 amino acids). The protein shares extensive homology with its paralog, *Hoxb-5* (Krumlauf et al., 1987). Their homeodomains are identical.

■ Expression

In situ: Starting on embryonic day 8 *Hoxa-5* mRNA is detected in the presumptive thoracic mesoderm. Sections from 12 day embryos indicated strong expression in the spinal cord with an anterior boundary of expression in the myelencephalon. The lung, ribs, and stomach represent the strongest signals at this stage. By embryonic day 18 the *in situ* results reported a significant decrease in detectable message levels in the thoracic mesodermal structures (Dony and Gruss 1987). Immunohistochemistry: Antibodies raised against peptide fragments detect *Hoxa-5*-like immunoreactive epitopes througout the 7.5 day old embryo, at 8.5 days of gestation, immunoreactivity was observed in all embryonic structures. With continuing development, the pattern of immunoreactivity became increasingly more restricted. Although strong immunoreactivity is observed in the spinal cord, and dorsal root ganglia. Additional labeling is seen in the lung and thymus as late as embryonic day 14. By day 17, the immunolabeling is most heavily localized to the spinal cord and ganglia, however, significant expression is still detectable in the lung, thymus and gut. In the adult, *Hoxa-5*-like immunoreactivity was observed in hippocampal granular neurons in the dentate gyrus, and purkinje cells of the cerebellum (Odenwald et al., 1987 and Tani et al., 1989). Northern analysis: Additional less abundant *Hoxa-5* transcripts of ~4 Kb, ~8 Kb, and ~9 Kb are also observed when poly (A)$^+$RNA from adult and embryonic tissues are examined by Northern analysis (Garbern et al., 1989). *In vitro*: In F9 embryonal carcinoma cells, the steady-state message level of *Hoxa-5* increases after treatment with retinoic acid (Murphy et al., 1988). *Hoxa-5* is also expressed in primary mouse embryo fibroblasts and NIH 3T3 fibroblasts but not mouse L-cells. Hoxa-5 protein is found in the nuclei of mitotically active or, at least, noncontact-inhibited cells but when contact-inhibited, Hoxa-5 protein is detected only in the cytoplasm. Transgenic mice: The spatial regulation of the *Hoxa-5* region-specific enhancer has been studied using DNA fragments from the 5′ regulatory region fused to different reporter genes (Zakany et al., 1988 and Tuggle et al., 1990).

■ Genetics, function

Hoxa-5 is a phosphoprotein which binds to DNA in a sequence specific manner. The consensus binding motif is

CPyPyNATTAT/GPy, with the core ATTA box essential for *in vitro* binding (Odenwald et al., 1989).

■ References

Colberg-Poley, A.M., Voss, S.D., Chowdhury, K. and Gruss, P. (1985). Clustered homeo boxes are differentially expressed during murine development. Cell 43, 39-45.

Dony, C. and Gruss, P. (1987). Specific expression of the Hox 1.3 homeo box gene in murine embryonic structures originating from or induced by the mesoderm. EMBO J. 6, 2965-2975.

Duboule, D., Baron, A., Mähl, P. and Galliot, B. (1986). A new homeo-box is present in overlapping cosmid clones which define the mouse Hox-1 locus. EMBO J. 5, 1973-1980.

Fibi, M., Zink, B., Kessel, M., Colberg-Poley, A.M., Labeit, S., Lehrach, H. and Gruss, P. (1988). Coding sequence and expression of the homeo box gene Hox-1.3. Development 102, 349-359.

Garbern, J., Odenwald, W.F., Tournier-Lasserve, E. and Lazzarini, R.A. (1989). Analysis of transcription of the murine homeobox gene Hox 1.3. In: Cell to Cell Signals in Mammalian Development, (S.W. de Laat, et al., eds.) NATO ASI Series, H26, 63-73, Springer-Verlag, Berlin.

Harvey, R.P., Tabin, C.J. and Melton, D.A. (1986). Embryoinic expression and nuclear localizationof *Xenopus* homeobox (Xhox) gene products. EMBO J. 5, 1237-1244.

Krumlauf, R., Holland, P.W.H., McVey, J.H. and Hogan, B.L.M. (1987). Developmental and spatial patterns of expression of the mouse homeobox gene, Hox 2.1. Development 99, 603-617.

Murphy, S.P., Garbern, J., Odenwald, W.F., Lazzarini, R.A. and Linney, E. (1988). Differential expression of the homeobox gene Hox 1.3 in F9 embryonal carcinoma cells. PNAS 85, 5587-5591.

Odenwald, W.F., Taylor, C.F., Palmer-Hill, F.J., Friedrich Jr. V., Tani, M. and Lazzarini, R.A. (1987). Expression of a homeo domain protein in noncontact-inhibited cultured cells and postmitotic neurons. Genes Dev. 1, 482-496.

Odenwald, W.F., Garbern, J., Arnheiter, H., Tournier-Lasserve, E. and Lazzarini, R.A. (1989).The Hox-1.3 homeo box protein is a sequence-specific DNA-binding phosphoprotein. Genes Dev. 3, 158-172.

Mavilio, F., Simeone, A., Giampolo, A., Faiella, A., Zapavigna, V., Acampora, D., Poiana, G., Russo, G., Peschle, C. and Boncinelli, E. (1986). Differential and stage-related expression in embryonic tissues of a new human homeobox gene. Nature 324, 664-668.

Tani, M., Odenwald, W.F., Lazzarini, R.A. and Friedrich Jr. V.L. (1989). Progressive restriction in the distribution of Hox-1.3 homeodomain protein during embryogenesis. J. Neurosci. Res. 24, 457-469.

Tournier-Lasserve, E., Odenwald, W.F., Garbern, J., Trojanowski, J. and Lazzarini, R.A. (1989). Remarkable intron and exon sequence conservation in human and mouse homeobox Hox 1.3 genes. Molec. and Cell. Biol. 9, 2273-2278.

Tuggle, C.K., Zakany, J., Cianetti, L., Peschle, C. and Nguyen-Huu, M.C. (1990). Region-specific enhancers near two mammalian homeo box genes define adjacent rostrocaudal domains in the central nervous system. Genes Dev. 4, 180-189.

Zakany, J., Tuggle, C.K., Patel, M.D. and Nguyen-Huu, M.C. (1988). Spatial regulation of homeobox gene fusions in the embryonic central nervous system of transgenic mice. Neuron 1, 679-691.

Hoxa-6 [m]

Species: *Mus musculus*
Chromosomal location: Chromosome 6
Other names: *Hox-1.2, m5*
Cognate genes: *HOXA6*
Type of homeobox: *Antp*-type
Accession number: M11988

■ Origin and description

The murine *Hoxa-6* gene was first identified on phages isolated with the *Drosophila Antp* Box as a member of the *HOXA* cluster (Colberg-Poley et al., 1985; Duboule et al., 1986). No cDNAs were reported up to now.

■ Expression

Expression during mouse embryogenesis was studied by *in situ* analysis (Dressler and Gruss, 1989; Toth et al., 1987). Major expression domains are the somites/prevertebrae (anterior boundary prevertebra 8 (Dressler and Gruss, 1989), the posterior myelencephalon and the cervical CNS.

■ References

Colberg-Poley, M.A., Voss, S.D.K.C., Stewart, C.L., Wagner, E.F. and Gruss, P. (1985). Clustered homeoboxes are differentially expressed during murine development. Cell 43, 39-45.

Dressler, G.R. and Gruss, P. (1989). Anterior boundaries of Hox gene expression in mesoderm-derived structures correlate with the linear gene order along the chromosome. Differentiation 41, 193-201.

Duboule, D., Baron, A., Mähl, P. and Galliot, B. (1986). A new homeo-box is present in overlapping cosmid clones which define the mouse Hox-1 locus. EMBO J. 5, 1973-1980.

Toth, L.E., Slawin, K.L., Pintar, J.E. and Nguyen-Huu, M.C. (1987). Region-specific expression of mouse homeobox genes in the embryonic mesoderm and central nervous system. Proc. Natl. Acad. Sci. USA 84, 6790-6794.

Hoxa-7 [m]

Species: *Mus musculus*
Chromosomal location: Chromosome 6
Other names: *Hox-1.1, m6*
Cognate genes: *HOXA7*
Type of homeobox: *Antp*-type
Accession number: X02399, M17192

■ Origin and description

The *Hoxa-7* gene was cloned in a low stringency screen with a *Drosophila Antp* Box probe (Colberg-Poley et al., 1985b). Mapping to the *HOXA* cluster (Colberg-Poley et al., 1985a), mapping to chromosome 6 (Bucan et al., 1986), a polyclonal mouse serum (Kessel et al., 1987), a monoclonal antibody (Schulze et al., 1987), and the cDNA cloning (Kessel et al., 1987; Balling et al., 1989) are reported.

■ Expression

The expression was studied by *in situ* analysis (Mahon et al., 1988; Dressler and Gruss, 1989; Kessel and Gruss, 1991). The gene is expressed at least from day 8 of gestation onwards, major expression domains are the neural tube, spinal ganglia (anterior boundary ganglion 5), somites/prevertebrae (anterior boundary prevertebra 10), and lung. RNA was also observed in adult tissues (lung, testes) and everal cell lines (Colberg-Poley et al., 1985; Fibi et al., 1988). Retinoic acid induces expression in teratocarcinoma cells, and anteriorizes the *Hoxa-7* expression domain, if embryos are exposed during gastrulation (Kessel and Gruss, 1991).

■ Function

The *Hoxa-7* gene was expressed in transgenic mice under control of a β-actin promoter. Ubiquitous expression led to craniofacial abnormalities including cleft palate, malformation of external ears and open eyes at birth (Balling et al., 1989). Posterior transformations were observed at the craniocervical transition, including manifestation of a proatlas and association of a vertebral body with the atlas (Kessel et al., 1990). Expression of the *Hoxa-7* gene was studied using fusions of upstream and downstream sequences with the lacZ gene (Püschel et al., 1990; Püschel et al., 1991). Major aspects of the endogenous *Hoxa-7* pattern could be reproduced. The *Hoxa-7* gene was disrupted in ES cells, however these cells were not transmitted to the germline (Zimmer and Gruss, 1989).

■ References

Acampora, D., D'Esposito, M., Faiell, A., Pannese, M., Migliaccio, E., Morelli, F., Stornaiuolo, A., Simeone, A. and Boncinelli, E. (1989). The human Hox gene family. Nucleic Acid Res. 24, 10385-10402.

Balling, R., Mutter, G., Gruss, P. and Kessel, M. (1989). Craniofacial abnormalities induced by ectopic expression of the homeobox gene *Hox-1.1* in transgenic mice. Cell 58, 337-347.

Bucan, M., Yang-Feng, T., Colberg-Poley, A.M., Wolgemuth, D.J., Guenet, J.-L., Francke, U. and Lehrach, H. (1986). Genetic and cytogenetic localization of the homeobox containing genes on mouse chromosome 6 and human chromosome 7. EMBO J. 5, 2899-2905.

Colberg-Poley, M.A., Voss, S.D.K.C., Stewart, C.L., Wagner, E.F. and Gruss, P. (1985a). Clustered homeoboxes are differentially expressed during murine development. Cell 43, 39-45.

Colberg-Poley, A.M., Voss, S.D., Chowdhury, K. and Gruss, P. (1985b). Structural analysis of murine genes containing homeobox sequences and their expression in embryonal carcinoma cells. Nature 314, 713-718.

Condie, B.G. and Harland, R. (1987). Posterior expression of a homeobox gene in early *Xenopus* embryos. Development 101, 93-105.

Dressler, G.R. and Gruss, P. (1989). Anterior boundaries of Hox gene expression in mesoderm-derived structures correlate with the linear gene order along the chromosome. Differentiation 41, 193-201.

Fibi, M.B.Z., Kessel, M., Colberg-Poley, A.M., Labeit, S., Lehrach, H. and Gruss, P. (1988). Coding sequence and expression of the homeobox containing gene *Hox-1.3*. Development 102, 349-359.

Kessel, M., Balling, R. and Gruss, P. (1990). Variations of cervical vertebrae after expression of a *Hox-1.1* transgene in mice. Cell 61, 301-308.

Kessel, M. and Gruss, P. (1991). Homeotic transformations of murine vertebrae and concomitant alteration of Hox codes induced by retinoic acid. Cell 67, 89-104.

Kessel, M., Schulze, F., Fibi, M. and Gruss, P. (1987). Primary structure and nuclear localization of a murine homeodomain protein. Proc. Natl. Acad. Sci. USA 84, 5306-5310.

Mahon, K.A., Westphal, H. and Gruss, P. (1988). Expression of homeobox gene *Hox-1.1* during mouse embryogenesis. Development (Supplement) 102, 187-195.

Püschel, A.W., Balling, R. and Gruss, P. (1990). Position-specific activity of the *Hox-1.1* promotor in transgenic mice. Development 108, 435-442.

Püschel, A.W., Balling, R. and Gruss, P. (1991). Separate elements cause lineage restriction and specify boundaries of *Hox-1.1* expression. Development 112, 279-287.

Schulze, F., Chowdhury, K., Zimmer, A., Drescher, U. and Gruss, P. (1987). The murine homeobox gene product, *Hox-1.1* protein, is growth controlled and associated with chromatin. Differentiation 36, 130-137.

Zimmer, A. and Gruss, P. (1989). Production of chimaeric mice containing embryonic stem (ES) cells carrying a homeobox Hox-1.1 allele mutated by homologous recombination. Nature 338, 150-153.

Hoxa-7   [x]

Species: *Xenopus laevis*
Chromosomal location: -
Other names: *XlHbox 3, Xhox 36*
Cognate genes: -
Type of homeobox: *Antp*-type
Accession numbers: X07103, M24752

■ Origin and description

This gene was isolated from a gastrula cDNA library using *Antp* homeobox probes (Condie and Harland, 1987; Fritz and De Robertis, 1988). It encodes a 209 amino acid protein with a homeodomain that shares 60 out of 62 amino acids with Antp (Condie and Harland, 1987). Embryos contain several kinds of transcripts, including noncoding and unspliced transcripts which are more abundant than the homeodomain protein coding mRNA (Condie et al., 1990).

■ Expression

XlHbox 3 mRNA levels peak at the neurula stage with lower levels in gastrula and tadpole (Condie and Harland, 1987; Fritz and De Robertis, 1988). Based on Northern analysis of dissected embryos, expression is maximal posteriorly, and the mRNA is detected in ectoderm and mesoderm but not endoderm (Condie and Harland, 1987). Exogastrulated embryos also express message in mesoderm and ectoderm. Treatment with lithium chloride, which anteriorizes embryos, decreases message levels, while UV treatment has no effect (Condie and Harland, 1987).

■ References

Condie, B. G. and Harland, R. M. (1987). Posterior expression of a homeobox gene in early *Xenopus* embryos. Development 35, 206-211.
Condie, B. G., Hemmati Brivanlou, A. and Harland, R. M. (1990). Most of the homeobox-containing Xhox 36 transcripts in early *Xenopus* embryos cannot encode a homeodomain protein. Mol. Cell. Biol. 10, 3376-3385.
Fritz, A. and De Robertis, E. M. (1988). *Xenopus* homeobox-containing cDNAs expressed in early development. Nucl. Acids Res. 16, 1453-1469.

Hoxa-9 [c]

Species: *Gallus gallus* (chicken)
Chromosomal location: -
Other names: *Chox-1.7*
Cognate genes: *HOXA9, Hoxa-9*
Type of homeobox: *Antp*-type, related to the *Drosophila Abd-B* gene
Accession number: -

■ Origin and description

The chicken *Hoxa-9* gene was isolated during a walk on the *HOXA* complex (Sasaki et al. 1992 and Kuroiwa et al., unpublished).

■ Expression

The expression of the chicken *Hoxa-9* gene was already observed in the 2-day old embryo and continued at the earliest to 4-week old chick. Transcripts show a graded distribution along rostro-caudal axis in spinal cord and has the highest concentration at the caudal side (Sasaki et al., 1992).

■ Function

An *E. coli* produced full length Hoxa-9 protein binds to NP sequence *in vitro* with nearly the same affinity as the Hoxa-4 protein (Sasaki et al., 1992). In transient transfection experiment, this protein strongly activates transcription from globin basal promoter through the artificial binding site NP6 (Sasaki et al., 1992).

■ References

Sasaki, H., Yamamoto, M. and Kuroiwa, A. (1992). Cell type dependent transcription regulation by chick homeodomain proteins. Mech. of Dev. 37, 25–36.

Hoxa-9 [m]

Species: *Mus musculus*
Chromosomal location: Chromosome 6
Other names: *Hox-1.7*
Cognate genes: *HOXA9*
Type of homeobox: *Antp*-type, related to *Abd-B*
Accession number: M28449

■ Origin and description

The murine *Hoxa-9* homeobox was isolated from a F9 teratocarcinoma cell cDNA library using human Hu-1 and Hu-2 homeobox probes (Rubin et al., 1987). The gene was mapped to mouse chromosome 6 (Rubin et al., 1987), its linkage to the *HOXA* cluster was recently proven by analysis of cosmid clones (unpublished).

■ Expression

The *Hoxa-9* gene is expressed in the posterior spinal cord (Rubin et al, 1987).

■ References

Rubin, M.R., King, W., Toth, L.E., Sawcuk, I.S., Levine, M., D'Eustachio, P. and Nguyen-Huu, M.C. (1987). Murine Hox-1.7 Homeo-box gene: Cloning chromosomal location and expression. Mol. Cell. Biol. 7, 3836-3841, correction p. 5593.

Hoxa-10 [c]

Species: *Gallus gallus* (chicken)
Chromosomal location: -
Other names: *Chox-1.8*
Cognate genes: *HOXA10, Hoxa-10*
Type of homeobox: *Antp*-type, related to the *Drosophila Abd-B* gene
Accession number: -

■ Origin and description

This gene was isolated during a walk on the chicken *HOXA* complex (Yokouchi et al., 1991 and Kuroiwa et al., unpublished).

■ Expression

The expression of the *Hoxa-10* gene was studied by *in situ* hybridization during limb development (Yokouchi et al., 1991).

■ References

Yokouchi, Y., Sasaki, H. and Kuroiwa, A. (1991). Homeobox gene expression correlated with the bifurcation process of limb cartilage development. Nature 353, 443-445.

Hoxa-11 [c]

Species: *Gallus gallus* (chicken)
Chromosomal location: -
Other names: *Chox-1.9*
Cognate genes: *HOXA11, Hoxa-11*
Type of homeobox: *Antp*-type, related to the *Drosophila Abd-B* gene
Accession number: -

■ Origin and description

The *Hoxa-11* gene was isolated during a walk on the chicken *HOXA* complex (Yokouchi et al., 1991 and Kuroiwa et al., unpublished).

■ Expression

The expression of this gene was studied by *in situ* hybridization during limb development (Yokouchi et al., 1991). The expression was first observed in the mesenchy-mal cells of both fore and hind limb buds at stage 19. The *Hoxa-11* expression at the distal region decrease and disappeared whereas *Hoxa-13* expression get prominent at this region. Finally, the *Hoxa-11* expression domain correspond to the zeugopod region.

■ References

Yokouchi, Y., Sasaki, H. and Kuroiwa, A. (1991). Homeobox gene expression correlated with the bifurcation process of limb cartilage development. Nature 353, 443-445.

Hoxa-13 [c]

Species: *Gallus gallus* (chicken)
Chromosomal location: -
Other names: *Chox-1.10*
Cognate genes: *HOXA13, Hoxa-13*
Type of homeobox: *Antp*-type, related to the *Drosophila Abd-B* gene
Accession number: -

■ Origin and description

This gene was isolated during a walk on the chicken *HOXA* complex (Yokouchi et al., 1991 and Kuroiwa et al., unpublished). This gene is the most 5'-located member of this complex and is a paralog of *Hoxd-13*.

■ Expression

The expression of the chicken *Hoxa-13* gene was studied by *in situ* hybridization during limb development (Yokouchi et al., 1991). The expression was first observed in the mesenchymal cells in a small area of the posterior-distal region of both fore and hind limb buds at stage 22. Its expression domain extends anteriorly and occupies the distal most region at a later stage. The *Hoxa-13* expression domain correspond to the primordial autopod region.

■ References

Yokouchi, Y., Sasaki, H. and Kuroiwa, A. (1991). Homeobox gene expression correlated with the bifurcation process of limb cartilage development. Nature 353, 443-445.

Hoxb-1 [m]

Species: *Mus musculus*
Chromosomal location: Chromosome 11
Other names: *Hox-2.9*
Cognate genes: *HOXB1*
Type of homeobox: *Antp*-type, related to the *Drosophila labial* gene
Accession number: X53063

■ Origin and description

This gene was isolated by cross-homology with *Drosophila Zen,* as part of a cosmid clone which contained the *Hoxb-3* and *Hoxb-2* genes (Rubock et al., 1990) and also by cDNA screening with a *Hoxa-1* probe (Fainsod et al., 1986; Murphy and Hill, 1991). It encodes a member of the murine *labial*-like subfamily together with *Hoxa-1* (Baron et al., 1987) and *Hoxd-1* (Hunt et al., 1991a, Frohman et al. 1990). The gene is the most 3'-located member of the *HOXB* cluster. *Hoxb-1* encodes a divergent *Antp*-like homeobox related to that of *Drosophila labial* (83% identical; Mlodzik et al., 1988). Unlike *Hoxa-1*, differentially spliced transcripts are not observed.

■ Expression

By Northern analysis, a 2.1 kb transcript of very low abundance is observed from E10.5-12.5 (Frohman et al., 1990). This result is similar to that observed for *Hoxa-1* but dissimilar to that observed for other *Hoxb* genes, expression of which reach maximal levels at E14.5 and persist through birth (Graham et al., 1989). *Hoxb-1* expression is first observed by *in situ* hybridization at E7.5 in the primitive streak (Frohman et al., 1990); Murphy and Hill, 1991). Shortly thereafter, E7.75-E8 transcripts are observed from the posterior end of the embryo to a region located within the still morphologically featureless hindbrain, first in mesoderm, and then in neuroectoderm (Frohman et al., 1990), to a level identical to that of *Hoxa-1* (Murphy and Hill, 1991). By E8-8.5, it becomes apparent that the anterior boundary to which *Hoxb-1* is expressed lies at the rhombomere (r) 3/r4 boundary (Frohman et al., 1990, Murphy and Hill, 1991; Hunt e al., 1991). Expression also becomes observed in branchial arch gut endoderm, surface ectoderm, lateral mesoderm to the level of the second branchial arch (Frohman et al., 1990), and neural crest derived from rhombomere 4 (Wilkinson et al., 1989, Frohman et al., 1990, Murphy and Hill, 1991; Hunt et al., 1991). By E8.5, *Hoxb-1* expression in the neural tube posterior to rhombomere 4 is extinguished, leaving a stripe of *Hoxb-1* expression solely in rhombomere 4 (Murphy et al., 1989; Wilkinson et al., 1989; Frohman et al., 1990; Hunt et al., 1991). This expression persists through E10.5 and is then extinguished. Expression is also observed at E9.5, but not subsequently, in branchial arch gut endoderm, surface ectoderm, and lateral mesoderm to the level of the third branchial arch (Frohman et al., 1990). Altered expression patterns for *Hoxb-1* have been observed in two experimental settings: After retinoic acid treatment (Morriss-Kay et al., 1991) and in the neuromeric segmentation mutant *kreisler* (Frohman et al., submitted). Transgenic experiments have shown that different regulatory regions are required to generate the early versus the r4 restricted domains of expression (Guthrie et al., 1992). Grafting experiments in the chicken have shown that the region which will eventually become r4 has a preprogrammed ability before rhombomere formation to initiate the proper *Hoxb-1* patterns when placed in ectopic locations (Guthrie et al., 1992).

■ References

Baron, A., Featherstone, M. S., Hill, R. E., Hall, A., Galliot, B. and Duboule, D. (1987). Hox-1.6: a mouse homeobox-containing gene member of the Hox-1 complex. EMBO J. 6, 2977-2986.

Frohman, M.A., Boyle, M. and Martin, G.R. (1990). Isolation of the mouse *Hox-2.9* gene: analysis of embryonic expression suggests that positional information along the anterior-posterior axis is specified by mesoderm. Development 110, 589-608.

Frohman, M.A., Martin, G.R., Cordes, S.P. and Barsh, G.S. (1993). Altered rhombomere-specific gene expression in the mouse segmentation mutant, kreisler. Development 117, 925–936.

Graham, A., Papalopulu, N. and Krumlauf, R. (1989). The murine and *Drosophila* homeobox gene complexes have common features of organization and expression. Cell 57, 367-378.

Guthrie, S., Muchamore, I., Kuriowa, A., Krumlauf, R. and Lumsden, A. (1991). Rhombomere transpositions in the chick embryo hindbrain reveal neuroectodermal autonomy of Hox-2.9 expression and segment phenotype. Nature 356, 157-159.

Hunt, P., Gulisano, M., Cook, M., Sham, M., Faiella, A., Wilkinson, D., Boncinelli, E. and Krumlauf, R. (1991a). A distinct *Hox* code for the branchial region of the head. Nature 353, 861-864.

Mlodzik, M., Fjose, A. and Gehring, W.H. (1988). Molecular structure and spatial expression of a homeobox gene from the *labial* region of the *Antennapedia*-complex. EMBO J. 7, 2569-2578.

Murphy, P., Davidson, D.R. and Hill, R.E. (1989). Segment-specific expression of a homeobox-containing gene in the mouse hindbrain. Nature 341, 156-159.

Murphy, P. and Hill, R.E. (1991). Expression of the mouse *labial*-like homeobox-containing genes, Hox-2.9 and Hox1.6, during segmentation of the hindbrain. Development 111, 61-74.

Moriss-Kay, G.M., Murphy, P., Hill, R.E. and Davidson, D.R. (1991). Effects of retinoic acid excess on expression of Hox-2.9 and Krox-20 and on morphological segmentation in the hindbrain of mouse embryos EMBO J. 10, 2985-2995.

Rubock, M., Larin, Z., Cook, M., Papalopulu, N., Krumlauf, N. and Lehrach, H. (1990). A yeast artificial chromosome containing the mouse homeobox cluster Hox-2. Proc. Natn. Acad. Sci. USA 87, 4751-4755.

Wilkinson, D.G., Bhatt, S., Cook, M., Boncinelli, E. and Krumlauf, R. (1989). Segmental expression of Hox-2 homeobox-containing genes in the developing mouse hindbrain. Nature 341, 405-409.

Hoxb-2 [m]

Species: *Mus musculus*
Chromosomal location: Chromosome 11, band 11D
Other names: *Hox-2.8*
Cognate genes: *HOXB2*
Type of homeobox: *Antp*-type, related to the *Drosophila proscipedia* gene
Accession number: -

■ Origin and description

This gene was first isolated as part of a yeast artificial chromosome clone containing most of the *HOXB* complex (Rubock et al., 1990). *Hoxb-2* was located at the 3' end of the clone and probes from this region were also used to isolate a cosmid containing *Hoxb-3, b-2* and *b-1* (Rubock et al., 1990). It is oriented in the same transcriptional orientation as other members of the *HOXB* complex. The sequence of the gene demonstrated that it had a high degree of identity with the *Drosophila pb* gene (Rubock et al., 1990; Cribbs et al., 1992) in multiple regions of the proteins and was a true *pb* homologue. The gene is 96% identical to its human homologue *HOXB2* (Acampora et al., 1989). One paralogue has been found in human (*HOXA2*) (Acampora et al., 1989) and mouse (*Hoxa-2*) (Hunt et al., 1991a).

■ Expression:

Hoxb-2 is the most anteriorly expressed gene of the *HOXB* cluster extending to the rhombomere 2/3 junction, basioccipital bones, second branchial arch and cranial ganglia gVII/VIII (Wilkinson et al., 1989; Hunt et al., 1991a, b; Sham et al., 1992). The anterior boundaries of expression are established in early presomite embryos at 7.5 dpc (Hunt et al., 1991a). At 8.5 dpc additional high levels of expression are generated in rhombomeres 3,4 and 5. This gene is rapidly induced by retinoic acid in mouse F9 cells (Papalopulu et al., 1991) and human N-tera2 cells (Simeone et al., 1990).

■ References

Acampora, D., D'Esposito, M., Faiella, A., Pannese, M., Migliaccio, E., Morelli, F., Stornaiuolo, A., Nigro, V., Simeone, A. and Boncinelli, E. (1989). The human HOX gene family. N.A.R. 17, 10385-10402.

Cribbs, D., Pultz, M., Johnson, D., Mazzulla, M. and Kaufman, T. (1992). Structural complexity and evolutionary conservation of the *Drosophila* homeotic gene *proboscipedia*. EMBO J. 11, 1437-1450.

Hunt, P., Gulisano, M., Cook, M., Sham, M., Faiella, A., Wilkinson, D., Boncinelli, E. and Krumlauf, R. (1991a). A distinct *Hox* code for the branchial region of the head. Nature 353, 861-864.

Hunt, P., Wilkinson, D. and Krumlauf, R. (1991b). Patterning the vertebrate head: murine Hox 2 genes mark distinct subpopulations of premigratory and migrating neural crest. Development 112, 43-51.

Papalopulu, N., Lovell-Badge, R. and Krumlauf, R. (1991). The expression of murine *Hox-2* genes is dependent on the differentiation pathway and displays collinear sensitivity to retinoic acid in F9 cells and *Xenopus* embryos. N.A.R. 19, 5497-5506.

Rubock, M., Larin, Z., Cook, M., Papalopulu, N., Krumlauf, N. and Lehrach, H. (1990). A yeast artificial chromosome containing the mouse homeobox cluster Hox-2. Proc. Natl. Acad. Sci. USA 87, 4751-4755.

Sham, M.-H., Hunt, P., Nonchev, S., Papalopulu, N., Graham, A., Boncinelli, E. and Krumlauf, R. (1992). Analysis of the murine *Hoxb-3* gene: conserved alternative transcripts with differential distributions in the nervous system and the potential for shared regulatory regions. EMBO J. 11, 1825-1836.

Simeone, A., Acampora, D., Arcioni, L., Andrews, P. W., Boncinelli, E. and Mavilio, F. (1990). Sequential activation of HOX2 homeobox genes by retinoic acid in human embryonal carcinoma cells. Nature 346, 763-766.

Wilkinson, D., Bhatt, S., Cook, M., Boncinelli, E. and Krumlauf, R. (1989). Segmental expression of hox 2 homeobox-containing genes in the developing mouse hindbrain. Nature 341, 405-409.

Hoxb-3 [m]

Species: *Mus musculus*
Chromosomal location: Chromosome 11, band 11D
Other names: *Hox-2.7*
Cognate genes: *HOXB3*
Type of homeobox: *Antp*-type, has intermediate identity to *Dfd, Zen* and *Pb*.
Accession number: -

■ Origin and description

This gene was identified as part of the *HOXB* complex because it was present on a phage clone and cosmid that contained *Hoxb-4* and *Hoxb-5* (Graham et al., 1988). The sequence showed it was organized in the same transcriptional orientation as other *Hoxb* genes and was related to the *Hoxa-3* and *Hoxd-3* genes in other complexes (Fainsod et al., 1987; Lonai et al., 1987). Multiple transcripts for this gene are present in embryos and comparison of the human (Acampora et al., 1989) and mouse (Sham et al., 1992) genes shows that there are at least two promoter 25 kb apart that encode identical proteins with different 5' untranslated regions. The distal promoter maps nearly 30kb upstream of the ATG with an important regulatory domain of the adjacent *Hoxb-4* gene (Sham et al., 1992). The position of this distal promoter is conserved in human, mouse and chicken *HOXB* clusters. Expression analysis shows that there are at least two other promoters which produce other types of transcripts, some of which may not contain a homeodomain (Sham et al., 1992). One type of *Hoxb-3* transcript has an alternative open reading frame which could encode a protein related to mammalian ATPases. The predicted protein is much larger than other *Hox* proteins in have a large domain of the carboxy terminal side of the homeodomain. There are no clear *Drosophila* homologues to this gene or protein although it is in the same relative position as *zen* in the *HOM-C* complex, and is often mistakenly referred to as a *Zen* or *Pb* homologue. This may represent a unique vertebrate paralogous group or a *Drosophila* gene which has been lost during evolution.

■ Expression

Hoxb-3 follows the colinear patterns of expression in the complex and is more anterior than *Hoxb-4*. The anterior boundaries of expression map to the rhombomere 4/5 junction, third branchial arch and posterior, first cervical vertebrae, and cranial ganglia gVX/XI (Graham et al., 1988, 1989; Wilkinson et al., 1989; Hunt et al., 1991b; Sham et al., 1992). The anterior boundaries of the *Hoxb-3/a-3/d-3* paralogous group are identical in the hindbrain and branchial arches except that there are high levels of *Hoxb-3/a-3* and not *Hoxd-3* in rhombomere 5 (Hunt et al., 1991a). The transcription pattern of the gene is very complex and the various transcript having different spatial distributions in the nervous system (Sham et al., 1992). Some transcripts containing the *Hoxb-3* homeobox extend to an anterior boundary at rhombomere 2/3 typical of the adjacent 3′ gene in the complex *Hoxb-2*, suggesting that *in vivo* some transcripts are influenced by regulatory regions from the adjacent gene in the cluster. Experiments in transgenic mice with *Hoxb-3/lacZ* fusion constructs have shown that the gene and flanking regions generate expression domains typical of the *Hoxb-2* gene indicating that there are regulatory regions in the cluster adjacent to the gene capable of regulating both *Hoxb-3* and *Hoxb-2*.

■ References

Acampora, D., D'Esposito, M., Faiella, A., Pannese, M., Migliaccio, E., Morelli, F., Stornaiuolo, A., Nigro, V., Simeone, A. and Boncinelli, E. (1989). The human HOX gene family. N.A.R. 17, 10385-10402.

Fainsod, A., Awgulewitsch, A. and Ruddle, F. H. (1987). Expression of the murine homeo box gene Hox 1.5 during embryogenesis. Devl Biol. 124, 125-133.

Graham, A., Papalopulu, N., Lorimer, J., McVey, J., Tuddenham, E. and Krumlauf, R. (1988). Characterization of a murine homeo box gene, *Hox 2.6*, related to the drosophila deformed gene. Genes Dev. 2, 1424-1438.

Graham, A., Papalopulu, N. and Krumlauf, R. (1989). The murine and *Drosophila* homeobox clusters have common features of organisation and expression. Cell 57, 367-378.

Hunt, P., Gulisano, M., Cook, M., Sham, M., Faiella, A., Wilkinson, D., Boncinelli, E. and Krumlauf, R. (1991a). A distinct *Hox* code for the branchial region of the head. Nature 353, 861-864.

Hunt, P., Wilkinson, D. and Krumlauf, R. (1991b). Patterning the vertebrate head: murine Hox 2 genes mark distinct subpopulations of premigratory and migrating neural crest. Development 112, 43-51.

Lonai, P., Arman, E., Czosnek, H., Ruddle, F. H. and Blatt, C. (1987). New murine homeoboxes: structure, chromosomal assignment, and differential expression in adult erythropoiesis. DNA. 6, 409-418.

Sham, M.-H., Hunt, P., Nonchev, S., Papalopulu, N., Graham, A., Boncinelli, E. and Krumlauf, R. (1992). Analysis of the murine *Hoxb-3* gene: conserved alternative transcripts with differential distributions in the nervous system and the potential for shared regulatory regions. EMBO J. 11, 1825-1836.

Wilkinson, D., Bhatt, S., Cook, M., Boncinelli, E. and Krumlauf, R. (1989). Segmental expression of hox 2 homeobox-containing genes in the developing mouse hindbrain. Nature 341, 405-409.

Hoxb-3 [x]

Species: *Xenopus laevis*
Chromosomal location: -
Other names: *XlHox2.7*
Cognate genes: *HOXB3*
Type of homeobox: *Antp*-type
Accession number: M91588

■ Origin and description

This gene was isolated during the cloning of the *Xenopus laevis HOXB* complex. Its position within the complex, as well as sequence data of the homeobox and of region upstream and downstream of the homeobox allowed to define it as *Hoxb-3*. This gene spans, like in human about 25 kb. A cDNA has been isolated.

■ Expression

The expression of this gene has been studied from egg to stage 35 of the Xenopus development. Expression starts at stage 11 (early gastrulation) and is found in areas posterior to the otic vesicle. These results are in correlation to what was expected by the position of this gene in the complex.

Hoxb-4 [m]

Species: *Mus musculus*
Chromosomal location: Chromosome 11, band 11D
Other names: *Hox-2.6*
Cognate genes: *HOXB4*
Type of homeobox: *Antp*-type, related to the *Drosophila Deformed* gene
Accession number: -

■ Origin and description

This gene was isolated by walking in a 3′ direction from clones containing the *Hoxb-5* gene (Graham et al., 1988). Sequence analysis showed that it was a mouse member of the *Dfd* family based on the high degree of sequence identity and colinear arrangements of the identity, both within the homeodomain and other regions (particularly the N-terminus) of the *Drosophila Dfd* protein (Graham et al., 1988). This gene has paralogues, *Hoxa-4, d-4* and *c-4* (Featherstone et al., 1988; Acampora et al., 1989; Galliot et al., 1989), and was the first example of the a paralogous group with members in each of the four complexes. Sequence comparisons of this group provided the clear initial evidence that mouse clusters were related by duplication and divergence from a common ancestor (Featherstone et al., 1988; Graham et al., 1988; Acampora et al., 1989; Galliot et al., 1989) and that this sequence identity extended to other vertebrate (Harvey et al., 1986; Regulski et al., 1987) and arthropods suggesting an ancient common ancestral cluster for both vertebrates and invertebrate homeobox complexes (Duboule and Dollé, 1989; Graham et al., 1989).

■ Expression

This gene is activated in response to retinoic acid in F9 cells and encodes multiple transcripts in cells and embryonic tissues (Graham et al., 1988; Papalopulu et al., 1991). The gene has an anterior boundary of expression established at 8.5 dpc in the hindbrain which corresponds to the rhombomere r6/7 junction (Wilkinson et al., 1989), and is spatially restricted in the neural crest (fourth branchial arach and posterior), cranial ganglia and surface ectoderm in the branchial region of the head (Hunt et al., 1991a,c). The boundary in paraxial mesoderm corresponds to the second cervical vertebrae and in the cranial ganglia to gX/XI (Hunt et al., 1991b). Expression is more posterior than the 3′ adjacent gene *Hoxb-3* and more anterior than the 5′ adjacent gene *Hoxb-5* in all tissues examined (Graham et al., 1988, 1989; Wilkinson et al., 1989). The major domains of *Hoxb-4* expression have been reconstructed using *lacZ* fusion constructs in transgenic mice

(Hunt et al., 1991b). Transgenic experiments have also identified at least three regulatory regions within and flanking the gene which function as spatially-specific enhancers capable of conferring restricted expression from heterologous promoters (Hunt et al., 1991b). The *Hoxb-4/lacZ* mice also show that in addition to typical expression in gut, kidney, gonads, PNS, blood vessels and lung, the gene is expressed in placodes of developing skin. A direct comparison of expression with two mouse paralogues show that some boundaries and domains of expression, such as the branchial arches and rhombomereic restrictions, are identical but that there are differences in the level, sites and timing of expression (Gaunt et al., 1989; Hunt et al., 1991a). The gene has been mutated in ES cells (Hasty et al., 1991).

■ References

Acampora, D., D'Esposito, M., Faiella, A., Pannese, M., Migliaccio, E., Morelli, F., Stornaiuolo, A., Nigro, V., Simeone, A. and Boncinelli, E. (1989). The human HOX gene family. N.A.R. 17, 10385-10402.

Duboule, D. and Dollé, P. (1989). The structural and functinal organization of the murine HOX gene family resembles that of *Drosophila* homeotic genes. EMBO J. 8, 1497-1505.

Featherstone, M. S., Baron, A., Gaunt, S. J., Mattei, M. G. and Duboule, D. (1988). Hox-5.1 defines a homeobox-containing gene locus on mouse chromosome 2. Proc. Natl. Acad. Sci. USA 85, 4760-4764.

Galliot, B., Dollé, P., Vigneron, M., Featherstone, M. S., Baron, A. and Duboule, D. (1989). The mouse Hox-1.4 gene: primary structure, evidence for promoter activity and expression during development. Development. 107, 343-359.

Gaunt, S. J., Krumlauf, R. and Duboule, D. (1989). Mouse homeogenes within a subfamily, Hox-1.4, -2.6 and -5.1, display similar anteroposterior domains of expression in the embryo, but show stage- and tissue-dependent differences in their regulation. Development. 107, 131-141.

Graham, A., Papalopulu, N., Lorimer, J., McVey, J., Tuddenham, E. and Krumlauf, R. (1988). Characterization of a murine homeo box gene, *Hox 2.6*, related to the *Drosophila* deformed gene. Genes Dev. 2, 1424-1438.

Graham, A., Papalopulu, N. and Krumlauf, R. (1989). The murine and *Drosophila* homeobox clusters have common features of organisation and expression. Cell 57, 367-378.

Harvey, R. P., Tabin, C. J. and Melton, D. A. (1986). Embryonic expression and nuclear localization of Xenopus homeobox (Xhox) gene products. EMBO. J. 5, 1237-1244.

Hasty, P., Ramirez-Solis, R., Krumlauf, R. and Bradley, A. (1991). Introduction of a subtle mutation into the *Hox-2.6* locus in embryonic stem cells. Nature 350, 243-246.

Hunt, P., Gulisano, M., Cook, M., Sham, M., Faiella, A., Wilkinson, D., Boncinelli, E. and Krumlauf, R. (1991a). A distinct *Hox* code for the branchial region of the head. Nature 353, 861-864.

Hunt, P., Whiting, J., Nonchev, S., Sham, M., Marshall, H., Graham, A., Cook, M., Allemann, R., Rigby, P., Gulisano, M., Faiella, A., Boncinelli, E. and Krumlauf, R. (1991b). The branchial *Hox* code and its implications for gene regulation, patterning of the nervous system and head evolution. Development 113 (Supplement 2), 63-77.

Hunt, P., Wilkinson, D. and Krumlauf, R. (1991c). Patterning the vertebrate head: murine Hox 2 genes mark distinct subpopulations of premigratory and migrating neural crest. Development 112, 43-51.

Papalopulu, N., Lovell-Badge, R., and Krumlauf, R. (1991). The expression of murine *Hox-2* genes is dependent on the differentiation pathway and displays collinear sensitivity to retinoic acid in F9 cells and *Xenopus* embryos. N.A.R. 19, 5497-5506.

Regulski, M., McGinnis, N., Chadwick, R. and McGinnis, W. (1987). Developmental and molecular analysis of *Deformed*: A homeotic gene controlling *Drosophila* head development. EMBO J. 6, 767-777.

Wilkinson, D., Bhatt, S., Cook, M., Boncinelli, E. and Krumlauf, R. (1989). Segmental expression of hox 2 homeobox-containing genes in the developing mouse hindbrain. Nature 341, 405-409.

Hoxb-4 [c]

Species: *Gallus gallus* (chicken)
Chromosomal location: -
Other names: *Chox-2.6, Chox-Z*
Cognate genes: *HOXB4, Hoxb-4*
Type of homeobox: *Antp*-type, related to the *Drosophila Dfd* gene
Accession number: X17612

■ Origin and description

The chicken *Hoxb-4* gene was isolated during a walk on the *HOXB* complex and identified as the *Deformed* type homeobox gene in this cluster (Sasaki et al., 1990a,b; Sasaki et al., 1992 and Kuroiwa et al., unpublished). The length of the *Hoxb-4* mRNA is 2.1 kB and encodes a homeodomain containing protein which consists of 245 amino acids. An amino acid sequence is 66% identical to the mouse protein (Sasaki et al., 1990a).

■ Expression

The expression of *Hoxb-4* was already observed in the 2-day old embryo and continued at the earliest to 4-week old chick. Transcripts showed a graded distribution along the rostro-caudal axis in spinal cord and has the highest concentration at the rostral side. The expression was observed in lung, kidney and testis but not in liver (Sasaki et al., 1992).

■ Function

In transient transfection experiment, the Hoxb-4 protein activated the transcription from the globin basal promoter through the artificial binding site NP6 (Sasaki et al., 1991).

■ References

Sasaki, H. and Kuroiwa, A. (1990a). The nucleotide sequence of the cDNA encoding a chicken Deformed family homeobox

gene, *Chox-Z*. Nucl. Acids Res. 18, 184.

Sasaki, H., Yokoyama, E. and Kuroiwa, A. (1990b). Specific DNA binding of the two chicken Deformed family homeodomain proteins, *Chox-1.4* and *Chox-a*. Nucl. Acids Res. 18, 1739-1747.

Sasaki, H., Yamamoto, M. and Kuroiwa, A. (1992). Cell type dependent transcription regulation by chick homeodomain proteins. Mech. Devl. 37, 25–36.

Hoxb-4 [x]

Species: *Xenopus laevis*
Chromosomal location: -
Other names: *Xhox-1A*
Cognate genes: *Hoxb-4; HOXB4*
Type of homeobox: *Antp*-type, related to the *Drosophila* gene *Dfd*
Accession number: XLHOXAB2

■ Origin and description

This gene lies on a genomic clone (*Xhox-1*) which was isolated from a *Xenopus* genomic library using a *Drosophila* Ubx homeobox probe (1). The *Xhox-1A* gene lies downstream (12.5 kb) of another homeobox gene (*Xhox-1B*). Both genes are likely to be the *Xenopus* homologues of the murine *HOXB* locus, the *Hoxb-4* and *Hoxb-5* genes respectively, thus representing a section of a larger homeobox cluster. A near full length cDNA clone has been isolated coding for the Xhox-1A protein (1). There are similarities between the Xhox-1A protein and the *Drosophila* deformed protein within an outside of the homeodomain, indicating a common evolutionary relationship (2). The Xhox-1A sequence contains the pentapeptide, Val Tyr Pro Trp Met, common to many other members of the *Hox* gene family (3). In Xenopus oocytes, the Xhox-1A protein has been shown to be karyophilic, with the karyophilic signal lying within, or C-terminal, to the homeodomain (1).

■ Expression

By Northern analysis, the *Xhox-1A* gene expresses a single predominant mRNA of 1.6 kb in length (1). A low level of maternally produced transcripts are found in ovary and eggs, but these are degraded after fertilization. Zygotic transcription begins at stage 11.5, midway through gastrulation, peaks during neurulation and persists into tadpole stages. In dissection experiments, expression is absent in the head area, present in both dorsal and ventral trunk (including the somites), and is absent in endoderm (4).

■ Function

The function of *Xhox-1A* in *Xenopus* development has

been approached by overexpression studies (4). An excess of synthetic *Xhox-1A* mRNA was introduced into fertilized *Xenopus* embryos at the two cell stage by microinjection. The mRNA degrades at a constant rate, but during gastrulation and neurulation there is a 20-100 fold excess of synthetic over endogenous mRNA which is distributed broadly over the embryo. The synthetic mRNA has been shown to be translated in *Xenopus* oocytes (1). The predominant phenotype is a lateral kinking in the injected half of the embryo, with concomitant dysplasia of somitic muscle. Mesoderm in this region has differentiated normally into muscle, but somite organization has been disrupted, in some cases severely. While the data does not provide evidence that *Xhox-1A* plays a direct role in somite formation, this patterning event is clearly sensitive to an inappropriate (spatial, temporal or level of) expression of this homeobox gene.

■ References

Fritz, A. and De Robertis, E.M. (1988). *Xenopus* homeobox-containing cDNAs expressed in early development. Nucleic Acids Res. 16, 1453-1468.

Harvey, R.P. and Melton, D.A. (1988). Microinjection of synthetic Xhox-1A homeobox mRNA disrupts somite formation in developing *Xenopus* embryos. Cell 53, 687-697.

Harvey, R.P., Tabin, C.J. and Melton, D.A. (1986). Embryonic expression and nuclear localization of *Xenopus* homeobox (Xhox) gene products. EMBO J. 5, 1237-1244.

Regulski, M., McGinnis, N., Chadwick, R. and McGinnis, W. (1987). Developmental and molecular analysis of deformed, a homeotic gene controlling *Drosophila* head development. EMBO J. 6, 767-777.

Hoxb-5 [m]

Species: *Mus musculus*

Chromosomal location: Chromosome 11, band 11D

Other names: *Hox-2.1, H24.1, Mu-1*

Cognate genes: *HOXB5*

Type of homeobox: *Antp*-type, related to the *Drosophila Scr* gene

Accession number: -

■ Origin and description

This gene was isolated by several groups screening a mouse genomic library with a *Drosophila Antennapedia* probe (Hart et al., 1985; Hauser et al., 1985; Jackson et al., 1985) and was localized to chromosome 11 (Hart et al., 1985; Munke et al., 1986; Nadeau et al., 1989). Its position in the middle of the *HOXB* complex was determined by the characterization of phage, cosmid and YAC clones from the region (Hart et al., 1985; Graham et al., 1988;

Rubock et al., 1990). This gene has two paralogues, *Hoxa-5* and *Hoxc-5* and is most closely related to the *Scr* gene. A sequence comparison with the zebrafish homologue (Njølstad and Fjose, 1988; Njølstad et al., 1988) indicated that over 90% of the entire protein coding sequences were identical and was one of the first suggestions that the genes would be highly conserved in all vertebrates.

■ Expression

The gene is active in some teratocarcinoma stem cells, but has been shown to be actively induced in cells induced to differentiate with retinoic acid (Hauser et al., 1985; Deschamps et al., 1987; Krumlauf et al., 1987; Papalopulu et al., 1991). Evidence also suggests that the response of this gene to RA may occur at both the transcriptional and post-transcriptional level (Deschamps et al., 1987; Papalopulu et al., 1991). In embryos the gene produced multiple transcripts which varied in a tissue specific manner (Krumlauf et al., 1987; Graham et al., 1989). *In situ* hybridisation and antibody staining experiments show that it is expressed in presomite embryos in posterior regions of the embryo and moves progressively more anterior to reach its anterior boundaries of expression in mesoderm and neuroectoderm at 8.25 dpc (Holland and Hogan, 1988; Wilkinson et al., 1989; Hunt et al., 1991; Wall et al., 1992). This was one of the first vertebrate genes cloned and shown to be expressed in many embryonic tissues and at more than one axial level. It was the first vertebrate gene found in mesodermal organs such as lung, kidney, gonads (Jackson et al., 1985; Krumlauf et al., 1987; Holland and Hogan, 1988). The anterior boundary of expression maps to the third cervical vertebrae. Patterns in the nervous system vary with time (Holland and Hogan, 1988; Wilkinson et al., 1989; Graham et al., 1991) and the gene marks the boundary between the hindbrain and spinal cord (Wilkinson et al., 1989). Like all genes from the *HOXB* complex the patterns of expression in the CNS undergo a sharp D-V restriction which mirror the birth of major classes of neurons (Graham et al., 1991). To date no loss or gain of function experiments on the gene have been reported in mouse embryos.

■ References

Deschamps, J., De-Laaf, R., Joosen, L., Meijlink, F. and Destree, O. (1987). Abundant expression of homeobox genes in mouse embryonal carcinoma cells correlates with chemically induced differentiation. Proc. Natl. Acad. Sci. U.S.A. 84, 1304-1308.

Graham, A., Papalopulu, N., Lorimer, J., Mcvey, J., Tuddenham, E. and Krumlauf, R. (1988). Characterization of a murine homeobox gene, *Hox 2.6*, related to the *drosophila* deformed gene. Genes Dev. 2, 1424-1438.

Graham, A., Papalopulu, N. and Krumlauf, R. (1989). The murine and *Drosophila* homeobox clusters have common features of organisation and expression. Cell 57, 367-378.

Graham, A., Maden, M. and Krumlauf, R. (1991). The murine Hox-2 genes display dynamic dorsoventral patterns of expression during central nervous system development. Development 112, 255-264.

Hart, C., Awgulewitsch, A., Fainsod, A., McGinnis, W. and Ruddle, F. (1985). Homeobox gene complex on mouse chromosome 11:

molecular cloning, expression in embryogenesis and homology to a human homeobox locus. Cell 43, 9-18.

Hauser, C., Joyner, A., Klein, R., Learned, T., Martin, G. and Tjian, R. (1985). Expression of homologous homeobox containing genes in differentiated human teratocarcinoma cells and mouse embryos. Cell 43, 19-28.

Holland, P.W. and Hogan, B.L. (1988). Spatially restricted patterns of expression of the homeobox-containing gene *Hox 2.1*. during mouse embryogenesis. Development 102, 159-174.

Hunt, P., Wilkinson, D. and Krumlauf, R. (1991). Patterning the vertebrate head: murine Hox 2 genes mark distinct subpopulations of premigratory and migrating neural crest. Development 112, 43-51.

Jackson, I., Schofield, P. and Hogan, B. (1985). A mouse homeobox gene is expressed during embryogenesis and in adult kidney. Nature 317, 745-748.

Krumlauf, R., Holland, P., McVey J. and Hogan, B. (1987). Developmental and spatial patterns of expression of the mouse homeobox gene,*Hox 2.1*. Development 99, 603-617.

Munke, M., Cox, D.R., Jackson, I.J., Hogan, B.L. and Francke, U. (1986). The murine Hox-2 cluster of homeo box containing genes maps distal on chromosome 11 near the tail-short (Ts) locus. Cytogenet. Cell. Genet. 42, 236-240.

Nadeau, J.H., Berger, F.G., Cox, D.R., Crosby, J.L., Davisson, M.T., Ferrara, D., Fuchs, E., Hart, C., Hunihan, L. and Lalley, P.A. et al. (1989). A family of type I keratin genes and the homeobox-2 gene complex are closely linked to the rex locus on mouse chromosome 11. Genomics. 5, 454-462.

Njølstad, P.R. and Fjose, A. (1988). In situ hybridization patterns of zebrafish homeobox genes homologous to Hox-2.1 and En-2 of mouse. Biochem. Biophys. Res. Commun. 157, 426-432.

Njølstad, P.R,. Molven, A. and Fjose, A. (1988). A zebrafish homologue of the murine *Hox-2.1* gene. FEBS. Lett. 230, 25-30.

Papalopulu, N., Lovell-Badge, R. and Krumlauf, R. (1991). The expression of murine *Hox-2* genes is dependent on the differentiation pathway and displays collinear sensitivity to retinoic acid in F9 cells and *Xenopus* embryos. N.A.R. 19, 5497-5506.

Rubock, M., Larin, Z., Cook, M., Papalopulu, N., Krumlauf, N. and Lehrach, H. (1990). A yeast artificial chromosome containing the mouse homeobox cluster Hox-2. Proc. Natn. Acad. Sci. USA 87, 4751-4755.

Wall, N., Jones, C., Hogan, B. and Wright, C. (1992). Expression and modification of *Hox-2.1* protein in mouse embryos. Mech. Dev. 37, 111-120.

Wilkinson, D., Bhatt, S., Cook, M., Boncinelli, E. and Krumlauf, R. (1989). Segmental expression of hox 2 homeobox-containing genes in the developing mouse hindbrain. Nature 341, 405-409.

Hoxb-5 [x]

Species: *Xenopus laevis*
Chromosomal location: -
Other names: *Xhox-1B, XlHbox4*
Cognate genes: *Hoxa-5, HOXA5*
Type of homeobox: *Antp*-type, related to the *Drosophila Scr* gene
Accession numbers: XLHOXAB1, XLHBOX4

■ Origin and description

This gene lies within a genomic clone (*Xhox-1*) which was isolated from a *Xenopus* genomic library using a *Drosophila Ubx* probe (1). The *Xhox-1B* gene lies upstream (12.5 kb) of another homeobox gene (*Xhox-1A*). Both genes are likely to be the *Xenopus* homologues of the murine *HOXB* locus, the *Hoxb-5* and *Hoxb-4* genes respectively, thus representing a section of a larger homeobox cluster. A partial length cDNA clone has been isolated (2).

■ Expression

By Northern analysis, *Xhox-1B* is expressed beginning in the mid-gastrula and peaking in the late neurula or tail-bud stage (1,2). A single transcript of 1.5-1.6 kb is detected.

■ References

Fritz, A. and De Robertis, E.M. (1988). *Xenopus* homeobox-containing cDNAs expressed in early development. Nucleic Acids Res. 16, 1453-1469.

Harvey, R.P., Tabin, C.J. and Melton, D.A. (1986). Embryonic expression and nuclear localization of *Xenopus* homeobox (Xhox) gene products. EMBO J. 5, 1237-1244.

Hoxb-5 [zf]

Species: *Brachydanio rerio* (zebrafish)
Chromosomal location: -
Other names: *ZF-21*
Cognate genes: *HOXB5*
Type of homeobox: *Antp*-type
Accession number: -

■ Origin and description

The *Hoxb-5* gene was cloned by cross-hybridization with the *Antp* and *Scr* homeobox sequences (Eiken et al., 1987; Njølstad et al., 1988a,b). A genomic lambda clone (C21) was isolated which contains both the *Hoxb-5* and the *Hoxb-6* gene (Njølstad et al., 1988a,b; Njølstad et al., 1990). A cDNA clone derived from the *Hoxb-5* gene, which contains the entire protein coding region, was sequenced. The predicted protein includes 275 amino acid residues and contains a homeodomain that is completely identical to the corresponding part of the murine *Hoxb-5* protein. The cognate zebrafish and mouse proteins are 81% identical.

■ Expression

The developmental expression of the *Hoxb-5* gene was studied by Northern blotting (Njølstad et al., 1988a,b) and *in situ* hybridization to tissue sections from hatching larvae (Njølstad and Fjose, 1988). A transcript of 2.3 kb was detected in 12 h embryos and older developmental stages. Transcripts derived from the *Hoxb-5* gene were detected in the posterior part of the hindbrain of hatching larvae. The zebrafish embryonic expression pattern has not been analyzed by *in situ* hybridization.

■ References

Eiken, H.G., Njølstad, P.R., Molven, A. and Fjose, A. (1987). A zebrafish homeobox-containing gene with embryonic transcription. Biochem. Biophys. Res. Commun. 149, 1165-1171.

Njølstad, P.R. and Fjose, A. (1988). *In situ* hybridization patterns of zebrafish homeobox genes homologous to *Hox*-2.1 and *En*-2 of mouse. Biochem. Biophys. Res. Commun. 157, 426-432.

Njølstad, P.R., Molven, A. and Fjose, A. (1988a). A zebrafish homologue of the murine *Hox*-2.1 gene. FEBS Lett. 230, 25-30.

Njølstad, P.R., Molven, A., Hordvik, I., Apold, J. and Fjose, A. (1988b). Primary structure, developmentally regulated expression and potential duplication of the zebrafish homeobox gene ZF-21. Nucl. Acids Res. 19, 9097-9111.

Njølstad, P.R., Molven, A., Apold, J. and Fjose, A. (1990). The zebrafish homeobox gene *Hoxb-6*: transcription unit, potential regulatory regions and *in situ* localization of transcripts. EMBO J. 9, 515-524.

Hoxb-6 [x]

Species: *Xenopus laevis*
Chromosomal location: -
Other names: *XlHox2.2*
Cognate genes: *HOXB6*
Type of homeobox: *Antp*-type
Accession number: M91587

■ Origin and description

This gene was isolated during the cloning of the *Xenopus laevis* HOXB complex. Its position within the complex, as well as sequence data (partial homeobox and the entire coding region downstream of the homeobox) allowed to define it as *Hoxb-6*.

Hoxb-6 [zf]

Species: *Brachydanio rerio* (zebrafish)
Chromosomal location: -
Other names: -
Cognate genes: *HOXB6*
Type of homeobox: *Antp*-type
Accession number: X17266

■ Origin and description

The *Hoxb-6* gene was cloned by cross-hybridization with the *Antp* and *Scr* homeobox sequences (Eiken et al., 1987; Njølstad et al., 1988a). A genomic lambda EMBL3 clone (C21) was isolated which contains both the *Hoxb-6* and the *Hoxb-5* gene (Njølstad et al., 1988a,b; Njølstad et al., 1990). Determination of the genomic sequence revealed that *Hoxb-6* encodes a putative protein of 228 amino acids. Compared to the murine *Hoxb-6* protein, the homeodomain and the entire amino acid sequences are 98.3% and 70.2% identical, respectively.

■ Expression

The expression of the *Hoxb-6* gene was studied by Northern blotting and *in situ* hybridization to tissue sections from hatching *larvae* (Njølstad et al., 1990). A transcript of about 1.4 kb was first detected at the 12 h stage. By *in situ* hybridization analysis, transcripts were mainly observed in the spinal cord and hindbrain. The anterior limit of expression is located in the posterior region of the hindbrain, but it has not been accurately mapped.

■ References

Eiken, H.G., Njølstad, P.R., Molven, A. and Fjose, A. (1987). A zebrafish homeobox-containing gene with embryonic transcription. Biochem. Biophys. Res. Commun. 149, 1165-1171.

Njølstad, P.R., Molven, A. and Fjose, A. (1988a). A zebrafish homologue of the murine *Hoxb-5* gene. FEBS Lett. 230, 25-30.

Njølstad, P.R., Molven, A., Hordvik, I., Apold, J. and Fjose, A. (1988b). Primary structure, developmentally regulated expression and potential duplication of the zebrafish homeobox gene ZF-21. Nucl. Acids Res. 19, 9097-9111.

Njølstad, P.R., Molven, A., Apold, J. and Fjose, A. (1990). The zebrafish homeobox gene *hox*-2.2: transcription unit, potential regulatory regions and *in situ* localization of transcripts. EMBO J. 9, 515-524.

Hoxb-7 [x]

Species: *Xenopus laevis*
Chromosomal location: -
Other names: *XlHbox2, MM3*
Cognate genes: *Hoxb-7, HOXB7*
Type of Homeobox: *Antp*-type
Accession numbers: X06592; X06593

■ Origin and description

This gene was isolated initially by screening a genomic library with a mixture of *Drosophila* homeobox probes (*Antp*, *ftz* and *Ubx*). Later on two related types of cDNA were isolated (Fritz and De Robertis, 1988), which correspond to the two duplicated forms of this gene (Fritz et al., 1989). Most *Xenopus laevis* genes are duplicated due to an ancestral tetraploidization of the genome. The gene encodes both homeobox-containing and homeobox-less transcripts (Wright et al., 1987). Similar alternative spliced homeobox-less transcripts lacking the homeobox are also found in *Hoxa-1* and *bicoid*. The exon upstream of the homeobox has an intriguing sequence similarity with the yeast mating-type protein a1 (score significant to 4.8 standard deviations, Wright et al., 1988). The non-coding 5′ leader sequence has very strong sequence conservation with paralogous genes (Bürglin et al., 1987), and has an inhibitory effect on the translation of *XlHbox 2* mRNA (Fritz and De Robertis, 1988). The recent cloning of seven genes of the *Xenopus HOXB* complex (Leroy and De Robertis, 1992) permits us to rename this gene as *Xenopus Hoxb-7*.

■ Expression

By Northern blot analysis this gene is strongly expressed in ovary, is absent in eggs and accumulates again starting at gastrulation. We have found no evidence of protein accumulation in pre-midblastula embryos or in the oocyte themselves. Therefore we cannot eliminate the possibility that the expression mRNA reported in Northern blots of oocytes (Müller et al., 1984) was due to contaminating follicle cells. (In mouse *Hoxb-7* is strongly expressed in macrophages). *XlHbox 2* is expressed in dissected *Xenopus* embryos according to the rules of colinearity and is induced by RA in embryos (Leroy and De Robertis, 1992) according to its position in the *HOXB* complex (reviewed by Boncinelli et al., 1991).

■ References

Boncinelli, E., Simeone, A., Acampora, D. and Mavilio, F. (1991). HOX gene activation by retinoic acid. Trends Genet. 7, 329-334.

Bürglin, T.R., Wright, C.V.E. and De Robertis, E.M. (1987). Translational control in homeobox mRNAs? Nature 330, 701-702.

Fritz, A. and De Robertis, E.M. (1988). *Xenopus laevis* homeobox-containing cDNAs expressed during early development. Nucleic Acids Res. 16, 1453-1469.

Fritz, A.F., Wright, C.V.E., Jegalian, B.G., and De Robertis, E.M. (1989). Duplicated Homeobox Genes in *Xenopus*. Developmental Biol. 131, 584-588.

Leroy, P. and De Robertis, E.M. (1992). Colinearity in the expression of genes from the *Xenopus laevis* Hox 2 complex: effects of lithium chloride and retinoic acid. Submitted.

Muller, M., Carrasco, A.E., and De Robertis, E.M., (1984). Isolation of a maternally expressed *Xenopus* homeotic-like gene. Cell 39, 157-162.

Wright, C.V.E., Cho, K.Y., Fritz, A., Burgling, T.R. and De Robertis, E.M. (1987). A *Xenopus laevis* gene encodes both homeobox-containing and homeobox-less transcripts. The EMBO J. 6, 4083-4094.

Hoxb-8 [x]

Species: *Xenopus laevis*
Chromosomal location: -
Other names: *XlHbox7*
Cognate genes: *Hoxb-8*, *HOXB8*
Type of homeobox: *Antp*-type
Accession number: M83947

■ Origin and description

This gene was isolated from a λEMBL4 library of genomic DNA using probes from the *XlHbox 2* gene. Its homeobox shares 51 out of 60 amino acids with *Antp* (De Robertis et al., 1989). *Hoxb-8* is located at the next-to-most 5′ position within the *Xenopus HOXB* complex (Leroy and De Robertis, manuscript submitted).

■ Expression

A polyclonal antibody has been obtained (Bittner, unpublished results). Expression of this gene is as predicted from its position in the complex (Bittner and Cho, unpublished results).

■ Reference

De Robertis, E.M., Bürglin, T.R., Fritz, A., Wright, C.V.E., Jegalian, B., Schnegelsberg, P., Bittner, D., Morita, E., Oliver, G. and Cho, K.W.Y. (1989). Families of vertebrate homeodomain proteins. In DNA-Protein Interactions in Transcription, Jay Gralla, ed. Alan R. Liss, Inc., New York, pp. 107-115.

Hoxb-9 [x]

Species: *Xenopus laevis*
Chromosomal location: -
Other names: *XlHbox6*
Cognate genes: *Hoxb-9*, *HOXB9*
Type of homeobox: *Abdominal-B*-type
Accession number: M17447

■ Origin and description

This gene was originally isolated by screening a *Xenopus* gastrula (stage 12) cDNA library with a mixed homeobox probe (Fritz and De Robertis, 1988). It exists in two versions, and gives two bands in genomic Southerns; this is due to the fact that *Xenopus laevis* is pseudotetraploid and has genomically duplicated copies of this gene (Fritz et al., 1989). *Xenopus Hoxb-9* is located at the 5′ most position within the *HOXB* complex (Leroy and De Robertis, 1992), and contains a homeodomain which is more similar to the homeodomain in *Drosophila Abdominal-B* than that of *Antennapedia* (Wright et al., 1990). In addition *Hoxb-9* contains a short poly-glutamine (opa or M-repeat) which has been found in many developmentally important genes in *Drosophila*, especially homeobox-containing genes (Sharpe et al., 1987).

■ Expression

In accordance with the rules of colinearity displayed by the Hox genes, *Xenopus Hoxb-9* is expressed in the posterior within the spinal cord and the underlying mesoderm (Sharpe et al., 1987). The mRNA is first detected at stage 13; it is approximately 1.8 kb on a Northern blot of polyA+ mRNA and it is estimated that there would be approximately 80 transcripts/cell. Between stages 13 and 20 there is a 5-fold increase in the accumulation of transcript which decreases to a tenth of this level by stage 41. Studies of conjugates prepared by combining ectoderm from early gastrula with dorsal mesoderm from mid-gastrula have shown that *Hoxb-9* is activated by neural induction (Sharpe et al., 1987). *Xenopus* Hoxb-9 protein is expressed in a posterior domain of the body encompassing the spinal cord and the underlying lateral plate mesoderm (Wright et al., 1990). The anterior border is just posterior to the hindbrain/spinal cord junction and extends throughout the spinal cord to the tip of the tail. Even though transcripts are detected in late gastrula embryos, the protein does not appear until the late neurula (stages 19-20). Hoxc-9 protein expression is biphasic. Just after the time that the neural tube closes, between stages 20 and 28, the levels of Hox 2.5 protein peak. It is expressed in almost all the neurons of the spinal cord including the proliferating and post-mitotic neurons. Expression continues through stage 35, however it becomes more restricted to the ependymal layer of dividing neurons as the outer neurons begin differentiating. The second phase begins around stages 48-49, when another burst of Hoxb-9 becomes apparent as two condensations of nuclei at the cervical and lumbar enlargements of the spinal cord; these regions will innervate the developing limbs. These condensations of Hoxb-9 expression also correlate with neurons which are proliferating. In *Xenopus* animal cap (AC) explants, *Hoxb-9* (*XlHbox 6*) expression is selectively activated by basic fibroblast growth factor (bFGF) (Cho and De Robertis, 1990). This corroborates the finding that peptide growth factors can provide anteroposterior positional information during early embryonic development. Expression of *Hoxb-9* in AC explants is potentiated by combinations of peptide growth factors and retinoic acid (RA); in contrast, RA alone cannot induce *Hoxb-9* expression in AC explants (Cho and De Robertis, 1990). In another study of axis formation, *Xenopus Hoxb-9* has been used as a posterior marker to analyze the effect of RA on gene expression in intact embryos and in dorsal ectoderm isolated from mid-gastrula embryos. It was found that an increased expression of *Hoxb-9* correlates with a decrease in anterior neurectodermal gene expression (Sive et al., 1990). This suggests that RA may alter cell lineages by changing the fate of dorsal ectoderm after its primary induction. In more developed explants from neurula-stage embryos *Xenopus Hoxb-9* has been used as a posterior marker for studying regional neural induction. At physiological concentrations RA acts as a signal which localizes the expression of *Hoxb-9* to its normal regional expression in the whole embryo (Sharpe, 1991). The expression of *Xenopus Hoxb-9* in comparison to other genes of the *Xenopus HOXB* complex has been studied (Leroy and De Robertis, 1992; Papalopulu et al., 1990; Durston and Boncinelli, personal communication). *Xenopus Hoxb-9* follows the rules of colinearity displayed by the Hox genes; it is located in the most 5' position of the *HOXB* complex, it is spatially expressed more posteriorly, and temporally expressed later than *Hoxb-3* which is located in a 3' position of the complex (Leroy and De Robertis, 1992). The expression of *Xenopus* Hoxb-9 at early neurula and its localization to the posterior nervous system have made it a useful molecular marker for neural induction. Analysis of conjugates prepared by combining mesoderm from stage 11 gastrula with either dorsal or ventral ectoderm from stage 10-1/2 embryos showed that dorsal ectoderm is biased to respond to neural induction; conjugates prepared with dorsal ectoderm, but not with ventral ectoderm, express *Hoxb-9* and N-CAM (Sharpe et al., 1987). In another study *Xenopus Hoxb-9* was used as a posterior neural marker to examine different factors which might contribute to regional neural induction (Sharpe and Gurdon, 1990). The experiments showed that the underlying dorsal mesoderm from late gastrula (stage 12-1/2) can be physically divided into two regions with different inducing capacity. Only the posterior region induced *Hoxb-9*.

■ Function

The function of the *Xenopus Hoxb-9* gene (*XlHbox 6*) was studied by overexpressing the protein in AC and then assaying its effect by transplantation into the blastocoel of host gastrulae. Expression of *XlHbox 6* is able to change the axial character of growth factor-treated AC explants when they are transplanted into host gastrulae; activin normally induced heads, but when *XlHbox 6* is overexpressed only tails are produced (Cho et al., 1991). When these animal caps expressing *XlHbox 6* (not treated with growth factors) are assayed by transplantation into the blastocoel of host gastrulae, they can produce tail-like structures, suggesting that homeodomain proteins are part of the biochemical pathway which generates the body axis (Cho et al., 1991). When blastomeres of the 32-cell stage embryo which normally give rise to dorso-anterior structures are injected with *Xenopus Hoxb-9* mRNA and colloidal gold lineage tracer, it was found that the progeny exhibited alterations to their normal cell fate (Niehrs and De Robertis, 1991). These changes were due to changes in cell migration during gastrulation and changes in the differentiation within the anterior central nervous system (CNS). Neurons could not form normally in the brain, but could participate in spinal cord differentiation (Niehrs and De Robertis, 1991).

References

Cho, K.W.Y. and De Robertis, E.M. (1990). Differential activation of *Xenopus* homeobox genes by mesoderm-inducing growth factors and retinoic acid. Genes Dev. 4, 1910-1916.

Cho, K.W.Y., Morita, E.A., Wright, C.V.E. and De Robertis, E.M. (1991). Overexpression of a homeodomain protein confers axis-forming activity to uncommitted *Xenopus* embryonic cells. Cell 65, 55-64.

Fritz, A. and De Robertis, E.M. (1988). *Xenopus* homeobox-containing cDNAs expressed in early development. Nucleic Acids Res. 16, 1453-1469.

Fritz, A., Cho, K.W.Y., Wright, C.V.E., Jegalian, B.G. and De Robertis, E.M. (1989). Duplicated homeobox genes in Xenopus. Dev. Biol.131, 584-588.

Jones, F.S., Prediger, E.A., Bittner, D.A., De Robertis, E.M. and Edelman, G.M. (1992). Cell adhesion molecules as targets for Hox genes: N-CAM transcription is modulated by co-transfection of Hox 2.5 and 2.4. Proc. Natl. Acad. Sci. USA, 89, 2086–2090.

Leroy, P. and De Robertis, E.M. (1992). Colinearity in the expression of genes from the *Xenopus laevis* Hox 2 complex: effects of lithium chloride and retinoic acid. Submitted for publication.

Niehrs, C. and De Robertis, E.M. (1991). Ectopic expression of a homeobox gene changes cell fate in *Xenopus* embryos in a position-specific manner. EMBO J. 10, 3621-3629.

Papalopulu, N., Hunt, P., Wilkinson, D., Graham, A. and Krumlauf, R. (1990). Hox-2 homeobox genes and retinoic acid: potential roles in patterning the vertebrate nervous system. Advances in Neural Regeneration Research, 291-307.

Sharpe, C.R., Fritz, A., De Robertis, E.M. and Gurdon, J.B. (1987). A homeobox-containing marker of posterior neural differentiation shows the importance of predetermination in neural induction. Cell 50, 749-758.

Sharpe, C.R. and Gurdon, J.B. (1990). The induction of anterior and posterior neural genes in *Xenopus laevis*. Development 109, 765-774.

Sharpe, C.R. (1991). Retinoic acid can mimic endogenous signals involved in transformation of the *Xenopus* nervous system. Neuron 7, 239-247.

Sive, H.L., Draper, B.W., Harland, R.M. and Weintraub, H. (1990). Identification of a retinoic acid-sensitive period during primary axis formation in *Xenopus laevis*. Genes Dev. 4, 932-943.

Wright, C.V.E., Morita, E.A., Wilkin, D.J. and De Robertis, E.M. (1990). The *Xenopus* XlHbox 6 homeoprotein, a marker of posterior neural induction, is expressed in proliferating neurons. Development 109, 225-234.

Hoxc-4 [m]

Species: *Mus musculus*

Chromosomal location: Chromosome 15

Other names: -

Cognate genes: *HOXC4*

Type of homeobox: *Antp*-type, related to *Drosophila Deformed* gene

Accession number: -

■ Origin and description

Hoxc-4 was identified from a chromosome walk 3' from *Hoxc-5*. The *Hoxc-4* homeobox lies 18 Kb 3' of *Hoxc-5* and belongs to the *Hoxa-4*, *Hoxb-4*, *Hoxd-4* subfamily (paralogues). Full length cDNA sequence of *Hoxc-4* has been determined (Geada et al., 1992).

■ Expression

The embryonic expression of *Hoxc-4* confirms it as a *Hoxa-4*, *Hoxb-4* and *Hoxd-4* paralogue having an anterior pre-vertebral boundary at pv4 and an anterior spinal cord boundary in the hindbrain. *Hoxc-4* does however show a major deviation from its paralogues in that the anterior limit spinal cord expression does not correspond to the boundary between rhombomeres 6 and 7 as *Hoxa-4*, *Hoxb-4* and *Hoxd-4*.. The *Hoxc-4* boundary is posterior to this and corresponds to the boundary between rhombomere 7 and 8 (Geada et al., 1992). *Hoxc-4* expression is more anterior than *Hoxc-5* and *Hoxc-6* confirming the colinearity of the anterior *Hoxc* genes.

■ References

Geada, A.M.C., Gaunt, S.J., Azzawi, M., Shimeld, S.M., Pearce, J. and Sharpe, P.T. (1992). Sequence and expression of the mouse Hox-3.5 gene. Development 116, 497-506.

Hoxc-5 [m]

Species: *Mus musculus*

Chromosomal location: Chromosome 15

Other names: *Hox-3.4*, *Hox-6.2*

Cognate genes: *HOXC5*

Type of homeobox: *Antp*-type, related to *Drosophila Scr* gene

Accession number: -

■ Origin and description

This gene was isolated through its close linkage with *Hoxc-6* (Sharpe et al. 1988). It is located 5 Kb 3' to *Hoxc-6* and 18 Kb 5' to *Hoxc-4*, and contains a homeobox related to the sex combs reduced homeotic gene. Paralogue genes on other clusters are *Hoxb-5* and *Hoxa-5* (Gaunt et al., 1990). A number of different *Hoxc-5* cDNA's have been isolated many of which appear unspliced or derived from sequence 5' to the transcription start. Putative full coding sequence has been deduced from cDNA and genomic sequence by comparison with human *HOXC5* and Zebrafish *ZF-25* sequences (Geada et al. in preparation).

■ Expression

The expression of *Hoxc-5* confirms it belongs to the *Hoxa-5/b-5* sub-family, the anterior boundary in the central nervous system at day 12.5 is just posterior to the myelen-cephalon which is almost identical to *Hoxb-5* and slightly more anterior to *Hoxa-5*. Expression is throughout the entire length of the spinal cord, and is localized to the central and ventral regions of the mantle - a pattern similar to that observed for *Hoxc-6* and *Hoxc-4*. Highest expression in the prevertebral column is over the posterior cervical and anterior thoracic prevertebrae, extending from pv6 to pv17 with a maximum between pv8-14. Low level expression is detectable in all prevertebrae anterior to pv6. The anterior prevertebral boundary of expression is similar to *Hoxa-5* and *Hoxb-5* *Hoxc-5* is weakly expressed in the outer mesodermal components of the lung (*Hoxa-5* and *Hoxb-5* show intense expression). Little expression is detected in the trachea but transcripts are clearly visible in the lining of the epithelium of the oesophagus; whereas neither *Hoxa-5* nor *Hoxb-5* transcripts are detectable in the oesophagus. *Hoxc-5* is expressed in the dorsal stomach wall, mesonephric and metanephric kidneys, mesodermal components of the testis.

■ References

Gaunt, S.J., Coletta, P.L., Pravtcheva, D. and Sharpe, P.T. (1990). Mouse *Hoxc-5*: homeobox sequence and embryonic expression patterns compared with other members of the *Hox* gene network. Development 109, 329-339.
Sharpe, P.T., Miller, J.R., Evans, E.P., Burtershaw, M.D. and Gaunt, S.J. (1988). Isolation and expression of a new mouse homeobox gene. Development 102, 347-407.

Hoxc-5 [x]

Species: *Xenopus laevis*
Chromosomal location: -
Other names: *XlHbox5*
Cognate genes: *Hoxc-5, HOXC5*
Type of homeobox: *Antp*-type
Accession number: X07105

■ Origin and description

This gene was isolated from a gastrula cDNA library by screening with Antp homeobox probes (Fritz and De Robertis, 1988). The homeodomain differs by only one amino acid from that of murine Hox 3.4 (Gaunt et al., 1990). In *Xenopus*, *XlHbox5* is linked to *XlHbox 1* (*Hoxc-6*) as in mammals (Leroy and De Robertis, submitted).

■ Expression

A 1.9 kb message is detected at late neurula and tailbud stages (Fritz and De Robertis, 1988).

■ References

Fritz, A. and De Robertis, E. M. (1988). *Xenopus* homeobox-containing cDNAs expressed in early development. Nucl. Acids Res. 16, 1453-1469.
Gaunt, S. J., Coletta, P. L., Pravtcheva, D. and Sharpe, P. T. (1990). Mouse Hox-3.4: homeobox sequences and embryonic expression patterns compared with other members of the *Hox* gene network. Development 109, 329-339.

Hoxc-5 [zf]

Species: *Brachydanio rerio* (zebrafish)
Chromosomal location: -
Other names: *ZF-25*
Cognate genes: *HOXC5*
Type of homeobox: *Antp*-type
Accession number: -

■ Origin and description

The *Hoxc-5* gene was cloned by cross-hybridization with the *Antp* and *Scr* homeobox sequences (Eiken et al., 1987). A genomic lambda EMBL3 clone (C25) was isolated which contains the *zf-25* homeobox sequence. The same homeobox is present in another clone (C26) where an additional homeobox sequence (*zf-61*) is also located (Njølstad et al., 1990). These two homeoboxes are closely related to the corresponding regions of the murine *Hoxc-5* and *Hoxc-6* genes, respectively. Sequencing of the entire protein coding region of *zf-25* has now confirmed that this is a cognate gene of *Hoxc-5/HOXC5* (Ericson et al., 1992).

■ Expression

The expression of the *Hoxc-5* has been analyzed by Northern blotting and *in situ* hybridization to embryonic and larval tissue sections (Eiken et al., 1987; Ericson et al., 1992). Transcripts (1.4 kb and 2.1 kb) were first detected at the 15 h stage and a much higher level of both transcripts was observed in 48 h embryos (Eiken et al., 1987). Also by *in situ* hybridization a relatively low level of transcripts was detected in tissue sections of 18 h embryos (Ericson et al., 1992). Stronger hybridization signals were observed in hatching larvae, but in both cases specific signals were detected only in the central nervous system. An anterior limit of expression is located near the posterior end of the hindbrain.

■ References

Eiken, H.G., Njølstad, P.R., Molven, A. and Fjose, A. (1987). A zebrafish homeobox-containing gene with embryonic transcription. Biochem. Biophys. Res. Commun. 149, 1165-1171.
Ericson, J.U., Krauss, S. and Fjose, A. (1993). Genomic sequence and embryonic expression pattern of the zebrafish homeobox

gene *Hoxc-5*. Int. J. Dev. Biol. 37, 263–272.
Njølstad, P.R., Molven, A., Apold, J. and Fjose, A. (1990). The
zebrafish homeobox gene *hox-2.2*: transcription unit, potential
regulatory regions and *in situ* localization of transcripts. EMBO
J. 9, 515-524.

Hoxc-6 [m]

Species: *Mus musculus*
Chromosomal location: Chromosome 15
Other names: *Hox-3.3, Hox-6.1*
Cognate genes: *HOXC6*
Type of homeobox: *Antp*-type, most
related to *Drosophila Antp*
Accession numbers: X16510, X16511

■ Origin and description

The *Hoxc-6* gene was isolated from an adult mouse kidney
cDNA library using a probe containing the *Hoxa-3* homeobox (Sharpe et al., 1988).This gene is located between
Hoxc-5 and *Hoxc-8* in the *HOXC* complex and is a member
of the paralogue group including *Hoxa-6* and *Hoxb-6*.
Two transcripts containing the *Hoxc-6* homeobox are produced from two independent promoters. A typical *Hox*
transcript is produced from the PRII promoter immediately 5' to the locus and consists of two exons separated
by a 700bp intron, with the homeobox located in the second exon (Coletta et al., 1991a). The PRI promoter is found
some 9 kb 5' of the rest of the gene and includes a third
exon which is spliced into the first exon of the PRII transcript, creating the PRI transcript (Shimeld et al., 1991).
This splice removes the presumptive start codon of the
PRII transcript and consequently translation of the PRI
message would produce a protein truncated by 82 amino
acids at the amino terminus.

■ Expression

The combined expression of both transcripts in embryos
has been studied by *in situ* hybridization (Sharpe et al.,
1988). At 8.25 days p.c. expression was seen in the presomitic mesoderm up to about the position of somite 6.
Expression extended posteriorly from here into the allantois but was not seen in any other extraembryonic tissues.
At 12.5 days p.c. labelling was seen in the spinal cord up
to the hindbrain. At this stage prevertebra (pv) 7 was
weakly labelled, strong labelling was observed over pv 9
to 16 and weaker labelling persisted over the rest of the
posterior pv. Of the internal organs, metanephric kidney,
gonad, mesodermal lung components, stomach and some
intestine showed hybridization. Antibodies raised to the
Xenopus cognate of *Hoxc-6* have been used to study the
distribution of both PRI and PRII homeoproteins in day 12

and 13 p.c. mouse embryos (Oliver et al., 1988). The PRI
protein was found to be expressed caudally from pv 1 and
in the CNS up to the hindbrain. The PRII protein was seen
more posteriorly in the CNS and in pv 9 to 16. The apparent confliction between *in situ* hybridization and antibody data has not been resolved. An area where mRNA
and protein localization matches is in the developing limb
buds (Coletta et al., 1991b). At 10.5 days p.c. expression in
the hind limb bud is confined to the ectoderm but in the
fore limb bud the mesoderm also expresses *Hoxc-6*.
Expression here appears to form an anterior/posterior gradient towards the proximal region of the limb bud.

■ References

Coletta, P.L., Shimeld, S.M., Chaudhuri, C., Muller, U., Clarke, J.P.
and Sharpe, P.T. (1991a). Characterization of the murine *Hox*-
3.3 gene and its promoter. Mechs. of Devel. 35, 129-142.
Coletta, P.L., Shimeld,S.M., Clarke, J.P. and Sharpe, P.T. (1991b).
The murine *Hox*-3.3 gene. In: Development and regeneration
of the nervous system. Chapmann and Hall, London.
Oliver, G., Wright, C.V.E., Hardwicke, J. and De Robertis, E.M.
(1988). Differential antero-posterior expression of two proteins
encoded by a homeobox gene in *Xenopus* and mouse embryos.
EMBO J. 7, 3199-3209.
Sharpe, P.T., Miller, J.R., Evans, E.P., Burtenshaw, M.D. and
Gaunt, S.J. (1988). Isolation and expression of a new mouse
homeobox gene. Development 102, 397-407.
Shimeld, S.M., Coletta, P.L., Geada, A.M.C. and Sharpe, P.T. In
preparation.

Hoxc-6 [x]

Species: *Xenopus laevis*
Chromosomal location: -
Other names: *Xlhbox1, AC1, Xeb 1*
Cognate names: *Hoxc-6, HOXC6, NvHbox 1*
Type of homeobox: *Antp*-type
Accession numbers: X12499; X12500

■ Origin and description

This gene was isolated by screening a genomic library with
an *Antennapedia* homeobox probe (Carrasco et al., 1984);
it was the first homeobox gene isolated from vertebrates
and provided an evolutionary jump from *Drosophila* to
higher organisms. Two types of cDNAs were isolated (Fritz
and De Robertis, 1988) which define two transcripts
derived from two promoters (PRI and PRII) separated by 9
kb in the genome. This alternative transcription leads to
the production of mRNAs coding for a short (transcripts
from the distal PRI) and a long protein (PRII transcripts).
These proteins share an identical homeodomain and
neighboring regions, with the long protein having an 82
aa extension at the amino terminus (Cho et al., 1988).
Both proteins have the same DNA-binding specificity, but

could have different transcriptional activation properties (Cho et al., 1988). A similar situation exists in the *Drosophila Abdominal-B* gene, in which the short protein acts as a repressor. In humans the cognate gene (called *HOXC6* or *Hoxc-6*) has a similar organization, and the distal PRI promoter has been shown to act as a master promoter which directs transcripts of the downstream genes *Hoxc-5* and *Hoxc-4* as well (Boncinelli et al., 1991). *XlHbox 1* has recently been found to be located in the genome next to *XlHbox 5*, a homologue of *Hoxc-5* (P. Leroy and De Robertis, submitted). This permits to designate this gene as *Xenopus Hoxc-6* in future.

■ Expression

In *Xenopus* expression has been analyzed by *in situ* hybridization (Carrasco and Malacinski, 1987) and by the generation of polyclonal antibodies specific for the long protein or common to both long and short proteins (Oliver et al., 1988a). The short protein starts to be expressed more anteriorly (in the hindbrain, at the level of the IX and X, or nodose, ganglion) than the long protein, which is expressed in a narrow band at the level of the anterior spinal cord (Oliver et al., 1988a). A band comprising neural crest and lateral mesoderm also expresses the protein and shares the same anterior and posterior borders with the CNS in *Xenopus* (De Robertis et al., 1989). This raises the question of whether this alignment of expression arises through an exchange of positional information between the mesoderm and neuroectoderm, or from two independent positional systems. The anti-XlHbox 1 antibodies crossreact with a wide variety of other vertebrates such as mouse (Oliver et al., 1988a), zebrafish (Molven et al., 1990) and chick (Oliver et al., 1990). In the mouse the expression of the vast majority of *Antennapedia*-type homeobox genes is more anterior in the CNS than in the mesoderm. The issue of whether there is a single or multiple mechanism for activating homeobox genes along the A-P axis of the CNS and the mesoderm remains unresolved. XlHbox 1 antibodies revealed a gradient of homeodomain protein, maximal in the anterior and proximal region in the mesoderm of the forelimb bud in *Xenopus*, mouse (Oliver et al. 1988b) and chick (Oliver et al., 1990) embryos. It is also present in the pectoral fin bud of the zebrafish, where the protein can be seen to demarcate a circular field of cells at early stages, before the limb bud is formed (Molven et al., 1990). In the developing chick feather bud XlHbox 1 antigen also forms a circular field, followed by a gradient in the anterior (Chuong et al., 1990). This has led to the view that homeobox genes may demarcate the classical morphogenetic gradient-fields that determine organ formation (reviewed by De Robertis et al., 1991).

■ Function

The function of *XlHbox 1* has been analyzed by microinjection of antibodies and of synthetic mRNAs. Injection of antibodies leads to defects in neural crest derivatives (Cho et al., 1988), and to morphological transformation of the anterior spinal cord into hindbrain structures (Wright et al., 1989). Injection of long protein mRNA disrupted segmentation and tissue organization. Injection of short protein mRNA led to spinal cord transformations similar, but more extensive, than those caused by the antibodies, suggesting that the short protein acts as an inhibitor (Wright et al., 1989). In animal cap explants *XlHbox 1* expression is activated by mesoderm induction and activin is a better inducer than FGF (Cho and De Robertis, 1990).

■ References

Boncinelli, E., Simeone, A., Acampora, D. and Malvilio, F. (1991). HOX gene activation by retinoic acid. Trends Genet. 7, 329-334.

Carrasco, A.E., McGinnis, W., Gehring, W.J. and De Robertis, E.M. (1984). Cloning of a *Xenopus laevis* gene expressed during early embryogenesis that codes for a peptide region homologous to *Drosophila* homeotic genes. Cell 37, 409-414.

Carrasco, A.E. and Malacinski, G.M. (1987). Localization of *Xenopus* Homeobox gene transcripts during embryogenesis and in the adult Nervous System. Dev. Biol. 121, 69-81.

Cho, K.W.Y., Goetz, J., Wright, C.V.E., Fritz, A., Hardwicke, J. and De Robertis, E.M. (1988). Differential utilization of the same reading frame in a *Xenopus* homeobox gene encodes two related proteins sharing the same DNA-binding specificity. EMBO J. 7, 2139-2149.

Cho, K.W.Y. and De Robertis, E.M. (1990). Differential activation of *Xenopus* homeobox genes by mesoderm-inducing growth factors and retinoic acid. Genes and Development 4, 1910-1916.

Chuong, C-M., Oliver, G., Ting, S.A. and De Robertis, E.M. (1990). Gradients of homeoproteins in developing feather buds. Development 110, 1021-1030.

De Robertis, E.M., Oliver, G. and Wright C.V.E. (1989). Determination of axial polarity in the vertebrate embryo: homeodomain proteins and homeogenetic induction. Cell 57, 189-191.

De Robertis, E.M., Morita, E. and Cho, K.W.Y. (1991). Gradient-fields and homeobox genes. Development 112, 669-678.

Fritz, A. and De Robertis, E.M. (1988). *Xenopus laevis* homeobox-containing cDNAs expressed during early development. Nucleic Acids Res. 16, 1453-1469.

Molven, A., Wright, C.V.E., Bremiller, R., De Robertis, E.M. and Kimmel, C.B. (1990). Expression of a homeobox gene product in normal and mutant zebrafish embryos: evolution of the tetrapod body plan. Development 109, 279-288.

Oliver, G., Wright, C.V.E., Hardwicke, J. and De Robertis, E.M. (1988a). Differential antero-posterior expression of two proteins encoded by a homeobox gene in *Xenopus* and mouse embryos. EMBO J. 7, 3199-3209.

Oliver, G., Wright, C.V.E., Hardwicke, J. and De Robertis, E.M. (1988b). A gradient of homeodomain protein in developing forelimbs of *Xenopus* and mouse embryos. Cell 55, 1017-1024.

Oliver, G., De Robertis, E.M., Wolpert, L. and Tickle, C. (1990). Expression of a homeobox gene in the chick wing bud following application of retinoic acid and grafts of polarizing region tissue. EMBO J. 9, 3093-3099.

Savard, P., Gates, P.B. and Brockes, J.P. (1988). Position dependent expression of a homeo box gene transcript in relation to amphibian limb regeneration. EMBO J. 7, 4275-4282.

Wright, C.V.E., Cho, K.W.Y., Hardwicke, J., Collins, R.H. and De Robertis, E.M. (1989). Interference with function of a homeobox gene in *Xenopus* embryos causes malformations of the anterior cervical spinal cord. Cell 59, 81-93.

Species: *Brachydanio rerio* (zebrafish)
Chromosomal location: -
Other names: *ZF-61*
Cognate genes: *HOXC6*
Type of homeobox: *Antp*-type
Accession number: X17267

■ Origin and description

The *Hoxc-6* gene was cloned by cross-hybridization with the *Antp* and *Scr* homeobox sequences (Eiken et al., 1987; Njølstad et al., 1990). A genomic lambda EMBL3 clone (C26) was isolated which contains two homeobox sequences, zf-25 (Eiken et al., 1987) and zf-61 (Njølstad et al., 1990). The zf-61 homeodomain is 98.3% identical to the corresponding part of the murine Hoxc-6 protein. Similarly, the homeodomains of zf-25, *Hoxc-5* (mouse) and *HOXC5* (human) are completely identical. Sequencing of the entire protein coding region of zf-25 has now confirmed that this is a cognate gene of *Hoxc-5/HOXC5* (Ericson et al., 1992). On this basis, it can be assumed that zf-61 is a direct homologue of the murine *Hoxc-6* gene.

■ Expression

The expression of the *Hoxc-6* gene has not been investigated, but *in situ* hybridization analyses of the closely linked *Hoxc-5* gene (Ericson et al., 1992) have demonstrated the presence of transcripts in the embryonic central nervous system. An anterior limit of expression was observed in the posterior region of the hindbrain.

■ References

Eiken, H.G., Njølstad, P.R., Molven, A. and Fjose, A. (1987). A zebrafish homeobox-containing gene with embryonic transcription. Biochem. Biophys. Res. Commun. 149, 1165-1171.
Ericson, J.U., Krauss, S. and Fjose, A. (1993). Genomic sequence and embryonic expression pattern of the zebrafish homeobox gene *hox-3.4*. Int. J. Dev. Biol. 37, 263–272.
Njølstad, P.R., Molven, A., Apold, J. and Fjose, A. (1990). The zebrafish homeobox gene *hox-2.2*: transcription unit, potential regulatory regions and *in situ* localization of transcripts. EMBO J. 9, 515-524.

Species: *Mus musculus*
Chromosomal location: Chromosome 15F1-3
Other names: *Hox-3.1*
Cognate genes: *HOXC8*
Type of homeobox: *Antp*-class
Accession number: M35603

■ Origin and description

The *Hoxc-8* gene was isolated by screening mouse genomic libraries with probes derived from *Drosophila Antp* and *Ubx* homeobox sequences (Awgulewitsch et al., 1986; Breier et al., 1986, 1988). A cDNA clone was isolated by screening mouse embryonic cDNA libraries with genomic sequences (Breier et al., 1988; Le Mouellic et al., 1988; Awgulewitsch et al., 1990). The sequence of the entire *Hoxc-8* genomic locus containing *Hoxc-9* and *Hoxc-8* intergenic regions has been determined (Awgulewitsch et al., 1990). The *Hoxc-8* gene was mapped to chromosome 15 (Awgulewitsch et al., 1986; Breier et al., 1986; Rabin et al., 1986) and was shown to be located towards the 5' of the homeobox gene cluster. The paralogous genes on the other clusters are *Hoxb-8* and *Hoxd-8*.

■ Expression

The expression of *Hoxc-8* has been studied by various groups (Awgulewitsch et al., 1986; Utset et al., 1987; Breier et al., 1988; Le Mouellic et al., 1988; Gaunt, 1988; Gaunt et al., 1988; Holland and Hogan, 1988; Awgulewitsch and Jacobs, 1990). The transcripts of *Hoxc-8* are detected as early as embryonic day p.c. (Le Mouellic et al., 1988; Gaunt, 1988). At this stage of development, the expression appears to be limited to the allantois. At subsequent stages of 8-9.5 dpc, the expression extends to most tissues of the embryo including mesodermal and neuroectodermal layers, in the region posterior to 13th somite. At 10th d. dpc, the *Hoxc-8* gene expression shifts rostrally and is seen posterior to the 5th somite. The transcripts begin to become spatially restricted within more anterior regions of the neural tube (Le Mouellic et al., 1988), showing highest levels in ventral subregions (Le Mouellic et al., 1988; Breier et al., 1988). This expression pattern is maintained at later stages of development (Awgulewitsch et al., 1986; Utset et al., 1987; Breier et al., 1988; Le Mouellic et al., 1988; Gaunt, 1988; Gaunt et al., 1988; Holland and Hogan, 1988; Awgulewitsch and Jacobs, 1990). At 12.5 dpc, the anterior boundary of the *Hoxc-8* transcript is posterior to the third cervical vertebra. *Hoxc-8* transcripts accumulate asymmetrically in the spinal cord along the dorsoventral axis. This gradient has been thought to reflect the expression of *Hoxc-8* gene in the motor neurons or in neuroblasts predetermined to form motor neurons. *Hoxc-8* transcripts are also detected in dorsal root ganglia and the sympathetic ganglia (Breier et al., 1988; Awgulewitsch and Jacobs, 1990). *Hoxc-8* is also expressed in sclerotome derived pre-vertebral condensations in the thoracic region at this stage of development. In the newborn and adult mouse, the expression of *Hoxc-8* is observed in the spinal cord and the kidney (Awgulewitsch et al., 1986; Utset et al., 1987). The expression of *Hoxc-8* is also observed in adult skin (Bieberich et al., 1991).

The 5' region of the *Hoxc-8* gene has been analyzed in transgenic mice for the regulatory elements that can confer *Hoxc-8* expression pattern on heterologous reporter gene (Bieberich et al., 1990). A 5 kb DNA fragment from the 5' region of the Hoxc-8 gene is sufficient to confer the *Hoxc-8* pattern of expression in 8.5 to 9.5 dpc embryos. However, in later stage embryos, the expression pattern of the reporter gene diverges from the endogenous pattern. This reporter gene is expressed in spinal ganglia in an inappropriately higher level compared to *Hoxc-8* gene expression. However, inclusion of an additional 2 kb of upstream sequences suppresses this aberrant expression in the developing dorsal root ganglia (Bieberich et al., 1990).

■ Function

The function of the *Hoxc-8* gene was studied by producing null mutant mice (Le Mouellic et al., 1992) in which *Hoxc-8* coding sequence was replaced with the lacZ gene via homologous recombination in ES cells (Le Mouellic et al., 1990). The homozygous mice were born alive, but most of them died within a few days. Anterior transformation of several skeletal segments was observed in the trunk regions of the homozygotes. The most striking feature was the appearance of a 14th pair of ribs on the 1st lumbar vertebra, which suggested homeotic transformation of 1st lumbar vertebra into the anterior thoracic segment. The skeletal elements of the 12th rib resembled those of the 11th rib. Additional sternabra was observed between the 6th and 7th ribs in many mutant mice and the 8th pair of ribs were found attached to the sternum. A few surviving homozygous adult mice exhibited certain neurological disorders which included abnormal reflexes and wavering movements. The fingers of the forelimb were found to be clenched with poor grasping capabilities. Spasmodic contraction of the hind limbs or back muscles were also observed in some of these mice. The pattern of β-galactosidase activity was identical in both heterozygotes and homozygotes and reflected faithfully the *Hoxc-8* expression pattern, thereby suggesting that the mutation modified the identity rather than the position of embryonic cells that would normally express *Hoxc-8*. Stable cell lines expressing *Hoxc-8* gene have been isolated and analyzed for potential target genes (Violette et al., 1992). Transcription of the gene encoding β-amyloid precursor protein, which is associated with the neurological disorder, Alzheimer disease, has been shown to be repressed by the *Hoxc-8* gene product.

■ References

Awgulewitsch, A., Utset, M.F., Hart, C.P., McGinnis, W. and Ruddle, F.H. (1986). Spatial restriction in expression of a mouse homeobox locus within the central nervous system. Nature 320, 328-335.

Awgulewitsch, A. and Jacobs, D. (1990). Differential expression of Hox 3.1 protein in subregions of the embryonic and adult spinal cord. Development 108, 411-420.

Awgulewitsch, A., Bieberich, C., Bogarad, L., Shashikant, C. and Ruddle, F.H. (1990). Structural analysis of the Hox-3.1 transcription unit and the Hoxc-9-Hox-3.1 intergenic region. Proc. Natl. Acad. Sci. USA 87, 6428-6432.

Bieberich, C.J., Utset, M.F., Awgulewitsch, A. and Ruddle, F.H. (1990). Evidence for positive and negative regulation of the Hox-3.1 gene. Proc. Natl. Acad. Sci. USA 87, 8462-8466.

Bieberich, C.J., Ruddle, F.H. and Stenn, K.S. (1992). Differential expression of the Hox3.1 gene in adult mouse skin. Annals NY Acad. Sci. 642, 346-354.

Breier, G., Bucan, M., Francke, U., Colberg-Poley, A.M. and Gruss, P. (1986). Sequential expression of murine homeobox genes during F9 EC cell differentiation. EMBO J. 5, 2209-2215.

Breier, G., Dressler, G.R. and Gruss, P. (1988). Primary structure and developmental expression pattern of Hox3.1, a member of the murine Hox3 complex gene cluster. EMBO J. 7, 1329-1336.

Gaunt, S.J. (1988). Mouse homeobox gene transcripts occupy different but overlapping domains in embryonic germ layers and organs: a comparison of Hox-3.1 and Hox-1.5. Development 103, 135-144.

Gaunt, S.J., Sharpe, P.T. and Duboule D. (1988). Spatially restricted domain of homeo gene transcripts in mouse embryos: relation to a segmented body plan. Development 104, 169-179.

Holland, P.W.H. and Hogan, B.L.M. (1988). Spatially restricted patterns of expression of the homeobox-containing gene Hox 2.1 during embryogenesis. Development 102, 159-174.

Le Mouellic, H., Condamine, H. and Brûlet, P. (1988). Pattern of transcription of the homeo gene Hox-3.1 in the mouse embryo. Genes Dev. 2, 125-135.

Le Mouellic, H., Lallemand, Y. and Brûlet, P. (1990). Targeted replacement of the homeobox gene Hox-3.1 by the Escherichia coli lacZ in mouse chimeric embryos. Proc. Natl. Acad. Sci. USA 87, 4712-4716.

Le Mouellic, H., Lallemand, Y. and Brûlet, P. (1992). Homeosis in the mouse induced by a null mutation in the Hox-3.1 gene. Cell 69, 251-264.

Rabin, M., Ferguson-Smith, A., Hart, C.P. and Ruddle, F.H. (1986). Cognate homeobox loci mapped on homologous human and mouse chromosomes. Proc. Natl. Acad. Sci. USA 83, 9104-9108.

Utset, M.F., Awgulewitsch, A., Ruddle, F.H. and McGinnis, W. (1987). Region-specific expression of two mouse homeobox genes. Science 235, 1379-1382.

Violette, S.M., Shashikant, C., Salbaum, M.J., Belting, H.G., Wang, J.C.H. and Ruddle, F.H. (1992). Repression of the β-amyloid gene in a Hox-3.1 producing cell line. Proc. Natl. Acad. Sci. USA 89, 3805-3809.

Hoxc-9 [m]

Species: *Mus musculus*
Chromosomal location: Chromosome 15
Other names: *Hox-3.2*
Cognate genes: *HOXC9*
Type of homeobox: *Antp*-type, related to *Abd-B*
Accession number: X07647

■ Origin and description

The murine *Hoxc-9* homeobox was first identified on phages near the *Hoxc-8* gene, therefore it is linked to the *HOXC* cluster on chromosome 15 (Breier et al., 1988). A cDNA with the complete reading frame is reported (Erselius et al., 1990).

■ Expression

Expression during mouse embryogenesis was studied by *in situ* analysis in day 8.5 - day 14.5 embryos (Erselius et al., 1990). Major expression domains lie in the posterior part of the embryos, in particular the somites/prevertebrae (anterior boundary prevertebra T10), the neural tube (sharp boundary at the level of prevertebra 10), the hindlimb buds, and the cortex of the developing kidney.

■ References

Breier, G., Dressler, G.R. and Gruss, P. (1988). Primary structure and developmental expression pattern of *Hox-3.1*, a member of the murine Hox-3 homeobox gene cluster. EMBO J. 7, 1329-1336.

Erselius, J.R., Goulding, M.D. and Gruss, P. (1990). Structure and expression pattern of the murine *Hox-3.2* gene. Development 110, 629-642.

Hoxc-10 [m]

Species: *Mus musculus*
Chromosomal location: Chromosome 15
Other names: *Hox-3.6*
Cognate genes: *HOXC10*
Type of homeobox: Related to the *Drosophila Abd-B* gene
Accession number: X63507

■ Origin and description

This gene was isolated from cloned genomic mouse DNA contained in the cosmid clone CosMoEA (Awgulewitsch et al., 1990) which also includes the *Hoxc-9* and *Hoxc-8* genes. The homeobox sequence was isolated and mapped by means of cross hybridization under reduced stringency conditions using a *Hoxc-9* homeobox-containing probe (Peterson et al., 1991). Sequencing of 6514 bp of genomic region and partial cDNAs provided the sequence of the homeobox and the presumptive *Hoxc-10* transcription unit. Sequence analysis of the *Hoxc-10* homeobox identified it as a member of the *Abd-B*-like subfamily (Izpisúa-Belmonte et al, 1991). Translation of the deduced *Hoxc-10* coding region revealed a predicted protein containing 342 amino acid residues. The *Hoxc-10* homeobox was mapped approximately 13 kb upstream of the *Hoxc-9* homeobox.

■ Expression

Northern blot analysis revealed a single transcript of approximately 2.3 kb in adult ovary and spinal cord, and in embryos ranging from developmental stages 8.5-12.5 days *post coitus* (p.c.). Production of the transcript appears to begin at around 8.5 days p.c. and continues throughout adulthood in several tissues, primarily in the spinal cord. *In situ* analysis revealed that *Hoxc-10* has an extreme posterior expression pattern in 12.5 days p.c. embryos with an anterior expression boundary localized at the level of the 13th thoracic vertebra. *Hoxc-10* also exhibits spacially restricted expression in the hindlimb bud, where its transcripts are localized in an anterio-proximal subregion in contrast to the posterio-distal domain reported for its cognate gene *Hoxd-10* (Dollé and Duboule, 1989).

■ References

Awgulewitsch, A., Bieberich, C., Bogarad, L., Shashikant, C. and Ruddle, F.H. (1990). Structural analysis of the *Hoxc-8* transcription unit and the *Hox-3.2-Hoxc-8* intergenic region. Proc. Natl. Acad. Sci. USA 87, 6428-6432.

Dollé, P. and Duboule, D. (1989). Two gene members of the murine HOX-5 complex show regional and cell-type specific expression in developing limbs and gonads. EMBO J. 8, 1507-1515.

Izpisúa-Belmonte, J.C., Falkenstein, H., Dollé, P., Renucci, A. and Duboule, D. (1991). Murine genes related to the *Drosophila AbdB* homeotic gene are sequentially expressed during development of the posterior part of the body. EMBO J. 10, 2279-2289.

Peterson, R.L, Jacobs, D.F. and Awgulewitsch, A. (1991). *Hoxc-10*: isolation and characterization of a new murine homeobox gene located in the 5′ region of the *Hox*-3 cluster. Mech. Dev. 37, 151-166.

Hoxd-1 [m]

Species: *Mus musculus*
Chromosomal location: Chromosome 2
Other names: *Hox-4.9*
Cognate genes: *HOXD1*
Type of homeobox: *Antp*-type, related to the *Drosophila labial* gene
Accession number: X60034

■ Origin and description

The gene was isolated by cross-homology with *Hoxb-1* and *Hoxa-1* using PCR and encodes the third and probably last member of the murine *labial*-like subfamily (Frohman and Martin, 1992). The gene is thought to be the most 3'-located member of the *HOXD* cluster. *Hoxd-1* encodes a divergent *Antp*-like homeobox related to that of *Drosophila labial* (82% identical; Mlodzik et al., 1988). Unlike *Hoxa-1*, differentially spliced transcripts are not observed. Sequence comparisons suggest that this gene is the cognate of the *Xllab* [*Xhoxlab1*] Xenopus gene (Blumberg et al., 1991; Sive and Cheng, 1991).

■ Expression

By Northern analysis, a 2.0 kb transcript of very low abundance is observed from E10.5-16.5 (Frohman and Martin, 1992). This result is dissimilar to that observed for *Hoxa-1* and *Hoxb-1*, since those genes are expressed at much higher abundance at E10.5 than *Hoxd-1* and are not detected after E12.5 (LaRosa and Gudas, 1988; Frohman et al., 1990). *Hoxd-1* expression is first observed by *in situ* hybridization at E7.5 in the posterior (or ventral) part of the primitive streak (Frohman and Martin, 1992). At E8.5, transcripts are observed from the posterior end of the embryo to the hindbrain (Frohman and Martin, 1992), although the strongest expression is detected at the posterior end in the primitive streak (Hunt et al., 1991; Frohman and Martin, 1992). From E7.5-E9.5, *Hoxd-1* is detected anterior of the primitive streak only in lateral mesoderm (Frohman and Martin, 1992). Unlike *Hoxa-1* and *Hoxb-1*, it is not expressed in the neural tube, in surface ectoderm, in paraxial mesoderm, in cranial neural crest, or branchial arch endoderm. From E9.5-E11.5, *Hoxd-1* expression is observed transiently in trunk neural crest just after it emigrates from the tips of the neural tube and in part of the dermatomes nearby. Using PCR, Murtha et al. (1991) have suggested that *Hoxd-1* may be expressed in the forebrain at E13.5.

■ References

Blumberg, B., Wright, C.V., De Robertis, E.M. and Cho, K.W. (1991). Organizer-specific homeobox genes in *Xenopus laevis* embryos. Science 253, 194-196.

Frohman, M.A. and Martin, G.R. (1992). Isolation and spatial analysis of *Hoxd-1*, a new murine *labial*-like gene, reveals that *labial* subfamily members are expressed similarly only in early anteroposterior axis formation. Mech. of Dev. 38, 55–68.

Frohman, M.A., Boyle, M. and Martin, G.R. (1990). Isolation of the mouse *Hoxb-1* gene: analysis of embryonic expression suggests that positional information along the anterior-posterior axis is specified by mesoderm. Development 110, 589-608.

Hunt, P., Gulisano, M., Cook, M., Sham, M.H. and Krumlauf, R. (1991). A distinct *Hox* code for the branchial region of the vertebrate head. Nature 353, 861-864.

LaRosa, G.J. and Gudas, L.J. (1988). Early retinoic acid-induced F9 teratocarcinoma stem cell gene ERA-1: alternate splicing creates transcripts for a homeobox-containing protein and one lacking the homeobox. Mol. Cell Biol. 8, 3906-3917.

Mlodzik, M., Fjose, A. and Gehring, W.H. (1988). Molecular structure and spatial expression of a homeobox gene from the *labial* region of the Antennapedia-complex. EMBO J. 7, 2569-2578.

Murtha, M., Leckman, J.F. and Ruddle, F.H. (1991). Detection of homeobox genes in development and evolution. PNAS 88, 10711-10715.

Sive, H.L. and Cheng, P.F. (1991). Retinoic acid perturbs the expression of Xhox.lab genes and alters mesodermal determination in *Xenopus laevis*. Genes and Dev. 5, 1321-1332.

Hoxd-1 [c]

Species: *Gallus gallus* (chicken)
Chromosomal location: -
Other names: *Chox-4.9, Chox-1, Chox-10*
Cognate genes: *HOXD1, Hoxd-1*
Type of homeobox: Related to the *Drosophila labial* gene
Accession number: M23611

■ Origin and description

This gene was cloned in two independent library screens, the first one a screen of a chicken genomic library with a *Drosophila Antp* gene and the second with a mixed homeobox probe, its characterization is described in Rangini et al. (1989). Sequencing of 1029 bp of the genomic region provided the sequence of the homeobox and the flanking regions. Sequence analysis revealed a new type of homeodomain which represents a new subfamily for which no other members have been reported. Translation of the sequenced genomic region revealed that the region codes for one contiguous open reading frame that includes the homeodomain. The *Hoxd-1* homeodomain belongs to the *labial* family (Mlodzik et al., 1988) together with a number of homeodomains from several organisms such as the human *HOXA1*, *HOXB1* and *HOXD1* (Simeone et al., 1991), the murine *Hoxa-1* and *Hoxb-1* (Baron et al., 1987; Frohman et al., 1990), the chicken *Ghox-lab* (Sundin et al.,

1990), the *Xenopus Xlab* (Blumberg et al., 1991) and the *C. elegans ceh-13* (Schaller et al., 1990) genes. The allocation of this gene to the *HOXD* complex is based on comparison of phage clones (Kuroiwa, A., personal communication) and its preliminary spatial pattern of expression (Martin, G., personal communication).

■ Expression

Like most other vertebrate members of this homeobox gene family, also *Hoxd-1* gives very weak signals in northern analysis. Northern analysis was performed on RNA from embryos from early head fold stages to 5 days of incubation. Three weak transcripts of 2.1, 5.7 and 6.7 kb can be detected. The 2.1 kb transcript is most abundant in early neurulation stages while the level of the other two transcripts increases later in development and remains constant until day 5. Preliminary *in situ* analysis suggests that the this gene is expressed in the somite and as it differentiates it becomes restricted to the dermomyotome.

■ References

Baron, A., Featherstone, M.S., Hill, R.E., Hall, A., Galliot, B. and Duboule, D. (1987). *Hox-1.6*: a mouse homeobox containing gene member of the *Hox-1* complex. EMBO J. 6, 2977-2986.

Blumberg, B., Wright, C.V.E., De Robertis, E.M. and Cho, K.W.Y. (1991). Organizer-specific homeobox genes in *Xenopus laevis* embryos. Science 253, 194-196.

Frohman, M.A., Boyle, M. and Martin, G.R. (1990). Isolation of the mouse Hox 2.9 gene; analysis of embryonic expression suggests that positional information along the anterior-posterior axis is specified by mesoderm. Development 110, 589-607.

Mlodzik, M., Fjose, A. and Gehring, W.J. (1988). Molecular structure and spatial expression of a homeobox gene from the *labial* region of the Antennapedia-complex. EMBO J. 7, 2569-2578.

Rangini, Z., Frumkin, A., Shani, G., Guttmann, M., Eyal-Giladi, H., Gruenbaum, Y. and Fainsod, A. (1989). The chicken homeobox genes Chox 1 and 3: Cloning, sequencing and expression during embryogenesis. Gene 76, 61-74.

Schaller, D., Wittmann, C., Spicher, A., Müller, F. and Tobler, H. (1990). Cloning and analysis of three new homeobox genes from the nematode *Caenorhabditis elegans*. Nucleic Acids Res. 18, 2033-2036.

Simeone, A., Acampora, D., Nigro, V., Faiella, A., D'Esposito, M., Stornaiuolo, A., Mavilio, F. and Boncinelli, E. (1991). Differential regulation by retinoic acid of the homeobox genes of the four HOX loci in human embryonal carcinoma cells. Mech. Dev. 33, 215-228.

Sundin, O.H., Busse, H.G., Rogers, M.B., Gudas, L.J. and Eichele, G. (1990). Region-specific expression in early chick and mouse embryos of Ghox-lab and Hox 1.6, vertebrate homeobox-containing genes related to *Drosophila labial*. Development 108, 47-58.

Hoxd-1 [x]

Species: *Xenopus laevis*
Chromosomal location: -
Other names: *Xhox.lab1, Xhox.labA, Xlab1*
Cognate genes: -
Type of homeobox: *Antp*-type, related to *Drosophila labial*
Accession number: M63873

■ Origin and description

Isolated by homology to the chicken *Ghox.lab* homeodomain (Sundin et al., 1990; Sive and Cheng, 1991). It is a member of the *labial* gene family, however it does not obviously correspond to members previously isolated from other species.

■ Expression

Expressed from stage 10 (late blastula) to stage 25 (tailbud), with RNA levels peaking in late gastrula and early neurula. By dissection of embryos, RNA is predominantly localized in dorsal mesoderm and ectoderm, particularly in posterior (trunk and tail) regions in early neurula. Strongly induced by exogenous retinoic acid (RA) during gastrula and neurula stages (up to 50-fold in dorsal ectoderm and mesoderm).

■ Function

Unknown. May be partly responsible for mediating the axial truncation caused by RA, since it becomes induced in anterior regions concomitant with anterior truncation, after RA application (Durston et al., 1989; Sive et al., 1990).

■ References

Blumberg, B., Wright, C.V.E., de Robertis, E.M. and Cho, K.W.Y. (1991). Organizer-specific homeobox genes in *Xenopus laevis*. Science 253, 194-196.

Durston, A.J., Timmermans, J.P.M., Hage, W.J., Hendriks, H.F.J., de Vries, N.J., Heideveld, M. and Nieuwkoop, P.D. (1989). Retinoic acid causes an anteroposterior transformation in the developing central nervous system. Nature 340, 140-144.

Sive, H.L. and Cheng, P.F. (1991). Retinoic acid perturbs the expression of *Xhox.lab* genes and alters mesodermal determination in *Xenopus laevis*. Genes Dev. 5, 1321-1332.

Sive, H.L., Draper, B.W., Harland, R.M. and Weintraub, H. (1990). Identification of a retinoic acid-sensitive period during primary axis formation in *Xenopus laevis*. Genes Dev. 4, 932-942.

Sundin, O.H., Busse, H.G., Rogers, M.B., Gudas, L.J. and Eichele, G. (1990). Region-specific expression in early chick and mouse embryos of *Ghox.lab* and Hox1.6, vertebrate homeobox-containing genes related to *Drosophila labial*. Development 108, 47-58.

Hoxd-3 [m]

Species: *Mus musculus*
Chromosomal location: Chromosome 2
Other names: *Hox-4.1, CH23; p23*
Cognate genes: *HOXD3*
Type of homeobox: *Antp*-type, related to *Drosophila Zen* and *Pb*
Accession number: -

■ Origin and description

The gene was isolated through its homology to *Hoxa-3* from a BALB/c genomic library (Lonai et al., 1987). Closely related to *Hoxa-3* and *Hoxb-3*, *Hoxd-3* was mistakenly associated with chromosome 12 of the mouse. Its mapping to chromosome 2 was shown by hybrid cell analysis (Pravcheva et al., 1989) and its linkage to the *HOXD* complex by jumping library and RFLP analysis (Stubbs et al., 1990). Chromosomal walk in a γ phage library demonstrated it to be 10-13 kb downstream from *Hoxd-4* in the same transcriptional orientation (Stubbs et al., 1990; Lonai et al., unpublished). *Hoxd-3* is the penultimate locus at the 3' end of the murine *HOXD* cluster.

■ Expression

Its border of expression is close to the third rhombomer (Hunt et al., 1991), it is closely similar to that of *Hoxa-3* and *Hoxb-3* (unpublished).

■ References

Hunt, P., Gulisano, M., Cook, M., Sham, M.-H., Faiella, A., Wilkinson, D., Boncinelli, E. and Krumlauf, R. (1991). A distinct Hox code for the branchial region of the vertebrate head. Nature 353, 861-864.

Lonai, P., Arman, E., Czosnek, H., Ruddle, F.H. and Blatt, C. (1987). New murine homeoboxes; structure, chromosomal assignment and differential expression in the embryo and in adult erythropoiesis. DNA 6, 409-415.

Pravcheva, M., Newman, M., Hunihan, L., Lonai, P. and Ruddle, F. (1989). Chromosome assignment of the murine *Hox-4.1* gene. Genomics 5, 541-545.

Stubbs, L., Poustka, A., Baron, A., Lehrach, H., Lonai, P. and Duboule, D. (1990). The murine genes *Hox-5.1* and *Hox-4.1* belong to the same complex on chromosome 2. Genomics 7, 422-427.

Hoxd-4 [m]

Species: *Mus musculus*
Chromosomal location: Chromosome 2, band D
Other names: *Hox-5.1, Hox-4.2*
Cognate genes: *HOXD4*
Type of homeobox: *Antp*-type, related to *Drosophila Deformed* gene
Accession number: J03770; X65950

■ Origin and description

Cross-hybridization to a probe covering the *Hoxa-4* (*Hoxa-4*) allowed the initial cloning of a cDNA molecule corresponding to *Hoxd-4* (Featherstone et al., 1988). As for its paralogous genes, *Hoxa-4*, *Hoxb-4* and *Hoxc-4*, *Hoxd-4* is related to the *Drosophila Deformed* (*Dfd*) homeotic gene (Regulski et al., 1987). Human *HOXD4* can substitute for Dfd for activation of the *Dfd* transcription unit, thus provoking *Dfd*-like homeotic transformations (McGinnis et al., 1990). Murine *Hoxd-4* is associated with autoregulatory (Pöpperl and Featherstone, 1992) and retinoic acid responsive elements (Pöpperl and Featherstone, 1993). The autoregulatory element maps to a spatial enhancer region defined in the human *HOXD4* gene (Tuggle et al., 1990). A 500 bp region of the human gene, spanning the autoregulatory element, is able to direct reporter gene expression to cells within the *Dfd* expression domain (Malicki et al., 1992).

■ Expression

The anterior border of expression of *Hoxd-4* lies between rhombomeres 6 and 7 in the developing hindbrain (Gaunt et al., 1989; Wilkinson et al., 1989), and at the second cervical prevertebra in the prevertebral column (Featherstone et al., 1988; Gaunt et al., 1989). *Hoxd-4* and *Hoxb-4* transcripts are detected at slightly more anterior positions than those of *Hoxa-4* (Gaunt et al., 1989). In many tissues at days 8.5 and 9.5, *Hoxd-4* transcripts are more abundant than those of *Hoxa-4* and *Hoxb-4* (Gaunt et al., 1989). This situation is reversed by 12.5 days, except in the meso- and metanephric kidneys where *Hoxd-4* shows the highest expression of these three genes (Gaunt et al., 1989). *Hoxd-4* transcripts are also detected throughout the limb bud mesenchyme (Dollé et al., 1989), in the gonads, thyroid, thymus, and floor of the pharynx (Gaunt et al., 1989). In the 12.5 day stomach, *Hoxd-4* is expressed in the neural-crest derived parasympathetic ganglia of the enteric plexus (Gaunt et al., 1989).

■ Function

HOXD4 protein is able to act through the *Dfd* autoregulatory enhancer (McGinnis et al., 1990), suggesting a strong conservation of function. To date, *Hoxd-4* null mutants have not been reported and so specific conclusions cannot yet be drawn concerning the role of this gene in axial specification. However, ectopic expression of *Hoxd-4* in the more anterior *Hoxa-1* expression domain results in posterior homeotic transformations of the occipital bones of the skull (Lufkin et al., 1992). These results support a role in vertebral morphogenesis consistent with normal expression in the vertebral column. In tissue culture, HOXD-4 acts as a transcriptional activator (Lufkin et al., 1992; Pöpperl and Featherstone, 1992).

■ References

Dollé, P., Izpisúa-Belmonte, J.-C., Falkenstein, H., Renucci, A. and Duboule, D. (1989). Coordinate expression of the murine Hox-5 complex homeobox-containing genes during limb pattern formation. Nature 342, 767-772.

Featherstone, M.S., Baron, A., Gaunt, S.J., Mattei, M.-G. and Duboule, D. (1988). Hox-5.1 defines a homeobox-containing gene locus on mouse chromosome 2. Proc. Natl. Acad. Sci. USA 85, 4760-4764.

Gaunt, S.J., Krumlauf, R. and Duboule, D. (1989). Mouse homeogenes within a subfamily, Hoxa-4, -2.6 and -5.1 display similar anteroposterior domains of expression in the embryo, but show stage- and tissue-dependent differences in their regulation. Development 107, 131-141.

Lufkin, T., Mark, M., Hart, C.P., Dollé, P., LeMeur, M. and Chambon, P. (1992). Homeotic transformation of the occipital bones of the skull by ectopic expression of a homeobox gene. Nature 359, 835-841.

Malicki, J., Cianetti, L., Peschle, C. and McGinnis, W. (1992). A human HOXD4 regulatory element provides head-specific expression in Drosophila embryos. Nature 358, 345-347.

McGinnis, N., Kuziora, M.A. and McGinnis, W. (1990). Human Hox-4.2 and Drosophila Deformed encode similar regulatory specificities in Drosophila embryos and larvae. Cell 63, 969-976.

Pöpperl, H. and Featherstone, M.S. (1992). An autoregulatory element of the murine Hox-4.2 gene. EMBO J. 11, 3673-3680.

Pöpperl, H. and Featherstone, M.S. (1993). Identification of a retinoic acid response element upstream of the murine Hoxd-4 gene. Mol. Cell Biol. 13, 257–265.

Regulski, M., McGinnis, N., Chadwick, R. and McGinnis, W. (1987). Developmental and molecular analysis of Deformed: a homeotic gene controlling Drosophila head development. EMBO J. 6, 767-777.

Tuggle, C.K., Zakany, J., Chianetti, L., Peschle, C. and Nguyen-Huu, M.C. (1990). Region-specific enhancers near two mammalian homeobox genes define adjacent rostrocaudal domains in the central nervous system. Genes Dev. 4, 180-189.

Wilkinson, D.G., Bhatt, S., Cook, M., Boncinelli, E. and Krumlauf, R. (1989). Segmental expression of Hox-2 homeobox-containing genes in the devleoping mouse hindbrain. Nature, 341, 405-409.

Hoxd-4 [c]

Species: *Gallus gallus* (chicken)
Chromosomal location: -
Other names: *Chox-4.2, Chox-a*
Cognate genes: *HOXD4, Hoxd-4*
Type of homeobox: *Antp*-type, related to the *Drosophila Dfd* gene
Accession number: *X17246*

■ Origin and description

The chicken *Hoxd-4* gene was isolated during a walk on the *HOXD* complex (Sasaki et al., 1990; Sasaki et al., 1992 and Kuroiwa et al., unpublished). The lengths of the mRNAs are 1.3 Kb and 2.8 Kb and the encoded homeodomain containing protein has a MW of ca. 26,960 D. The protein is 74% identical to the mouse protein (Sasaki et al., 1990a).

■ Expression

The expression of *Hoxd-4* was already observed in the 2-day old embryo and continued at the earliest to 4-week old chick. Transcripts showed a graded distribution along rostro-caudal axis in spinal cord with the highest concentration at the rostral side. The expression was observed in kidney but not in liver, lung and testis (Sasaki et al., 1992).

■ Function

An *E. coli* produced full length Hoxd-4 protein binds to 5' upstream region of the *Hoxa-4* and *Hoxd-4* genes *in vitro* (Sasaki et al., 1990). In transient transfection experiment, this protein weakly activated the transcription from globin basal promoter through the artificial binding site NP6 (Sasaki et al., 1992).

■ References

A. Kuroiwa, pers. com.

Hoxd-8 [m]

Species: *Mus musculus*
Chromosomal location: Chromosome 2, bands D, E
Other names: *Hox-4.3, Hox-5.4*
Cognate genes: *HOXD8*
Type of homeobox: *Antp*-type
Accession number: -

Origin and description

The *Hoxd-8* gene was cloned during a walk on the *HOXD* complex by cross-hybridization with the *Hoxa-1* and *Hoxd-12* homeobox probes (Izpisúa-Belmonte et al., 1990; Duboule et al., 1990). This gene was originally not detected using probes derived from neighbour genes but was further isolated after the description of the cognate human gene *HOXD8* (Oliver et al., 1989). *Hoxd-8* belong to the group 8 with the two paralogous genes *Hoxb-8* and *Hoxc-8* (Izpisúa-Belmonte et al., 1990; Sadoul and Featherstone, 1991). No counterpart has been described on the *HOXA* complex so far.

Expression

During development, *Hoxd-8* is expressed in structures of either mesodermal or ectodermal origins located or derived from below a precise cranio-caudal level that correspond to the prevertebra 18, in the mesoderm and to the posterior hindbrain, in the neuro-ectoderm. In addition, *Hoxd-8* is strongly expressed in the metanephric kidneys, the genital ridges and, like other *Hoxd* genes, in the developing limb buds. In 12.5 day old limbs, expression is restricted =/50to proximal blastemas (Izpisúa-Belmonte et al., 1990).

References

Duboule, D., Boncinelli, E., DeRobertis, E., Featherstone, M.S., Lonai, P., Oliver, G. and Ruddle, F. (1990). An update of mouse and human HOX genes nomenclature. Genomics 7, 458-459.

Izpisúa-Belmonte, J.C., Dollé, P., Renucci, A., Zappavigna, V., Falkenstein, H. and Duboule, D. (1990). Primary structure and embryonic expression pattern of the mouse Hox-4.3 homeobox gene. Development 110, 733-746.

Sadoul, R. and Featherstone, M.S. (1991). Sequence analysis of the homeobox containing exon of the murine Hox-4.3 homeogene. Bioch.Biophys. Acta 1089, 259-261.

Hoxd-9 [m]

Species: *Mus musculus*

Chromosomal location: Chromosome 2, bands D, E

Other names: *Hox-4.4, Hox-5.2*

Cognate genes: *HOXD9*

Type of homeobox: *Antp*-type, related to *Drosophila Abd-B*

Accession number: X62669

Origin and description

Hoxd-9 was isolated during a walk on the *HOXD* complex (Duboule and Dollé, 1989; Duboule, 1990). This gene is member of the group 9 and is therefore the most 3' located gene on the *HOXD* complex which is related to the *Drosophila* homeotic gene *AbdominalB*. It is one of the few Hox genes for which paralogous genes exist on the three other complexes. As all *AbdB*-related genes, *Hoxd-9* does not have the conserved pentapeptide upstream the homeobox. Rather, it harbors the *AbdB* consensus Trp residue at position -7 from the homeobox (Izpisúa-Belmonte et al., 1991).

Expression

Hoxd-9 mRNAs are 2.6 and 2.9 Kb large and are encoded by two exons and the protein contains 339 amino acids (Renucci et al., 1992). The expression of this gene was studied by using in situ hybridization. In fetuses, transcripts were observed in mesodermal and ectodermal derivatives, located below the level of the first lumbar prevertebra (20-21), in the mesoderm, and below a level corresponding to the 17-18th prevertebrae in the neuro-ectoderm (Duboule and Dollé, 1989; Dollé et al., 1991a). Expression is first detected around gestational day 8, at a stage where other *AbdB*-related genes are not yet expressed (Izpisúa-Belmonte et al., 1991). Later, *Hoxd-9* is strongly expressed in the developing uro-genital system (Dollé and Duboule, 1989; Dollé et al., 1991b). In addition to the trunk, *Hoxd-9* expression is detected at high level during limb development (Dollé and Duboule, 1989; Oliver et al., 1989). Transcripts are present in the lateral plate mesoderm at the budding stage (Dollé and Duboule, 1989; Dollé et al., 1989) and remain homogenously distributed in almost the entire limb buds until about day 10. Later, the expression is progressively restricted to the periphery of the chondrogenic blastema (the perichondrium), with a higher concentration in more proximal areas of the limbs (e.g. in the humeral blastema; Dollé et al., 1989). The regulation of *Hoxd-9* expression has been studied by using a transgenic approach. The insertion of a marker gene (lacZ) in frame in a large genomic piece of DNA demonstrated that most (but not all) of the regulatory elements required for the correct expression of this gene are located in a rather close vicinity (Renucci et al., 1992). In *in vitro* cotransfection experiments, the HOXD9 protein was shown to have transcriptional activating properties (Zappavigna et al., 1991; Arcioni et al., 1992).

References

Arcioni, L., Simeone, A., Guazzi, S., Zappavigna, V., Boncinelli, E. and Mavilio, F. (1992). The upstream region of the human homeobox gene HOXC5 is a target for regulation by retinoic acid and HOX proteins. EMBO J. 11, 265-278.

Dollé, P. and Duboule, D. (1989). Two genes members of the murine HOX-5 complex show regional and cell-type specific expression in developing limbs and gonads. EMBO J. 8, 1507-1515.

Dollé, P., Izpisúa-Belmonte, J.-C., Falkentsein, H., Renucci, A. and Duboule, D. (1989). Coordinate expression of the murine HOX-5 homeobox containing gene during limb pattern formation. Nature 342, 767-772.

Dollé, P., Izpisúa-Belmonte, J-C., Boncinelli, E. and Duboule, D.

(1991). The Hox-4.8 gene is localised at the 5′ end of the HOX-4 complex and is expressed at the posterior extremity of the body during development. Mech. Dev. 36, 3-14.

Dollé, P., Izpisúa-Belmonte, J.-C., Tickle, C., Brown, J. and Duboule, D. (1991). Hox-4 genes and the morphogenesis of mammalian genitalia. Genes Dev. 5, 1767-1776.

Duboule D., Boncinelli E., DeRobertis E., Featherstone M.S., Lonai P., Oliver G. and Ruddle, F. (1990). An update of mouse and human HOX genes nomenclature. Genomics 7, 458-459.

Duboule, D. and Dollé P. (1989). The structural and functional organization of the murine HOX gene family resembles that of *Drosophila* homeotic genes. EMBO J. 8 1497-1505.

Izpisúa-Belmonte, J.C., Falkenstein, H., Dollé, P., Renucci, A. and Duboule D. (1991). Murine genes related to the Drosophila AbdB homeotic gene are sequentially expressed during development of the posterior part of the body. EMBO J. 10, 2279-2289.

Oliver, G., Sidell, N., Fiske, W., Heinjmann, C., Mohandas, T., Sparkes, R.S. and DeRobertis, E. M. (1989). Complementary homeoprotein gradients in developing limb buds. Genes Dev. 3, 641-650.

Renucci, A., Zappavigna, V., Zakany, J., Izpisúa-Belmonte, J-C., Bürki, K. and Duboule, D. (1992). Comparison of mouse and human Hox-4 complexes defines conserved promoter sequences involved in the regulation of the Hox-4.4 gene. EMBO J. 11, 1459-1468.

Zappavigna, V., Renucci, A., Izpisúa-Belmonte, J.-C., Urier, G., Peschle, C. and Duboule, D. (1991). HOX4 genes encode transcription factors with potential auto- and cross-regulatory capacities. EMBO J. 10, 4177-4187.

Hoxd-10 [m]

Species: *Mus musculus*

Chromosomal location: Chromosome 2, bands D, E

Other names: *Hox-4.5, Hox-5.3*

Cognate genes: *HOXD10*

Type of homeobox: *Antp*-type, related to *Drosophila Abd-B*

Accession number: X62669

■ Origin and description

Hoxd-10 was isolated during a walk on the *HOXD* complex (Duboule and Dollé, 1989; Duboule et al., 1990). This gene is member of the group 10 and is thus related to the *Drosophila* homeotic gene *AbdominalB*. Paralogous genes exist on the *HOXA* and *HOXC* complexes. As all *AbdB*-related genes, *Hoxd-10* does not have the conserved pentapeptide upstream the homeobox. As member of group 10 (like for groups 9 and 12), this *AbdB*-related gene does have the consensus Trp residue at position -7 from the homeobox (Izpisúa-Belmonte et al., 1991).

■ Expression

Hoxd-10 mRNAs are 2.6 and 3.3 Kb large and are encoded by two exons and the protein contains 341 amino acids (Renucci et al., 1992). The expression of this gene was studied by using in situ hybridization. In fetuses, transcripts were observed in mesodermal and ectodermal derivatives, located below the level of the third lumbar vertebrae (22-23), in the mesoderm, and below a level corresponding to the 21st-22nd vertebrae in the neuro-ectoderm (Duboule and Dollé, 1989; Dollé et al., 1991a). Expression is first detected around gestational day 8.5, soon after the onset of expression of the *Hoxd-9* gene but before *Hoxd-11* (Izpisúa-Belmonte et al., 1991). Later, *Hoxd-10* is strongly expressed in the developing uro-genital system (Dollé and Duboule, 1989; Dollé et al., 1991b). In addition to the trunk, *Hoxd-10* expression is detected at high level during limb development (Dollé and Duboule, 1989). Transcripts are not present in the forelimb lateral plate mesoderm at the budding stage but are detected in the forelimbs where they remain homogenously distributed in practically all the developing limbs except, however, in the most antero-proximal part, until about day 10 (Dollé and Duboule, 1989; Dollé et al., 1989). Later, the expression is progressively restricted to the periphery of the condensed material (the perichondrium), with a higher concentration in more proximo-central areas of the limbs (e.g. in the humeral and radius-ulna blastema; Dollé et al., 1989). In *in vitro* co-transfection experiments, the HOXD10 protein was shown to have transcriptional activating properties (Zappavigna et al., 1991; Arcioni et al., 1992).

■ References

Arcioni, L., Simeone, A., Guazzi, S., Zappavigna, V., Boncinelli, E. and Mavilio, F. (1992). The upstream region f the human homeobox gene HOXC4 is a target for regulation by retinoic acid and HOX proteins. EMBO J. 11, 265-278.

Dollé, P. and Duboule, D. (1989). Two genes members of the murine HOX-5 complex show regional and cell-type specific expression in developing limbs and gonads. EMBO J. 8, 1507-1515.

Dollé, P., Izpisúa-Belmonte, J.-C., Falkentsein, H., Renucci, A. and Duboule, D. (1989). Coordinate expression of the murine HOX-5 homeobox containing gene during limb pattern formation. Nature 342, 767-772.

Dollé, P., Izpisúa-Belmonte, J-C., Boncinelli, E. and Duboule, D. (1991a). The Hoxd-13 gene is localised at the 5′ end of the HOX-4 complex and is expressed at the posterior extremity of the body during development. Mech. Dev. 36, 3-14.

Dollé, P., Izpisúa-Belmonte, J.-C., Tickle, C., Brown, J. and Duboule, D. (1991b). Hox-4 genes and the morphogenesis of mammalian genitalia. Genes Dev. 5, 1767-1776.

Duboule, D. and Dollé P. (1989). The structural and functional organization of the murine HOX gene family resembles that of *Drosophila* homeotic genes. EMBO J. 8, 1497-1505.

Duboule, D., Boncinelli, E., DeRobertis E., Featherstone, M.S., Lonai P., Oliver G. and Ruddle, F. (1990). An update of mouse and human HOX genes nomenclature. Genomics 7, 458-459.

Izpisúa-Belmonte, J.C., Falkenstein, H., Dollé, P., Renucci, A. and Duboule D. (1991). Murine genes related to the *Drosophila* *AbdB* homeotic gene are sequentially expressed during development of the posterior part of the body. EMBO J. 10, 2279-2289.

Renucci, A., Zappavigna, V., Zakany, J., Izpisúa-Belmonte, J-C., Bürki, K. and Duboule, D. (1992). Comparison of mouse and human Hox-4 complexes defines conserved promoter sequences involved in the regulation of the Hoxd-9 gene. EMBO J. 11, 1459-1468.

Zappavigna, V., Renucci, A., Izpisúa-Belmonte, J.-C., Urier, G., Peschle, C. and Duboule, D. (1991). HOX4 genes encode transcription factors with potential auto- and cross-regulatory capacities. EMBO J. 10, 4177-4187.

Hoxd-11 [m]

Species: *Mus musculus*

Chromosomal location: Chromosome 2, bands D, E

Other names: *Hox-4.6, Hox-5.5*

Cognate genes: *HOXD11*

Type of homeobox: *Antp*-type, related to *Drosophila Abd-B*

Accession number: X48849

■ Origin and description

Hoxd-11 was isolated during a walk on the *HOXD* complex (Izpisúa-Belmonte et al., 1991; Duboule, 1990). This gene is a member of the group 11 and is thus related to the *Drosophila* homeotic gene *Abdominal-B*. Paralogous genes exist on the *HOXA* and *HOXC* complexes. As all *AbdB*-related genes, *Hoxd-11* does not have the conserved pentapeptide upstream the homeobox. As a member of group 11 (like for genes in the group 13) this *Abd-B*-related gene does not have the consensus Trp residue at position -7 from the homeobox which is found in *Abd-B*-related genes from groups 9, 10 and 12 (Izpisúa-Belmonte et al., 1991).

■ Expression

The HOXD11 protein contains 323 amino acids encoded by two exons. The first exon contains large stretches of monotonic amino acids (poly-Gly and poly-Ala). The expression of this gene was studied by using in situ hybridization. In fetuses, transcripts were observed in mesodermal and ectodermal derivatives, located below the level of pre-vertebrae 26 to 27, in the mesoderm, and below a level corresponding to pre-vertebrae 24 to 26 in the neuro-ectoderm (Izpisúa-Belmonte et al., 1991; Dollé et al., 1991a). Expression is first detected around late gestational day 8, early day 9, soon after the onset of expression of the *Hoxd-10* gene but before *Hoxd-12* (Izpisúa-Belmonte et al., 1991). Later, *Hoxd-11* is strongly expressed in the developing uro-genital system (Dollé et al., 1991b). In addition to the trunk, *Hoxd-11* expression is detected at high level during limb development (Dollé et al., 1989). Transcripts are not present in the forelimb lateral plate mesoderm at the budding stage but are detected in the forelimbs where they remain homogenously distributed specifically in postaxial mesoderm (Dollé et al., 1989). Later, the expression domain extends anteriorly, in the most distal regions, and concomitantly, becomes restricted to the periphery of the condensed material (the perichondrium), with a higher concentration in more central-distal areas of the limbs (the radius-ulna blastema as well as more distal ones such as for the wrist bones or the metacarpal; Dollé et al., 1989). The regulation of *Hoxd-11* was studied by a transgenic approach using the lacZ reporter gene. Several regulatory elements, located either in 5' or in 3' from the coding region seem to be required for the expression of the transgene in a pattern that resembles that of the endogenous gene. However, a correct domain in limbs was not observed (Gérard et al., 1993)

■ Function

The function of the Hoxd-11 protein was investigated in two systems. Using a retroviral vector, the mouse protein was expressed at an ectopic position, in the pre-axial mesoderm of the developing chicken limbs. Infected chicken thus expressed a "posterior" protein at a more "anterior" position. The resulting phenotype is complex but the morphology of digit one now resembles, with a high incidence, that of digit two (it bears an additional phalange). This changes in morphologies somehow resemble homeotic transformations occurring in *Drosophila* (Morgan et al., 1992). The same protein was expressed, upon heat shock, in P-element mediated transformed *Drosophila* embryos, at the time of expression of the endogenous homeotic genes. The phenotype obtained indicates that anterior segments are transformed towards the identities of more posterior segments (Bachiller et al., 1993).

■ References

Bachiller, D., Macias, A., Duboule, D. and Morata, G. (1994). Homeotic transformations in *Drosophila* produced by over-expression of vertebrate *AbdB*-related genes. EMBO J., in press.

Dollé, P., Izpisúa-Belmonte, J.-C., Falkentein, H., Renucci, A. and Duboule, D. (1989). Coordinate expression of the murine HOX-5 homeobox containing gene during limb pattern formation. Nature 342, 767-772.

Dollé, P., Izpisúa-Belmonte, J.-C., Boncinelli, E. and Duboule, D. (1991a). The Hox-4.8 gene is localised at the 5' end of the HOX-4 complex and is expressed at the posterior extremity of the body during development. Mech. Dev. 36, 3-14.

Dollé, P., Izpisúa-Belmonte, J.-C., Tickle, C., Brown, J. and Duboule, D. (1991b). Hox-4 genes and the morphogenesis of mammalian genitalia. Genes Dev. 5, 1767-1776.

Duboule D., Boncinelli E., DeRobertis E., Featherstone M.S., Lonai P., Oliver G. and Ruddle, F. (1990). An update of mouse and human HOX genes nomenclature. Genomics 7, 458-459.

Gérard, M., Duboule, D. and Zákány, J. (1993). *In vivo* mapping of the mouse Hoxd-11 control elements. EMBO J. 12, 3539–3550.

Izpisúa-Belmonte, J.C., Falkenstein, H., Dollé, P., Renucci, A. and

Duboule D. (1991). Murine genes related to the *Drosophila AbdB* homeotic gene are sequentially expressed during development of the posterior part of the body. EMBO J. 10, 2279-2289.

Morgan , B. A., Izpisúa-Belmonte, J.-C., Duboule, D. and Tabin, C. J. (1992). Targeted misexpression of Hox-4.6 in the avian limb bud causes apparent homeotic transformations. Nature 358, 236-239.

Hoxd-12   [m]

Species: *Mus musculus*

Chromosomal location: Chromosome 2, bands D, E

Other names: *Hox-4.7, Hox-5.6*

Cognate genes: *HOXD12*

Type of homeobox: *Antp*-type, related to *Drosophila Abd-B*

Accession number: X58848

■ Origin and description

Hoxd-12 was isolated during a walk on the *HOXD* complex (Izpisúa-Belmonte et al., 1991; Duboule, 1990). This gene is a member of the group 12 and is thus related to the *Drosophila* homeotic gene *Abdominal-B*. A paralogous gene exists only on the *HOXC* complex (*Hoxc-12*). As all *Abd-B*-related genes, *Hoxd-12* does not have the conserved pentapeptide upstream the homeobox. As a member of group 12 (like for genes in the groups 9 and 10) this *Abd-B*-related gene does have the consensus Trp residue at position -7 from the homeobox (Izpisúa-Belmonte et al., 1991).

■ Expression

The HOXD12 protein contains 275 amino acids encoded by two exons separated by an unusually short intronic sequence (131bp). The expression of this gene was studied by using in situ hybridization. In fetuses, transcripts were observed in mesodermal and ectodermal derivatives, located below the level of pre-vertebrae 30 to 31, in the mesoderm, and below a level corresponding to pre-vertebrae 27 to 28 in the neuro-ectoderm (Izpisúa-Belmonte et al., 1991; Dollé et al., 1991a). Expression is first detected around early gestational day 9, soon after the onset of expression of the *Hoxd-11* gene (Izpisúa-Belmonte et al., 1991). Later, *Hoxd-12* is strongly expressed in the developing uro-genital system where it persists until very late fetal stages (Dollé et al., 1991b). In addition to the trunk, *Hoxd-12* expression is detected at high level during limb development (Dollé et al., 1989). Transcripts are not present in both forelimb and hindlimb lateral plate mesoderm at the budding stage but are detected in the

posterior mesoderm of both forelimbs and hindlimbs (Dollé et al., 1989). Later, the expression domain extends anteriorly, in the most distal regions, so that *Hoxd-12* transcripts are basically restricted to the autopods. Concomitantly, the signal becomes restricted to the periphery of the condensed material (the perichondrium), with a higher concentration around the distal blastemas of the limbs (e.g the metacarpal blastemas as well as the phalanges; Dollé et al., 1989; Duboule et al., unpublished). At this late stage of limb development, the expression pattern of *Hoxd-12* is similar to that of *Hoxd-13* and there are no obvious differences detected between forelimbs and hindlimbs. In birds, however, slight differences between fore- and hindlimbs are observed at a comparable stage (Mackem and Mahon, 1992).

■ References

Dollé, P., Izpisúa-Belmonte, J.-C., Falkentsein, H., Renucci, A. and Duboule, D. (1989). Coordinate expression of the murine HOX-5 homeobox containing gene during limb pattern formation. Nature 342, 767-772.

Dollé, P., Izpisúa-Belmonte, J.-C., Boncinelli, E. and Duboule, D. (1991a). The Hox-4.8 gene is localised at the 5′ end of the HOX-4 complex and is expressed at the posterior extremity of the body during development. Mech. Dev. 36, 3-14.

Dollé, P., Izpisúa-Belmonte, J.-C., Tickle, C., Brown, J. and Duboule, D. (1991b). Hox-4 genes and the morphogenesis of mammalian genitalia. Genes Dev. 5, 1767-1776.

Duboule D., Boncinelli E., DeRobertis E., Featherstone, M.S., Lonai, P., Oliver, G. and Ruddle, F. (1990). An update of mouse and human HOX genes nomenclature. Genomics 7, 458-459.

Izpisúa-Belmonte, J.-C., Falkenstein, H., Dollé, P., Renucci, A. and Duboule D. (1991). Murine genes related to the *Drosophila AbdB* homeotic gene are sequentially expressed during development of the posterior part of the body. EMBO J. 10, 2279-2289.

Mackem, S. and Mahon, K. A. (1991). GHox-4.7: A chick homeobox gene expressed primarily in limb buds with limb-type differences in expression. Development 112, 791-806.

Hoxd-13   [m]

Species: *Mus musculus*

Chromosomal location: Chromosome 2, bands D, E

Other names: *Hox-4.8*

Cognate genes: *HOXD13*

Type of homeobox: *Antp*-type, related to *Drosophila Abd-B* gene

Accession number: -

■ Origin and description

The *Hoxd-13* gene was cloned during a walk on the *HOXD* complex (Duboule et al., 1990) by cross-hybridisation with a cognate human probe (Dollé et al., 1991a). This gene is the most 5'-located member of the *HOXD* complex as judged by the presence of the *Evx-2* gene about 8 Kb upstream (Bastian et al., 1992). The *Hoxd-13* homeobox is rather divergent from the *Antp* prototype and can be classified as related to that of the *Abd-B* homeotic gene. However, the *Hoxd-13* homeobox is more divergent, with respect to *Abd-B*, that that of other *Abd-B*-related *Hoxd* genes (from *Hoxd-9* to *Hoxd-12*; Izpisúa-Belmonte et al., 1991; Dollé et al., 1991a).

■ Expression

The expression of the *Hoxd-13* gene was studied by in situ hybridisation during prenatal development. Its expression pattern correlates well with its extreme 5' position within the *HOXD* complex as it is the most posteriorly expressed *Hox* gene described so far. It is expressed starting at midday 9 p.c. in posterior regions of the lateral plate mesoderm, in the condensations of the sclerotome as well as in the neural tube. The anterior limit of expression is detected at the level of prevertebra 30-32 (Dollé et al., 1991a). It is also strongly expressed in the developing genital tubercle and in its derivatives (e.g. the penis) until after birth (Dollé et al., 1991b and in posterior parts of the genito-urinary tractus. In contrast to most Hox genes, *Hoxd-13* is not expressed in the fetal kidneys. In addition to the trunk, *Hoxd-13* expression is detected at high level during limb development (Dollé et al., 1991a). Transcripts are not present in both forelimb and hindlimb lateral plate mesoderm at the budding stage but are detected in the posterior mesoderm of both forelimbs and hindlimbs. Later, the expression domain extends anteriorly, in the most distal regions, so that *Hoxd-13* transcripts are found in most of the distal parts of the autopods. Concomitantly, the signal becomes restricted to the periphery of the condensed material (the perichondrium), with a higher concentration around the distal blastemas of the limbs (e.g in the future metacarpals and phalanges; Dollé et al., 1991a; Duboule, unpublished).

■ Function

The function of the *Hoxd-13* gene was addressed by the production of an insertional mutant via homologous recombination in ES cells. Homozygote mutant mice are fully viable and have defects in the skeletal elements along all three axes, i.e. in the penis, the vertebral column and in the limbs. In adult limbs, a reduction (or deletion) of bony elements is observed in the areas that express *Hoxd-13* during chondrogenesis and ossification, i.e. in all digits of the fore- and hindlimbs (Dollé et al., 1993).

■ References

Dollé, P., Izpisúa-Belmonte, J.-C., Boncinelli, E. and Duboule, D. (1991a). The Hox-4.8 gene is localised at the 5' end of the HOX-4 complex and is expressed at the posterior extremity of the body during development. Mech. Dev. 36, 3-14.
Dollé, P., Izpisúa-Belmonte, J.-C., Tickle, C., Brown, J. and

Duboule, D. (1991b). Hox-4 genes and the morphogenesis of mammalian genitalia. Genes and Dev. 5, 1767-1776.
Dollé, P., Dierich, A., Schimmang, T., Schuhbaur, B., LeMeur, M., Chambon, P. and Duboule D. (1993). Cell 75, 431-441.
Duboule, D., Boncinelli, E., DeRobertis, E., Featherstone, M.S., Lonai, P., Oliver, G. and Ruddle, F. (1990). An update of mouse and human HOX genes nomenclature. Genomics 7, 458-459.
Izpisúa-Belmonte, J.C., Falkenstein, H., Dollé, P., Renucci, A. and Duboule, D. (1991). Murine genes related to the *Drosophila AbdB* homeotic gene are sequentially expressed during development of the posterior part of the body. EMBO J. 10, 2279-2289.

Hx1Af [ar]

Species: *Artemia franciscana*
Chromosomal location: -
Other names: -
Cognate genes: -
Type of homeobox: *Antp*-type
Accession number: X70079

■ Origin and description

Cloned by PCR with degenerate homeobox primers, followed by inverse PCR of 0.5 kb fragment from genomic DNA. Amino acid conservation in the homeodomain identifies this gene as a close relative to the *Antp* class but not a distinct homologue of any known gene.

■ References

Averof, M. and Akam, M. (1993). Hom/Hox genes in a crustacean; implication for the origin of insect and crustacean body plans. Current Biology 3, 73–78.

Hx1Sg [Sg]

Species: *Schistocerca gregaria* (grasshopper)
Chromosomal location: -
Other names: *Sg Dax*
Cognate genes: -
Type of homeobox: *Antp*-type
Accession number: X73982

■ Origin and description

Cloned by low stringency screening of a *Schistocerca gregaria* genomic library using *Drosophila* homeobox probes.

Matches *Antp* class consensus, but sequence within and flanking the homeobox does not identify it as an orthologue of any specific *Drosophila* gene.

■ Expression

Expressed from 10% embryogenesis in the posterior of the germ disc and growing germ band. Expressed later (from 22% embryogenesis) in a subset of neural precursor cells that closely resembles the neural expression of *fushi-tarazu* in *Drosophila*.

■ References

Akam, M., Dawson, I. and Tear, G. (1988). Homoeotic genes and the control of segment identity. Development 104, 123-133.
Dawes, R., Dawson, I., Tear, G. and Akam, M. in preparation.

Kn1 [zm]

Species: *Zea mays*

Chromosomal location: long arm of chromosome 1

Other names: -

Cognate genes: -

Type of homeobox: -

Accession number: X61308

■ Origin and description

Knotted-1 (*Kn1*) was cloned by transposon tagging in maize (Hake et al., 1989). The *Kn1* homeodomain (Vollbrecht et al., 1991) is more similar to that of human RL (Nourse et al., 1990; Kamps et al., 1990) and yeast MATPi (Kelly et al., 1988).

■ Expression

Kn1 was discovered by dominant mutations that affect leaf development (Freeling and Hake, 1985). Clonal analysis identified the inner cell layer of the leaf as the site of mutant gene action (Sinha and Hake, 1990). The wildtype protein is most abundantly expressed in the meristematic tissue of the plant.

■ Genetics, function

All the mutations described are dominant. One is a tandem duplication (Veit et al., 1990), others are caused by insertions in introns (Hake et al., 1989).

■ Remarks

Other homeobox genes have been isolated using the *Kn1* homeobox as a hybridization probe in maize (Vollbrecht et al., 1991) and other plant species. They all contain an extended region of conservation 5' to the homeodomain.

■ References

Freeling, M. and Hake, S. (1985). Developmental genetics of mutants that specify Knotted leaves in maize. Genetics 111, 617-634.
Hake, S., Vollbrecht, E. and Freeling, M. (1989). Cloning Knotted, the dominant morphological mutant in maize using *Ds2* as a transposon tag. EMBO J. 8, 15-22.
Kamps, M.P., Murre, C., Sun, X.-H. and Baltimore, D. (1990). A new homeobox gene contributes the DNA binding domain of the t (1;19) translocation protein in Pre-B ALL. Cell 60, 547-555.
Kelly, M., Burke, J., Smith, M., Klar, A. and Beach, D. (1988). Four mating-type genes control sexual differentiation in the fission yeast. EMBO J. 7, 1537-1547.
Nourse, J., Mellentin, J.D., Galili, N., Wilkinson, J., Stanbridge, E., Smith, S.D. and Cleary, M.L. (1990). Chromosomal translocation t (1;19) results in synthesis of a homeobox fusion mRNA that codes for a potential chimeric transcription factor. Cell 60, 535-545.
Sinha, N. and Hake, S. (1990). Mutant characters of Knotted maize leaves are determined in the innermost tissue layers. Dev. Biol. 141, 203-210.
Veit, B., Vollbrecht, E., Mathern, J. and Hake, S. (1990). A tandem duplication causes the *Kn1-O* allele of Knotted, a dominant morphological mutant of maize. Genetics 125, 623-631.
Vollbrecht, E., Veit, B., Sinha, N. and Hake, S. (1991). The developmental gene Knotted-1 is a member of a maize homeobox gene family. Nature, 350, 241-243.

lab [d]

Species: *Drosophila melanogaster*

Chromosomal location: 84A

Other names: *labial*

Cognate genes: mammalian group 1 *Hox* genes, *XHoxlab-2*, *GHox-lab*

Type of homeobox: *Antp*-class, *lab*-subclass

Accession number: -

■ Origin and description

labial (*lab*) is the most proximal gene of the *Antennapedia*-Complex (ANT-C, Wakimoto and Kaufman, 1981). *lab* was isolated by cross-hybridization to the *Ubx* homeobox (Mlodzik et al., 1988) and independently by a genomic walk from more distal located sequences within the ANT-C (Diederich et al., 1989). The *lab* homeo domain is 67% identical to the Antp homeo domain and it is split by an intron after position 44.

Expression

labial is first expressed at the blastoderm stage in a stripe just anterior to the *Dfd* expression domain in the anlagen of the presumptive intercalary segment and in the most posterior somatic cells. Throughout embryogenesis *lab* expression is restricted to derivatives of the ectodermal and endodermal germ layers. Expression in the ectodermal layer is confined to procephalic structures along the lateral and ventral areas anterior to the cephalic furrow. When segmentation becomes visible, *lab* expression is restricted to the antero-ventral regions of the procephalic lobe. Later it is also detected in the dorsal ridge regions of the head segments. In the endoderm *labial* is expressed in both the anterior and posterior midgut primordia. After fusion of the two primordia *lab* is restricted to a discrete stripe of expression about seven cells in width (Diederich et al., 1989; Mlodzik et al., 1988).

Function

labial appears to be required for normal head involution. The cells that express *lab* in the procephalic lobes are at critical positions with respect to the cellular movements that underly the process of head involution. As a secondary consequence of this failure the salivary glands, the H-piece and the ventral arm of the cephalopharyngeal apparatus are missing. In the adult, as judged from hypomorphs, the maxillary palp and vibrissae region are deleted or disrupted and dorsally the posterior head appears transformed to thoracic like structures (Merrill et al., 1989).

References

Diederich, R.J., Merrill, V.K.L., Pultz, M.A. and Kaufman, T.C. (1989). Isolation, structure and expression of labial, a homeotic gene of the *Antennapedia* complex involved in *Drosophila* head development. Genes Dev. 3, 399-414.

Merrill, V.K.L., Diederich, R.J., Turner, F.R. and Kaufman, T.C. (1989). A genetic and developmental analysis of mutations in *labial*, a gene necessary for proper head formation in *Drosophila melanogaster*. Dev. Biol. 135, 376-391.

Mlodzik, M., Fjose, A. and Gehring, W.J. (1988). Molecular structure and spatial expression of a homeobox gene from the *labial* region of the Antennapedia-complex. EMBO J. 7, 2569-2578.

Wakimoto, B.T. and Kaufman, T.C. (1981). Analysis of larval segmentation in lethal genotypes associated with the *Antennapedia* gene complex in *Drosophila*. Dev. Biol. 81, 51-64.

LH-2 [m,r]

Species: *Mus musculus, Rattus norvegicus*

Chromosomal location: Mouse chromosome 2

Other names: *PB36*

Cognate genes: -

Type of homeobox: LIM-class, related to *Drosophila apterous* homeobox

Accession number: -

Origin and description

A 300bp cDNA sequence of this gene was isolated from murine pre-B cell lines by subtractive hybridization (Yancopoulos et al., 1990). The subtractive cDNA was used as a probe to screen a rat brain cDNA library (Furley et al., 1990) and a mouse pre-B cell cDNA library as described (Yancopoulos et al., 1990) for full length cDNAs. The rat and mouse *LH-2* cDNA are more than 98% identical (Xu et al., submitted). The homeodomain of the LH-2 protein is related to that of other LIM/Homeodomain proteins, most strikingly with that of the *Drosophila apterous* protein (Cohen et al., 1992).

Expression

LH-2 is expressed in most pre-B cell lines and several pre-T cell lines tested, it was not expressed in B and T cell lines representing the later stages of lymphocyte development. *LH-2* was not expressed in any fibroblast, epithelial, macrophage cell lines tested (Xu et al., submitted). Developmental expression of *LH-2* transcript in murine tissues was examined by northern blotting as well as *in situ* hybridization (Xu et al., submitted). *LH-2* transcripts were abundant in the brain with high level present throughout embryonic development and lower levels postnatally. Lower levels of *LH-2* transcripts were also detected in embryonic spinal cord, fetal liver. *LH-2* transcripts were not detected in postnatal spleen, thymus. No expression of *LH-2* was detected in heart, lung, muscle, intestine and kidney. The highest levels of *LH-2* expression in the developing central nervous system were found in the forebrain although expression of *LH-2* was also observed at low level in the midbrain, very low level in hindbrain and spinal cord.

Function

The novel expression pattern and structural characteristics of *LH-2* gene suggest that it encodes a transcriptional regulatory protein involved in the control of cell differentiation in the developing lymphoid and neural cell types. We have defined the DNA binding site of the homeodomain of *LH-2* protein (Xu et al., in preparation) and generated *LH-2* transgenic mouse.

References

Yanopulos, G.D., Oltz, E.M., Rathbun, G., Berman, J.E., Smith, R.K., Lansford, R.D., Rothman, P., Okada, A., Lee, G., Morrow, M., Kaplan, K. and Alt, F. (1990). Coordinately regulated genes that are expressed in discrete stages of B-cell development. Proc. Natl. Acad. Sci. U.S.A. 87, 5759-5763.

Furley, A.M., Morton, S.B., Manalo, D., Karagogeos, D., Dodd, J. and Jessell, T.M. (1990). The axonal glycoprotein TAG-1 is an immunoglobulin superfamily member with neurite outgrowth-promoting activity. Cell 61, 157-170.

Cohen, B., McGuffin, M.E., Pfeifle, C., Segal, D. and Cohen, S.M. (1992). Apterous, a gene required for imaginal disc development in *Drosophila* encodes a member of the LIM family of developmental regulatory proteins. Genes Dev. 6, 715-729.

Xu, Y., Baldassare, M., Fisher, P., Rathbun, G., Oltz, E., Yanopoulos, G.D., Jessell, T.M. and Alt, F. (1992). LH-2: A novel LIM/Homeodomain gene expressed in lymphocytes and in the developing central nervous system. Proc. Natl. Acad. Sci. USA, in press.

Lin-11   [ce]

Species: *Caenorhabditis elegans* (N2)
Chromosomal location: I
Other names: -
Cognate genes: -
Type of homeobox: LIM class
Accession number: X54355

■ Origin and description

lin-11 was first isolated genetically (Trent et al., 1983). Cloning of the gene showed that it contains a homeodomain similar to *mec*-3 and two copies of a cysteine-rich motif termed the LIM-motif (Freyd et al., 1990). The LIM-motif of *lin*-11 has been shown to be an iron-sulfur- and zinc-containing metallodomain (Li et al., 1991).

■ Expression

A single transcript of about 1.7 kb has been observed (Freyd et al., 1990).

■ Genetics, function

Freyd et al., 1983: "The gene *lin*-11 functions in the asymmetric division of the 2° vulval blast cells, which generate part of the hermaphrodite vulva (Ferguson and Horvitz, 1985; Ferguson et al., 1987). Whereas in wild-type animals each of the two 2° cells divides to produce daughter cells that express different fates, in *lin*-11 mutant animals these cells produce daughters that express the same fate (Ferguson et al., 1987). Animals mutant for *lin*-11 are unable to form a functional vulva and therefore cannot lay eggs (Ferguson and Horvitz, 1985). In addition, *lin*-11 males have abnormalities in tail morphology and animals of both sexes are slightly uncoordinated (Ferguson and Horvitz, 1985), suggesting that *lin*-11 has additional functions in *C. elegans* development."

References

Ferguson, E.L. and Horvitz, H.R. (1985). Identification and characterization of 22 genes that affect the vulval cell lineages of the nematode Caenorhabditis elegans. Genetics 110, 17-72.

Ferguson, E.L, Sternberg, P.W. and Horvitz, H.R. (1987). A genetic pathway for the specification of the vulval cell lineages of *Caenorhabditis elegans*. Nature 326, 259-267.

Freyd, C., Kim, S. and Horvitz, R.H. (1990). Novel cysteine-rich motif and homeodomain in the product of the Caenorhabditis elegans cell lineage gene *lin*-11. Nature 344, 876-879.

Li, P.M., Reichert, J., Freyd, G., Horvitz, H.R. and Walsh, C.T. (1991). The LIM region of a presumptive *Caenorhabditis elegans* transcription factor is an iron-sulfur- and zinc-containing metallodomain. Proc. Natl. Acad. Sci. USA 88, 9210-9213.

Trent, C., Tsung, N. and Horvitz, H.R. (1983). Egg-laying defective mutants of the nematode *Caenorhabditis elegans*. Genetics 104, 619-647.

Lox1   [Hm]

Species: *Hirudo medicinalis* (leech)
Chromosomal location: -
Other names: -
Cognate genes: -
Type of homeobox: *Antp*-type
Accession number: -

■ Origin and description

The *Lox1* homeobox was originally cloned from a genomic library by low stringency hybridization with *Scr*. A partial cDNA clone of this gene was later isolated and a probe derived from this cDNA was used to identify a genomic fragment containing the putative translation start. The *Lox1* homeobox is related to *Antp*-like homeoboxes of the ANT-C of *Drosophila* (*Dfd* -47/60 amino acid identity-, *Scr* -49/60-, *ftz* -44/60- and *Antp* -48/60-) and to vertebrate genes like human *HOX2A*, murine *Hoxa-5* frog *Xlhbox4* and zebrafish *ZF-21* (49/60 amino acid identity with each one of these four genes).

■ Expression

Lox1 is detected by antibody staining in a repeated set of neurons in every segmental ganglion of the leech embryonic central nervous system. The *Lox1* neurons appear in bilaterally symmetrical positions that are nearly identical from segment to segment. There is no detectable *Lox1* staining between 4 and 8 days of embryonic development. Expression is first detected late on day 8, when ganglia are forming, and then maintained until the end of embryonic development. The pattern of *Lox1* is therefore reminiscent of the late expression of fly segmentation genes like *ftz*. A *Lox1* mRNA is detected in adult leeches by Northern blot hybridization and RNase protection assays.

■ References

Baumgarten, M.J. Homeobox genes in the leech. Ph. D. Thesis, Columbia University (1990).
Baumgarten, M.J. and Macagno, E.R. (1986). Expression of homeobox-containing genes in the leech *Hirudo medicinalis*. Soc. Neurosci. Abstr. 12, 1122.

Lox2 [Hm]

Species: *Hirudo medicinalis* (leech)
Chromosomal location: -
Other names: -
Cognate genes: *Ubx; abdA*
Type of homeobox: *Antp*-type
Accession number: X17566, codename HIRLOX2G, EMBL HMLOX2G

■ Origin and description

Lox2 was cloned from a genomic library by low stringency hybridization with *Scr*. The DNA fragment isolated corresponds to the 3'-terminal exon of the gene. It contains a 118 amino acid open reading frame that includes a homeobox with high homology to several *Antp*-like homeobox genes. In addition, the homology between *Lox2* and *Ubx* was found to extend 11 amino acids beyond the C-terminus of the homeodomain, in a region that is also conserved between *Ubx* and *abdA*. This short sequence is absent from other homeobox genes and therefore identifies *Lox2* as an *Ubx/abdA* homolog.

■ Expression

In situ hybridization of leech embryos with *Lox2* shows that this gene is expressed in the posterior two-thirds of the developing leech central nervous system in 7-14 day old embryos. The strongest expression is found in midbody ganglia 7 to 21, with somewhat weaker expression in the tail ganglia. In addition, two small clusters of cells express *Lox2* in the posterior margin of midbody ganglion 6, representing the anterior boundary of *Lox2* expression. Weaker signal is also detectable on the remainder of the germinal plate during these stages. Northern blots show that *Lox2* is also expressed in adult leeches.

■ References

Wysocka-Diller, J.W., Aisemberg, G.O., Baumgarten, M., Levine, M. and Macagno, E.R. (1989). Characterization of a homologue of bithorax-complex genes in the leech Hirudo medicinalis. Nature 341, 760-763.

Lox3A, Lox3B, Lox3C, Lox3D [Hm]

Species: *Hirudo medicinalis* (leech)
Chromosomal location: -
Other names: -
Cognate genes: *Htr-A2*
Type of homeobox: *XlHbox8*-type
Accession number: -

■ Origin and description

The three linked homeoboxes *Lox3A-C* were isolated from a genomic library by low stringency hybridization with *Scr*. A closely related fourth member of this family (*Lox3D*) was obtained from an adult cDNA library. In the region of overlap, the four coding sequences were found to be almost identical (96-98% homology at the amino acid level). *Lox3A-D* are very homologous to the gene *Htr-A2* from the leech *Helobdella miserialis* (55/60 amino acid identity, plus short stretches of homology in regions flanking the homeoboxes) and to the *Xenopus* gene *XlH*box8 (52/60).

■ Expression

One or more members of the *Lox3* family are expressed in the gut primordium of 11-day old embryos. *In situ* hybridization detects transverse *Lox3* stripes in midbody segments 3-13 and a diffuse, non-striped pattern posteriorly to midbody segment 13. Postembryonic juvenile leeches exhibit *Lox3* expression in what appears to be the gut epithelium. This is an agreement with *Xenopus* *XlH*box8 expression in endoderm-derived structures. *Lox3* expression has also been detected in embryos, juveniles and adults by northern blot and RNase protection. Towards the end of embryogenesis, *Lox3* expression switches from a 1.7 kb transcript predominant in 11-day old embryos to a 0.8 kb mRNA with higher levels in juveniles.

■ References

Wysocka-Diller, J.W., Aisemberg, G.O.and Macagno, E.R. Manuscript in preparation.

mab-5 [ce]

Species: *Caenorhabditis elegans* (N2)
Chromosomal location: III
Other names: -
Cognate genes: -
Type of homeobox: *Antp*-type
Accession number: M22751

■ Origin and description

mab-5 (male abnormal) has been first isolated genetically (Hodgkin, 1983). Subsequent cloning of the gene showed that it encodes a homeodomain gene of the *Antp*-type and has a hexapeptide upstream of the homeodomain (Costa et al., 1988). This gene is part of the *C. elegans* homeobox cluster (Bürglin et al., 1991; Kenyon and Wang, 1991).

■ Expression

In situ hybridization of L1, L2 and early L3 animals showed that *mab-5* is expressed in the posterior body region of both hermaphrodites and males. Hybridization has been observed in the region of the gonad up to the anus, with the strongest staining towards the posterior (Costa et al., 1988). Analysis of *mab-5-lacZ* fusions in transgenic animals shows weak expression in QL before division, and strong expression in the descendants. No expression is observed in QR and its descendants (Salser and Kenyon, 1992). The *mab-5-lacZ* fusion is first expressed at approximately 180 minutes (halfway through gastrulation) in the posterior region (Cowing and Kenyon, 1992). Some of the staining blast cells are P7 - P12 and V6.

■ Genetics, Function

mab-5 null mutations alter the fates of many ectodermal and mesodermal cells. For most of these cells, *mab-5* mutations appear to cause discrete transformations in cell fate such that cells undergo sequences of division, differentiation, and/or migration characteristic of other cells (Kenyon, 1986, see figure 1 for summary). The affected cells are all located in the posterior body region, and the transformations cause most of the cells to adopt fates specific for their more anterior homologues (Kenyon, 1986). Mosaic analysis indicates that *mab*-5 gene activity is cell-autonomous (Kenyon, 1986). A *mab*-5 gain-of-function mutation (*lin*-21) causes transformations opposite to those of *mab*-5 null alleles in several tissues (Hedgecock et al., 1987; Waring and Kenyon, 1991; Salser and Kenyon, 1992). A well studied case are the sister neuroblasts QL and QR (reviewed in Hedgecock et al., 1987): in wild-type QR and its descendants migrate anteriorly while QL and its descendants migrate posteriorly. In *mab*-5 null mutants,

QL first moves towards the posterior, but its daughters turn around and move to the anterior. In the *mab-5* gain-of-function mutation the QR daughters turn around and move towards the posterior. Inappropriate expression of *mab-5* under heat-shock control causes the descendants of QR to migrate to the posterior (*mab-5* is normally not expressed in QR descendants, Salser and Kenyon, 1992).

■ References

Bürglin, T.R., Ruvkun, G., Coulson, A., Hawkins, N.C., McGhee, J.D., Schaller, D., Wittmann, C., Müller, F. and Waterston, R.H. (1991). Nematode homeobox cluster. Nature 351, 703-703.

Costa, M., Weir, M., Coulson, A., Sulston, J. and Kenyon, C. (1988). Posterior pattern formation in *C. elegans* involves position-specific expression of a gene containing a homeobox. Cell 55, 747-756.

Cowing, D.W. and Kenyon, C. (1992) Expression of the homeotic gene *mab-5* during *Caenorhabditis elegans* embryogenesis Development 116, 484–490.

Hedgecock, E.M., Culotti, J.G., Hall, D.H. and Stern, B.D. (1987). Genetics of cell and axon migrations in *Caenorhabditis elegans*. Development 100, 365-382.

Hodgkin, J. (1983). Male phenotypes and mating efficiency in *Caenorhabditis elegans*. Genetics 103, 43-64.

Kenyon, C. (1986). A gene involved in the development of the posterior body region of *C. elegans*. Cell 46, 477-487.

Kenyon, C. and Wang, B. (1991). A cluster of *Antennapedia*-class homeobox genes in a nonsegmented animal. Science 253, 516-517.

Salser, S.J. and Kenyon, C. (1992). Activation of a *C. elegans Antennapedia* homologue in migrating cells controls their direction of migration. Nature 355, 255-258.

Waring, D.A. and Kenyon, C. (1991). Regulation of cellular responsiveness to inductive signals in the developing *C. elegans* nervous system. Nature 350, 712-715.

mec-3 [ce]

Species: *Caenorhabditis elegans* (N2)
Chromosomal location: IV
Other names: -
Cognate genes: -
Type of homeobox: LIM-class
Accession number: M20224

■ Origin and description

mec-3 mutations were first identified in screens for touch-insensitive mutants (Chalfie and Sulston, 1981). Subsequent cloning showed that it was a homeobox containing gene (Way and Chalfie, 1988), and it was later recognized to contain two copies of the LIM-motif (Freyd et al., 1990). Sequence comparison to *C. vulgarensis mec-3* identified conserved sequence elements and identified the 5' exons of *mec-3* correctly (Way et al., 1991).

Expression

mec-3 is expressed in several sensory neurons, as assayed by expression of a *mec-3*-lacZ fusion in transgenic animals (Way and Chalfie, 1989). These cells are the six touch receptor neurons (PLML, PLMR, ALML, ALMR, AVM, PVM), which mediate the response to gentle touch, and the FLPL/R and PVDL/R neurons. The PVDs mediate response to harsh mechanical stimuli, and the FLPs have an ultra-structure suggestive of a mechanoreceptor (Way and Chalfie, 1989). *unc*-86 is necessary for all *mec-3*-lacZ expression, while *mec-3* activity itself appears to be required for maintained expression of the *mec-3*-lacZ fusion (Way and Chalfie, 1989), indicating that *mec-3* is autoregulatory. Sequences important for the regulation of *mec-3* have been mapped to the promoter region, and MEC-3 and UNC-86 protein footprint in this promotor (Way et al., 1991, Xue et al., 1992).

Genetics, function

The *mec-3* gene is essential for proper differentiation of the set of six touch receptor neurons. In mutants lacking *mec-3* activity, the touch receptors express none of their unique differentiated features and appear to be transformed into other types of neurons (Way and Chalfie, 1988). *mec-3* is also needed for PVD function: The PVD neurons no longer mediate a response to harsh mechanical stimuli in *mec-3* mutants (Way and Chalfie, 1989).

References

Chalfie, M. and Sulston, J. (1981). Developmental genetics of mechanosensory neurons of *Caenorhabditis elegans* . Dev. Biol. 82, 358-370.

Freyd, G., Kim, S. and Horvitz, R.H. (1990). Novel cysteine-rich motif and homeodomain in the product of the *Caenorhabditis elegans* cell lineage gene *lin*-11. Nature 344, 876-879.

Way, J.C. and Chalfie, M. (1988). *mec*-3, a homeobox-containing gene that specifies differentiation of the touch receptor neurons in *C. elegans*. Cell 54, 5-16.

Way, J.C. and Chalfie, M. (1989). The *mec*-3 gene of *Caenorhabditis elegans* requires its own product for maintained expression and is expressed in three neuronal cell types. Genes Dev. 3, 1823-1833.

Way, J.C., Wang, L., Run, J.-Q., Wang, A. (1991). The *mec*-3 gene contains *cis*-acting elements mediating positive and negative regulation in cells produced by asymmetric cell division in *Caenorhabditis. elegans*. Genes Dev. 5, 2199-2211.

Xue, D., Finney, M., Ruvkun, G. and Chalfie, M. (1992). Regulation of the *mec*-3 gene by the *C. elegans* homeoproteins UNC-86 and MEC-3. EMBO J. 11, 4969-4979.

MHox [m]

Species: *Mus musculus*
Chromosomal location: Distal region of chromosome 1
Other names: *PHox; PHox 1; K-2*
Cognate genes: *PHox-1* (human)
Type of homeobox: *paired*-type
Accession number: L06502

Origin and description

The cDNA was originally cloned from a gt11 expression library constructed from differentiated C_2C_{12} myogenic cell mRNA by virtue of its binding affinity to a DNA element found in the mouse muscle creatine kinase enhancer (Cserjesi et al., 1992). Sequence analysis indicated the central region of the longest open reading frame had a very high degree of homology with the homeobox of *S8* (Opstelten et al., 1991). The homeoboxes of both *MHox* and *S8* are most similar to the *paired* class of homeoboxes but both contain the amino acid substitution of a glutamine for a serine within the DNA binding helix. By rescreening of the library with the original clone, we obtained a number of clones encoding an alternate carboxyl end. Comparison with the genomic clone, has shown these cDNA clones are products of an alternate splicing event (Martin, personal communication). This alternate splice variant has recently been reported by Kern et al. (1992). The human homologue of *MHox* has also recently been cloned by Grueneberg et al. (1992).

Expression

Developmental distribution was examined by *in situ* hybridization using a *MHox* specific probe. The message is expressed primarily in mesodermally derived cells of the embryo with the highest levels of expression in appendicular structures. By 9 d.p.c., high levels of expression are seen in visceral arches. During development, the highest levels of expression are found in developing frontonasal processes and limbs. High levels of expression are also found in the dermamyotome of the somites, the dorsal aorta, dermis, diaphragm, and in the region of condensing cartilage. Low levels of expression are found in most areas containing mesenchyme cells. There is no expression in the central nervous system or viscera during embryogenesis. In adult animals, *MHox* RNA is present at low levels in heart, skeletal muscle, vascular smooth muscle and uterus. Approximately equal ratios of the two alternatively spliced species are found in adult tissues and cultured cell lines examined to date (Cserjesi, personal communication). Kem et al. (1992) indicate that the transcript encoding the larger protein is the predominant species during embryonic development.

References

Cserjesi, P., Lilly, B., Bryson, L., Wang, Y., Sassoon, D.A., Olson, E.N. (1992). MHox: a mesodermally restricted homeodomain protein that binds an essential site in the creatine kinase enhancer. Development 115, 1087-1101.

Grueneberg, D.A., Natesan, S., Alexandre, C., Gilman, M.Z. (1992). Human and *Drosophila* homeodomain proteins that enhance the DNA-binding activity of serum response factor. Science 257, 1089-1095.

Kern, M.J., Witte, D.P., Valerius, M.T., Aronow, B.J., Potter, S.S. (1992). A novel murine homeobox gene isolated by a tissue specific PCR cloning strategy. Nucl. Acid. Res. 20, 5189-5195.

Opstelten, D.-J., Vogels, R., Robert, B., Kalkhoven, E., Zwartkruis, de L., Destree, O.H., Deschamps, J., Lawson, K.A. and Meijlink, F. (1991). The mouse homeobox gene, S8, is expressed during embryogenesis predominantly in mesenchyme. Mech. Dev. 34, 29-42.

Mix-1 [x]

Species: *Xenopus laevis*
Chromosomal location: -
Other names: -
Cognate genes: -
Type of homeobox: *paired*-type
Accession number: -

■ Origin and description

Isolated as rapidly inducible by mesoderm inducers (Activin A).

■ Expression

As far as known, *Mix*-1 is only expressed during late blastula and gastrula stages. RNA is localized in vegetal hemisphere, primarily in marginal zone.

■ References

Rosa, F. (1989). *Mix*.1, a homeobox mRNA inducible by mesoderm inducers, is expressed in the presumptive endodermal cells of Xenopus embryos. Cell 57, 965-974.

Mocut [m]

Species: *Mus musculus*
Chromosomal location: 5G
Other names: -
Cognate genes: *CDP, Clox*
Type of homeobox: Related to the *Drosophila cut* gene
Accession number: -

■ Origin and description

An incomplete cDNA was cloned by screening an expression library from mouse N2a neuroblastoma cells with a homeodomain-binding sequence identified in the promoter region of the NCAM gene (Hirsch et al., 1991). Rescreening the same and a commercial mouse brain library with the original clone extended the cDNA sequence up to a putative initiation methionine. The composite sequence contains an open reading frame of 3993 nt. It contains a homeobox of the type found in the *Drosophila cut* gene (Blochlinger et al., 1988) and three internal repeats which are also conserved in the *Drosophila* gene. Likely species homologues are human *CDP* (Neufeld et al., 1992) (first described on the basis of functional criteria by Barberis et al., 1987) and dog *Clox* (Andres et al., 1992).

■ Expression

On RNA from N2a cells, a *Mocut* probe detects transcripts of 6 and 8 kb. On the basis of Northern blot analysis, *Mocut* is expressed in the mouse embryo from 7.5 day p.c. onwards and in adult brain, heart, muscle, intestine, thymus, spleen and adrenal medulla.

■ References

Andres, V., Nadal-Ginard, B. and Mahdavi, V. (1992). *Clox*, a mammalian homeobox gene related to *Drosophila cut* encodes DNA-binding regulatory proteins differentially expressed during development. Development 116, 321-334.

Barberis, A., Superti-Furga, B. and Busslinger, M. (1987). Mutually exclusive interaction of the CCAAT-binding factor and a displacement protein with overlapping sequences of a histone gene promoter. Cell 50, 347-359.

Blochlinger, K., Bodmer, R., Jack, J., Jan, L.Y. and Jan, Y.N. (1988). Transformation of sensory organ identity by ectopic expression of *Cut* in *Drosophila*. Nature 333, 629-630.

Hirsch, M.R., Valarché, I., Deagostini-Bazin, H., Pernelle, C., Joliot, A. and Goridis, C. (1991). An upstream regulatory element of the NCAM promoter contains a binding site for homeodomains. FEBS Lett. 287, 197-202.

Neufeld, E.J., Skalnik, D.G., Lievens, P.M.J. and Orkin, S.H. (1992). Human CCAAT displacement protein is homologous to the *Drosophila* homeoprotein *cut*. Nature Genetics 1, 50-55.

Mox-1 [m]

Species: *Mus musculus*
Chromosomal location: 11
Other names: -
Cognate genes: -
Type of homeobox: *Mox*-type
Accession number: Z15103

■ Origin and description

Mox-1 was isolated from a murine 8.5 days post coitum (d.p.c.) cDNA library using a partial, homeobox-containing cDNA probe from the frog endoderm specific gene *XlHbox 8*. The homeodomain of *Mox-1* is of the *Antp* family but defines a novel subclass together with a related gene, *Mox-2*. The genomic locus for *Mox-1* has been isolated from C3H and 129 murine genomic libraries and is comprised of three exons. A full characterization of *Mox-1* and *Mox-2* is described in Candia et al. (1992). PCR clones for a *Xenopus* homolog of *Mox-1* (*XMox-1*) have been isolated and will assist in obtaining full length cDNAs.

■ Expression

Northern analysis of *Mox-1* reveals a single 2.1 kb transcript beginning in pools of 6.5-7.0 d.p.c. mouse embryos and continues through 9.5 d.p.c., the latest stage examined by northerns. *In situ* hybridization detects abundant message for *Mox-1* at approximately 7.25 d.p.c., before a head fold is seen. The onset of *Mox-1* expression is concomitant with, or may even precede, sequential deployment of the clustered *Hox* genes. Expression at 7.25 d.p.c. is restricted to the mesodermal germ layer posterior to the presumptive cardiac mesoderm, and thus contrasts the genes of the *Hox* clusters, which are expressed in neuroectoderm and mesoderm. At 8.5 d.p.c. expression of *Mox-1* is in the entire somite, lateral plate, and intermediate mesoderm, as well as cranial mesoderm anterior to the somites (somitomeric mesoderm), and in posterior presomitic mesoderm equivalent to the length of about 2 somites. The anterior limit of *Mox-1* expression is coincident with that of *Hoxb-1*, *Motch*, and fibroblast growth factor receptor 1 (Conlon, personal communication). Expression of *Mox-1* continues through 15.5 d.p.c. in mesodermal/somitic tissues and in a number of neural crest derived foci at specific muscle attachment sites in the head. Because transcript levels of *Mox-1* decrease with gestation and the differentiation state of tissues, expression of *Mox-1* after 15.5 d.p.c. has been difficult to detect. Immunohistochemical studies with polyclonal antibodies directed against a *Mox-1* fusion protein have confirmed the *in situ* analyses. Details and further characterization can be found in Candia et al. (1992).

■ References

Candia, A.F., Hu, J., Crosby, J., Lalley, P.A., Noden, D., Nadeau, J.H. and Wright, C.V.E. (1992). *Mox-1* and *Mox-2* define a novel homeobox gene subfamily and are differentially expressed during early mesodermal patterning in mouse embryos. Development 116, 1123–1136.

Mox-2 [m]

Species: *Mus musculus*
Chromosomal location: 12
Other names: -
Cognate genes: -
Type of homeobox: *Mox*-type
Accession number: Z16406

■ Origin and description

Mox-2 was isolated from a murine 8.5 days post coitum (d.p.c.) cDNA library using a partial, homeobox-containing probe for the frog gene *XlHbox 8*. The homeodomain of *Mox-2* is of the Antp family but defines a novel subclass together with another gene, *Mox-1*. The genomic locus for *Mox-2* has been isolated from C3H and 129 murine genomic libraries and is comprised of three exons. Characterization of *Mox-2* (and *Mox-1*) is described in Candia et al. (1992). The genomic locus and a partial cDNA for the Xenopus homolog of *Mox-2* (*XMox-2*) has been isolated. A rat homolog of *Mox-2* had recently been cloned from an adult vascular smooth muscle library (Walsh, personal communication).

■ Expression

In situ analysis shows that *Mox-2* is expressed at low levels, beginning at 8.0-8.5 d.p.c., in the entire epithelial somite. Expression in other embryonic or extraembryonic tissues is not detected. By 9.0 d.p.c. expression of *Mox-2* had become restricted to the sclerotome of the somite. Expression continues in sclerotomally derived tissues through 10.5 d.p.c. at which point detection of *Mox-2* transcripts becomes difficult. By comparing *in situ* analyses, *Mox-2* is expressed at lower levels than *Mox-1*, the other member of this mesoderm-specific gene family.

■ References

Candia, A.F., Hu, J., Crosby, J., Lalley, P.A., Noden, D., Nadeau, J.H. and Wright, C.V.E. (1992). *Mox-1* and *Mox-2* define a novel homeobox gene subfamily and are differentially expressed during early mesodermal patterning in mouse embryos. Development 116, 1123–1136.

msh [Hv]

Species: *Chlorohydra viridissima (C.v)*
Chromosomal location: -
Other names: -
Cognate genes: -
Type of homeobox: *msh*-type
Accession number: X64629

■ Origin and description

This gene was isolated from a cDNA library by cross hybridization with a 50 guessmer encoding the third helix of the *Antp* homeodomain according to the hydra codon usage.

■ Expression

msh expression, measured by quantitative PCR, does not significantly vary during head regeneration of a *C.v.* multiheaded mutant.

■ References

Schummer, M., Scheurlen, I., Schaller, C. and Galliot, B. (1992). HOM/HOX homeobox genes are present in hydra (*Chlorohydra viridissima*) and are differentially expressed during regeneration. EMBO J. 11, 1815-1823.

msh [Ci]

Species: *Ciona intestinalis* (ascidian)
Chromosomal location: -
Other names: -
Cognate genes: -
Type of homeobox: *msh*-type
Accession number: M38581

■ Origin and description

Homeobox region only, cloned from genomic DNA via PCR using degenerate primers for *msh*-like genes (Holland, 1991).

■ References

Holland, P.W.H. (1991). Cloning and evolutionary analysis of *msh*-like homeobox genes from mouse, zebrafish and ascidian. Gene 98, 253-257.

msh [d]

Species: *Drosophila melanogaster*
Chromosomal location: 99B
Other names: -
Cognate genes: -
Type of homeobox: *NK*-type
Accession number: -

■ Origin and description

This gene was isolated together with *engrailed* (*en*) (Fjose et al., 1985) by cross-homology with the *Ultrabithorax* (*Ubx*) homeobox (McGinnis et al., 1984). It is most closely related to the *NK*-class of homeoboxes (Kim and Nierenberg, 1989; see also *bap* and *Nkx*) and *msx* (Robert et al., 1989).

■ Expression

The transcripts of the gene are present throughout embryogenesis and exhibit a complex type of spatial expression pattern. During early gastrulation stages, transcripts accumulate as seven irregular patches located on the lateral sides of the embryo. At the early germ band elongation stage, cells between the seven patches become labelled, giving rise to a very characteristic symetrical "sledge-shaped" labeling in the lateral ectoderm. At full germ band extension, a complex periodic pattern is seen, including expression in a subset of neuroblasts, in some mesodermal cells and in addition in the stomodeal and proctodeal regions. In late embryos, the msh transcripts are detected in the pharyngeal, the somatic and the visceral musculatures and also in the CNS.

■ References

Fjose, A., McGinnis, W.J. and Gehring, W.J. (1985). Isolation of a homeobox containing gene from the engrailed region of *Drosophila* and the spatial distribution of its transcripts. Nature 313, 284-289.

Kim, Y. and Nierenberg, M. (1989). *Drosophila* NK-homeobox genes. Proc. Natl. Acad. Sci. U.S.A. 86, 7716-7720.

McGinnis, W., Garber, R.L., Wirz, J., Kuroiwa, A. and Gehring, W.J. (1984). A homologous protein-coding sequence in *Drosophila* homeotic genes and its conservation in other metazoans. Cell 37, 403-408.

Robert, B., Sassoon, D., Jacq, B., Gehring, W. and Buckingham, M. (1989). Hox-7, a mouse homeobox gene with a novel pattern of expression during embryogenesis. EMBO J. 8, 91-100.

msh [dO]

Species: *Drosophila melanogaster* (Oregon R)
Chromosomal location: -
Other names: -
Cognate genes: -
Type of homeobox: *msh*-type
Accession number: M38582

■ Origin and description

Homeobox region only, cloned from genomic DNA via PCR using degenerate primers for *msh*-like genes (Holland, 1991).

■ References

Holland, P.W.H. (1991). Cloning and evolutionary analysis of *msh*-like homeobox genes from mouse, zebrafish and ascidian. Gene 98, 253-257.

msh A [zf]

Species: *Brachydanio rerio* (zebrafish)
Chromosomal location: -
Other names: -
Cognate genes: -
Type of homeobox: *msh*-type
Accession number: M38578

■ Origin and description

Homeobox region only, cloned from genomic DNA via PCR using degenerate primers for *msh*-like genes (Holland, 1991).

■ References

Holland, P.W.H. (1991). Cloning and evolutionary analysis of *msh*-like homeobox genes from mouse, zebrafish and ascidian. Gene 98, 253-257.

msh B [zf]

Species: *Brachydanio rerio* (zebrafish)
Chromosomal location: -
Other names: -
Cognate genes: -
Type of homeobox: *msh*-type
Accession number: M38579

■ Origin and description

Homeobox region only, cloned from genomic DNA via PCR using degenerate primers for *msh*-like genes (Holland, 1991).

■ References

Holland, P.W.H. (1991). Cloning and evolutionary analysis of *msh*-like homeobox genes from mouse, zebrafish and ascidian. Gene 98, 253-257.

msh C [zf]

Species: *Brachydanio rerio* (zebrafish)
Chromosomal location: -
Other names: -
Cognate genes: -
Type of homeobox: *msh*-like
Accession number: M38580

■ Origin and description

Homeobox region only, cloned from genomic DNA via PCR using degenerate primers for *msh*-like genes (Holland, 1991).

■ References

Holland, P.W.H. (1991). Cloning and evolutionary analysis of *msh*-like homeobox genes from mouse, zebrafish and ascidian. Gene 98, 253-257.

msh-2 [d]

Species: *Drosophila melanogaster*
Chromosomal location: 93E1-3 (chromosome 3R)
Other names: *tinman, NK-4*
Cognate genes: -
Type of homeobox: -
Accession number: -

■ Origin and description

This gene had been isolated by PCR using degenerate primers derived from the homeobox of the *cut* gene (Bodmer et al., 1990), and independently by cross-homology to homeobox-derived oligonucleotides (Kim and Nirenberg,1989) and by chromosome walking (M. Frasch, unpublished). The *tinman* homeodomain is rather divergent from other homeodomains (30%-50%), the closest relatives are homeodomains of genes expressed in the mesoderm (59% amino acid homology to the *Drosophila NK-2* and *NK-3* genes). The *tinman* gene is located about 7 kb proximal to *NK-3* on the right arm of chromosome 3.

■ Expression

tinman expression starts at late blastoderm stage in the mesoderm primordia of the body segments and a group of cells near the future stomodeum. Expression in the body segments is dependent on *twist*, a zygotic determinant of mesoderm. At mid-germband extended stage the mesoderm extends dorsally along the body wall and *tinman* expression becomes restricted to the internal dorsal regions - the presumptive visceral mesoderm. Soon thereafter expression becomes further restricted to the dorsal margin of the mesoderm - the heart precursor cells - where expression persists at least until the heart is differentiated (Bodmer et al., 1990).

■ Function

The analysis of deficiencies of the 93 region which includes the *tinman* gene showed no heart and visceral mesoderm formation (Bodmer et al., 1990). Recently, a small deficiency and a point mutant for *tinman* (Df(3R)GC14 and EC40, respectively, Mohler and Pardue, 1984) have been analyzed (R. Bodmer, unpublished). EC40 has a stop codon in the recognition helix of the homeodomain and its muscle phenotype is indistinguishable from Df(3R)GC14. Such *tinman* null mutant embryos do not form an embryonic heart and no evidence for the appearance of cardiac precursor cells has been found. In addition, *tinman* mutants form visceral muscles and their precursors in the midgut region, and the characteristic midgut constrictions are absent (but visceral muscles around the hindgut appear normal). Somatic muscle formation appears relatively unaffected and minor abnormalities may be due to midgut malformations. This suggests that even though *tinman* is expressed initially in all the mesodermal cells of the body segments *tinman* is predominantly required for the specification of the heart and visceral primorida. It thus appears that *tinman* is involved in a first major subdivision of the mesoderm, namely the specification of the visceral mesoderm (including the precursors for the heart) as opposed to the somatic mesoderm.

■ References

Bodmer, R., Jan, L.Y. and Jan, Y.N. (1990). A new homeobox-containing gene, *msh-2*, is transiently expressed early during mesoderm formation in *Drosophila*. Development 110, 661-669.

Kim, S. and Nirenberg, M. (1989). *Drosophila* NK-homeobox genes. PNAS 86, 7716-7720.

Mohler, J. and Pardue, M.-L. (1984). Mutational analysis of the region surrounding the 93D heat shock locus of *Drosophila melanogaster*. Genetics 106, 249-265.

Species: *Gallus gallus* (chicken)
Chromosomal location: -
Other names: *Ghox*-7
Cognate genes: *msx-1*, *Xhox-7.1*
Type of homeobox: *msh*-type
Accession number: X61922

■ Origin and description

This cDNA was isolated from a total chick cDNA library (stages 19-23; Robert et al. 1991). It is the chick homolog to the mouse *msx-1* (Robert et al. 1989; Hill et al. 1989), according to its sequence and to its pattern of expression during development.

■ Expression

In the chick, *Msx-1* is expressed in the roof plate of the neural tube, and in the neural crest, the visceral arches and the fronto-nasal territories; in the endocardial cushion, that will form the atrio-ventricular valves, *Msx-1* is expressed in the endothelium and in the mesenchymal cells that derive from it; expression takes place in the lateral mesoderm. In the limb bud, expression is observed in the apical ectodermal ridge from the time it appears; in the mesenchyme, transcript distribution is originally widespread throughout the limb bud, but becomes restricted at stage 20 to the region which underlies the ridge. In general, *Msx-1* is expressed in regions of interaction between ectoderm and mesoderm that will result in the formation of a progress zone, like the limb bud, the mandibular arch and the genital eminence.

■ References

Hill, R.E., Jones, P.F., Rees, A.R., Sime, C.M., Justice, M.J., Copeland, N.G., Jenkins, N.A., Graham, E. and Davidson, D.R. (1989). A new family of mouse homeobox-containing genes: molecular structure, chromosomal location, and developmental expression of *Hox*-7.1. Genes Dev. 3, 26-37.

Robert, B., Sassoon, D., Jacq, B., Gehring, W. and Buckingham, M. (1989). *Hox*-7, a mouse homeobox gene with a novel pattern of expression during embryogenesis. EMBO J. 8, 91-100.

Robert, B., Lyons, G., Simandl, B.K., Kuroiwa, A. and Buckingham, M. (1991). The apical ectodermal ridge regulates *Hox*-7 and *Hox*-8 gene expression in developing chick limb bud. Genes Dev. 5, 2363–2374.

msx-1   [m]

Species: *Mus musculus*
Chromosomal location: -
Other names: *Hox-7.1, Hox-7*
Cognate genes: -
Type of homeobox: *msh*-type
Accession number: -

■ Origin and description

This gene was isolated by crosshybridization with the Drosophila *msh* gene and was shown to be expressed in areas of apical growth during fetal development (Hill et al., 1989; Robert et al., 1989; Monaghan et al., 1991). The homeobox was also cloned from genomic DNA via PCR using degenerate primers for *msh*-like genes (Holland, 1991).

■ References

Hill, R.E., Jones, P.F., Rees, A.R., Sime, C.M., Justice, M.J., Copeland, N.G., Jenkins, N.A., Graham, E. and Davidson, D.R. (1989). A new family of mouse homeobox-containing genes: molecular structure, chromosomal location, and developmental expression of *Hox*-7.1. Genes Dev. 3, 26-37.

Holland, P.W.H. (1991). Cloning and evolutionary analysis of *msh*-like homeobox genes from mouse, zebrafish and ascidian. Gene 98, 253-257.

Monaghan, A.P., Davidson, D.D., Sime, C., Graham, E., Baldock, R., Bhattacharya, S.S. and Hill, R.E. (1991). The *msh*-like homeobox genes define domains in the developing eye. Development 112, 1053-1061.

Robert, B., Sassoon, D., Jacq, B., Gehring, W. and Buckingham, M. (1989). *Hox*-7, a mouse homeobox gene with a novel pattern of expression during embryogenesis. EMBO J. 8, 91-100.

msx-2   [c]

Species: *Gallus gallus*
Chromosomal location: -
Other names: *Hox-8, Ghox-8,*
Cognate genes: *msx-1, Quox-7, Xhox-7.1*
Type of homeobox: *msh*-type
Accession number: X62541

■ Origin and description

This cDNA was isolated from a total chick cDNA library (stages 19-23; Robert et al. 1991). It is the chick homolog to the mouse *Msx-2* (Davidson et al. 1991; Monaghan et al. 1991), according to its sequence and to its pattern of expression during development.

■ Expression

In the chick, *Msx-2* is expressed in the roof plate of the neural tube, in its most external side, and in the neural crest, the visceral arches and the fronto-nasal territories. In the endocardial cushion, that will form the atrio-ventricular valves, *Msx-2* is expressed in the myocardial part of the cushion. Expression is high in the lateral mesoderm. In the limb bud, *Msx-2* expression is similar to that of *Hox*-7 except that it is prominent in the apical ridge, and weaker in the underlying mesenchyme. Furthermore, *Msx-2* transcripts do not show a uniform antero-posterior distribution in the mesoderm, their concentration being highest on the anterior side, as if the gene was influenced by the polarizing zone. *Msx-2* is expressed at many places where ecto-mesodermal interactions result in the formation of a progress zone.

■ References

Davidson, D.R., Crawley, A., Hill, R.E. and Tickle, C. (1991). Position-dependent expression of two related homeobox genes in developing vertebrate limbs. Nature 352, 429-431.

Monaghan, A.P., Davidson, D.D., Sime, C., Graham, E., Baldock, R., Bhattacharya, S.S. and Hill, R.E. (1991). The *msh*-like homeobox genes define domains in the developing eye. Development 112, 1053-1061.

Robert, B., Lyons, G., Simandl, B.K., Kuroiwa, A. and Buckingham, M. (1991). The apical ectodermal ridge regulates *Hox*-7 and *Hox*-8 gene expression in developing chick limb bud. Genes Dev. 5, 2363–2374.

msx-2   [m]

Species: *Mus musculus*
Chromosomal location: -
Other names: *Hox-8.1, Hox-8*
Cognate genes: *msx-2, Quox 7*
Type of homeobox: *msh*-type
Accession number: M38576

■ Origin and description

msx-2 was isolated using the *msh* probe (Hill et al., 1989) or by PCR (Holland, 1991). Expression was repeated during development of the eyes and teeth (Mackenzie et al., 1992; Monaghan et al., 1991) as well as in other placed of epithelial-mesenchymal interactions.

■ References

Hill, R.E., Jones, P.F., Rees, A.R., Sime, C.M., Justice, M.J., Copeland, N.G., Jenkins, N.A., Graham, E. and Davidson, D.R. (1989). A new family of mouse homeobox-containing genes: molecular structure, chromosomal location, and developmental expression of Hox-7.1. Genes Dev. 3, 26-37.

Holland, P.W.H. (1991). Cloning and evolutionary analysis of msh-like homeobox genes from mouse, zebrafish and ascidian. Gene 98, 253-257.

MacKenzie, A., Ferguson, M. W. J. and Sharpe, P. T. (1992). Expression patterns of the homeobox gene Hox-8 in the mouse embryo suggest a role in specifying tooth initiation and shape. Development 115, 403-420.

Monaghan, A.P., Davidson, D.D., Sime, C., Graham, E., Baldock, R., Bhattacharya, S.S. and Hill, R.E. (1991). The msh-like homeobox genes define domains in the developing eye. Development 112, 1053-1061.

msx-2 [qu]

Species: *Coturnix coturnix japonica* (quail)

Chromosomal location: -

Other names: *Quox-8; Quox-7*

Cognate genes: *msx-2; Ghox-8*

Type of homeobox: *msh*-type

Accession number: M37164

■ Origin and description

This gene was isolated from a cDNA library prepared from embryonic quail limb bud, by cross-hybridization with a fragment of a mouse related gene: *msx-1* (Takahashi et Le Douarin, 1990). The sequence comparison with *msx-1* (Robert et al., 1989; Hill et al., 1989) revealed a well conserved domain containing the homeobox and its 5′ and 3′ flanking regions. The homeobox shows remarkable homology with the msh homeobox of *Drosophila*. Comparison with new members of the *msh*-like family, reveals that this quail gene is in fact the homolog of the murine gene *msx-2* (Monaghan et al., 1991) and the chick gene *Ghox 8* (Coelho et al., 1991).

■ Expression

The expression of *msx-2* has been studied by *in situ* hybridization during embryonic development, from 8 somites to 5 days of incubation (E5). In the trunk region, *msx-2* is transcribed in the neural folds before the closure of the neural tube, then in a subpopulation of migrating neural crest cells. Then, and up to E5, the expression is localized in the very medio-dorsal part of the neural tube, the overlying mesenchyme and ectoderm, along the entire length of the spinal cord. *msx-2* is also expressed in the limb buds and the somatopleura derivatives. In the cephalic region, at E3, the branchial arches, which are neural crest derivatives, are strongly positive for *msx-2* expression in their medio-ventral part. The expression in the mandibular arches has been studied in an organ-culture *in vitro* system; it has been shown to be strictly dependent upon the presence of the epithelium. The mandibular epithelium can be successfully replaced by others but not all epithelia. *msx-2* expression seems well correlated with normal later differentiation of the skeletal mandibular derivatives; cartilage and membrane bone (Takahashi and Le Douarin, 1991).

■ References

Coelho, C., Sumoy, L., Rodgers, B., Davidson, D., Hill, R., Upholt, W. and Kosher, R. (1991). Expression of the chicken homeobox-containing gene GHox 8 during embryonic chick limb development. Mech. Dev. 34, 143-154.

Hill, R., Jones, P., Rees, A., Sime, C., Justice, M., Copeland, N., Jenkins, N., Graham, E. and Davidson, D. (1989). A new family of mouse homeo box-containing genes: molecular structure, chromosomal location, and developmental expression of Hox-7.1. Genes Dev. 3, 26-37.

Monaghan, A., Davidson, D., Sime, C., Graham, E., Baldock, R., Bhattacharya, S. and Hill, R. (1991). The msh-like homeobox genes define domains in the developing vertebrate eye. Development 112, 1053-1061.

Robert, B., Sassoon, D., Jacq, B., Gehring, W. and Buckingham, M. (1989). Hox 7, a mouse homeobox with a novel pattern of expression during embryogenesis. EMBO J. 8, 91-100.

Takahashi, Y. and Le Douarin, N. (1990). cDNA cloning of a quail homeobox gene and its expression in neural crest-derived mesenchyme and lateral plate mesoderm. Proc. Natl. Sci. USA 87, 7482-7486.

Takahashi, Y., Bontoux, M. and Le Douarin, N. (1991). Epithelio-mesenchymal interactions are critical for Quox 7 expression and membrane bone differentiation in the neural crest derived mandibular mesenchyme. EMBO J. 10, 2387-2393.

msx-3 [m]

Species: *Mus musculus*

Chromosomal location: -

Other names: *Msh3*

Cognate genes: -

Type of homeobox: *msh*-type

Accession number: M38577

■ Origin and description

Homeobox region only, cloned from genomic DNA via PCR using degenerate primers for *msh*-like genes (Holland, 1991).

■ Reference

Holland, P.W.H. (1991). Cloning and evolutionary analysis of msh-like homeobox genes from mouse, zebrafish and ascidian. Gene 98, 253-257.

nkch4 [d]

Species: *Drosophila melanogaster*
Chromosomal location: right arm of the chromosome 3 (93D9-E2 region)
Other names: -
Cognate genes: -
Type of homeobox: Related to the human *HOX11* (*tcl3*) oncogene
Accession number: Z11704

■ Origin and description

This gene was isolated from *Drosophila* genomic library screened with a degenerate oligonucleotide probe derived from the microsequence of the GEBF-1 protein (Jagla et al., 1992). nkch4 maps in the proximal region of the *NK*-homeobox gene cluster, and like *NK-1* contains an intron between codons for the homeodomain amino acid residues 44 (Glu) and 45 (Val). Third helix of the *nkch4* homeodomain differs from all other homeodomains at several conserved residues known to be important for DNA recognition.

■ Expression

Low levels of *nkch4* transcripts were detected during late stages of embryogenesis as well as in third instar larvae and pupae. In late embryos *nkch4* is expressed in the developing central nervous system (CNS).

■ References

Jagla, K., Georgel, Ph., Bellard, F., Dretzen, G. and Bellard, M. (1993). nkch4, a novel homeobox gene from the *Drosophila* 93E region. Gene 127, 165–171.

Nkx-2.2 [m]

Species: *Mus musculus*
Chromosomal location: Chromosome 2 (Band 2H)
Other names: -
Cognate genes: -
Type of homeobox: *NK-2*-type
Accession number: -

■ Origin and description

This gene was isolated by low stringency hybridization with the homeobox of *TTF-1* (Guazzi et al., 1989). It is a member of the murine *Nkx-2* family, since it contains both a *NK2* type homeobox (Kim and Nirenberg, 1989) and an additional conserved region, encoding for a 17 amino acids long peptide (Price et al., 1992).

■ Expression

Nkx-2.2 expression begins at day 9 p.c. in mouse development. At day 10 p.c. *Nkx-2.2* is expressed in the developing diencephalon. It is also expressed in the hindbrain and spinal cord, in motorneurons directly adjacent to the floor plate. At day 12.5 p.c., expression is detected in a narrow band of cells, between the dorsal and ventral thalami, and in the hypothalamus. In these areas, *Nkx-2.2* expression domains are adjacents and partially overlap those of *Dlx-1* and *TTF-1* (Price et al., 1992).

■ References

Guazzi, S., Price, M., De Felice, M., Damante, G., Mattei, M.-G. and Di Lauro, R. (1990). Thyroid nuclear factor 1 (TTF-1) contains a homeodomain and displays a novel DNA binding specificity. EMBO J. 9, 3631-3639.

Kim, Y. and Niremberg, M. (1989). *Drosophila* NK-homeobox genes. Proc. Natl. Acad. Sci. USA 86, 7716-7720.

Price, M., Lazzaro, D., Pohl, T., Mattei, M.-G., Rüther, U., Olivo, J.-C., Duboule, D. and Di Lauro, R. (1992). Regional expression of the homeobox gene Nkx-2.2 in the developing mammalian forebrain. Neuron 8, 1-20.

Nkx-2.3 [m]

Species: *Mus musculus*
Chromosomal location: Chromosome 12
Other names: -
Cognate genes: -
Type of homeobox: *NK-2*-type
Accession number: -

■ Origin and description

This gene was isolated by cross-homology with the *Nkx-2.1* (*TTF-1*) homeobox (Guazzi et al., 1990). It is a member of the *Nkx-2* class of vertebrate genes (Price et al., 1991)

■ Expression

The *Nkx-2.3* gene is expressed during fetal development, at mid gestation, in the mesoderm of the gut. It is also found at the periphery of the branchial arches and in the tooth germs (P. Dollé, personal communication).

■ References

Price, M., Lazzaro, D., Pohl, T., Mattei, M.-G., Ruether, U., Olivo, J.-C., Duboule, D. and DiLauro, R. (1992). Regional expression of the homeobox gene *Nkx-2.2* in the developing mammalian forebrain. Neuron, 8, 241-255.

Nkx-2.4 [m]

Species: *Mus musculus*
Chromosomal location: -
Other names: -
Cognate genes: -
Type of homeobox: *NK-2*-type
Accession number: -

■ Origin and description

This gene was isolated by cross-homology with the *Nkx-2.1* (*TTF-1*) homeobox (Guazzi et al., 1990). It is a member of the *Nkx-2* class of vertebrate genes (Price et al., 1991)

■ Expression

The *Nkx-2.4* gene is expressed exclusively in the diencephalon, from about day 10 of development. Its expression becomes progressively restricted to a domain in the hypothalamic region. This domain is located at a posterodorsal position, slightly above the future neurohypophysis, in the presumptive mammillary area (P. Dollé et al., in preparation).

■ References

Price, M., Lazzaro, D., Pohl, T., Mattei, M.-G., Ruether, U., Olivo, J.-C., Duboule, D. and DiLauro, R. (1992). Regional expression of the homeobox gene *Nkx-2.2* in the developing mammalian forebrain. Neuron 8, 241-255.

NvHbox 1 [Nv]

Species: *Notophthalmus viridescens* (red-spotted newt)
Chromosomal location: -
Other names: -
Cognate genes: *Xlhbox1*, *Hoxc-6*
Type of homeobox: *Antp*-type
Accession number: -

■ Origin and description

Identified by screening a newt limb blastema cDNA library with a mixture of *Drosophila* and *Xenopus* homeobox probes. Subsequently a newt genomic clone was also isolated and mapped. The amino acid sequence conservation with *Xenopus*, mouse and human homologs is very high, both in the box and outside it.

■ Expression

Detectable by Northern blotting in newt embryos, adult newt forelimb and hindlimb and their regeneration blastemas, but not in adult tail, liver, heart or spleen. A proximal forelimb blastema expresses at approximately 3-fold higher level than a distal. When animals were injected with doses of retinoic acid that proximalize a distal blastema the level of expression did not change by the mid-bud stage of the blastema. Expression was not detectable by Northern analysis of RNA from *Xenopus* forelimb mesenchyme.

■ Remarks

It is interesting that the gene is expressed in the adult limb of the newt, a regenerative urodele amphibian, and not adult *Xenopus*, nor possibly at the protein level in chick or mouse where it is apparently turned off in the limb bud after the developmental phase of expression (see *Xlhbox1*).

■ References

Savard, P, Gates, P.B. and Brockes, J.P. (1988). Position dependent expression of a homeobox gene transcript in relation to amphibian limb regeneration. EMBO J. 7, 4275-4282.

NvHbox 2 [Nv]

Species: *Notophthalmus viridescens* (red-spotted newt)
Chromosomal location: -
Other names: -
Cognate genes: related to mammalian group 11 *Hox* genes
Type of homeobox: *Antp*-type
Accession number: -

■ Origin and description

Identified by screening a newt limb blastema cDNA library with a degenerate oligonucleotide probe. The homeobox is identical to mouse *Hoxc-11* and there is significant sequence identity with *c-11* outside the homeobox, although not enough at this point to be sure that it is the newt homolog.

■ Expression

Expression in any adult tissue is not detectable by RNAse protection analysis. After amputation of the limb or tail, it is expressed by mesenchymal blastemal cells (the progenitor cells of the regenerate), but not by the limb wound epidermis. A proximal blastema (mid humerus) expresses at 3-5 fold higher levels than a distal blastema (mid radius/ulna). When animals were injected with doses of retinoic acid that proximalize a distal blastema, the level of expression in a distal blastema did not change at two time points analyzed.

■ Remarks

The lack of response to retinoic acid is surprising in view of the responsiveness of the putative homolog *Hoxc-11* in the chick limb bud. Limb regeneration (and development) in urodeles shows a number of differences from limb development in the chick. The gene is a clear example of one that is turned on in the adult after amputation of regenerative structures.

■ References

Brown, R. and Brockes, J.P. (1991). Identification and expression of a regeneration-specific homeobox gene in the newt limb blastema. Development 111, 489-496.

Oct-1 [m]

Species: *Mus musculus*
Chromosomal location: Chromosome 1
Other names: *OTF-1, OBP100, NFIII, NF-A1*
Cognate genes: -
Type of homeobox: POU-type
Accession number: -

■ Origin and description

cDNAs of human *Oct-1* were isolated by screening phage expression libraries with the DNA binding motif, ATG-CAAAT (Sturm et al., 1988). *Oct-1* is a member of the POU gene family, which was defined by the sequence homology of *Pit-1/GHF-1*, *Oct-1*, *Oct-2* and *Unc-86* and thus contains a homeobox that is specific for the POU gene family (Herr et al., 1988).

■ Expression

The expression pattern of the *Oct-1* gene in the developing and adult mouse has not been investigated by *in situ* hybridization. However, as determined by the electrophoretic mobility shift assay, *Oct-1* appears to be expressed ubiquitously (Schöler et al., 1989; Schreiber et al., 1990).

■ Function

The *Oct-1* protein (also referred to as OTF-1, OBP100, NFIII and NF-A1) binds to the so-called octamer sequence motif ATGCAAAT (for review: Kemler et al., 1990; Schöler, 1991). When placed in the context of an RNA polymerase II small nuclear RNA promoter, the octamer motif is active in cells containing no other *Oct*-factors in addition to *Oct-1* (Tanaka et al., 1988 and references therein). However, the activation of an octamer motif placed in RNA polymerase II mRNA-encoding promoters appears to require the presence of other *Oct*-factors. For example, the expression of the immunoglobulin gene promoters correlates with the expression pattern of *Oct-2* (for review: Kemmler et al., 1990).

■ References

Herr, W., Sturm, R.A., Clerc, R.G., Corcoran, L.M., Baltimore, D., Sharp, P.A., Ingraham, H.A., Rosenfeld, M.G., Finney, M., Ruvkun, G. and Horvitz, H.R. (1988). The POU domain: a large conserved region in the mammalian *pit*-1, *oct*-1, *oct*-2 and *Caenorhabditis elegans unc*-86 gene products. Genes and Dev. 2, 1513-1516.

Schöler, H.R. (1991). Octamania: The POU factors in murine development. Trends Genet. 7, 323-329.

Schöler, H.R., Hatzopoulos, A.K., Balling, R., Suzuki, N. and Gruss, P. (1989A). A family of octamer-specific proteins present during mouse embryogenesis: evidence for germline-specific expression of an *Oct*-factor. EMBO J. 8, 2543-2550.

Schreiber, E., Harshman, K., Kemler, I., Malipiero, U., Schaffner, W. and Fontana, A. (1990). Astrocytes and glioblAstoma cells express novel octamer-DNA binding proteins distinct from the ubiquitous *Oct*-1 and B cell type *Oct*-2 proteins. Nucleic Acids Res. 18, 5495-5503.

Sturm, R.A., Das, G. and Herr, W. (1988). The ubiquitous octamer binding protein Oct-1 contains a POU domain with a homeobox subdomain. Genes Dev. 2, 1582-1599.

Tanaka, M., Grossniklaus, U., Herr, W. and Hernandez, N. (1988). Activation of the U2 snRNA promoter by the octamer motif defines a new class o RNA polymerase II enhancer elements. Genes Dev. 2, 1764-1778.

Oct-2 [m]

Species: *Mus musculus*
Chromosomal location: Chromosome 7
Other names: *OTF-2, NF-A2*
Cognate genes: -
Type of homeobox: POU-type
Accession number: -

■ Origin and description

cDNAs of human *Oct-2* were isolated after sequencing purified *Oct-2* protein (Scheidereit et al, 1988) and by screening phage expression libraries with the DNA binding motif, ATGCAAAT (Clerc et al., 1988; Müller-Immerglück et al., 1988; Staudt et al., 1988). Murine genomic sequences and differential splicing products were isolated using human *Oct-2* probes (Hatzopoulos et al., 1990). *Oct-2* is a member of the POU gene family, which was defined by the sequence homology of *Pit-1/GHF-1*, *Oct-1*, *Oct-2* and *Unc-86* and thus contains a homeobox that is specific for the POU gene family (Herr et al., 1988).

■ Expression

The *Oct-2* gene is expressed during murine embryogenesis in the neural tube and the entire brain, except for the telencephalon (Hatzopoulos et al., 1990; Schöler, 1991). In the adult mouse it is expressed in lymphoid cells, the nervous system, intestine, testis and kidney (Schreiber et al., 1990; Hatzopoulos et al., 1990). However, except for lymphoid cells there is no evidence that *Oct-2* mRNA is translated into *Oct-2* protein (Schreiber et al., 1990).

■ Function

The *Oct-2* protein (also referred to as OTF-2, and NF-A2) is a trans-activator activating transcription factors via the octamer motif ATGCAAAT. This regulatory element is found in a number of B cell specific genes (for review: Kemler et al., 1990).

■ References

Clerc, R.G., Corcoran, L.M., LeBowitz, J.H., Baltimore, D. and Sharp, P.A. (1988). The B-cell-specific *Oct-2* protein contains POU box- and homeo box-type domains. Genes Dev. 2, 1570-1581.

Hatzopoulos, A.K., Stoykova, A.S., Erselius, J.R., Golding, M., Neuman, T. and Gruss, P. (1990). Structure and expression of the mouse *Oct*2a and *Oct*2b, two differentially spliced products of the same gene. Development 109, 349-362.

Herr, W., Sturm, R.A., Clerc, R.G., Corcoran, L.M., Baltimore D., Sharp, P.A., Ingraham, H.A., Rosenfeld, M.G., Finney, M., Ruvkun, G. and Horvitz, H.R. (1988). The POU domain: a large conserved region in the mammalian *pit*-1, *oct*-1, *oct*-2 and *Caenorhabditis elegans unc*-86 gene products. Genes and Dev. 2, 1513-1516.

Kemler, I. and Schaffner, W. (1990). Octamer transcription factors and the cell type-specificity of immunoglobulin gene expression. FASEB J. 4, 1444-1449.

Müller-Immerglück, M.M., Ruppert, S., Schaffner, W. and Matthias, P. (1988). A cloned transcription factor stimulates transcription from lymphoid-specific promoters in non-B cells. Nature 336, 544-551.

Scheidereit, C., Cromlish, J.A., Gerster, T., Kawakami, K., Balmaceda, C.-G., Currie, R.A. and Roeder, R.G. (1988). A human lymphoid-specific transcription factor that activates immunoglobulin genes is a homeobox protein. Nature 336, 551-557.

Schöler, H.R. (1991). Octamania: The POU factors in murine development. Trends Genet. 7, 323-329.

Schreiber, E., Harshman, K., Kemler, I., Malipiero, U., Schaffner, W. and Fontana, A. (1990). Astrocytes and globlAstoma cells express novel octamer-DNA binding proteins distinct from the ubiquitous *Oct*-1 and B cell type *Oct*-2 proteins. Nucleic Acids Res. 18, 5495-5503.

Staudt, L.M., Clerc, R.G., Singh, H., LeBowitz, H., Sharp, P.A. and Baltimore, D. (1988). Cloning of a lymphoid-specific cDNA encoding a protein binding the regulatory octamer DNA motif. Science 241, 577-580.

Oct-2   [h]

Species: *Homo sapiens*
Chromosomal location: Chromosome 19
Other names: *OTF-2, NF-A2*
Cognate genes: -
Type of homeobox: POU-homeobox, most related to *oct-1*
Accession numbers: M36542, M36653, M36718, X53468

■ Origin and description

The full-length cDNA was isolated with synthetic oligonucleotides directed against tryptic peptide sequences of the purified protein (Scheidereit et al., 1988) and with the South-Western screening method using the DNA binding site (Clerk et al., 1988; Ko et al., 1988; Müller et al., 1988; Staudt et al., 1988). It encodes a reading frame of 55 kD containing POU-specific box and POU-homeobox as well as a potential leucine zipper (of unknown function). The cloned cDNAs differ for a differentially spliced miniexon, absent in Scheidereit et al., 1988, and for a C-terminally altered product, due to differential splicing (Clerc et al., 1988). Transactivation domains have been mapped by cotransfection in HeLa cells (Gerster et al., 1990; Müller-Immerglück et al., 1990; Tanaka and Herr, 1990).

■ Expression

Oct-2 mRNA is generally found in B-Lymphocytes, albeit at different levels but also glial cells and some T cells express *oct-2* (Clerc et al., 1988; Müller et al., 1988; Scheidereit et al., 1988; Staudt et al., 1988). For rodents, *oct-2* expression has been detected in the suprachiasmatic nucleus, medial mammillary nucleus, granule and red nucleus in the adult rat nervous system and is widely spread in the embryonic rat nervous system (He et al., 1989).

■ Function

Oct-2 is considered to be one of the major transcriptional regulatory proteins for immunoglobulin gene promoters and the heavy chain enhancer, as well as for further yet unidentified cellular genes.

■ References

Clerc, R.G., Corcoran, L.M., LeBowitz, J.H., Baltimore, D. and Sharp, P.A. (1988). The B-cell-specific Oct-2 protein contains POU box- and homeo box-type domains. Genes Dev. 2, 1570-1581.

Gerster, T., Balmaceda, C.-G. and Roeder, R.G. (1990). The cell type-specific octamer transcription factor OTF-2 has two domains required for the activation of transcription. EMBO J. 9, 1635-1643.

He, X., Treacy, M.N., Simmons, D.M., Ingraham, H.A., Swanson, L.W. and Rosenfeld, M.G. (1989). Expession of a large family of POU-domain regulatory genes in mammalian brain development. Nature 340, 35-42.

Ko, H.-S., Fast, P., McBride, W. and Staudt, L.M. (1988). A human protein specific for the immunoglobulin octamer DNA motif contains a functional homeobox domain. Cell 55, 135-144.

Müller, M.M., Ruppert, S., Schaffner, W. and Matthias, P. (1988). A cloned octamer transcription factor stimulates transcription from lymphoid-specific promoters in non-B-cells. Nature 336, 544-551.

Müller-Immerglück, M.M., Schaffner, W. and Matthias, P. (1990). Transcription factor Oct-2A contains functionally redundant activation domains and works selectively from a promoter but not from a remote enhancer position in non-lymphoid (HeLa) cells. EMBO J. 9, 1625-1634.

Scheidereit, C., Cromlish, J.A., Gerster, T., Kawakami, K., Currie, R., Balmaceda, C.-G. and Roeder, R.G. (1988). A human lymphoid specific transcription factor that activates immunoglobulin genes is a homeobox protein. Nature 336, 551-557.

Staudt, L.M., Clerc, R.G., Singh, H., Lebowitz, J.H., Sharp, P.A. and Baltimore, D. (1988). Cloning of a lymphoid-specific cDNA encoding a protein binding the regulatory octamer DNA motif. Science 241, 577-580.

Tanaka, M. and Herr, W. (1990). Differential transcriptional activation by oct-1 and oct-2: Interdependent activation domains induce oct-2 phosphorylation. Cell 60, 375-386.

Oct-4 [m]

Species: *Mus musculus*

Chromosomal location: Chromosome 17, in the MHC between Q and T

Other names: *Oct-3*, *NF-A3*

Cognate genes: -

Type of homeobox: POU-type

Accession number: -

■ Origin and description

The cDNA was isolated from embryonic carcinoma cDNA libraries by cross-homology with the *Oct-2* POU region (Okamoto et al., 1990; Rosner et al., 1990; Schöler et al., 1990A). It is located at the distal end of the murine *t*-com-plex in the *MHC* between *Q* and *T* (Schöler et al., 1990B; Yeom et al., 1991). In addition to *Oct-4*, two other very early embryonic genes map in this general region, namely t^{12} and *Ped* (Yeom et al, 1991). Rat, cat, mink and human reveal cross-hybridizing fragments in an interspecies hybridization analysis, whereas other vertebrates such as chicken, frog and fish are negative, as well as invertebrates (Yeom et al., 1991). *Oct-4* is a member of the POU gene family, which was defined by the sequence homology of *Pit-1/GHF-1*, *Oct-1*, *Oct-2* and *Unc-86* (Herr et al., 1988). Thus, *Oct-4* contains a homeobox that is specific for the POU gene family (Schöler, 1991).

■ Expression

The expression of the *Oct-4* was studied using the electrophoretic mobility shift assay (Schöler et al., 1989) and after its cloning by *in situ* hybridization (Rosner et al., 1990; Schöler et al., 1990B; Yeom, 1991). The expression pattern of *Oct-4* correlates with an undifferentiated cell phenotype. *Oct-4* expression is detected in the totipotent and pluripotent stem cells of the pregastrulation embryo and is downregulated during differentiation of the cells, eventually becoming confined to the germline lineage. In embryonic stem cells and embryonal carcinoma cell lines, *Oct-4* expression is downregulated when the cells are induced to differentiate by retinoic acid.

■ Function

The function of *Oct-4* (also referred to as *NF-A3*, and *Oct-3*) was studied in fertilized oocytes by antisense *Oct-4* oligonucleotides (Rosner et al., 1991). Microinjection resulted in a loss of *Oct-4* mRNA and as a consequence in an inhibition of DNA synthesis and arrest of the embryo at the one-cell stage. Several lines of evidence indicate that *Oct-4* might regulate DNA replication in the one-cell stage (Rosner et al., 1991). In addition to its possible role in DNA replication, *Oct-4* protein activates transcription via the octamer motif ATGCAAAT. Since target genes of *Oct-4* still have to be determined, the octamer motif is only one possible DNA motif. Possible candidates for target genes are *Hoxa-5* and the *hst/k-FGF* proto-oncogene (Schöler, 1991).

■ References

Herr, W., Sturm, R.A., Clerc, R.G., Corcoran, L.M., Baltimore, D., Sharp, P.A., Ingraham, H.A., Rosenfeld, M.G., Finney, M., Ruvkun, G. and Horvitz, H.R. (1988). The POU domain: a large conserved region in the mammalian *pit*-1, *oct*-1, *oct*-2 and *Caenorhabditis elegans unc*-86 gene products. Genes and Dev. 2, 1513-1516.

Okamoto, K., Okazawa, H., Okuda, A., Sakai, M., Muramatsu, M. and Hamada, H. (1990). A novel octamer binding transcription factor is differentially expressed in mouse embryonic cells. Cell 60, 461-472.

Rosner, M.H., Vigano, M.A., Ozato, K., Timmons, P.M., Poirier, F., Rigby, P.W.J. and Staudt, L.M. (1990). A POU-domain transcription factor in early stem cells and germ cells of the mammalian embryo. Nature 345, 686-692.

Rosner, M.H., De Santo, R.J., Arnheiter, H. and Staudt, L.M. (1991). *Oct-3* is a maternal factor required for the first mouse embryonic division. Cell 64, 1103-1110.

Schöler, H.R., Hatzopoulos, A.K., Balling, R., Suzuki, N. and Gruss, P. (1989A). A family of octamer-specific proteins present during mouse embryogenesis: evidence for germline-specific expression of an *Oct*-factor. EMBO J. 8, 2543-2550.

Schöler, H.R., Ruppert, S., Suzuki, N., Chowdhury, K. and Gruss. P. (1990A). New type of POU domain in germline-specific protein *Oct-4*. Nature 344, 435-439.

Schöler, H.R., Dressler, G.R., Balling, R., Rohdewohld, H. and Gruss, P. (1990B). *Oct-4*: a germline-specific transcription factor mapping to the mouse t-complex. EMBO J. 9, 2185-2195.

Schöler, H.R. (1991). Octamania: The POU factors in murine development. Trends Genet. 7, 323.

Yeom, Y.I., Ha, H.-S., Balling, R., Schöler, H.R. and Artzt, K. (1991). Structure, expression and chromosomal location of the *Oct-4* gene. MOD, 35, 171–180.

Oct-6 [m]

Species: *Mus musculus*

Chromosomal location: Chromosome 4

Other names: *SCIP* (rat); *Tst-1* (rat)

Cognate genes: -

Type of homeobox: POU-type

Accession number: -

■ Origin and description

The *Oct-6* cDNA was isolated from embryonic carcinoma cDNA libraries by cross-homology with the *Oct-2* POU region (Meijer et al., 1990; Suzuki et al., 1990). *Oct-6* has also been cloned from rat, where it is termed *SCIP* (for suppressed, cAMP-inducible POU; Monuki et al., 1990; 1991) or *Tst-1* (because of its isolation from a rat testis cDNA library; He et al., 1990). *Oct-6* is located on mouse chromosome 4 and contains a homeobox that is specific for the POU gene family (Schöler, 1991). The POU gene family was defined by the sequence homology of *Pit-1/GHF-1*, *Oct-1*, *Oct-2* and *Unc-86* (Herr et al., 1988).

■ Expression

The expression of the *Oct-6* gene in the mouse was studied first using the electrophoretic mobility shift assay (Schöler et al., 1989) and after its cloning by *in situ* hybridization (Suzuki et al., 1990). *Oct-6* is expressed at early embryonic stages and later in specific neurons of the developing and adult brain and also in testis (Suzuki et al., 1990). In rat expression has been specified postnatally for developing Schwann cells (Monuki et al., 1990). In embryonic stem cells and embryonal carcinoma cell lines, *Oct-6* expression is downregulated when the cells are induced to differentiate by retinoic acid (Schöler et al., 1989; Suzuki et al., 1990).

■ Function

Oct-6 activates transcription via the octamer motif ATG-CAAAT. Because *Oct-6* is transiently expressed during the period of rapid cell division separating the premyelinating and myelinating phases of Schwann cell differentiation, it may play a role in the proliferation of these cells (Monuki et al., 1990). Possible candidates for target genes are the myelin-specific genes myelin basic protein (MBP) and protein zero (Po) (Monuki et al., 1990; 1991). Possible target genes in the early embryo and in testis were not yet determined.

■ References

He, X., Treacy, M.N., Simmons, D.M., Ingraham, H.A., Swanson, L.W. and Rosenfeld, M.G. (1989). Expression of a large family of POU-domain regulatory genes in mammalian brain development. Nature 340, 35-42.

Herr, W., Sturm, R.A., Clerc, R.G., Corcoran, L.M., Baltimore, D., Sharp, P.A., Ingraham, H.A., Rosenfeld, M.G., Finney, M., Ruvkun, G. and Horvitz, H.R. (1988). The POU domain: a large conserved region in the mammalian *pit-1*, *oct-1*, *oct-2* and *Caenorhabditis elegans unc-86* gene products. Genes and Dev. 2, 1513-1516.

Meijer, D., Graus, A., Kraay, R., Langeveld, A., Mulder, M.P. and Grosveld, G. (1990). The octamer binding factor *Oct6*: cDNA cloning and expression in early embryonic cells. Nucleic Acid Res. 18, 7357-7365.

Monuki, E.S., Weinmaster, G., Kuhn, R. and Lemke, G. (1989). SCIP: A glial POU domain gene regulated by cAMP. Neuron 2, 783-793.

Monuki, E.S., Weinmaster, G., Trapp, B.D. and Lemke, G. (1990). Expression and activity of the POU transcription factor SCIP. Science 249, 1300-1303.

Schöler, H.R., Hatzopoulos, A.K., Balling, R., Suzuki, N. and Gruss, P. (1989). A family of octamer-specific proteins present during mouse embryogenesis: evidence for germline-specific expression of an Oct-factor. EMBO J. 8, 2543-2550.

Schöler, H.R. (1991). Octamania: The POU factors in murine development. Trends Genet. 7, 323-329.

Suzuki, N., Rohdewohld, H., Neuman, T., Gruss, P. and Schöler, H.R. (1990). *Oct-6*: a POU transcription factor expressed in embryonal stem cells and in the developing brain. EMBO J. 9, 3723.

otd [d]

Species: *Drosophila melanogaster*

Chromosomal location: X chromosome (polytene location 8A1)

Other names: *orthodenticle, ocelliless (oc)*

Cognate genes: -

Type of homeobox: related to the *paired-*class

Accession number: -

Origin and description

otd mutations were first identified by their effect on embryonic pattern formation (Wieschaus et al., 1984). In addition, *otd* mutations are allelic to *ocelliless* mutations, which had been previously identified (Bedichek, 1934). Molecular isolation of the *otd* gene showed that it contained a homeobox most related to those of the *paired* class (Finkelstein et al., 1990a). *otd* contains an unusual amino acid at position 9 of the "recognition" helix (lysine), and its binding specificity resembles that of the *Drosophila bicoid* protein (M. Simpson, C. Desplan, unpublished observation).

Expression

In the blastoderm embryo, *otd* is expressed in a circumferential stripe in the presumptive head region. During later embryonic and larval stages, the *otd* protein continues to be expressed in the head, at the ventral midline, and several other locations. *otd* is expressed in several regions of specific imaginal discs, particularly in the region of the eye-antennal disc that gives rise to the ocelli of the adult head.

Function

Null *otd* mutations result in embryonic lethality. Mutant embryos are defective in anterior head development (Wieschaus et al., 1984; Finkelstein and Perrimon, 1990b). In addition, *otd* mutations cause defects in the development of the medial region of the embryo, in both the CNS and the epidermis. *otd* is specifically required for the differentiation of identified medial neurons necessary for axonal guidance. ocelliless mutations are a second class of lesions in the *otd* gene. These mutations result in viable adult flies lacking the ocellar visual structures located in the medial region of the adult head. *otd* is required to specify the region of the eye-antennal imaginal disc which gives rise to the ocelli.

References

Bedichek, S. (1934). Dros. Inf. Serv. 2, 9.

Finkelstein, R., Smouse, D., Capaci, T.M., Spradling, A.C. and Perrimon, N. (1990a). The orthodenticle gene encodes a novel homeodomain protein involved in the development of the *Drosophila* nervous system and ocellar visual structures. Genes and Develop. 4, 1516-1527.

Finkelstein, R. and Perrimon, N. (1990a). The orthodenticle gene is regulated by bicoid and torso and specifies *Drosophila* head development. Nature 346, 485-488.

Wieschaus, E., Nusslein-Volhard, C. and Jurgens, G. (1984). Mutations affecting the pattern of the larval cuticle in *Drosophila melanogaster*. III. Zygotic loci on the X chromosome and the fourth chromosome. Roux Arch. devl. Biol. 193, 296-307.

OTX1 [h]

Species: *Homo sapiens*
Chromosomal location: -
Other names: -
Cognate genes: *Otx-1*
Type of homeobox: *otd*-type
Accession number: -

Origin and description

This gene was isolated by cross-homology with the *Drosophila otd* homeobox (Simeone et al., 1992). A cDNA library prepared from 8-week human embryos was screened at low stringency conditions with a short *otd* genomic sequence including the homeobox (Finkelstein et al., 1990).

Expression

Otx-1 has been found expressed in basal telencephalon, diencephalon and mesencephalon but not in spinal cord. It is also expressed in dorsal telencephalon and regions of metencephalon. From E9.5 onward, its expression clearly marks the posterior boundary of mesencephalon (Simeone et al., 1992). *Otx-1* genes is also expressed in developing sense organs: in E12.5 olfactory epithelium as well as in the respiratory epithelium of nasal cavities, in the developing ear and in the developing eye.

References

Finkelstein, R., Smouse, D., Capaci, T.M., Spradling, A.C. and Perrimon, N. (1990a). The orthodenticle gene encodes a novel homeodomain protein involved in the development of the *Drosophila* nervous system and ocellar visual structures. Genes and Develop. 4, 1516-1527.

Simeone, A., Acampora, D., Gulisano, M., Stornaiuolo, A. and Boncinelli, E. (1992b). Nested expression domains of four homeobox genes in developing rostral brain. Nature 358, 687-690.

OTX2 [h]

Species: *Homo sapiens*
Chromosomal location: -
Other names: -
Cognate genes: *Otx-2*
Type of homeobox: *otd*-type
Accession number: -

■ Origin and description

This gene was isolated by cross-homology with the *Drosophila otd* homeobox (Simeone et al., 1992). A cDNA library prepared from 8-week human embryos was screened at low stringency conditions with a short *otd* genomic sequence including the homeobox (Finkelstein et al., 1990). Two classes of homologous cDNA clones, termed *OTX1* and *OTX2*, were found.

■ Expression

Otx-2 is first expressed at the epiblast stage during mouse development. Later on it is expressed in basal telencephalon, diencephalon and mesencephalon but not in spinal cord. It has a specific localization in choroid plexuses, both in lateral ventriculi and myelencephalon. It is also expressed in various regions of diencephalon, namely in the epithalamus, dorsal thalamus and mammillary region of posterior hypothalamus; in mesencephalic regions of tectum and tegmentum. From E9.5 onward, its expression clearly marks the posterior boundary of mesencephalon (Simeone et al., 1992). *Otx-2* genes is also expressed in developing sense organs: in E12.5 olfactory epithelium as well as in the respiratory epithelium of nasal cavities, in the developing ear and in the developing eye.

■ References

Finkelstein, R., Smouse, D., Capaci, T.M., Spradling, A.C. and Perrimon, N. (1990a). The orthodenticle gene encodes a novel homeodomain protein involved in the development of the *Drosophila* nervous system and ocellar visual structures. Genes and Develop. 4, 1516-1527.

Simeone, A., Acampora, D., Gulisano, M., Stornaiuolo, A. and Boncinelli, E. (1992b). Nested expression domains of four homeobox genes in developing rostral brain. Nature 358, 687-690.

pal-1 [ce]

Species: *Caenorhabditis elegans* (N2)
Chromosomal location: III
Other names: *ceh-3*
Cognate genes: -
Type of homeobox: related to *cad*
Accession number: X57140, X62782

■ Origin and description

Isolated using highly degenerate oligonucleotides for the most conserved region of helix 3 (Bürglin et al., 1989). Subsequently it was shown that the genetically identified gene *pal-1* (Waring and Kenyon, 1990) can be rescued with *ceh-3* (Waring and Kenyon, 1991). The highest similarity is with the *caudal*-class genes.

■ Genetics, function

pal-1 was mainly analyzed in male animals, since they have easily observable tail structures. In wild-type animals the laterally symmetrical pair of V6 cells that lie in the posterior body region undergo a characteristic pattern of cell division that produce male rays, which are neuronal structures. In *pal-1* animals, the V6 cells are homeotically transformed such that they produce cell lineages characteristic for more anteriorly located cells (V1-V4) that produce epidermal seam cells (Waring and Kenyon, 1990). Mosaic analysis showed that *pal-1* acts within the V6 cells (Waring and Kenyon, 1991). Two other phenotypes, a low frequency of embryonic lethality, and variable bulges in the body are associated with both *pal-1* alleles analyzed (Waring and Kenyon, 1990). These alleles are probably not complete loss-of-function mutations (L. Edgar, personal communication). *pal-1* appears to act upstream of *mab-5*, since both *mab-5* loss-of-function, and *mab-5* gain-of-function mutations are epistatic to *pal-1* (Waring and Kenyon, 1990, 1991).

■ References

Bürglin, T.R., Finney, M., Coulson, A. and Ruvkun, G. (1989). *Caenorhabditis elegans* has scores of homeobox-containing genes. Nature 341, 239-243.

Waring, D.A. and Kenyon, C. (1990). Selective silencing of cell communication influences anteroposterior pattern formation in *C. elegans*. Cell 60, 123-131.

Waring, D.A. and Kenyon, C. (1991). Regulation of cellular responsiveness to inductive signals in the developing *C. elegans* nervous system. Nature 350, 712-715.

Pax-3 [m]

Species: *Mus musculus*
Chromosomal location: Chromosome 1
Other names: -
Cognate genes: *HuP2*/WS1
Type of homeobox: *paired*-type
Accession number: X59358

■ Origin and description

This gene was isolated through hybridization with the *Pax-1* (Deutsch et al., 1988) paired box (Dressler et al., 1988). It contains a paired box and a paired-type homeodomain and encodes the octapeptide found between paired domain and homeodomain in most paired domain containing proteins (Goulding et al., 1991; Burri et al., 1989). *Pax-3* is a member of the *Pax* gene family (for review see Deutsch and Gruss, 1991) and was mapped to chromosome 1 near *Villin* (Walther et al., 1991). Two transcripts are detected by Northern hybridization (Goulding et al., 1991). The sequence and the structure of *Pax-7* (Jostes et al., 1991) and *Pax-3* are highly similar, suggesting that these genes arose via duplication from a common ancestral gene.

■ Expression

Pax-3 expression is confined to embryogenesis between day 8.5 p.c. to day 15 p.c. In the neural tube, expression begins just prior to neural tube closure. At day 14 of embryogenesis, expression is no longer detected in the anterior neural tube. Transcripts are detected in the dorsal part of the ventricular layer of the neural tube in the alar and the roof plate. In early embryogenesis *Pax-3* is expressed in all brain areas, from day 11 p.c. onwards expression is no longer detectable in the telencephalon. *Pax-3* transcripts are also present in neural crest derivatives, particularly in spinal ganglia and cephalic neural crest, including the nasal process and structures derived from the first and second branchial arches. In addition, expression is detected in the segmented mesoderm in the dermomyotome from day 8.5 to day 11 of embryogenesis and in the undifferentiated mesenchyme of the limb buds from day 10 to day 11 p.c. (Goulding et al., 1991). Note that the expression patterns of *Pax-3* and *Pax-7* (Jostes et al., 1991) are very similar. *Pax-3* is not expressed in undifferentiated F9 cells or P19 cells. Differentiation of F9 cells into parietal endoderm and of P19 cells into neuroectoderm-like cells resulted in the expression of two *Pax-3* transcripts (Goulding et al., 1991).

■ Function

Pax-3 was mapped near *Villin*, close to a known developmental mutant, *splotch* (*Sp*) (Walther et al., 1991). Homozygous *splotch* mice show a variety of phenotypic alterations in structures in which *Pax-3* is expressed (Auerbach, 1954; Moase and Trasler, 1989; Franz, 1990; Goulding et al., 1991). Two allels of splotch, Sp^r and Sp^{2H} show mutations of the *Pax-3* gene. In Sp^r, *Pax-3* is deleted and in Sp^{2H} 32bp are missing resulting in premature termination of the open reading frame in the homeobox (Epstein et al., 1991). Hence, *Pax-3* is important for proper development of CNS structures and neural crest derived structures.

The human Waardenburg's syndrome I was mapped to a region on human chromosome 2 (Ishikiriyama et al., 1989) which is syntenic to the region of mouse chromosome 1 where *Pax-3* is localized. Subsequently, the human homologue of *Pax-3*, *HuP2*, was found to map to this region. Analysis of *HuP2* in affected individuals of different families showed that *HuP2* is mutated in Waardenburg's syndrome patients (Tassabehji et al., 1992; Baldwin et al., 1992).

■ References

Auerbach, R. (1954). Analysis of the developmental effects of a lethal mutation in the house mouse. J. Exp. Zool. 127, 3905-329.

Baldwin, C.T., Hoth, C.F., Amos, J.A., da-Silva, E.O. and Milunski, A. (1992). An exonic mutation in the HuP2 paired domain gene causes Waaredenburg's syndrome. Nature 355, 637-638.

Burri, M., Tromvoukis, Y., Bopp, D., Frigerio, G. and Noll, M. (1989). Conservation of the paired domain in metazoans and its structure in three isolated human genes. EMBO J. 8, 1183-1190.

Deutsch, U., Dressler, G.R. and Gruss, P. (1988). Pax-1, a member of a paired box homologous murine gene family, is expressed in segmented structures during development. Cell 53, 617-625.

Deutsch, U. and Gruss, P. (1991). Murine paired domain proteins as regulatory factors of embryonic development. Sem. Dev. Biol. 2, 413–424.

Dressler, G.R., Deutsch, U., Balling, R., Simon, D., Guénet, J.-L. and Gruss, P. (1988). Murine genes with homology to Drosophila segmentation genes. Development 104 (Supplement), 181-186.

Epstein, D.J., Vekemans, M. and Gros, P. (1991). splotch (Sp^{2H}), a mutation affecting development of the mouse neural tube, shows a deletion within the paired homeodomain of Pax-3. Cell 67, 767-774.

Franz, T. (1990). Defective ensheatment of motoric nerves in the Splotch mutant mouse. Acta Anat. 138, 246-253.

Goulding, M.D., Chalepkis, G., Deutsch, U., Erselius, J.R. and Gruss, P. (1991). Pax-3, a novel murine DNA binding protein expressed during early neurogenesis. EMBO J. 10, 1135-1147.

Ishikiriyama, S., Tonoki, H., Shibuya, Y., Chin, S., Harada, N., Abe, K., Niikawa, N. (1989). Waardenburg syndrome Type I in a child with de novo inversion (2) (q35q37.3). Am. J. Med. Genet, 33, 505-507.

Jostes, B., Walther, C. and Gruss, P. (1991). The murine paired box gene, Pax7, is expressed specificall during the development of the nervous and muscular system. Mech. Dev. 33, 27-38.

Moase, C.E. and Trasler, D.G. (1989). Spinal ganglia reduction in the Splotch-delayed mouse neural tube defect mutant. Teratology 40, 67-75.

Tassabehji, M., Read, A.P., Newton, V.E., Harris, R., Balling, R., Gruss, P. and Strachan, T. (1992). Waardenburg's syndrome patients have mutations in the human homologue of the Pax-3 paired box gene. Nature 355, 635-636.

Walther, C., Guénet, J.-L., Simon, D., Deutsch, U., Jostes, B., Goulding, M., Plachov, D., Balling, R. and Gruss, P. (1991). Pax: A murine multigene family of paired box containing genes. Genomics 11, 424-434.

Pax-4 [m]

Species: *Mus musculus*
Chromosomal location: Chromosome 6
Other names: -
Cognate genes: -
Type of homeobox: *paired*-type
Accession number: -

■ Origin and description

This gene was isolated (Walther et al., 1991) by hybridization with the paired boxes of *Pax-1* (Deutsch et al., 1988), *Pax-3* (Dressler et al., 1988; Goulding et al., 1991) and *gooseberry-distal* (Bopp et al., 1986; Côté et al., 1987).

Pax-4 is a member of the *Pax* gene family (for review see Deutsch and Gruss, 1991) and maps to chromosome 4 (Walther et al., 1991). It contains a rather divergent paired box (Walther et al., 1991) and a *paired*-type homeobox (Asano and Gruss, unpublished).

■ References

Bopp, D., Burri, M., Baumgartner, S., Frigerio, G. and Noll, M. (1986). Conservation of a large protein domain in the segmentation gene *paired* and in functionally related genes of *Drosophila*. Cell 47, 1033-1040.

Côté, S., Preiss, A., Haller, J., Schuh, R., Kienlin, A., Seifert, E. and Jäckle, H. (1987). The *gooseberry-zipper* region of *Drosophila*: five genes encode different spatially restricted transcripts in the embryo. EMBO J. 6, 2793-2801.

Deutsch, U. and Gruss, P. (1991). Murine paired domain proteins as regulatory factors of embryonic development. Sem. Dev. Biol. 2, 413–424.

Deutsch, U., Dressler, G.R. and Gruss, P. (1988). Pax-1, a member of a paired box homologous murine gene family, is expressed in segmented structures during development. Cell 53, 617-625.

Dressler, G.R., Deutsch, U., Balling, R., Simon, D., Guénet, J.-L. and Gruss, P. (1988). Murine genes with homology to *Drosophila* segmentation genes. Development 104 (Supplement), 181-186.

Goulding, M.D., Chalepkis, G., Deutsch, U., Erselius, J.R. and Gruss, P. (1991). *Pax-3*, a novel murine DNA binding protein expressed during early neurogenesis. EMBO J. 10, 1135-1147.

Walther, C., Guénet, J.-L., Simon, D., Deutsch, U., Jostes, B., Goulding, M., Plachov, D., Balling, R. and Gruss, P. (1991). Pax: A murine multigene family of paired box containing genes. Genomics 1, 424-434.

Pax-6 [m]

Species: *Mus musculus*

Chromosomal location: Chromosome 2

Other names: -

Cognate genes: *pax*[zf-a], *AN*

Type of homeobox: *Paired*-type

Accession number: -

■ Origin and description

This gene was isolated (Walther et al., 1991) by hybridization with the *Pax-1* (Deutsch et al., 1988), *Pax-3* (Dressler et al., 1988; Goulding et al., 1991) and gooseberry-distal (Bopp et al., 1986; Côté et al., 1987) paired box. It is a member of the *Pax* gene family (for review see Deutsch and Gruss, 1991) and was mapped to chromosome 2 (Walther et al., 1991). It contains a rather divergent paired box as well as a rather divergent homeobox of the paired-type. *Pax-6*, like *paired*, does not encode an octapeptide usually found between paired domain and homeodomain (Burri et al., 1989). So far, two differentially spliced transcripts are known, one contains an insertion of 42bp in a highly conserved region of the paired box (Walther and Gruss, 1991).

■ Expression

Expression is mainly confined to the ectoderm and starts around day 8 of embryogenesis in the forebrain and the hindbrain. In the neural tube *Pax-6* is first expressed in a broad band of cells in the transverse plane of the neural tube, being absent only from the most dorsal and ventral cells at day 8.5 p.c. In later stages *Pax-6* transcripts are confined mainly to the ventricular layer of the basal plate, sparing the floorplate and cells directly adjacent to it. In the brain, *Pax-6* is expressed in the telencephalon, diencephalon, and rhombencephalon sparing the roof of the mesencephalon in all developmental stages examined. Expression in the metencephalon/cerebellum begins around day 15 p.c. and is still detected in the granular layer of the adult cerebellum. *Pax-6* transcripts are detected in all structures of the developing eye and nose and in the olfactory bulbs (Walther and Gruss, 1991).

■ Function

Pax-6 maps near the semidominant mutant *small eye* (*Sey*). Analysis of *Pax-6* in different alleles of this mutant revealed that *Pax-6* sequences are deleted in *Sey*[H], and that *Sey* and *Sey*[Neu] have point mutations in the *Pax-6* gene, resulting in a premature termination of the open reading frame before or after the homeobox, respectively (Hill et al., 1991). The heterozygote mutants of *Sey* and *Sey*[Neu] have smaller eyes (Roberts, 1967) and the homozygotes lack eyes, nose and the olfactory bulbs (Hogan et al., 1986, 1988). This is in concordance with *Pax-6* being involved in the regulation of the inductive events leading to the formation of eyes and nose.

A human cDNA, *AN*, was cloned and shown to be completely or partially deleted in two patients with Aniridia (Ton et al., 1991). *AN* represents the human homologue of *Pax-6*, suggesting that mutations in the human homolog of *Pax-6* cause the Aniridia syndrome.

■ References

Bopp, D., Burri, M, Baumgartner, S., Frigerio, G. and Noll, M. (1986). Conservation of a large protein domain in the segmentation gene *paired* and in functionally related genes of *Drosophila*. Cell 47, 1033-1040.

Burri, M., Tromvoukis, Y., Bopp, D., Frigerio, G. and Noll, M. (1989). Conservation of the paired domain in metazoans and its structure in three isolated human genes. EMBO J. 8, 1183-1190.

Côté, S., Preiss, A., Haller, J., Schuh, R., Kienlin, A., Seifert, E. and Jäckle, H. (1987). The *gooseberry-zipper* region of *Drosophila*: five genes encode different spatially restricted transcripts in the embryo. EMBO J. 6, 2793-2801.

Deutsch, U. and Gruss, P. (1991). Murine paired domain proteins as regulatory factors of embryonic development. Sem. Dev. Biol. 2, 413–424.

Deutsch, U., Dressler, G.R. and Gruss, P. (1988). Pax-1, a member of a paired box homologous murine gene family, is expressed in segmented structures during development. Cell 53, 617-625.

Dressler, G.R., Deutsch, U., Balling, R., Simon, D., Guénet, J.-L. and Gruss, P. (1988). Murine genes with homology to *Drosophila* segmentation genes. Development 104 (Supplement), 181-186.

Goulding, M.D., Chalepkis, G., Deutsch, U., Erselius, J.R. and Gruss, P. (1991). *Pax-3*, a novel murine DNA binding protein expressed during early neurogenesis. EMBO J. 10, 1135-1147.

Hogan, B.L.M., Horsburgh, G., Cohen, J., Hetherington, C.M., Fisher, G. and Lyon, M.F. (1986). *Small eyes* (Sey): a homozygous lethal mutation on chromosome 2 which affects the differentiation of both lens and nasal placodes in the mouse. Embryol. Exp. Morph. 97, 95-110.

Hogan, B.L.M., Hirst, E.M.A., Horsburgh, G. and Hetherington, C.M. (1988). *Small eye* (Sey): a mouse model for the genetic analysis of cranofacial abnormalities. Development 103 Supplement, 115-119.

Hill, R.E., Favor, J., Hogan, B.L.M., Ton, C.C.T., Saunders, G.F., Hanson, I.M., Prosser, J., Jordan, T., Hastie, N.D. and van Heyningen, V. (1991). Mouse *Small eye* results from mutations in a paired-like homeobox-containing gene. Nature 354, 522-525.

Roberts, R.C. (1967). *Small eyes*, a new dominant mutant in the mouse. Genet. Res. 9, 121-122.

Ton, C.C.T., Hirvonen, H., Miwa, H., Weil, M.M., Monaghan, P., Jordan, T., van Heyningen, V., Hastie, N.D., Meijers-Heijboer, H., Drechsler, M., Royer-Pokora, B., Collins, F., Swaroop, A., Strong, L.C. and Saunders, G.F. (1991). Positional cloning of a paired box- and homeobox-containing gene from the Aniridia region. Cell 67, 1059-1074.

Walther, C. and Gruss, P. (1991). *Pax-6*, a murine paired box gene, is expressed in the developing CNS. Development 113, 1435-1449.

Walther, C., Guénet, J.-L., Simon, D., Deutsch, U., Jostes, B., Goulding, M., Plachov, D., Balling, R. and Gruss, P. (1991). Pax: A murine multigene family of paired box containing genes. Genomics 11, 424-434.

Pax-6 [zf]

Species: *Brachydanio rerio* (zebrafish)
Chromosomal location: -
Other names: *pax* [zf-a]
Cognate genes: *Pax-6*
Type of homeobox: *prd*-type
Accession number: X61389

■ Origin and description

A zebrafish genomic fragment, *pax* [zf-d], was isolated by cross-hybridization to the paired box sequence of the murine *Pax-1* gene (Deutsch et al., 1988; Eiken, Villand and Fjose, unpublished). Four cDNA clones corresponding to the *pax-6* [zf-a] gene were isolated from lambda gt11 and ZAP-II cDNA libraries by cross-hybridization screening with the murine *Pax-1* and zebrafish pax [zf-d] paired box probes (Krauss et al., 1991a,b). One of the four cDNA clones (cZk3) contains the entire protein coding region of the zebrafish *pax-6* gene (Krauss et al., 1991b). In addition to a paired box sequence, the cDNA encodes a homeodomain which is about 60% identical to other known paired type homeodomains.

■ Expression

The developmental expression of *pax-6* has been studied by Northern blotting and *in situ* hybridization (Krauss et al., 1991a,b). A 3.0 kb transcript was detected in embryonic RNA and specific hybridization signals were observed within the eyes and two restricted regions of the central nervous system of 12-24 h embryos. At later developmental stages a more complex expression pattern was observed within the eyes and most of the central nervous system.

■ References

Deutsch, U., Dressler, G.R. and Gruss, P. (1988). *Pax-1*, a member of a paired box homologous murine gene family, is expressed in segmented structures during development. Cell 53, 617-625.

Krauss, S., Johansen, T., Korzh, V. and Fjose, A. (1991a). Expression pattern of zebrafish *pax* genes suggests a role in early brain regionalization. Nature 353, 267-270.

Krauss, S., Johansen, T., Korzh, V., Moens, U., Ericson, J.U. and Fjose, A. (1991). Zebrafish pax [zf-a]: a paired box-containing gene expressed in the neural tube. EMBO J. 10, 3609-3619.

Pax-7 [m]

Species: *Mus musculus*
Chromosomal location: Chromosome 4
Other names: -
Cognate genes: *HuP1*
Type of homeobox: *Paired*-type
Accession number: -

■ Origin and description

This gene was isolated by hybridization with the *Pax-1* (Deutsch et al., 1988), *Pax-3* (Dressler et al., 1988) and gooseberry-distal (Bopp et al., 1986; Côté et al., 1987) *paired* box (Jostes et al., 1991). It contains a paired box and a paired-type homeobox and encodes the octapeptide usually found between paired domain and homeodomain (Jostes et al., 1991; Burri et al., 1989). *Pax-7* is a member of the *Pax* gene family (for review see Deutsch and Gruss, 1991) and was mapped to chromosome 4 (Walther et al., 1992). The gene *HuP1* (Burri et al., 1989) most likely represents the human homolog of *Pax-7*. The sequence and the structure of *Pax-7* and *Pax-3* are highly similar, suggesting that these genes arose via duplication from a common anchestral gene.

■ Expression

Pax-7 is expressed in the ventricular zone of the neural tube beginning around day 9 of embryogenesis. Expression is confined to the ventricular zone of the alar plate of the neural tube, no expression is detected in the roof plate. Between day 8 to day 11 of embryogenesis *Pax-7* transcripts are detected in all brain areas. Thereafter, the anterior expression boundary is retracted to the diencephalon. *Pax-7* is also expressed in the olfactory epithelium. In addition to neuroectodermal structures, *Pax-7* transcripts are found in the dermamyotome of the differentiating somites and in later stages in the intercostal muscles, the skeletal muscles of the trunk, the shoulders and the forelimbs. Smooth muscles and dermatome-derived tissues do not show expression of *Pax-7* (Jostes et al., 1991). Note that the expression patterns of *Pax-7* and of *Pax-3* are very similar.

■ References

Bopp, D., Burri, M, Baumgartner, S., Frigerio, G. and Noll, M. (1986). Conservation of a large protein domain in the segmentation gene *paired* and in functionally related genes of *Drosophila*. Cell 47, 1033-1040.

Burri, M., Tromvoukis, Y., Bopp, D., Frigerio, G. and Noll, M. (1989). Conservation of the paired domain in metazoans and its structure in three isolated human genes. EMBO J. 8, 1183-1190.

Côté, S., Preiss, A., Haller, J., Schuh, R., Kienlin, A., Seifert, E. and Jäckle, H. (1987). The *gooseberry-zipper* region of *Drosophila*: five genes encode different spatially restricted transcripts in the embryo. EMBO J. 6, 2793-2801.

Deutsch, U. and Gruss, P. (1991). Murine paired domain proteins as regulatory factors of embryonic development. Sem. Dev. Biol. 2, 413–424.

Deutsch, U., Dressler, G.R. and Gruss, P. (1988). *Pax-1*, a member of a paired box homologous murine gene family, is expressed in segmented structures during development. Cell 53, 617-625.

Dressler, G.R., Deutsch, U., Balling, R., Simon, D., Guénet, J.-L. and Gruss, P. (1988). Murine genes with homology to *Drosophila* segmentation genes. Development 104 (Supplement), 181-186.

Goulding, M.D., Chalepkis, G., Deutsch, U., Erselius, J.R. and Gruss, P. (1991). *Pax-3*, a novel murine DNA binding protein expressed during early neurogenesis. EMBO J. 10, 1135-1147.

Jostes, B., Walther, C. and Gruss, P. (1991). The murine paired box gene, *Pax-7*, is expressed specifically during the development of the nervous and muscular system. Mech. Dev. 33, 27-38.

Walther, C., Guénet, J.-L., Simon, D., Deutsch, U., Jostes, B., Goulding, M., Plachov, D., Balling, R. and Gruss, P. (1991). Pax: A murine multigene family of paired box containing genes. Genomics 11, 424-434.

pb [d]

Species: *Drosophila melanogaster*
Chromosomal location: 84A4-5 (ANT-C)
Other names: *proboscipedia*
Cognate genes: *HOXB2*; *Hoxb-2*, *Hoxa-2*
Type of homeobox: *Antp*-class
Accession number: X63728 and X63729

■ Origin and description

The region encompassing *pb* was isolated in a chromosomal "walk" of the Ant-C (Scott et al., 1983; Pultz et al., 1988). A collection of *pb*-null mutations associated with chromosome rearrangements permitted *pb* to be delimited genetically; mRNA-coding sequences were subsequently localized within this region (Pultz et al, 1988). The homeodomain encoded in the *pb* protein is of the *Antp*-class, but is relatively diverged from the Antp/Scr/Ubx/abd-A homeodomains of the fruit fly. Helix 3 of the *pb*-class homeodomain (*pb* and its human and murine homologs, 95% identical in the homeodomain) is unique among the known homeotic proteins, possessing Val-47 in place of Ile. Alternative splicing of *pb* pre-mRNA adjacent to the homeobox may permit the synthesis of at least four different proteins, each with its homeodomain in an altered protein context (Cribbs et al., 1992a).

■ Expression and function

Null mutations of *pb* lead to an adult-specific homeotic transformation, of labial palps (mouthparts) to prothoracic legs; in contrast, partial loss-of-function mutations of *pb* result in a partial but quite distinct adult transformation of the labial palps, to antennal aristae. In both the cases of complete and partial loss-of-function the maxillary palps are modified, though in varying degrees, in a fashion suggesting a transformation toward antennal identity (Kaufman, 1978). In contrast to the adult homeotic transformations observed, the absence of *pb* function in an embryo appears to be without detectable consequence (Pultz et al., 1988). The action of *pb* is thus unique among the known homeotic genes in these two senses: *pb* is adult-specific, and complete versus partial loss of function leads to distinct developmental transformations. Despite the adult-specificity of requirements for *pb* function, *pb* mRNA and protein are accumulated in embryos as well as at the larval and pupal stages; in fact, maximal accumulation of mRNA is detected at about 6-8 hours of embryogenesis. Prominent localized embryonic *pb* protein expression is observed in the labial and maxillary segments of the gnathocephalon that give rise to head-specific structures of the larva and the adult, whereas later embryonic expression is found within the central nervous system (Pultz et al., 1988). Larval expression has likewise been detected within the CNS, as well as the labial imaginal disks that yield the adult labium (Randazzo et al., 1991). Rescue of the homeotic transformation due to the *pb*-condition was achieved by germline transformation, using a P element harboring a *pb* "minigene". This mini-gene is expressed in the labial disks and fully rescues the labial defects of *pb*-animals; however, expression is not detected in the CNS and these flies have corresponding behavioral deficiencies (Randazzo et al,, 1991). While *pb* expression is normally limited to the head apart from the small number of cells in the CNS, ectopic expression of *pb* in the thorax has been detected as a consequence of a mutation that fuses the *Antp* P1 promoter to *pb*. Expression of *pb* protein from the newly formed fusion gene causes dominant thoracic defects correlated to diminished *Antp* expression from the intact *Antp* copy on the homologous chromosome. Thus one potential target of the *pb* homeodomain protein in transcriptional regulation is *Antp* (Cribbs et al., 1992b).

■ References

Cribbs, D.L., Pultz, M.A., Johnson, D., Mazzulla, M. and Kaufman, T.C. (1992a). Structural complexity and evolutionary conservation of the *Drosophila* homeotic gene *proboscipedia*. EMBO J. 11, 1437-1449.

Cribbs, D.L, Pattatucci, A.M., Pultz, M.A. and Kaufman, T.C. (1992b). Ectopic expression of the *Drosophila* homeotic gene *proboscipedia* under the control of the *Antennapedia* P1 promoter causes dominant thoracic defects. Genetics 132, 699-711.

Kaufman, T.C. (1978). Cytogenetic analysis of chromosome 3 in *Drosophila melanogaster*. Isolation and characterization of four new alleles of the *proboscipedia* locus. Genetics 90, 579-596.

Pultz, M.A., Diederich, R., Cribbs, D.L. and Kaufman, T.C. (1988). The *proboscipedia* locus of the *Antennapedia* complex: a molecular and genetic analysis. Genes Dev. 2, 901-920.

Randazzo, F.M., Cribbs, D.L. and Kaufman, T.C. (1991). Rescue and regulation of *proboscipedia*: a homeotic gene of the *Antennapedia* complex. Development 113, 257-271.

PBX1 [h]

Species: *Homo sapiens*

Chromosomal location: 1q23

Other names: -

Cognate genes: -

Type of homeobox: Most similar to yeast MATa1

Accession numbers: M31170, M31222, B33061, A33061

■ Origin and description

This gene was isolated from the t(1;19)(q23;p13) chromosomal translocation breakpoint in acute pre-B cell leukemias. The *PBX1* gene spans over 80 kilobases of chromosome band 1q23; the t(1;19) breakpoints appear to fall within a large intron between exons 2 and 3. The *PBX1* gene codes for two predicted proteins of 46.5 (*Pbx1a*) and 38.4 (*Pbx1b*) kilodaltons that arise by differential splicing and differ in their carboxy terminal composition but otherwise are identical. The *Pbx1* homeodomain is most similar to MATal (35% identical) and, like MATa2, has a 3 amino acid insertion at/near helix 1.

■ Expression

RNA expressed in most fetal and adult human tissues and cell lines with the notable exception of lymphoid lineage cell lines which lack detectable *PBX1* RNA. Differential splice products of *PBX1* RNA are expressed tissue-specifically to result in alternative proteins of 46.5 (*Pbx1a*) and 38.4 (*Pbx1b*) kilodaltons.

■ Genetics, function

Widely expressed homeodomain protein suggested to play a generalized role in transcription control. Oncogenic activation occurs by protein fusion to the transactivation domain of helix-loop-helix protein E2A following t(1;19) chromosomal translocations in acute pre-B cell leukemias.

■ References

Hunger, S.P., Galili, N., Carroll, A.J., Crist, W.M., Link, M.P. and Cleary, M.L. (1991). The t(1;19)(q23;p13) results in consistent fusion of E2A and PBX1 coding sequences in acute lymphoblastic leukemias. Blood 77, 687-693.

Kamps, M.P., Murre, C., Sun, X.H. and Baltimore, D. (1990). A new homeobox gene contributes the DNA binding domain of the t(1;19) translocation protein in pre-B ALL. Cell 60, 547-555.

Monica, K., Galili, N., Nourse, J., Saltman, D. and Cleary, M.L. (1991). PBX2 and PBX3, new homeobox genes with extensive homology to the human proto-oncogene PBX1. Molec. Cell. Biol. 11, 6149-6157.

Nourse, J., Mellentin, J.D., Galili, N., Wilkinson, J., Stanbridge, E., Smith, S.D. and Cleary, M.L. (1990). Chromosomal translocation t(1;19) results in synthesis of a homeobox fusion mRNA that codes for a potential chimeric transcription factor. Cell 60, 535-545.

PBX2 [h]

Species: *Homo sapiens*

Chromosomal location: 3q22-23

Other names: -

Cognate genes: -

Type of homeobox: Most similar to yeast MATa1

Accession number: X59842

■ Origin and description

This gene was isolated by cross homology with the *PBX1* homeobox.

■ Expression

RNA expressed in most fetal and adult human tissues and in most human cell lines.

■ Genetics, function

Suggested to play a generalized role in transcription control based on the widespread expression of *PBX2* RNA in many human fetal and adult tissues.

■ References

Monica, K., Galili, N., Nourse, J., Saltman, D. and Cleary, M.L. (1991). PBX2 and PBX3, new homeobox genes with extensive homology to the human proto-oncogene PBX1. Molec. Cell. Biol. 11, 6149-6157.

PBX3 [h]

Species: *Homo sapiens*
Chromosomal location: 9q33-34
Other names: -
Cognate genes: -
Type of homeobox: Most similar to yeast MATa1
Accession number: X59841

■ Origin and description

This gene was isolated by cross homology with the *PBX1* homeobox.

■ Expression

RNA expressed in most fetal and adult human tissues and in most human cell lines. Differential splicing of *PBX3* RNA results in alternative proteins with predicted molecular weights of 47.1 (*Pbx3a*) and 38.8 (*Pbx3b*) kilodaltons, respectively, that differ in their carboxy-terminal amino acid composition.

■ Genetics, function

Widely expressed homeodomain protein suggested to play a generalized role in transcription control.

■ References

Monica, K., Galili, N., Nourse, J., Saltman, D. and Cleary, M.L. (1991). PBX2 and PBX3, new homeobox genes with extensive homology to the human proto-oncogene PBX1. Molec. Cell. Biol. 11, 6149-6157.

pdm-1 [d]

Species: *Drosophila melanogaster*
Chromosomal location: 33F
Other names: *dPOU-19/pdm1*
Cognate genes: -
Type of homeobox: *Oct-2* type
Accession number: -

■ Origin and description

Genomic DNA of *pdm-1* were isolated with a human *Oct-2* cDNA probe. Using this genomic DNA as a probe, we have isolated several cDNA clones that all belonged to the same class as judged by restriction mapping. The longest cDNA clone contained a 1806 bp open reading frame that has the potential to code for a 601 amino acid peptide. At its C-terminal end, the open reading frame codes for a POU domain in which 92% of the residues of the POU specific domain and 81% of the residues of the homeodomain are identical to human *Oct-2* (Affolter et al., 1993). Comparing our sequence with other recently published *Drosophila* POU box cDNAs revealed that we had characterized the transcripts of the same gene as the one named *pdm-1* (Billin et al., 1991; Lloyd and Sakonjo, 1991) or *dPOU-19* (Dick et al., 1991).

■ Expression

The *pdm-1* transcript is first detected shortly before cellularization as a band of expression extending from 20% to 50% egg length (0% egg length is located at the posterior pole). Upon cellularization, this band of expression resolves into two stronger stripes due to the disappearance of transcripts in the centre. During germ band extension, each of the two bands split again and slowly fade away during stage 8. Before full germ band extension strong segmentally repeated expression is observed in the region of the neuroectoderm. This pattern evolves in a quite dynamic fashion until at about stage 10, virtually all cells of the neurogenic region contain high levels of *pdm-1* transcripts. In stage 11, ectodermal staining has disappeared and transcripts can be detected in a subset of neuroblasts of the developing CNS. *pdm-1* transcripts can now also be seen in the developing anterior and posterior midgut primordias. The staining is quite uniform and extends throughout both primordias until they fuse during stage 12. Only slightly later, the uniform staining of the entire midgut endoderm is dramatically modified. Transcripts disappear from two regions, an anterior domain and a more centrally located domain. This distinct expression pattern in the endoderm is essentially maintained during stage 13 and 14 during which the midgut close up at the dorsal and ventral side. In stages 15 and 16, strong expression is still detectable in the midgut at the level of proventriculus, the first midgut constriction and from slightly anterior to the third constriction to the end of the posterior midgut. Expression in these stages is also apparent in certain cells in the head, the CNS and the PNS (see also Dick et al., 1991).

■ Regulation

Changes in gene expression (i.e. induction of lab) based on an induction process between the visceral mesoderm and the endoderm involving a cascade of interacting genes have previously been reported (Immerglück et al., 1990; Reuter et al., 1990). With respect to the modulation of *pdm-1* expression upon midgut fusion, we found that the repression of previous ubiquitous primordial expression in the central domain is dependent on the activity of *Ubx* and *dpp* in the adjacent VM, repression in the more anteriorly located domain is independent of known homeotic selector genes.

■ Genetics, function

pdm-1 is deleted in Df(2L) prd1.7. We did not find any alteration with respect to the expression of several segmentation genes (in addition to the prd defects). In addition, *pdm-1* appears not to be required for gut fusion and the formation of gut constrictions. No *pdm-1* mutants are available at the moment.

■ References

Affolter, M., Walldorf, U., Kloter, U., Schier, A. and Gehring, W.J. (1993). Regional repression of a *Drosophila* POU box gene in the endoderm involves interaction between germ layers. Development 117, 1199–1210.

Billin, A.N., Cockerill, K.A. and Poole, S.J. (1991). Isolation of a family of *Drosophila* POU domain genes expressed in early development. Mech. Dev. 34, 75-84.

Dick, T., Yang, Y., Yeo, S. and Chia, W. (1991). Two closely linked *Drosophila* POU domain genes are expressed in neuroblasts and sensory elements. Proc. Natl. Acad. Sci. USA, 88, 7645-7649.

Immerglück, K., Lawrence, P.A. and Bienz, M. (1990). Induction across germ layers in *Drosophila* mediated by a genetic cascade. Cell 62, 261-268.

Lloyd, A. and Sakonju, S. (1991). Characterization of two *Drosophila* POU domain genes, related to oct-1 and oct-2, and the regulation of their expression patterns. Mech. Dev. 36, 87-102.

Reuter, R., Panganiban, G.E.F., Hoffmann, F.M. and Scott, M.P. (1990). Homeotic genes regulate the spatial expression of putative growth factors in the visceral mesoderm of *Drosophila* embryos. Development 101, 1031-1040.

pdm-2   [d]

Species: *Drosophila melanogaster*
Chromosomal location: 33F
Other names: *DPOU-28*
Cognate genes: -
Type of homeobox: POU-domain
Accession number: M81958

■ Origin and description

This gene was isolated by PCR using degenerate oligos specific for the POU-specific region of the *Oct-1*, *Oct-2*, *Pit-1*, and *unc-86* genes (Billin et al. 1991; Dick et al. 1991). It is located at band 33F of the left arm of the second chromosome. Its POU domain is quite similar to the *Drosophila pdm-1* gene, also located at band 33F, but is unrelated outside of this. Among the mammalian genes, it is most closely related to the *Oct-1* POU domain.

■ Expression

On Northern blots, the *pdm-2* gene hybridizes to a 2.5 kb RNA with its highest expression in 3-6 hour embryos, with weaker expression in later embryos. Its expression pattern is very similar to that of the *pdm-1* gene. It is first expressed at cellular blastoderm in a wide band stretching from midway along the length of the embryo to ~20% from the posterior end, with an additional sharp arc of expression at the extreme anterior end. During early gastrulation, the broad band splits into two, and additional narrower stripes appear flanking the cephalic furrow and farther posteriorly. The anterior arc of expression moves somewhat dorsally. By early germ band extension, the gene is expressed weakly in a segmentally reiterated series of stripes along the length of the embryo (Billin et al. 1991; Dick et al. 1991). Shortly after this, the striped expression disappears, and a new, complex pattern of expression is seen in the neurogenic ectoderm. As with the *pdm-1* gene, in later embryos the *pdm-2* gene is expressed in a subset of cells of the central nervous system, and peripherally in cells likely to be part of peripheral sense organs (Billin and Poole, 1991; Dick et al. 1991).

■ Function

The function is unknown at present. From its expression pattern, it may be involved in segmentation of the embryo and in neurogenesis.

■ References

Billin, A., Cockerill, K. and Poole, S. (1991). Isolation of a family of *Drosophila* POU domain genes expressed in early development. Mech. Dev. 34, 75-84.

Dick, T., Yang, X., Yeo, S. and Chia, W. (1991). Two closely linked *Drosophila* POU domain genes are expressed in neuroblasts and sensory elements. Proc. Natl. Acad. Sci. USA, 88, 7645-7649.

Pem [m]

Species: *Mus musculus*
Chromosomal location: -
Other names: -
Cognate genes: -
Type of homeobox: related to *Prd* family
Accession number: -

■ Origin and description

The *Pem* cDNA clone was isolated by subtraction hybridization on the basis of differential expression between two T-cell lymphoma cell clones which differ in maturation and tumorigenic characteristics (MacLeod et a., 1990).

■ Expression

The *Pem* gene has an "oncofetal" pattern of expression. *Pem* mRNA is expressed in a variety of immortalized and malignant rodent cell lines from many different lineages, including lymphoid, myeloid, mesencymal, neuronal and epithelial (MacLeod et a., 19901, Wilkinson et al., 1990). In contrast, *Pem* mRNA is not detectably expressed in any adult tissue tested, with the exception of mouse (not rat) testis, which expresses low amounts of *Pem* transcripts. *Pem* mRNA is expressed at least as early as day 6 of mouse fetal development (MacLeod et al., 1990). Its expression is apparently confined to extra-embryonic tissue after day 9 p.c. *Pem* transcripts are present in undifferentiated F9 EC cells and ES stem cells. *Pem* mRNA accumulation is augmented when EC or ES cells are induced to differentiate (Sasaki et al., 1991). The mesencymal stem cell line, 10T$\frac{1}{2}$, expresses trace amounts of *Pem* mRNA. Following commitment of myoblasts, *Pem* mRNA levels in 10T$\frac{1}{2}$ cells are dramatically increased (Sasaki et al., 1991). The regulation of *Pem* gene expression has been examined in somatic cell hybrids formed between T cell lymphomas which differ in maturation status. These studies have shown that unlike many "classical T-cell genes", the *Pem* gene is not transcriptionally regulated by trans-acting inhibitor molecules (Wilkinson et al., 1991).

■ References

MacLeod, C.L., Fong, A.M., Seal, B.M., Walls, L. and Wilkinson, M.F. (1990). Isolation of novel cDNA clones from T-lymphoma cells: one encodes a putative multiple membrane spanning protein. Cell Growth & Diff. 1, 271-279.
Wilkinson, M.F., Kleeman, J., Richards, J. and MacLeod, C.L.

(1990). A novel onco-fetal gene expressed in a stage-specific manner in murine embryonic development. Develop. Biol. 141, 451-455.
Sasaki, A., Doskow, J., MacLeod, C.L., Rogers, M., Gudas, L. and Wilkinson, M.F. (1991). The oncofetal gene *Pem* encodes a homeodomain and is regulated in embryonic and pre-muscle stem cells. Mech. Dev. 34, 155-164.
Wilkinson, M.F., Doskow, J., von Borstell, II., R.C., Fong, A.M. and MacLeod, C.L. (1991). The expression of several T cell specific and novel genes are repressed by trans-acting factors in immature T lymphoma clones. J. Exp. Med. 174, 269-280.

PHox 2 [m]

Species: *Mus musculus*
Chromosomal location: 7E1-7F2
Other names: -
Cognate genes: -
Type of homeobox: Related to the *Drosophila paired* gene
Accession number: -

■ Origin and description

An incomplete cDNA was cloned by screening an expression library from mouse N2a neuroblastoma cells with a homeodomain-binding sequence identified in the promoter region of the *NCAM* gene (Hirsch et al., 1991). Rescreening the same library with the original clone yielded a full length cDNA. The composite sequence contains an open reading frame of 840 nt. It contains a homeobox encoding most of the residues diagnostic for a homeodomain of the *paired* family (Frigerio et al., 1986), except for the presence of a glutamine instead of a serine at position 9 of the recognition helix. The *Phox 2* homeobox shows the highest degree of sequence similarity with human *Phox 1* (Gruenberg et al., 1992), mouse *Mhox* (Cserjesi et al., 1992) and mouse *S8* (Opstelten et al., 1991).

■ Expression

On RNA from N2a cells, a *Phox 2* probe detects a major transcript of 1.7 kb. *In situ* hybridization shows strong expression of the *Phox 2* gene in the mouse embryo in sympathetic, parasympathetic and enteric ganglia and in the adrenal medulla. Weak expression in the CNS can be ascribed to the noradrenergic areas A1-A6 in the met- and myelencephalon.

■ References

Cserjesi, P., Lilly, B., Bryson, L., Wang, Y., Sassoon, D.A. and Olson, E.N. (1992). MHox: a mesodermally restricted homeodomain protein that binds an essential site in the muscle creatine kinase enhancer. Development 115, 1087-1101.

Frigenio, G., Burri, M., Bopp, D., Baumgartner, S. and Noll, M. (1986). Structure of the segmentation gene paired and the *Drosophila* PRD gene set as part of a gene network. Cell 47, 735-746.

Gruenberg, D.A., Natesan, S., Alexandre, C. and Gilman, M.Z. (1992). Human and *Drosophila* homeodomain proteins that enhance the DNA-binding activity of serum response factor. Science 257, 1089-1095.

Hirsch, M.R., Valarché, I., Deagostini-Bazin, H., Pernelle, C., Joliot, A. and Goridis, C. (1991). An upstream regulatory element of the NCAM promoter contains a binding site for homeodomains. FEBS Lett. 287, 197-202.

Opstelten, D.-J.E., Vogels, R., Robert, B., Kalkhoven, E., Zwartkruis, F., de Laaf, L., Destrée, O.H., Deschamps, J., Lawson, K.A. and Meijlink, F. (1991). The mouse homeobox gene *S8* is expressed during embryogenesis predominantly in mesenchyma. Mech. Dev. 34, 29-42.

PHO2   [Sac]

Species: *Saccharomyces cerevisiae* (S288C)
Chromosomal location: IV
Other names: *BAS2, GRF10*
Cognate genes: -
Type of homeobox: -
Accession numbers: X05062, M22259, M24613, X54293 (GG100-14D)

■ Origin and description

The gene was cloned independently several times either from strain S288C (Sengstag and Hinnen, 1987; Arndt et al., 1987; Berben et al., 1988), or strain GG100-14D (McCarthy et al., 1991). Of the three yeast homeodomain genes identified thus far, the one of *PHO2* (Bürglin, 1988; Berben et al., 1988) has the best match to the canonical 60 amino acid homeodomain. The homeodomain is located at position 77-136 in the 559 amino acid long protein. Deletion of the first 64 amino acids does not result in loss of DNA binding, while deletion of the first 111 amino acids results in loss of DNA binding (Arndt et al., 1991).

■ Genetics, function

PHO2 has been shown to be a regulator of several different genes. It acts synergistically with *BAS1* to provide the high basal level of *HIS4* transcription in the absence of amino acid or phosphate starvation (Arndt et al., 1987; Tice-Baldwin et al., 1989; Devlin et al., 1991). *PHO2* is also a positive regulator of the *PHO5* gene in the phosphate metabolism pathway (Oshima, 1982; Vogel et al., 1989). It has been shown that *PHO2* binds directly to sequences upstream of *PHO5* and *HIS4* (Arndt et al., 1987; Tice-Baldwin et al., 1989; Vogel et al., 1989; Devlin et al., 1991). Furthermore, it binds upstream of the *TRP4* gene and is involved in the modulation of the general control

response of *TRP4* expression under phosphate limitation (Braus et al., 1989). The binding sites on the *HIS4*, *PHO5* and *TRP4* promoters have been determined by footprinting (Tice-Baldwin et al., 1989; Vogel et al., 1989; Braus et al., 1989). A possible role of *PHO2* in the yeast life cycle has been suggested by Berben et al. (1988): Homozygous *PHO2* null dipoids fail to sporulate.

■ References

Arndt, K.T., Styles, C. and Fink, G.R. (1987). Multiple global regulators control HIS4 transcription in yeast. Science 237, 874-880.

Berben, G., Legrain, M. and Hilger, F. (1988). Studies on the structure, expression and function of the yeast regulatory gene PHO2. Gene 66, 307-312.

Braus, G., Mösch, H.-U., Vogel, K., Hinnen, A. and Hütter, R. (1989). Interpathway regulation of the TRP4 gene of yeast. EMBO J. 8, 939-945.

Bürglin, T.R. (1988). The yeast regulatory gene PHO2 encodes a homeodomain. Cell, 53, 339-340.

Develin, C., Tice-Baldwin, K., Shore, D. and Arndt, K.T. (1991). RAP1 is required for BAS1/BAS2- and GCN4-dependent transcription of the yeast HIS4 gene. Mol. Cell. Biol. 11, 3642-3651.

McCarthy, B.J., Creasy, C.L. and Bergman, L.W. (1991). Molecular analysis of a temperature sensitive allele of the PHO2 gene of *Saccharomyces cerevisiae*. Nucl. Acis Res. 19, 3463.

Oshima, Y. (1982). Regulation of phosphatases. Strathern, J.N., Jones, E.W. and Broach, J.R. (eds.) in Molecular Biology of the Yeast Saccharomyces. Cold Spring Harbor Laboratory Press, Cold Spring Harbor. 159-180.

Sengstag, C. and Hinnen, A. (1987). The sequence of the *Saccharomyces cerevisiae* gene PHO2 codes for a regulatory protein with unusual aminoacid composition. Nucl. Acids Res. 15, 233-246.

Tice-Baldwin, K., Fink, G.R. and Arndt, K.T. (1989). BAS1 has a Myb motif and activates HIS4 transcription only in combination with BAS2. Science 246, 931-935.

Vogel, K., Hörz, W. and Hinnen, A. (1989). The two positively acting regulatory proteins PHO2 and PHO4 physically interact with PHO5 upstream activation regions. Mol. Cell. Biol. 9, 2050-2057.

Pi   [Sp]

Species: *Schizosaccharomyces pombe*
Chromosomal location: -
Other names: -
Cognate genes: -
Type of homeobox: atypical
Accession number: XO7643

■ Origin and description

Kelly et al. (1988) describe the cloning of the mating-type region of fission yeast. The gene *Pi* in the *mat*2-P cassette is the only one of the four genes to display similarity to the homeobox genes. The homeodomain appears most closely related to the yeast α2 homeodomain, and probably - like α2 - has 3 extra amino acids between helix 1 and helix 2. In addition the *Pi* homeodomain is truncated at the carboxy-terminus.

■ Expression

Kelly et al. (1988): The transcription of *Pi* is strongly influenced by nutritional conditions. During vegetative growth of both the haploid and diploid strains the *Pi* transcript is undetectable and remains undetectable even upon entry into asexual stationary phase. Subsequent removal of the nitrogen source causes the level of *Pi* transcript to be dramatically elevated in both strains (neither conjugation nor sporulation occurs in nitrogen-rich medium).

■ Genetics, function

Kelly et al. (1988): The mating-type region of fission yeast consists of three components, *mat*1, *mat*2-P and *mat*3-M, each separated by 15 kb. Cell-type is determined by the alternate allele present at *mat*1, either P in an h⁺ or M in an h⁻ cell. *mat*2-P and *mat*3-M serve as donors of information that is transposed to *mat*1 during a switch of mating type. *Pi* is not required for the h⁺ or h⁻ mating phenotype, however it appears to be essential for meiosis and sporulation.

■ References

Kelly, M., Burke, J., Smith, M., Klar, A. and Beach, D. (1988). Four mating-type genes control sexual differentiation in the fission yeast. EMBO J. 7, 1537-1547.

prd [d]

Species: *Drosophila melanogaster*
Chromosomal location: Chromosome 2, 33C1
Other names: *paired*
Cognate genes: -
Type of homeobox: *prd*-type
Accession number: M14548 (DROPRD)

■ Origin and description

This pair-rule segmentation gene (Nüsslein-Volhard and Wieschaus, 1980) has been isolated by chromosomal walking in a region defined by the large deficiency *Df(2L)prd*[1.25] (Nüsslein-Volhard et al., 1984) and has been identified by differential transcript mapping and subsequent mapping of a 1.1 kb insertion in the *prd* transcript

(Kilchherr et al., 1986) of the X-ray induced mutant *prd*[2.45] (Tearle and Nüsslein-Volhard, 1987). The *prd* homeodomain is the prototype of a separate homeodomain class (Bopp et al., 1986; Frigerio et al., 1986). Additional members of this class are encoded by the segment-polarity locus *gooseberry* (*gsb*) (Nüsslein-Volhard and Wieschaus, 1980) which includes two genes most closely related to *prd* (Bopp et al., 1986; Baumgartner et al., 1987). Like *prd*, the two *gsb* proteins harbor at their amino terminal a paired-domain and a *prd*-type homeodomain. All three homeodomains of *prd* and *gsb* are extended at their amino terminal by 18 amino acids (Bopp et al., 1986).

■ Expression

The expression patterns during embryogenesis of *prd* transcripts (Kilchherr et al., 1986) and of *prd* protein (Gutjahr et al., 1993) have been described in detail. The most striking feature of these patterns is their transition, during blastoderm cellularization, from an initial, transiently expressed pair-rule pattern with a double-segment repeat to a segment-polarity pattern exhibiting a single-segment periodicity during blastoderm cellularization. Later stages, during and after germ band retraction, show specific expression in the head region as well as in the central nervous system (Gutjahr et al., 1993). The *prd* gene is regulated in a combinatorial manner by all other pair-rule gene products and is at the bottom of the gene regulatory cascade of pair-rule genes (Baumgartner and Noll, 1990) while its initial activation does not depend on any of the pair-rule genes (Gutjahr et al., 1993).

■ References

Baumgartner, S. and Noll, M. (1990). Network of interactions among pair-rule genes regulating *paired* expression during primordial segmentation of *Drosophila*. Mech. Dev. 33, 1-18.

Baumgartner, S., Bopp, D., Burri, M. and Noll, M. (1987). Structure of two genes at the *gooseberry* locus related to the *paired* gene and their spatial expression during *Drosophila* embryogenesis. Genes Dev. 1, 1247-1267.

Bopp, D., Burri, M., Baumgartner, S., Frigerio, G. and Noll, M. (1986). Conservation of a large protein domain in the segmentation gene *paired* and in functionally related genes of *Drosophila*. Cell 47, 1033-1040.

Frigerio, G., Burri, M., Bopp, D., Baumgartner, S. and Noll, M. (1986). Structure of the segmentation gene *paired* and the *Drosophila* PRD gene set as part of a gene network. Cell 47, 735-746.

Gutjahr, T., Frei, E. and Noll, M. (1993). Complex regulation of early *paired* expression: Initial activation by gap genes and pattern modulation by pair-rule genes. Development 117, 609–623.

Kilchherr, F., Baumgartner, S., Bopp, D., Frei, E. and Noll, M. (1986). Isolation of the *paired* gene of *Drosophila* and its spatial expression during early embryogenesis. Nature 321, 493-499.

Nüsslein-Volhard, C. and Wieschaus, E. (1980). Mutations affecting segment number and polarity in *Drosophila*. Nature 287, 795-801.

Nüsslein-Volhard, C., Wieschaus, E. and Kluding, H. (1984).

Mutations affecting the pattern of the larval cuticle in *Drosophila melanogaster*. I. Zygotic loci on the second chromosome. Roux's Arch. Dev. Biol. 193, 267-282.

Tearle, R. and Nüsslein-Volhard, C. (1987). Tübingen mutants and stock list. Dros. Inf. Serv. 66, 209-269.

Quox-1 [qu]

Species: *Coturnix coturnix japonica* (quail)
Chromosomal location: -
Other names: -
Cognate genes: -
Type of homeobox: *Antp*-type
Accession number: M59714

■ Origin and description

The *Quox-1* cDNA was isolated from a cDNA library of E5 quail spinal cord by cross-hybridization with the Antp homeobox (Xue et al., 1991). The open reading frame of *Quox-1* corresponds to a predicted protein of 242 amino acids. *Quox-1* shares extensive protein homology with the mouse Hox-1.1 in the homeodomain (100%) and 5' coding region (76%). Interestingly this homology does not extend 3' of the homeodomain. However, the COOH-terminal domain of the *Quox-1* protein has no significant homology with other known homeoproteins. Another feature of *Quox-1* is of interest: in the 3' translated region, a 12-bp sequence repeats perfectly a highly conserved sequence of the IV helix of homeodomain. The genomic DNA of *Quox-1* has also been cloned.

■ Expression

Expression of *Quox-1* gene was studied by *in situ* hybridization in quail embryos from E3 to E6. *Quox-1* gene is widely expressed in the developing central nervous system, including the forebrain. *Quox-1* is the first vertebrate homeobox gene to be described in which expression occurs throughout the rostral part of the central nervous system, more anteriorly than any Hox gene so far described. Outside the central nervous system, transcriptions of *Quox-1* were mainly detected in the neural crest-derived spinal ganglia, in the olfactory placode and in the endoderm-derived esophagus, trachea, liver and digestive organs, as well as in the perichondrium of vertebrae.

■ References

Zhi-Gang Xue, Gehring, W.J. and Le Douarin, N. (1991). Quox-1, a quail homeobox gene expressed in the embryonic central nervous system, including the forebrain. Proc. Natl. Acad. Sci. USA, 88, 2427-2431.

ro [d]

Species: *Drosophila melanogaster*
Chromosomal location: 97D5-7
Other names: *rough*
Cognate genes: -
Type of homeobox: -
Accession number: -

■ Origin and decsription

The *rough* mutation was discovered by Muller in 1913 (Lindsley and Grell, 1968). The gene has been isolated by P-element tagging (Tomlinson et al., 1988) and independently by cross homology to a homeobox consensus oligonucleotide (Saint et al., 1988). The *rough* homeodomain is 57 and 58 % identical to the homeo domains of *Antp* and*Scr*, respectively.

■ Expression

rough is expressed in the eye imaginal disc starting in the morphogenetic furrow (MF) and in the developing ommatidial preclusters. It is expressed only in the eye disc, consistent with its phenotype (see below). In the MF most nuclei of the precluster express *rough*, R2, R3, R4, R5, R8 and the mistery cells. Probably there are some nuclei in the basal region that do not show expression. The second tier of *rough* expression begins posterior to the MF (~6 hr of ommatidial development) when it becomes refined to 4 cells, R2, R3, R4 and R5. The expression is still detectable at the anterior margin in 40-hr pupal retinae and is then fading out gradually (Kimmel et al., 1990).

■ Function

The *rough* gene is required for ommatidial development and *rough*⁻ flies have ommatidia with too few photoreceptor cells. Analysis of somatic clones suggests that *rough* is required only in R2 and R5 (Tomlinson et al., 1988). Misexpression of the *rough* gene (Basler et al., 1990; Kimmel et al., 1990) and a phenotypic analysis using cell type specific markers and double mutant combinations (Heberlein et al., 1991) indicate that *rough* specifies the photoreceptor cell subtype R2/R5, which are required to induce other neighboring cells as photoreceptors. In *rough*⁻ ommatidia R2/R5 are transformed to the R3/R4 celltype (Heberlein et al., 1991).

■ References

Basler, K., Yen, D., Tomlinson, A. and Hafen, E. (1990). Reprogramming cell fate in the developing *Drosophila retina*: transformation of R7 cells by ectopic expression of *rough*. Genes and Dev. 4, 728-739.

Heberlein, U., Mlodzik, M. and Rubin, G.M. (1991). Cell-fate determination in the developing *Drosophila* eye: role of the *rough* gene. Development 112, 703-712.

Kimmel, B.E., Heberlein, U. and Rubin, G.M. (1990). The homeo domain protein rough is expressed in a subset of cells in the developing *Drosophila* eye where it can specify photoreceptor cell subtype. Genes and Dev. 4, 712-727.

Lindsley, D.L. and Grell, E.H. (1968). Genetic variations of *Drosophila melanogaster*. Carnegie Inst. Wash. Publ. No. 627.

Saint, R., Kalionis, B., Lockett, T.J. and Elizur, A. (1988). Pattern formation in the developing eye of *Drosophila melanogaster* is regulated by the homeobox gene, *rough*. Nature 334, 151-154.

Tomlinson, A., Kimmel, B.E. and Rubin, G.M. (1988). *rough*, a *Drosophila* hoemobox gene required in photoreceptors R2 and R5 for inductive interactions in the developing eye. Cell 55, 771-784.

SCIP [r]

Species: *Rattus norvegicus*
Chromosomal location: –
Other names: *Testes-1 (Tst-1)*
Cognate genes: *Oct-6* (mouse)
Type of homeobox: POU domain
Accession number: GenBank M35205

■ Origin and description

This gene was identified through screening of rat sciatic nerve and rat Schwann cell cDNA libraries with degenerate oligonucleotide probes corresponding to the POU-specific and POU-homeobox domains of *Oct-1*, *Oct-2*, and *Pit-1* (Monuki et al. 1989), and independently, by PCR of rat testes cDNA (He et al. 1989; N.B. This paper contains a sequence error in the SCIP POU domain). The complete mRNA encodes a protein of 451 amino acids (Monuki et al. 1990), with the POU domain located in the carboxy-terminal half of the protein. Nearly 50% of the amino acids upstream of the *SCIP* POU domain are either glycine or alanine, and the mRNA is exceptionally GC-rich (Monuki et al. 1990; He et al. 1991). In the rat, the *SCIP* gene is devoid of introns, and has other features of an expressed retroposon (Kuhn et al. 1991). The SCIP protein has been shown to be localized to the nucleus, and to be of a size (47 kD) predicted by the cloned cDNA (Kuhn et al. 1991).

■ Expression

SCIP/Oct-6 is widely expressed during early mammalian development, but in late embryonic and postnatal animals expression is largely confined to cells of the developing central and peripheral nervous systems and to the testes (Monuki et al. 1989; He et al. 1989; Monuki et al. 1990; Suzuki et al. 1990). Within the postnatal nervous systems, a large fraction of expressing cells appear to be progenitors to glial cells, and in particular, to myclinating glia (Monuki et al. 1989; 1990), although some expression in mature neurons cannot be excluded (He et al. 1989; Suzuki et al. 1990). In cultured peripheral glial cells (Schwann cells), *SCIP* expression is strongly potentiated by elevation of intracellular cyclic AMP. *SCIP* expression is highest in glial progenitors, and falls to much lower levels when these cells exit the cell cycle and differentiate into myelin-forming cells (Monuki et al. 1990; Bögler et al. 1992). When Schwann cells de-differentiate in response to nerve transection, *SCIP* expression is transiently re-elevated when these cells transiently re-enter the cell cycle (Monuki et al. 1990).

■ Function

SCIP is an octamer binding protein like *Oct-1*, *Oct-2*, and *Oct-3/4* (Suzuki et al. 1990; Kuhn et al. 1991). When transfected into cultured Schwann cells along with reporter constructs driven by the promoters of the major myelin-specific genes (e.g. those encoding the myelin-specific structural proteins P_O and MBP), SCIP acts as a strong transcriptional repressor of these genes (Monuki et al. 1990). In contrast, when co-transfected with reporter constructs driven by an octamer motif and a heterologous TATA box, the protein acts as a transcriptional activator (Suzuki et al. 1990). This simultaneous activation/repression demonstration has been made using the same protein concentration in the same cells. SCIP does not act as a transactivator of myelin-specific promoters when tested by transfection into heterologous non-glial (HeLa) cells. Recent studies employing anti-sense oligonucleotides have suggested that elevated *SCIP* expression is a necessary but not sufficient condition for Schwann cell division.

■ Remarks

SCIP pronounced "skip", for suppressed cyclic AMP-inducible POU).

■ References

Bögler, O., Kuhn, R., Entwistle, A., Monuki, E.S., Lemke, G. and Noble, M. (1992). Single cell analysis of SCIP protein expression in the oligodendrocyte-type-2 astrocyte lineage. J. Cell Biol. submitted.

He, X., Treacy, M.M., Simmons, D.M., Ingraham, H.A., Swanson, L.W. and Rosenfeld, M.G. (1989). Expression of a large family of POU domain regulatory genes in mammalian brain development. Nature 340, 35-42.

He, X., Gerrero, R., Simmons, D.M., Park, R.E., Lin, C.R., Swanson, L.W. and Rosenfeld, M.G. (1991). Tst-1, a member of the POU domain gene family, binds the promoter of the gene encoding the cell adhesion molecule P_O. Molec. Cell. Biol. 11, 1739-1744.

Kuhn, R., Monuki, E.S. and Lemke, G. (1991). The gene encoding the transcription factor SCIP has features of an expressed retroposon. Molec. Cell. Biol. 11, 4642-4650.

Monuki, E.S., Weinmaster, G., Kuhn, R. and Lemke, G. (1989). SCIP: A glial POU domain gene regulated by cyclic AMP. Neuron 3, 783-793.

Monuki, E.S., Kuhn, R., Weinmaster, G., Trapp, B.D. and Lemke, G. (1990). Expression and activity of the POU transcription factor SCIP. Science 249, 1300-1303.

Suzuki, N., Rohdewohld, H., Neuman, T., Gruss, P. and Schöler, H.R. (1990). Oct-6: a POU transcription factor expressed in embryonal stem cells and in the developing brain. EMBO J. 9, 3723-3732.

Scr [d]

Species: *Drosophila melanogaster*
Chromosomal location: Chromosome 3, map position 3-47.5
 Salivary gland chromosome band 84B1-2
Other names: *Sex combs reduced, Multiple sex combs* (*Msc*)
Cognate genes: -
Type of homeobox: *Antp*-class
Accession number: -

■ Origin and description

Sex combs reduced (*Scr*) was first isolated as an embryonic lethal mutant by Kaufman et al. (1980). Most *Scr* alleles are late embryonic lethals and show homeotic transformations of the first thoracic and the labial segments (Wakimoto and Kaufman, 1981; Sato et al., 1985) which are described below. The location of the *Scr* gene on the chromosomal DNA isolated by chromosomal walking was defined by mapping breakpoints of chromosomal rearrangements (Garber et al., 1983; Scott et al., 1983). Subsequently, the gene was cloned and partially sequenced (Kuroiwa et al., 1985; Le Motte et al., 1989). It consists of three exons approximately 0.5, 1.0 and 2.5 kb in length (listed in 5' to 3' order) separated by two introns of approximately 6 and 15 kb respectively, spanning 25 kb of genomic DNA. Breakpoint associated mutations extend considerably further, and some seem to map even beyond the next gene on the 5' side, which is *fushi tarazu* (*ftz*). In this context it may be important to realize that the *ftz* gene in *Drosophila melanogaster* is inverted relative to the one in *Drosophila hydei* (Maier et al., 1990; see *ftz*) and as a consequence of the inversion some of the *cis*-regulatory elements of *Scr* may have been relocated on the other side of the *ftz* gene. In any case the *Scr* gene (including the control regions) is considerably larger than the transcribed region alone. The cDNA of the entire *Scr* message of 4135 bp was sequenced (Le Motte et al., 1989). It encodes a protein of 413 amino acids. The homeobox is located in exon 3 and shows 82% sequence identity to that of *Antp*, whereas the amino acid sequence of the homeodomain is 92% identical. 5' to the homeodomain a YPWM box is found (as in many homeoproteins of the *Antp*-class) which has extended similarly to *Deformed* (*dfd*) IYPWMK..H and the first three amino acids at the 5' end are also identical to those of *Dfd* (if the second ATG is used for initiation of translation). In addition, the protein contains glycine repeats and M- or Opa-repeats (CAX) encoding Gln, Asn or Ser. Unexpectedly, a CAX repeat was also found in 3' untranslated trailer region.

■ Expression

The expression of *Scr* was studied by *in situ* hybridization (Kuroiwa et al., 1985; Martinez-Arias et al., 1987) and by antibody staining for protein localization (Riley et al., 1987; Le Motte et al., 1989; Mahaffey and Kaufman, 1987a and b; Mahaffey et al., 1989; Glicksman and Brower, 1988). *Scr* transcripts are first detected during gastrulation in a band of cells just behind the cephalic furrow, in the position of the second *engrailed* (*en*) stripe, which corresponds to PS2. Using a highly sensitive antibody, the *Scr* protein (SCR) can also be detected at this stage with a relatively short time lag (Le Motte et al., 1989). During the extended germband stage the SCR labelling moves ~6-8 cells posteriorly with respect to the cephalic furrow and then extends into PS3. There is a gradual shift from parasegmental to segmental expression in the labial and prothoracic segments. Besides the epidermis, the CNS becomes labelled in a single neuromere of the suboesophageal ganglion in PS2. During the extended germband stage there is a gap of expression in the epidermis between the areas of expression of *Scr* in PS2 and *Antp* in PS4. However, at germband retraction the gap disappears and the areas of expression of *Antp* and *Scr* abut. This is not the case for the CNS where the gap in PS3 persists. Following germband retraction a third tissue expresses SCR, the visceral mesoderm. SCR is not only detected in the anterior midgut, anterior to the first gut constriction, where ANTP is expressed, but also in the posterior midgut, behind the area of UBX expression. In third instar larvae *Scr* is expressed in the labial and humeral (dorsal prothoracic) imaginal discs, in the adepithelial (muscle-precursor) cells of the three pairs of leg discs, and in a small patch of cells in the antennal disc. In the CNS, SCR is detected in a narrow band of cells in the suboesophageal ganglion (Glicksman and Brower, 1988). In the adult fly, the same neuromere of the suboesophageal ganglion, which presumably corresponds to the labial segment is labelled by SCR antibodies (Mahaffey and Kaufman, 1987a).

■ Function

The function of *Scr* has been deduced from the phenotype of loss-of-function mutations. *Scr* null mutants die at the end of embryogenesis and show evidence of homeotic transformation of the labial and first thoracic segments, where the gene is most strongly expressed. The first thoracic segment (T1) is partially transformed into T2, which results in a reduction of the "beard", a cluster of denticles on the ventral side of the cuticle, which is characteristic for T1. The labial segment is transformed towards the maxillary segment and forms an additional antenno-maxillary complex of sensory organs. Thus, the two segments are transformed in opposite directions, the labial segment anteriorly, the prothoracic segment posteriorly. This is an exception to the rules established for the Bithorax and *Antennapedia* genes, in which loss-of-function mutants result in anterior transformations. The dominant gain-of-function mutants have an opposite effect with respect to T1, they transform T2 and T3 leg discs towards T1, which results in the formation of extra sex combs on T2 and T3. The known dominant gain-of-function mutants do not affect the embryo. However, an artificial dominant gain-of-function mutant, constructed by insertion of the cDNA into a heatshock vector and germline transformation, has a dramatic effect on embryogenesis. Heat induction at early embryonic stages leads to a failure of head involution and the homeotic transformation of T2 and T3 to T1 (Gibson et al., 1990). This clearly indicates that *Scr* specifies the prothoracic (T1) and the labial segments. The expression in the CNS and visceral mesoderm strongly suggests an important function in those tissues as well.

■ References

Garber, R.L., Kuroiwa, A. and Gehring, W.J. (1983). Genomic and cDNA clones of the homeotic locus *Antennapedia* in *Drosophila*. EMBO J. 2, 2027-2036.

Gibson, G., Schier, A., Le Motte, P. and Gehring W.J. (1990). The specification of *sex combs reduced* and *Antennapedia* are defined by a distinct portion of each protein that includes the homeodomain. Cell 62, 1087-1103.

Glicksman, M.A. and Brower, D.L. (1988). Expression of the *Sex combs reduced* protein in *Drosophila larvae*. Devel. Biol. 127, 113-118.

Kaufman, Lewis, R. and Wakimoto, B. (1980). Cytogenetic analysis of chromosome 3 in *Drosophila melanogaster*: the homeotic gene complex in polytene chromosome interval 84A-B. Genetics 94, 115-133.

Kaufman, T.C., Seeger, M.A. and Olsen, G. (1990). Molecular and genetic organization of the *Antennapedia* gene complex of *Drosophila melanogaster*. Adv. Genet. 27, 309-362.

Kuroiwa, A., Kloter, U., Baumgartner, P. and Gehring W.J. (1985). Cloning of the homeotic *Sex combs reduced* gene in *Drosophila* and *in situ* localization of its transcripts. EMBO J. 4, 3757-3764.

LeMotte, P., Kuroiwa, A., Fessler, L.I. and Gehring, W.J. (1989). The homeotic gene *Sex combs reduced* of *Drosophila*: gene structure and embryonic expression. EMBO J. 8, 219-227.

Mahaffey, J.W. and Kaufman, T.C. (1987a). Distribution of the *Sex combs reduced* gene products in *Drosophila melanogaster*. Genetics 117, 51-60.

Mahaffey, J.W. and Kaufman, T.C. (1987b). The homeotic genes of the *Antennapedia* complex and Bithorax complex of *Drosophila*.. Developmental genetics of higher organisms: a primer in developmental biology. Macmillan New York (ed. G.M. Malacinski) pp. 329-360.

Mahaffey, J.W., Diederich, R.J. and Kaufman, T.C. (1989). Novel patterns of homeotic protein accumulation in the head of the *Drosophila* embryo. Development 105, 167-174.

Maier, D., Preiss, A. and Powell, J.R. (1990). Regulation of the segmentation gene *fushi tarazu* has been functionally conserved in *Drosophila*. EMBO J. 9, 3957-3966.

Martinez-Arias, A., Ingham, P., Scott, M. and Akam, M. (1987). The spatial and temporal deployment of *Dfd* and *Scr* transcripts throughout development of *Drosophila*. Development 100, 673-683.

Riley, P.D., Carroll, S.B. and Scott, M.P. (1987). The expression and regulation of *Sex combs reduced* protein in *Drosophila* embryos. Genes and Dev. 1, 716-730.

Sato, T., Hayes, P.H. and Denell, R.E. (1985). Homeosis in *Drosophila*: Roles and spatial patterns of expression of the *Antennapedia* and *Sex combs reduced* loci in embryogenesis. Devel. Biol. 111, 171-192.

Scott, M.P., Weiner, A.J., Polisky, B.A, Hazelrigg, T.I., Pirrotta, V., Scalenghe, F. and Kaufman, T.C. (1983). The molecular organization of the *Antennapedia* complex of *Drosophila*. Cell 35, 763-776.

Wakimoto, B.T. and Kaufman, T.C. (1981). Analysis of larval segmentation in lethal genotypes associated with the *Antennapedia* gene complex in *Drosophila melanogaster*. Dev. Biol. 81, 51-64.

ScrAf [ar]

Species: *Artemia franciscana*
Chromosomal location: -
Other names: -
Cognate genes: *Scr*
Type of homeobox: *Antp*-class
Accession number: X70080

■ Origin and description

Cloned by PCR with degenerate homeobox primers, followed by inverse PCR of 1.6 kb fragment from genomic DNA. Amino acid conservation in the homeodomain and immediately flanking regions uniquely identifies this gene as a homologue of *Drosophila Scr*.

■ References

Averof, M. and Akam, M. (1993). Hom/Hox genes in a crustacean; implication for the origin of insect and crustacean body parts. Current biology 3, 73–78.

ScrSg [Sg]

Species: *Schistocerca gregaria* (grasshopper)
Chromosomal location: -
Other names: -
Cognate genes: *Scr*
Type of homeobox: *Antp*-type
Accession number: -

■ Origin and description

Cloned by low stringency screening of a *Schistocerca gregaria* genomic library using *Drosophila* homeobox probes.

Amino acid conservation extends both upstream and downstream of the homeodomain, and uniquely identifies this gene as a homologue of *Drosophila Scr*.

■ Expression

Analyzed only by *in situ* hybridization to sections of embryos at 45% development expressed in posterior maxillary, labial and part of first thoracic segments.

■ References

Akam, M., Dawson, I. and Tear, G. (1988). Homeotic genes and the control of segment identity. Development, 104 Supplement, 123-133.
Tear, G. (1990). D. Phil. thesis, University of Cambridge.

smox-1 [Sm]

Species: *Schistosoma mansoni*
Chromosomal location: -
Other names: -
Cognate genes: -
Type of homeobox: *Antp*-type
Accession number: M85306

■ Origin and description

This gene was isolated from a cDNA library of the cercarial stage of development screened with a degenerate oligomer encoding the peptide sequence KIWFQNRR (Webster and Mansour, 1992). In common with many other Antp-class genes, *smox-1* contains a YPWM sequence upstream of the homeodomain (reviewed in Scott et al., 1989).

■ References

Scott, M.P., Tamkun, J.W. and Hartzell III, G.W. (1989). The structure and function of the homeodomain. Biochim. Biophys. Acta 989, 25-48
Webster, P.J. and Mansour, T.E. (1992). Conserved classes of homeodomains in *Schistosoma mansoni*, an early bilateral metazoan. Mech. Dev. 38, 25-32.

smox-2 [Sm]

Species: *Schistosoma mansoni*
Chromosomal location: -
Other names: -
Cognate genes: -
Type of homeobox: *en*-type
Accession number: M85305

■ Origin and description

This gene was isolated from a cDNA library of unisex female worms screened with a degenerate oligomer encoding the peptide sequence KIWFQNRR (Webster and Mansour, 1992). The 524 amino acid deduced protein sequence contains a region of approximately 110 amino acids, including the homeodomain, which is highly conserved in all *engrailed* homologues (reviewed in Joyner and Hanks, 1992).

■ References

Joyner, A. and Hanks, M. (1992). The engrailed genes: evolution of function. Sem. Dev. Biol. 2, 435–445.
Webster, P.J. and Mansour, T.E. (1992). Conserved classes of homeodomains in *Schistosoma mansoni*, an early bilateral metazoan. Mech. Dev. 38, 25–32.

smox-3   [Sm]

Species: *Schistosoma mansoni*
Chromosomal location: -
Other names: -
Cognate genes: -
Type of homeobox: *paired*-type
Accession number: M85303

■ Origin and description

This gene was isolated from a cDNA library of the schistosomule stage of development screened with a degenerate oligomer encoding the peptide sequence KIWFQNRR (Webster and Mansour, 1992). It contains a paired-class homeodomain; the sequence does not extend sufficiently to determine if it also contains a paired box.

■ References

Webster, P.J. and Mansour, T.E. (1992). Conserved classes of homeodomains in *Schistosoma mansoni*, an early bilateral metazoan. Mech. Dev. 38, 25–32.

smox-4   [Sm]

Species: *Schistosoma mansoni*
Chromosomal location: -
Other names: -
Cognate genes: -
Type of homeobox: -
Accession number: M85302

■ Origin and description

This gene was isolated from a cDNA library of adult female worms screened with a degenerate oligomer encoding the peptide sequence KIWFQNRR (Webster and Mansour, 1992). It contains a homeodomain which is not closely homologous to any of the currently described classes of homeodomains.

■ References

Webster, P.J. and Mansour, T.E. (1992). Conserved classes of homeodomains in *Schistosoma mansoni*, an early bilateral metazoan. Mech. Dev. 38, 25–32.

smox-5   [Sm]

Species: *Schistosoma mansoni*
Chromosomal location: -
Other names: -
Cognate genes: -
Type of homeobox: -
Accession number: M85304

■ Origin and description

This gene was isolated from a cDNA library of unisex female worms screened with a degenerate oligomer encoding the peptide sequence KIWFQNRR (Webster and Mansour, 1992). It contains a homeodomain which is not closely homologous to any of the currently described classes of homeodomains.

■ References

Webster, P.J. and Mansour, T.E. (1992). Conserved classes of homeodomains in *Schistosoma mansoni*, an early bilateral metazoan. Mech. Dev. 38, 25–32.

smox-6   [Sm]

Species: *Schistosoma mansoni*
Chromosomal location: -
Other names: -
Cognate genes: -
Type of homeobox: -
Accession number: M85307

■ Origin and description

This gene was isolated from a cDNA library of the cercarial stage of development screened with a degenerate oligomer encoding the peptide sequence KIWFQNRR (Webster and Mansour, 1992). The clone does not extend sufficiently to determine the sequence of the entire homeodomain.

■ References

Webster, P.J. and Mansour, T.E. (1992). Conserved classes of homeodomains in *Schistosoma mansoni*, an early bilateral metazoan. Mech. Dev. 38, 25–32.

S8   [m]

Species: *Mus musculus*
Chromosomal location: Chromosome 2
Other names: -
Cognate genes: *CS8A*
Type of homeobox: related to *paired*
Accession number: X52875

■ Origin and description

A part of this gene was first isolated by Kongsuwan et al. in a screen of a spleen cDNA library with oligos designed to detect paired-type homeoboxes. The complete homeodomain was sequenced by Opstelten et al., (1991). Although the *S8* homeodomain is rather similar to the *paired* homeodomain, it contains a Q at position 50 unlike the S found in typical *paired* type homeodomains including those encoded by the some of the mouse Pax-genes. It is therefore expected to have a DNA binding specificity different from that of the paired homeodomains. There is circumstantial evidence for the existence of a second *S8*-like gene.

■ Expression

At 7.5 days p.c. the *S8* gene is expressed in all embryonic mesoderm except in and very near the primitive streak (Meijlink and Lawson, unpublished). In later stages expression remains mesoderm-specific, but a characteristic pattern develops with high local concentrations in craniofacial mesenchyme, the limbs, the genital tubercle and lower levels in the meninges, the heart, somites and sclerotomes all along the axis. No expression was found in endodermal derivatives, the central or peripheral neural system, the splanchnopleure, nor in epithelial layers covering sites of often very high mesenchymal expression. Many sites of high expression correlate with the location of migrated neural crest cells. Another correlation is that with sites where development of structures proceeds through reciprocal epithelio-mesenchymal interactions (Opstelten et al., 1991).

■ Genetics, function

S8 maps close to the stubby (*stb*) mutation on mouse chromosome 2.

■ References

Kongsuwan et al. (1988). Expression of multiple homeobox genes within diverse mammalian haemopoietic lineages. EMBO J. 7, 2131-2138.

Opstelten, D.-J., Vogels, R., Robert, B., Kalkhoven, E., Zwartkruis, F., De Laaf, L., Destrée, O.H., Deschamps, J.U., Lawson, K.A. and Meijlink, F. (1991). The mouse homeobox gene, S8, is expressed during embryogenesis predominantly in mesenchyme. Mech. Dev. 34, 29-42.

TgHbox1 [Tg]

Species: *Tripneustes gratilla* (sea urchin)
Chromosomal location: -
Other names: *HB1*
Cognate genes: -
Type of homeobox: *Antp*-class
Accession number: X13148

■ Origin and description

The *TgHbox1* homeobox was isolated from a genomic λ Charon 4 library by probing with *Drosophila Scr* sequences. This screening produced a single lambda clone with a 14.6 kb insert that contained the last exon of the *TgHbox1* gene (Dolecki et al., 1986). Subsequently, a 3.1 kb cDNA was isolated from an oligo (dT) primed cDNA library in λgtll using the genomic homeobox sequence as a probe (Angerer et al., 1989). The *TgHbox1* homeobox is a mainstream *Antp*-class homeobox and definitive cognates from other species have yet to be identified.

■ Expression

TgHbox1 is activated to produce a 6.9 kb messenger RNA beginning at early blastula stage. Transcription of this message increases through gastrula stage when a second 7.7 kb messenger RNA is detected. After gastrula stage transcription of the 6.9 kb transcript decreases several fold and the 7.7 kb messenger RNA is no longer detectable (Dolecki et al., 1986). *In situ* localization of expression shows that the messenger RNA is first activated in the dorsal ectoderm but becomes progressively restricted to the posterior mid-dorsal ectoderm overlying the vertex of the mature *pluteus larvae* (Angerer et al., 1989). *TgHbox1* is expressed extensively as a single 6.9 kb mRNA in many adult tissues. It is most abundant in small intestine, about twofold less abundant in large intestine, Aristotle's lantern, ovary and testis and a further twofold less abundant in coelomocyte poly(A)+RNA.

■ References

Angerer, L.M., Dolecki, G.J., Gagnon, M., Lum, R., Wang, G., Yang, Q., Humphreys, T. and Angerer, R.C. (1989). Progressively restricted expression of a homeobox gene within the aboral ectoderm of developing sea urchin embryos. Genes Dev. 3, 370-383.

Dolecki, G.J., Wannakrairoj, S., Lum, R., Wang, G., Riley, H.D., Carlos, R., Wang, A. and Humphreys, T. (1986). Stage-specific expression of a homeo box-containing gene in the non-segmented sea urchin embryo. EMBO J. 5, 925-930.

TgHbox3 [Tg]

Species: *Tripneustes gratilla* (sea urchin)
Chromosomal location: -
Other names: *HB3*
Cognate genes: -
Type of homeobox: *Antp*-class
Accession number: X13146

■ Origin and description

TgHbox3 was isolated from a 8.9 kb genomic insert in a λ charon 4 phage which had been selected with a *Drosophila Scr* probe. The *TgHbox3* homeobox is a mainstream *Antp*-class homeobox and definitive cognates from other species have yet to be identified.

■ Expression

Reaction of the *TgHbox3* probe to Northern transfers of RNA from various embryonic stages and adult tissues indicate that *TgHbox3* is expressed primarily in adult gonads in both ovary and testis. Very faint reactions were detected in adult small intestine and at gastrula stage in the embryos. In all tissues, the gene was expressed as a 5.7 kb mRNA.

■ Reference

Dolecki, G.J., Wang, G. and Humphreys, T. (1988). Stage- and tissue-specific expression of two homeobox genes in sea urchin embryos and adults. Nuc. Acid Res. 16, 11543-11558.

TgHbox4 [Tg]

Species: *Tripneustes gratilla* (sea urchin)
Chromosomal location: -
Other names: *HB4, Hbox4*
Cognate genes: *Abd-B*
Type of homeobox: *Antp*-class, related to *Abd-B*
Accession number: X13147

■ Origin and description

TgHbox4 was isolated from a genomic library in λ charon 4, using *Drosophila Scr* as a probe. The *TgHbox4* gene contains a homeobox related to the *Drosophila Abd-B* homeobox and its cognates in other species such as vertebrate group 9 *Hox* genes.

■ Expression

TgHbox4 is expressed as 4.4 kb and 3.7 kb messenger RNAs. The former appears at early blastula stage, persists through gastrula stage and then decreases, while the latter appears in late blastula stage and persists at high levels through pluteus stage. *TgHbox4* was expressed in most adult tissues as a 3.5 kb messenger RNA being very abundant in large intestine and small intestine and expressed in decreasing amounts in lantern ovary and testis. This message is barely detectable in coelomocytes (Dolecki et al., 1988).

■ Function

The *TgHbox4* homeodomain, produced as a bacterially expressed protein, binds to the 3' enhancer of the late L1 H2B gene. Antisera to the *TgHbox4* homeodomain blocks binding of the nuclear factor which recognizes this enhancer while antisera to other *Antennapedia* proteins do not block binding of this factor (Zhao et al., 1991).

■ Reference

Dolecki, G.J., Wang, G. and Humphreys, T. (1988). Stage- and tissue-specific expression of two homeobox genes in sea urchin embryos and adults. Nuc. Acid Res. 16, 11543-11558.

Zhao, A.Z., Vansant, G., Bell, J., Humphreys, T. and Maxson, R. (1991). Activation of the L1 late H2B histone gene in blastula-stage sea urchin embryos by an *Antennapedia*-class homeoprotein. Mech. Dev. 34, 21-28.

TgHbox5 [Tg]

Species: *Tripneustes gratilla* (sea urchin)
Chromosomal location: -
Other names: -
Cognate genes: *ceh-9*
Type of homeobox: *TgHbox5*
Accession number: -

■ Origin and description

TgHbox5 was isolated from a genomic library in λ charon 4, using *Drosophila Scr* as a probe. The *TgHbox5* homeodomain is very divergent from all known homeoboxes and *ceh-9* (Hawkins and McGhee, 1990) is the only known homologue. These two genes may together define a new class of homeoboxes.

■ Expression

Two major transcripts of *TgHbox5* appear during embryonic development. A 5.0 kb mRNA which is abundant at blastula stage and persists through pluteus stage and a 2.7 kb mRNA that appears at gastrula and pluteus stage. *TgHbox5* is expressed as four different mRNAs of 5.0 kb, 3.5 kb, 3.1 kb and 2.7 kb in various adult tissues. In the small intestine, the 5.0 kb is expressed most abundantly with decreasing amounts of the 3.5, 3.1 and 2.7 kb mRNAs. In the large intestine, the 3.5 and the 3.1 RNAs are expressed much more abundantly than the smaller and the larger RNA. The 5.0, 3.5 and 3.1 kb RNA transcripts in the Aristotle's lantern and ovary lanes are expressed faintly and at about equal levels. The expression of *TgHbox5* cannot be detected in testis and coelomocytes.

■ Reference

Hawkins, N.C. and McGhee, J.D. (1990). Homeobox containing genes in the nematode *Caenorhabditis elegans*. Nucleic Acids Res. 18, 6101-6106.
Wang, G.V.L., Dolecki, G.J., Carlos, R. and Humphreys, T. (1990). Characterization and expression of two sea urchin homeobox gene sequences. Develop. Genetics 11, 77-87.

TgHbox6 [Tg]

Species: *Tripneustes gratilla* (sea urchin)
Chromosomal location: -
Other names: -
Cognate genes: -
Type of homeobox: *Antp*-class
Accession number: -

■ Origin and description

TgHbox6 was isolated from a genomic library in λ charon 4, using *Drosophila Scr* as a probe. The *TgHbox6* homeobox is a mainstream *Antp*-class homeobox and definitive cognates from other species have yet to be identified.

■ Expression

Detectable reactions of *TgHbox6* to Northern transfers of RNA from embryonic stages were not obtained. *TgHbox6* messenger RNA was present in all adult tissues examined which include small and large intestines, Aristotle's lantern, ovary, testis, and coelomocytes. All tissues contained at 3.2 kb mRNA; ovary and testis contained in addition a 3.6 kb mRNA.

■ Reference

Wang, G.V.L., Dolecki, G.J., Carlos, R. and Humphreys, T. (1990). Characterization and expression of two sea urchin homeobox gene sequences. Dev. Genet. 11, 77-87.

TTF-1 [r,m,h]

Species: *Rattus norvegicus, Mus musculus, Homo sapiens*
Chromosomal location: Mouse chromosome 12 (bands C1-C3); human 14 (band 14q13)
Other names: *Nkx-2.1*
Cognate genes: -
Type of homeobox: Related to *Drosophila NK-2*
Accession number: X53858

■ Origin and description

TTF-1 (Thyroid Transcription Factor 1) was identified as a thyroid specific protein that binds to promoters expressed exclusively in thyroid follicular cells (Civitareale et al., 1989; Francis-Lang et al., 1992a). A degenerate oligonucleotide, derived from the partial amino acid sequence of a highly purified preparation of calf TTF-1, was used to clone the *TTF-1* cDNA from a thyroid cDNA library (Guazzi et al., 1990). *TTF-1* contains a homeodomain, highly homologous to that found in the *Drosophila NK-2* gene. The DNA binding specificity of this homeodomain is different from that of Antennapedia, even though both homeodomains have a very similar recognition helix, providing the first functional evidence for a role of other regions of homeodomains in DNA site selection (Damante and Di Lauro,1991). Another region of homology, downstream of the homeobox, is observed between the coding regions of *TTF-1* and *NK-2*. This region, encoding a 17 amino acids long peptide, as well as a homeodomain of the *NK-2* type, are also found in three other mouse genes (*Nkx-2* family, see also *Nkx-2.2, Nkx-2.3, Nkx-2.4*) (Price et al., 1992).

■ Expression

The expression of *TTF-1* was studied by *in situ* hybridization, Northern blots and immunohistochemistry in both prenatal development and in adult rats and mice (Lazzaro et al., 1991; Guazzi et al., 1990; Price et al., 1992). No discrepancies were observed between protein and mRNA data. *TTF-1* expression begins at day 10.5 p.c. of rat development in three different regions; the thyroid anlage, the lung bud and the developing brain. In both thyroid anlage and lung bud expression *TTF-1* is highest in the leading edge of the proliferating primordia. Expression is strictly confined to endodermal cells and is maintained throughout development and in adult life. In 4-week old rats, *TTF-1* is present in thyroid follicular cells and in the epithelium lining the bronchioles, but is not expressed in tracheal epithelium. In the CNS, *TTF-1* is expressed in the developing diencephalon (hypothalamus and posterior pituitary) and in the telencephalon (lateral and medial corpus striatum). Two other homeobox genes are expressed in these regions, *Nkx-2.2* and *Dlx-1* (Price et al., 1992). In 4-week old rats, expression of *TTF-1* in the CNS is greatly reduced (Price et al., 1992), except for the posterior pituitary, that maintains high concentrations of *TTF-1* (Lazzaro et al., 1991).

■ Function

TTF-1 is a transcriptional activator. Several, functionally relevant binding sites for TTF-1 are contained in the promoters of thyroglobulin (Civitareale et al., 1990) and thyroperoxidase (Francis-Lang et al., 1992a), two genes exclusively expressed in thyroid follicular cells. Expression of *TTF-1* in non-thyroid cells results in transcriptional activation of both promoters and of artificial promoters containing at least two binding sites (De Felice, Damante and Di Lauro, unpublished). The function of TTF-1 at early stages of thyroid differentiation is unknown, since thyroglobulin and thyroperoxidase gene transcription is activated only five days later (Lazzaro et al., 1991). Similarly, no information is available on the function of TTF-1 in lung and brain. A *paired* domain containing factor, *Pax-8*, is expressed in thyroid and binds to a site overlapping with one of the TTF-1 sites in the promoters of both thyroglobulin and TPO (Zannini, Francis-Lang and Di Lauro, submitted). The functional relevance of this interaction is not clear.

■ References

Civitareale, D., Lonigro, R., Sinclair, A.J. and Di Lauro, R. (1989). A thyroid specific nuclear protein essential for tissue-specific expression of the thyroglobulin promoter. EMBO J. 8, 2537-2542.

Damante, G. and Di Lauro, R. (1991). Several regions of Antennapedia and thyroid transcription factor 1 homeodomains contribute to DNA binding specificity. Proc. Natl. Acad. Sci. USA 88, 5388-5392.

Francis-Lang, H., Price, M., Polycarpou-Schwartz, M. and Di Lauro, R. (1992). Cell-type-specific expression of the rat thyroperoxidase promoter indicates common mechanisms for thyroid-specific gene expression. Mol. Cell. Biol. 12, 576-588.

Guazzi, S., Price, M., De Felice, M., Damante, G., Mattei, M.-G. and Di Lauro, R. (1990). Thyroid nuclear factor 1 (TTF-1) contains a homeodomain and displays a novel DNA binding specificity. EMBO J. 9, 3631-3639.

Lazzaro, D., Price, M., De Felice, M. and Di Lauro, R. (1991). The transcription factor TTF-1 is expressed at the onset of thyroid and lung morphogenesis and in restricted regions of the foetal brain. Development 113, 1093-1104.

Price, M., Lazzaro, D., Pohl, T., Mattei, M.-G., Rüther, U., Olivo, J.-C., Duboule, D. and Di Lauro, R. (1992). Regional expression of the homeobox gene Nkx-2.2 in the developing mammalian forebrain. Neuron 8, 1-20.

Ubx [d]

Species: *Drosophila melanogaster*
Chromosomal location: Right arm of chromosome 3, bands 89E1-2
Other names: *Ultrabithorax*
Cognate genes: -
Type of homeobox: *Antp*-type
Accession numbers: K01963, X05723, X05724, X05725, X05726, X05727, X05427, M24609

■ Origin and history

The *Ubx* gene (originally described by Hollander in 1934: see Lindsley and Grell, 1968) is one of the very first homeotic genes identified in *Drosophila*. *Ubx* was shown to belong to a locus containing different homeotic mutations that formed a "pseudo allelic series". This locus was named the bithorax complex by Ed Lewis who pioneered the field of developmental genetics in *Drosophila* (Lewis; 1951;1955, 1968, 1978). *Ubx* is the first homeotic gene that has been cloned (Bender et al., 1983). Finally *Ubx* is one of the genes at the basis of the discovery of the homeobox (McGinnis et al.,1984; Scott and Weiner, 1984).

■ Function

Ubx mutations are homozygous lethal. In the dead 1st instar larvae, parasegments 5 and 6 (PS5 and PS6) are transformed into copies of PS 4 (Lewis, 1951, 1955, 1978; Morata and Garcia-Bellido, 1976; Morata and Kerridge, 1981; Kerridge and Morata, 1982; Hayes et al., 1984; PS5 and 6 form the 3rd thoracic and 1st abdominal segments of an adult fly). Thus the function of the *Ubx*+ product is to provide proper identity of metameres in the posterior thorax and anterior abdomen of the adult fly. *Ubx* mutations are haplo-insufficient; heterozygous flies (*Ubx*/+), have slightly enlarged halteres on the 3rd thoracic segment. The *Ubx* gene is one of the three homeotic genes of the bithorax complex (Sanchez-Herrero et al., 1985; Tiong et al., 1985). The finding that parasegmental identity can be switched by a single mutation suggested that homeotic genes (selector genes) control developmental pathways by regulating subordinate target genes responsible for parasegment-specific morphogenesis (realisator genes: Garcia-Bellido, 1977). Several evidences support very strongly this hypothesis. 1. *Ubx* products contain a homeodomain motif (McGinnis et al.,1984; Scott and Weiner, 1984) which has sequence similarity to procaryotic and eucaryotic regulators (Laughon and Scott, 1984; Sherperd et al., 1984; reviewed in Gehring, 1987 and in Scott et al., 1989). 2. On the one hand, *Ubx* has been shown in embryos to control negatively the expression of *Antennapedia*, another homeotic product (Hafen et al., 1984; Caroll et al., 1986; Wirz et al., 1986, Boulet and Scott, 1988). On the other hand, *Ubx* has also been to act positively on its own expression in the visceral mesoderm of embryos (Bienz and Tremml, 1988). These negative and positive interactions are likely to be direct, since the *Ubx* product has bee shown to bind *in vitro* to the *Antennapedia* and to the *Ubx* promoter sequences (Beachy et al., 1988). In addition, transcriptional activation and repression by *Ultrabithorax* proteins has been reproduced in yeast (Samson et al., 1989) or tissue culture cells (Thali et al., 1988; Krassnow et al., 1989). 3. Cloning of DNA sequences that bind to the *Ubx* products revealed the presence of genes that are regulated by *Ubx* (Gould et al., 1990).

■ Expression

In 1981 Rob Saint and Michel Goldschmidt-Clermont in Dave Hogness laboratory discovered that the *Ubx* transcription unit covers 70kb of DNA (Hogness et al., 1985). 12 different transcripts can be generated by alternative splicing and polyadenylation. Translation of these mRNAs yield a family of six *Ubx* proteins characterized by constant amino- and carboxy-proximal regions of 247 and 99 amino acid residues (the homeobox sequence is encoded by the 3' end common exon). The members of this family are distinguishable by a short variable region that links the constant regions and consists of different combinations of three optional elements of 9, 17 and 17 residues (Beachy et al., 1985; O'Connors et al., 1988; Kornfeld et al., 1989). The spectrum of RNA products changes with time and tissue. It has bee demonstrated that the *Ubx* protein carrying the second micro-exon is not expressed in the nervous system (Weinzierl et al., 1987). However all different isoforms of the *Ubx* product are not essential since a mutation that eliminates the second micro-exon (and thus 4 isoforms) has no effect on fly viability and development (Busturia et al., 1990). The pattern of expression of the *Ubx* products during development has been extensively studied by *in situ* hybridization to RNA on frozen sections (Akam, 1983; Akam and Martinez-Arias, 1985) or antibody staining of whole embryos (White and Wilcox, 1984; Beachy et al., 1985). *Ubx* is expressed in parasegments affected by *Ubx* mutations. The pattern appears in PS5 in few cells of the epidermis and central nervous system (CNS). In PS6, nearly all cells of the epidermis and CNS express *Ubx* . There are mutations that affect specifically the identity of PS5 (the *abx/bx* mutations) or of PS6 (the *bxd/pbx* mutations; Lewis, 1978). The *abx/bx* mutations are spread over 28 kb of DNA within the large intron of the *Ubx* transcription unit. while the *bxd/pbx* mutations are distributed over a 40 kb region of DNA upstream of the promoter (Bender et al., 1983; Peifer and Bender, 1986). The *abx/bx* and *bxd/pbx* mutations define very large parasegment-specific enhancers that direct the very intricate patterns of expression of *Ubx* in PS5 and 6 respectively (White and Wilcox, 1985; White and Akam, 1985; Cabrera et al., 1985; Peifer et al., 1987; Little et al., 1990; Irvine et al., 1991). In the visceral mesoderm, *Ubx* is expressed in PS7 (Bienz and Tremml, 1988). The pattern of expression in PS7 and in more posterior parasegments decreases in intensity. This is due to repression by the two other homeotic genes of the BX-C, *abd-A* and *Abd-B* genes which are expressed in these parasegments (Struhl and White, 1985; Harding et al., 1985).

■ References

Akam, M.E. (1983). The location of the *Ultrabithorax* transcripts in *Drosophila* tissue sections. EMBO J. 2, 2075-2084.

Akam, M.E. and Martinez-Arias, A. (1985). The distribution of *Ultrabithorax* transcripts in *Drosophila* embryos. EMBO J. 4, 1689-1700.

Bender, W., Akam, M., Karch, F., Beachy, P.A., Peifer, M., Spierer, P., Lewis, E.B. and Hogness, D.S. (1983). Molecular genetics of the bithorax complex in *Drosophila*. Science 221, 23-29.

Beachy, P.A., Helfand, S.L. and Hogness, D.S. (1985). Segmental distribution of bithorax complex proteins during *Drosophila* development. Nature 313, 545-551.

Beachy, P.A., Krasnow, M.A., Gavis, E.R. and Hogness, D.S. (1988). An *Ultrabithorax* protein binds sequences near its own and the *Antennapedia* P1 promoter. Cell 55, 1069-1081.

Bienz, M. and Tremml, G. (1988). Domain of *Ultrabithorax* expression in *Drosophila* visceral mesoderm from autoregulation and exclusion. Nature 33, 576-578.

Boulet, A.M. and Scott, M.P. (1988). Control elements of the P2 promoter of the *Antennapedia* gene. Genes Dev. 2, 1600-1614.

Busturia, A., Vernos, I., Casanova, J. and Morata, G. (1990).

Different forms of *Ultrabithorax* proteins generated by alternate splicing are functionally equivalent. EMBO J. 9, 3551-3555.

Caroll, S.B., Laymon, R.A., McCutcheon, M.A., Riley, P.D. and Scott, M.P. (1986). The localization and regulation of *Antennapedia* protein expression in *Drosophila* embryos. Cell 47, 113-122.

Cabrera, C., Botas, J. and Garcia-Bellido, A. (1985). Distribution of *Ultrabithorax* proteins in mutants of *Drosophila* bithorax complex and its transregulatory genes. Nature 318, 569-571.

Garcia-Bellido, A. (1977). Homeotic and atavic mutations in insects. Am. Zool. 17, 613-629.

Gehring, W.J. (1987). Homeo boxes and the study of development. Science 236, 1245-1252.

Gould, A.P., Brookman, J.J., Strutt, D.I. and White, A.H. (1990). Targets of homeotic gene control in *Drosophila*. Nature 348, 308-312.

Hafen, E., Levine, M. and Gehring, W.J. (1984). Regulation of *Antennapedia* transcript distribution by the bithorax complex in *Drosophila*. Nature 307, 287-289.

Harding, K., Wedeen, C., McGinnis, W. and Levine, M. (1985). Spatially regulated expression of homeotic genes in *Drosophila*. *Science* 229,1236-1242.

Hayes, P.H., Sato, T. and Denell, R.E. (1984). Homeosis in *Drosophila*: The *Ultrabithorax* larval syndrome. Proc. Natl. Acad. Sci. USA 81, 545-549.

Hogness, D.S., Lipshitz, H.D., Beachy, P.A., Peattie, D.A., Saint, R.B., Goldschmidt-Clermont, M., Harte, P.J., Gavis, E.R. and Helfand, S.L. (1985). Regulation and products of the *Ubx* domain of the bithorax complex. Cold Spring Harb. Symp. Quant. Biol. 50, 181-194.

Irvine, K.D., Helfand, S.L. and Hogness, D.S. (1991). The large upstream control region of the homeotic gene *Ultrabithorax*. Development 111, 407-424.

Kerridge, S. and Morata, G. (1982). Developmental effects of some newly induced *Ultrabithorax* alleles of *Drosophila*. J. Embryol. Exp. Morph. 68, 211-234.

Kornfeld, K., Saint, R.B., Beachy, P.A., Harte, P.J., Peattie, D.A. and Hogness, D.S. (1989). Structure and expression of a family of *Ultrabithorax* mRNAs generated by alternate splicing and polyadenylation in *Drosophila*. Genes Dev. 3, 243-258.

Krasnow, M.A., Saffman, E.E., Kornfeld, K. and Hogness, D.S. (1989). Transcriptional activation and repression by *Ultrabithorax* proteins in cultured *Drosophila* cells. Cell 57, 1031-1043.

Laughon, A and Scott, M.P. (1984). Sequence of a *Drosophila* segmentation gene: protein structure homology with DNA-binding proteins. Nature 310, 25-31

Lewis, E. B. (1951). Pseudoallelism and gene evolution. Cold Spring Harb. Symp. Quant. Biol. 16, 159-174.

Lewis, E. B. (1955). Some aspects of position pseudoallelism. Amer. Nat. 89, 73-89.

Lewis, E. B. (1978). A gene complex controlling segmentation in *Drosophila*. Nature 276, 565-570.

Lindsley, D.L. and Grell, E.H. (1968). Genetic variation of *Drosophila melanogaster*. Carnegie Inst. Wash. Publ. 627.

Little, J., Byrd, C. and Brower, D. (1990). Effects of *abx, bx, pbx* mutations on expression of homeotic genes in *Drosophila* larvae. Genetics 124, 899-908.

McGinnis, W., Levine, M.S., Hafen, E., Kuroiwa, A. and Gehring, W. (1984). A conserved DNA sequence in homeotic genes of the *Drosophila Antennapedia* and bithorax complexes. Nature 308, 428-433.

Morata, G. and Garcia-Bellido, A. (1976). Developmental analysis of some mutants of the bithorax system of *Drosophila*. Wilhem Roux's Arch. 179, 125-143.

Morata, G. and Kerridge, S. (1981). Sequential functions of the bithorax complex of *Drosophila*. Nature 290, 778-781.

O'Connors, M.B., Binari, R., Perkins, L.A. and Bender, W. (1988). Alternative RNA products from the *Ultrabithorax* domain of the bithorax complex. EMBO J. 7, 435-445.

Peifer, M. and Bender, W. (1986). The anterobithorax and bithorax mutations of the bithorax complex. EMBO J. 5, 2293-2303.

Peifer, M., Karch, F. and Bender, W. (1987). The bithorax complex: control of segmental identity. Genes Dev. 1, 891-898.

Samson, M.L., Jackson-Grusby, L. and Brent, R. (1989). Gene activation and DNA binding by *Drosophila Ubx* and *abd-A* proteins. Cell 57, 1045-1052.

Sanchez-Herrero, E., Vernos, I., Marco, R. and Morata, G. (1985). Genetic organization of the *Drosophila* bithorax complex. Nature 313, 108-113.

Scott, M.P. and Weiner, A.J. (1984). Structural relationships among genes that control development: sequence homology between the *Antennapedia*, *Ultrabithorax*, and *fushi tarazu* loci of *Drosophila*. Proc. Natl. Acad. Sci. USA 81, 4115-4119.

Scott, M.P., Tamkun, J.W. and Harzell, G.W. (1989). The structure and function of the homeodomain. BBA Rev. Cancer 989, 25-48.

Sherperd, J.C.W., McGinnis, E., Carrasco, A.E., DeRobertis, E.M. and Gehring, W.J. (1984). Fly and frog homeo domains show homology with yeast mating type regulatory proteins. Nature 310, 70-71.

Struhl, G. and White, A.H. (1985). Regulation of the *Ultrabithorax* gene of *Drosophila* by other bithorax complex genes. Cell 43, 507-519.

Thali, M., Müller, M.M., DeLorenzi, M., Matthias, P. and Bienz, M. (1988). *Drosophila* homeotic genes encode transcriptional activators similar to mammalian OTF-2. Nature 336, 598-601.

Tiong, S.Y.K., Bone, L.M. and Whittle, J.R.S. (1985). Recessive lethal mutations within the bithorax complex in *Drosophila*. Mol. Gen. Genet. 200, 335-342.

Weinzierl, R., Axton, J.M., Ghysen, A. and Akam, M. (1987).*Ultrabithorax* mutations in constant and variable regions of the protein coding sequence. Genes Dev. 1, 386-397.

White, R.A.H. and Wilcox, M. (1984). Protein products of the bithorax complex in *Drosophila*. Cell 39, 163-171.

White, R.A.H. and Wilcox, M. (1985). Regulation of the distribution of *Ultrabithorax* proteins in *Drosophila*. Nature 318, 563-567.

White, R.A.H. and Akam, M.E. (1985). Contrabithorax mutations cause inappropriate expression of Ultrabithorax products in *Drosophila*. Nature 318, 567-569.

Wirz, J., Fessler, L.I. and Gehring, W.J. (1986). Localization of the *Antennapedia* protein in *Drosophila* embryos and imaginal discs. EMBO J. 5, 3327-3334.

UbxAf [ar]

Species: *Artemia franciscana*

Chromosomal location: -

Other names: -

Cognate genes: *Ubx*

Type of homeobox: *Antp*-class

Accession numbers: X70081

■ Origin and history

Cloned by PCR with degenerate homeobox primers, followed by inverse PCR of 1.3 kb fragment from genomic DNA. Amino acid conservation in the homeodomain and immediately flanking regions uniquely identifies this gene as a homologue of *Drosophila Ubx*.

■ References

Averof, M. and Akam, M. (1993). Hom/Hox genes in a crustacean; implication for the origin of insect and crustacean body plans. Current biology 3, 73–78.

unc-4 [ce]

Species: *Caenorhabditis elegans* (N2)

Chromosomal location: II

Other names: *ceh-4*

Cognate genes: -

Type of homeobox: *prd*-type

Accession number: -

■ Origin and description

unc-4 (uncoordinated) was one of the first genes defined by mutation in *C. elegans* (Brenner, 1974). *ceh-4* was isolated using highly degenerate oligonucleotides for the most conserved region of helix 3 (Bürglin et al., 1989). It was subsequently shown that the homeobox gene *ceh-4* is *unc-4* (Miller et al., 1991). The homeodomain belongs to a growing group of genes having homeodomains that are distantly similar to *prd*-homeodomains, but do not contain a paired domain and thus have been grouped separately into a *prd*-like class (Miller et al., 1991).

■ Expression

A single low abundance transcript of 1.2 kb has been observed (Miller et al., 1991).

■ Genetics, function

unc-4 mutants are unable to crawl backwards due to a specific neural wiring defect in their motor neuron circuit (Chalfie et al., 1985; White et al., 1991). EM reconstruction of the ventral nerve cord in an *unc-4* mutant (White et al., 1991) showed that certain motor neurons of the VA class accept synaptic input that is appropriate for VB motor neurons. This homeotic switch in the pattern of connectivity is restricted to the nine VA cells that have VB sisters in the motor neuron lineage (Miller et al., 1991). *unc-4* activity is necessary to distinguish the fates of VA and VB neuroblasts that arise as sister cells, but is not required to specify presynaptic input to VA motor neurons derived from other sublinages (Miller et al., 1991). Experiments with a temperature sensitive *unc-4* allele have established that the creation of synapses between VA motor neurons and particular interneurons depends upon *unc-4* activity at about the time that these connections are made (Miller et al., 1991).

■ References

Brenner, S. (1974). The genetics of *Caenorhabditis elegans*. Genetics 77, 71-94.

Bürglin, T.R., Finney, M., Coulson, A. and Ruvkun, G. (1989). *Caenorhabditis elegans* has scores of homeobox-containing genes. Nature 341, 239-243.

Chalfie, M., Sulston, J.E., White, J.G., Southgate, E., Thomson, J.N. and Brenner, S. (1985). The neural circuit for touch sensitivity in *Caenorhabditis elegans*. J. Neurosci. 5, 956-964.

Miller, D.M., Shen, M.M., Shamu, C.E., Bürglin, T.R., Ruvkun, G., Dubois, M.L., Ghee, M. and Wilson, L. (1992). *C. elegans unc-4* gene encodes a homeodomain protein that determines the pattern of synaptic input to specific motor neurons. Nature 355, 841-845.

White, J.G., Southgate, E. and Thomson, J.N. (1992). Mutations in the *Caenorhabditis elegans unc-4* gene alter the synaptic input to ventral cord motor neurons. Nature 355, 838-841.

unc-86 [ce]

Species: *Caenorhabditis elegans* (N2)
Chromosomal location: III
Other names: -
Cognate genes: *Brn-3*, T-POU
Type of homeobox: POU-IV
Accession number: M22363

■ Origin and description

unc-86 was first identified on the basis of a mutation that causes defects in movement and egg laying (Hodgkin, Horvitz and Brenner, 1978). The defective behaviour was traced to neural defects, many of which are caused by alterations in the cell lineages that generate the nervous system (Chalfie, Sulston and Horvitz, 1981). The gene was cloned by walking from linked DNA polymorphisms, and helped to define the POU family of homeodomain proteins (Finney, Ruvkun and Horvitz, 1988; Herr et al., 1988). A single transcript is made, encoding a protein of 38,000 daltons.

■ Expression

unc-86 protein expression has been followed using a specific antiserum, at the level of single cells (Finney and Ruvkun, 1990). It is expressed in a precise lineage-specific manner, in one of the two daughter cells of particular divisions, within 10-30 minutes after cell division. Expression is initiated in 22 distinct cell types; 13 of these differentiate into neurons, and 9 divide further, generating only neurons. By the adult stage *unc-86* is expressed in 57 neurons, nearly a fifth of the 302-neuron nervous system. These neurons are of many different functional classes, and are primarily found in the peripheral nervous system.

■ Function

Some mutations completely eliminate the production of *unc-86* protein (Finney and Ruvkun, 1990); from these null mutations the function can be inferred. In these cell lineages that can be easily assayed in the null mutants, the daughter cell that would normally express *unc-86* is transformed so that it behaves like its mother (Chalfie, Sulston and Horvitz, 1981; Finney and Ruvkun, 1990). Therefore *unc-86* is needed to make those daughter cells different from their mothers. In other cells *unc-86* does not influence the pattern of cell division, but is required for aspects of the neuron's differentiated phenotype (Finney and Ruvkun, 1990).

■ References

Chalfie, M., Horvitz, H.R. and Sulston, J. (1981). Mutations that lead to reiterations in the cell lineages of *C. elegans*. Cell 24, 59-69.

Finney, M., Ruvkun, G. and Horvitz, H.R. (1988). The *C. elegans* cell lineage and differentiation gene *unc-86* encodes a protein with a homeodomain and extended similarity to transcription factors. Cell 55, 757-769.

Finney, M. and Ruvkun, G. (1990). The *unc-86* gene product couples cell lineage and cell identity in *C. elegans*. Cell 63, 895-905.

Herr, W., Sturm, R.A., Clerc, R.G., Corcoran, L.M., Baltimore, D., Ingraham, H.A., Rosenfeld, M.G., Finney, M., Ruvkun, G. and Horvitz, H.R. (1988). The POU domain: a large conserved region in the mammalian *pit-1*, *oct-1*, *oct-2* and *Caenorhabditis elegans unc-86* gene products. Gen. Dev. 2, 1513-1516.

Hodgkin, J., Horvitz, H.R. and Brenner, S. (1979). Nondisjunction mutants of the nematode *Caenorhabditis elegans*. Genetics 91, 67-94.

vHNF1 [r,m,h]

Species: *Rattus sp.; Mus musc.; Homo sap.*
Chromosomal location: Mouse: 11C; Human: 17q12
Other names: *LF-B3, HNF-1beta, vAPF, TCF2*
Cognate genes: -
Type of homeobox: Extra-large *HNF1*-type
Accession numbers: Rat: X56546; Human: X58840

■ Origin and description

vHNF1 (variant Hepatic Nuclear Factor 1) was initially characterized as an *HNF1*-like DNA-binding activity, present in dedifferentiated variants of hepatoma cell lines as well as in somatic hybrids that fail to express most hepatic markers and where, in turn, *HNF1* undetectable (Baumhueter et al., 1988; Cereghini et al., 1988; Cereghini et al., 1990). cDNA clones for *vHNF1* were isolated on the basis of postulated sequence-homology with the DNA-binding domain of *HNF1* (Rey-Campos et al., 1991; De Simone et al., 1991; Mendel et al., 1991, Bach et al., 1991). *vHNF1* comprises a DNA-binding domain very similar to the *HNF1* tri-partite structure (an extra-large atypical homeodomain, a pseudo-POU-domain and a distal dimerization domain) and differs completely from *HNF1* in its activation domain. *vHNF1* can form homo- and hetero-dimers with *HNF1*, *in vitro* and *in vivo*.

■ Expression

Contrary to initial analysis, *vHNF1* was subsequently detected not only in dedifferentiated hepatoma cells but also, at low amounts, in liver and differentiated hepatoma cells. Its gene is expressed in yolk sac and fetal liver as well as in kidney and, at low levels, in developed liver, intestine and lung (Rey-Campos et al., 1991; De Simone et al., 1991; Mendel et al., 1991). *vHNF1* was also shown, contrary to *HNF1*, to be induced in F9 cells, after differentiation by retinoic acid (Kuo et al., 1991; De Simone et al., 1991; Cereghini et al., in preparation). In conclusion, *vHNF1* presents an expression pattern very similar to that of *HNF1*, although wider (e.g. lungs), and it appears earlier (e.g. in nephrons) during the epithelial differentiation programs (Ott et al., 1991; Lazzaro et al., 1992).

■ Function

Surprisingly, *vHNF1* was shown, in most studies, to be a potent activator of the transcription via an *HNF1* binding site (which is found in many regulatory regions of genes expressed specifically in liver: see Mendel and Crabtree, 1991; De Simone and Cortese, 1991; Rey-Campos and Yaniv, 1992, for reviews; Rey-Campos et al., 1991, De Simone et al., 1991). Nevertheless, the endogenous version of the factor, contained in those cells, fails to activate transcription in the same conditions. In the developing animal, the patterns of expression suggest a role for *vHNF1* early in the differentiation of the polarized epithelia, specially in kidney, but formal evidences for such a function are still lacking.

■ References

Bach, I., Mattei, M.G., Cereghini, S. and Yaniv, M. (1991). Two members of an HNF1 homeoprotein family are expressed in human liver. Nucl. Acids Res. 13, 3553-3559.

Baumhueter, S., Courtois, G. and Crabtree, G.R. (1988). A variant nuclear protein in dedifferentiated hepatoma cells binds to the same functional sequences in the b fibrinogen gene promoter as HNF1. EMBO J. 7, 2485-2493.

Cereghini, S., Blumenfeld, M. and Yaniv, M. (1988). A liver-specific factor essential for albumin transcription differs between differentiated and dedifferentiated rat hepatoma cells. Genes Dev. 2, 957-974.

Cereghini, S., Yaniv, M. and Cortese, R. (1990). Hepatocyte dedifferentiation and extinction is accompanied by a block in the synthesis of mRNA coding for the transcription factor HNF1/LFB1. EMBO J. 9, 2257-2263.

De Simone, V. and Cortese, R. (1991). Transcriptional regulation of liver-specific gene expression. Curr. Op. in Cell Biol. 3, 960-965.

De Simone, V., De Magistris, L., Lazzaro, D., Gerstner, J., Monaci, P., Nicosia, A. and Cortese, R. (1991). LFB3, a heterodimer-forming homeoprotein of the LFB1 family, is expressed in specialized epithelia. EMBO J. 10, 1435-1443.

Kuo, C.J., Mendel, D.B., Hansen, L.P. and Crabtree, G.R. (1991). Independent regulation of HNF1 alpha and HNF1 beta by retinoic acid in F9 teratocarcinoma cells. EMBO J. 10, 2231-2236.

Lazzaro, D., De Simone, V., De Magistris, L., Lehtonen, E. and Cortese, R. (1992). LFB1 and LFB3 homeoproteins are sequentially expressed during kidney development. Development 114, 469–479.

Mendel, D.B. and Crabtree, G.R. (1991). HNF1, a member of novel dimerizing homeodomain protein. J. Biol. Chem. 266, 677-680.

Mendel, D.B., Hansen, L.P., Graves, M.K., Conley, P.B. and Crabtree, G.R. (1991). HNF1 alpha and HNF1 beta (vHNF1) share dimerization and homeo domains, but not activation domains, and form heterodimers in vitro. Genes and Dev. 5, 1042-1056.

Ott, M.-O., Rey-Campos, J., Cereghini, S. and Yaniv, M. (1991). vHNF1 is expressed in epithelial cells of distinct embryonic origin during development and precedes HNF1 expression. Mech. Dev. 36, 47-58.

Rey-Campos, J., Chouard, T., Yaniv, M. and Cereghini, S. (1991). vHNF1 is a homeoprotein that activates transcription and forms heterodimers with HNF1. EMBO J. 10, 1445-1457.

Rey-Campos, J. and Yaniv, M. (1992). Regulation of albumin gene expression. In: Genetic intervention in diseases with unknown etiology. (T.O. Yoshida, ed.) Elsevier Science Publishers B.V. pp 1-14.

W26 [d]

Species: *Drosophila melanogaster*
Chromosomal location: 57B
Other names: -
Cognate genes: -
Type of homeobox: Related to *paired*
Accession number: -

■ Origin and description

This gene was isolated by cross-homology with the *caudal* (*cad*) homeobox (Mlodzik and Gehring, 1985; MacDonald and Struhl, 1986). It contains a homeobox most closely related to the *paired* (*prd*) gene (Kilcherr et al., 1986) and the *C. elegans unc-4* gene (Miller et al., 1992). W26 has two different transcripts which encode two proteins differing at the N-terminus and in a region C-terminal of the homeodomain.

■ Expression

The *W26* gene is first expressed in the proctodeum at stage 7 of embryogenesis when the germband is extending. This expression persists in the developing hindgut and in the anal pads until the end of embryogenesis. An additional expression is seen in the nervous system starting at stage 12 when the germband retracts and stays also until the end of embryogenesis. Using specific probes, it could be shown that the two different transcripts of the *W26* gene are expressed in different tissues, the 2.7 kb transcript in the nervous system and the 2.3 kb transcript in the proctodeum, hindgut and anal pads.

■ References

Kilcherr, F., Baumgartner, S., Bopp, D., Frei, E. and Noll, M. (1986). Isolation of the paired gene of *Drosophila* and its spatial expression during early embryogenesis. Nature 321, 493-499.

MacDonald, P.M. and Struhl, G. (1986). A molecular gradient in early *Drosophila* embryos and its role in specifying the body pattern. Nature 324, 537-545.

Miller, D.M., Shen, M.M., Shamu, C.E., Bürglin, T.R., Ruvkun, G., Dubois, M.L., Ghee, M. and Wilson, L. (1992). *C. elegans unc-4* gene encodes a homeodomain protein that determines the pattern of synaptic input to specific motor neurons. Nature 335, 841-845.

Mlodzik, M., Fjose, A. and Gehring, W.J. (1985). Isolation of caudal, a *Drosophila* homeobox containing gene with maternal expression, whose transcripts form a concentration gradient at the pre-blastoderm stage. EMBO J. 4, 2961-2969.

XANF-1 [x]

Species: *Xenopus laevis*
Chromosomal location: -
Other names: -
Cognate genes: -
Type of homeobox: Novel class, related to that of *HES-1*
Accession number: X60099

■ Origin and description

Sequence isolated from a differential cDNA library screening using a PCR-based method applied to a single neural plate of a late gastrula embryo. The name stands for "*Xenopus* anterior neural fold".

■ Expression

The *XANF-1* gene is expressed predominantly in the anterior part of the developing nervous system, after an initial homogenous distribution in early gastrula ectoderm.

■ References

Zaraiski, A. G., Lukyanov, S. A., Vasiliev, O. L., Smirnov, Y. V., Belyavsky, A. V. and Kasanskaya, O.V. (1992). A novel homeobox gene expressed in the anterior neural plate of the *Xenopus* embryo. Dev. Biol. 152, 373-382.

Xcad1 [x]

Species: *Xenopus laevis*
Chromosomal location: -
Other names: -
Cognate genes:
Type of homeobox: related to the *Drosophila caudal* gene
Accession number: M63874

■ Origin and description

Xcad1 was isolated from a dorsal blastopore lip cDNA library (Blumberg et al., 1991) using a 1024-fold degenerate oligonucleotide corresponding to a highly-conserved amino acid sequence, KIWF(Q/K)NRR found in the DNA-binding helix 3 of the homeodomain (Bürglin et al., 1989). *Xcad1* is related to the *Drosophila melanogaster* homeobox gene *caudal*, the vertebrate genes *cdx-1* (Duprey et al., 1988) and *CHox-cad* (Frumkin et al., 1991), and the *C. elegans* gene *Ceh-3* (T.R. Bürglin, personal communication).

■ Expression

The ~2.3 kb *Xcad1* mRNA is first detectable in the dorsal blastopore lip at approximately stage 10.5 (Blumberg et al., 1991). Steady state mRNA levels are maximal at stage 13, and decrease thereafter, reaching very low levels by stage 30.

■ References

Blumberg, B., Wright, C.V.E., De Robertis, E.M. and Cho, K.W.Y. (1991). Organizer-specific homeobox genes in *Xenopus laevis* embryos. Science 253, 194-196.

Bürglin, T.R., Finney, M., Coulson, A. and Ruvkun, G. (1989). *Caenorhabditis elegans* has scores of homeobox-containing genes. Nature 341, 239-243.

Duprey, P., Chowdhury, K., Dressler, G.R., Balling, R., Simon, D., Guenet, J.-L. and Gruss, P. (1988). A mouse gene homologous to the *Drosophila* gene caudal is expressed in epithelial cells from the embryonic intestine. Genes Dev. 2, 1647-1654.

Species: *Xenopus laevis*
Chromosomal location: -
Other names: -
Cognate genes: *cdx-1, CHox-cad, ceh-3, caudal*
Type of homeobox: related to the *Drosophila caudal* gene
Accession number: M63875

■ Origin and description

Xcad2 was isolated from a dorsal blastopore lip cDNA library (Blumberg et al., 1991) using a 1024-fold degenerate oligonucleotide corresponding to a highly-conserved amino acid sequence, KIWF(Q/K)NRR found in the DNA-binding helix 3 of the homeodomain (Bürglin et al., 1989). *Xcad2* is related to the *Drosophila melanogaster* homeobox gene *caudal*, the vertebrate genes *cdx-1* (Duprey et al., 1988), and *CHox-cad* (Frumkin et al., 1991), and the C. elegans gene *ceh-3* (T.R. Bürglin personal communication). Sequence comparisons among members of this gene family suggest that *Xcad2, cdx-1,* and *CHox-cad* correspond to the same gene in different species.

■ Expression

The ~4.5 kb *Xcad2* mRNA is first detectable in the dorsal blastopore lip at approximately stage 10.5 (Blumberg et al., 1991). Steady state mRNA levels are maximal at stage 13, and decrease thereafter, reaching very low levels by stage 30.

■ References

Blumberg, B., Wright, C.V.E., De Robertis, E.M. and Cho, K.W.Y. (1991). Organizer-specific homeobox genes in *Xenopus laevis* embryos. Science 253, 194-196.

Bürglin, T.R., Finney, M., Coulson, A. and Ruvkun, G. (1989). *Caenorhabditis elegans* has scores of homeobox-containing genes. Nature 341, 239-243.

Duprey, P., Chowdhury, K., Dressler, G.R., Balling, R., Simon, D., Guenet, J.-L. and Gruss, P. (1988). A mouse gene homologous to the *Drosophila* gene *caudal* is expressed in epithelial cells from the embryonic intestine. Genes Dev. 2, 1647-1654.

Species: *Xenopus laevis*
Chromosomal location: -
Other names: -
Cognate genes: *XCad1, XCad2, Cdx1, ceh-3, Chox-cad*
Type of homeobox: related to the *Drosophila cad* type
Accession number: -

■ Origin and description

This gene was isolated by homology to PCR generated homeobox sequences from *Xenopus laevis*.

■ Expression

Xcad3 is expressed at the beginning of gastrulation and maintains a constant level through the end of neurulation and then its transcript levels begin to decline. It is expressed equally in the dorsal and the ventral regions of the embryo and is localized to the posterior region of the anterior-posterior axis. *Xcad3* transcript is not present in the uninduced *Xenopus* animal cap and its expression is induced by the addition of either bFGF or activin A.

■ References

Frohman, M.A., Northrop, J.L. and Kimelman, D. Progressive activation of homeobox gene expression during vertebrate gastrulation. Submitted.

Species: *Xenopus laevis*
Chromosomal location: -
Other names: -
Cognate genes: *Gst-1, Gst-2, Chox-7*
Type of homeobox: novel type
Accession number: -

■ Origin and description

This gene was isolated by homology to PCR generated homeobox sequences from *Xenopus laevis*.

■ Expression

XGst-1 is expressed at the beginning of gastrulation and maintains a constant level through the end of neurulation and then its transcript levels begin to decline. It is expressed equally in the dorsal and the ventral regions of the embryo and is localized to the middle region of the anterior-posterior axis. *XGst-1* transcript is present in the uninduced *Xenopus* animal cap and its expression is induced several fold by the addition of either bFGF or activin A.

■ References

Frohman, M.A., Northrop, J.L. and Kimelman, D. Progressive activation of homeobox gene expression during vertebrate gastrulation. Submitted.

XlHbox 8 [x]

Species: *Xenopus laevis*
Chromosomal location: -
Other names: -
Cognate genes: -
Type of homeobox: divergent form of the *Antp* family
Accession number: X16849

■ Origin and description

This gene was cloned from a genomic *Xenopus laevis* library by screening with an *Antennapedia* homeobox probe (Wright et al., 1988).

■ Expression

The interest of *XlHbox 8* lies in the fact that it is exclusively expressed in endodermal cells. A polyclonal antibody probe detects the protein at the tailbud stage in a narrow band in the posterior foregut. This region corresponds to the posterior 2/3 of the duodenum. This protein is expressed initially in nuclei of the gut lining, and when the pancreatic anlagen are derived from this region the exocrine epithelium continues to express this gene, as does the duct system that connects the developing pancreas to the duodenum. The liver does not express *XlHbox 8*. Because differentiation of the epithelium of the digestive tract is instructed by induction from the mesoderm, *XlHbox 8* provides a useful probe to characterize mesodermal/endodermal interactions at the molecular level. *XlHbox 8* demarcates a very small band of the duodenum; presumably other related homeobox genes specify position in other regions of the digestive tract.

■ Reference

Wright, C.V.E., Schnegelsberg, P. and De Robertis, E.M. (1989). Xlhbox 8: A novel *Xenopus* homeo protein restricted to a narrow band of endoderm. Development 104, 787-794.

Xlim-1 [x]

Species: *Xenopus laevis*
Chromosomal location: -
Other names: -
Cognate genes: -
Type of homeobox: LIM class; related to *mec-3*, *lin-11* and *Isl-1*
Accession number: -

■ Origin and description

Isolated by PCR with consensus sequences from LIM class homeobox genes *mec-3, lin-11, ISL-1,* and *CEH-14.*

■ Expression

Expressed in late blastula/gastrula in the dorsal blastopore lip and dorsal mesoderm ("Spemann's organizer"). In later embryogenesis expressed in several locations. In the adult primarily expressed in the brain. Inducible in animal explants of blastula embryos by activin A and by retinoic acid, and synergistically by both.

■ References

Taira, M., Jamrich, M., Good, P.J. and Dawid, I.B. (1992). The LIM domain-containing homeobox gene Xlim-1 is expressed specifically in the organizer region of *Xenopus* gastrula embryos. Genes Dev. 6, 356–366.

Xlnr1-16 [x]

Species: *Xenopus laevis*
Chromosomal location: -
Other names: -
Cognate genes: -
Type of homeobox: POU specific
Accession number: X54681

■ Origin and description

This partial POU-box sequence was isolated by PCR using cDNA specific of neurula-stage embryos as template (Baltzinger et al. 1990). It is a member of the POU II class, as defined by He et al. 1989.

References

Baltzinger, M., Stiegler, P. and Remy, P. (1990). Cloning and sequencing of POU-boxes expressed in *Xenopus laevis* neurula embryos. Nucleic Acids Res. 18, 6131.

He, X., Treacy, M.N., Simmons, D.M., Ingraham, H.A., Swanson, L.W. and Rosenfeld, M.G. (1989). Expression of a large family of POU-domain regulatory genes in mammalian brain development. Nature 340, 35-42.

Xlnr1-19; Xlnr1-20; Xlnr1-22; Xlnr1-34 [x]

Species: *Xenopus laevis*
Chromosomal location: -
Other names: -
Cognate genes: -
Type of homeobox: POU specific
Accession numbers: X54682-X54685

Origin and description

These partial POU-box sequences were isolated by PCR using cDNA specific of neurula-stage embryos as template (Baltzinger et al. 1990). They are close variants and show the typical features of the POU III class described by He et al. 1989.

References

Baltzinger, M., Stiegler, P. and Remy, P. (1990). Cloning and sequencing of POU-boxes expressed in *Xenopus laevis* neurula embryos. Nucleic Acids Res. 18, 6131.

He, X., Treacy, M.N., Simmons, D.M., Ingraham, H.A., Swanson, L.W. and Rosenfeld, M.G. (1989). Expression of a large family of POU-domain regulatory genes in mammalian brain development. Nature 340, 35-42.

Xloct1-5; Xloct1-13; Xloct1-17; Xloct1-32; [x]

Species: *Xenopus laevis*
Chromosomal location: -
Other names: -
Cognate genes: *Oct1*
Type of homeobox: POU specific
Accession numbers: X54677-X54680

Origin and description

These partial POU-box sequences were isolated by PCR using cDNA specific of neurula-stage embryos as template (Baltzinger et al. 1990). The nucleotide sequence of *Xloct1-32* codes for the *Oct1* POU domain described by Smith and Old (1990). The three other sequences (17, 13 and 5) are close variants of clone 32. They all belong to the POU II class defined by He et al. 1989.

References

Baltzinger, M., Stiegler, P. and Remy, P. (1990). Cloning and sequencing of POU-boxes expressed in *Xenopus laevis* neurula embryos. Nucleic Acids Res. 18, 6131.

He, X., Treacy, M.N., Simmons, D.M., Ingraham, H.A., Swanson, L.W. and Rosenfeld, M.G. (1989). Expression of a large family of POU-domain regulatory genes in mammalian brain development. Nature 340, 35-42.

Smith, D.P. and Old, R.W. (1990). Nucleotide sequence of *Xenopus laevis* Oct-1 cDNA. Nucleic Acids Res. 18, 369.

XLPOU 1 [x]

Species: *Xenopus laevis*
Chromosomal location: -
Other names: -
Cognate genes: *Oct1*
Type of homeobox: POU (Class III)
Accession number: X59055

Origin and description

This cDNA was isolated by screening a *Xenopus* brain cDNA library using a PCR probe derived from degenerate oligonucleotides based on the flanking amino acids of the *Pit-1* POU domain and adult *Xenopus* brain mRNA as template.

Expression

Transcripts were observed in adult brain and skin. Transcripts were localized in the dorsoanterior portion of embryos, with the earliest detection at the neural plate stage.

Reference

Agarwal, V.R. and Sato, S.M. (1991). XLPOU 1 and XLPOU 2, two novel POU domain genes expressed in the dorsoanterior region of *Xenopus* embryos. Dev. Biol. 147, 363-373.

XLPOU 2 [x]

Species: *Xenopus laevis*
Chromosomal location: -
Other names: -
Cognate genes: -
Type of homeobox: POU (Class III)
Accession number: X59056

■ Origin and description

This cDNA was isolated by screening a *Xenopus* brain cDNA library using a PCR probe derived from degenerate oligonucleotides based on the flanking amino acids of the *Pit-1* POU domain and adult *Xenopus* brain mRNA as template.

■ Expression

Transcripts were observed in adult brain and kidney. Transcripts were localized in the dorsoanterior portion of the embryo with earliest detection at the neurula stage.

■ Reference

Agarwal, V.R. and Sato, S.M. (1991). XLPOU 1 and XLPOU 2, two novel POU domain genes expressed in the dorsoanterior region of *Xenopus* embryos. Dev. Biol. 147, 363-373.

zen [d]

Species: *Drosophila melanogaster*
Chromosomal location: 84A (ANT-C)
Other names: *zerknüllt; z1; z2, zen*-1
Cognate genes: -
Type of homeobox: *Antp*-class
Accession number: -

■ Origin and description

zen was isolated in a screen for clones that cross-hybridize with the homeobox sequence of the *Sex combs reduced* (Scr) homeotic gene (Doyle et al., 1986). It maps to the *Antennapedia* complex (ANT-C) which contains several homeobox genes. *zen* maps proximal to *bcd* and distal to *pb*. The *zen* region consists of two closely linked homeobox genes, designated *z1* and *z2*. These homeodomains share approximately 60% amino acid identity with the *Antp* class homeodomains, including those associated with *Antp*, *ftz* and *Ubx*. The *z1* and *z2* homeodomains share more homology with each other (75%) than either

one does with others in the *Antp* class. There is no homology outside of the homeodomain.

■ Expression

z1 and *z2* show very similar temporal and spatial patterns of expression (Rushlow et al., 1987). Transcripts can be detected as early as cell cycle 10 broadly distributed along dorsal regions of the embryo. Considered in cross section, transcripts encompass about 40% of the embryo's circumference, extending from the dorsal-most surface toward more lateral regions. During cellularization the pattern refines such that transcripts are lost from lateral and polar regions leaving a dorsal stripe of about 6 nuclei along the dorsal surface of the embryo. By gastrulation most of the transcripts are localized within the differentiating amnioserosa. However, transcripts are also detected in the presumptive optic lobe just anterior to the cephalic furrow. Transcripts (and protein products) diminish during the late stages of germ band elongation.

■ Function

zen is one of several genes involved in dorsal-ventral pattern formation. *zen*-mutants are embryonic lethal and show a ventralized phenotype as early as the onset of gastrulation. Dorsal most tissues such as the amnioserosa do not differentiate because the fate of these cells has been transformed to a more lateral fate. Thus gastrulation movements are affected: the dorsal folds (derived from amnioserosa tissue) never form and the germ band fails to extend along the dorsal surface. Thus the posterior midgut invagination does not occur in the right place and will never meet the anterior midgut. Moreover the cephalic fold extends from the normal lateral positions over the dorsal surface creating a separation of the trunk and head regions during later development. The initial dorsal on/ ventral off pattern of *zen* expression is controlled by the *dorsal* morphogen (Rushlow et al., 1987). *dorsal* represses *zen* in ventral regions. This repression is mediated by a silencer element in the distal promoter (Doyle et al., 1989). *dorsal* protein binds to sites in the element *in vitro* (Ip et al., 1989) and mutations in these sites cause derepression *in vivo* (Rushlow and Warrior, 1991). The refinement process of *zen* is two-fold: the dorsal expression is maintained at high levels during cellularization by *dpp*, *tld*, *scw* while the refinement during gastrulation is regulated by *sog* (Rushlow and Levine, 1989). P-transformation assays were done to determine which transcript, *z1*, *z2*, or both are required for *zen+* function (Rushlow et al., 1987). A genomic fragment containing the *z1* transcript, 1.6 Kb of 5' flanking DNA, and 1.5 Kb of 3' flank was shown to rescue the *zen*-phenotype when transformed lines were crossed into a *zen*-background. Genomic DNA that contained the *z2* transcript with 6 Kb of flanking sequences did not rescue the *zen*- phenotype. Furthermore, genetic studies showed that *z2*- embryos are viable (barely) and do not show a ventralized phenotype. This demonstrates that the *z1* gene is critical for *zen+* function.

■ References

Doyle, H., Harding, K., Hoey, T. and Levine, M. (1986). Transcripts encoded by a homeobox gene are restricted to dorsal tissue of *Drosophila* embryos. Nature 322, 76-79.

Doyle, H., Kraut, R. and Levine, M. (1989). Spatial regulation of *zerknüllt*: a dorsal-ventral patterning gene in *Drosophila*. Genes Dev. 3, 1518-1533.

Ip, Y., Kraut, R., Levine, M. and Rushlow, C. (1991). The *dorsal* morphogen is a sequence-specific DNA-binding protein that interacts with a long-range repression element in *Drosophila*. Cell 64, 439-446.

Rushlow, C., Doyle, H., Hoey, T. and Levine, M. (1987a). Molecular characterization of the *zerknüllt* region of the *Antennapedia* gene complex in *Drosophila*. Genes Dev. 1, 1268-1279.

Rushlow, C. and Warrior, R. (1992). The rel family of proteins. Bioessays 14, 89-95.

Rushlow, C., Frasch, M., Doyle, H. and Levine, M. (1987b). Maternal regulation of *zerknüllt*: a homeobox gene controlling differentiation of dorsal tissues in *Drosophila*. Nature 330, 583-586.

Rushlow, C. and Levine, M. (1989). The role of the *zerknüllt* gene in dorsal-ventral pattern formation in *Drosophila*. In: Advances in Genetics, Genetic Regulatory Hierarchies in Development. T. Wright and J. Scandalios, eds. (San Diego: Academic Press, Inc.).

Rushlow, C., Han, K., Manley, J. and Levine, M. (1989). The graded distribution of the *dorsal* morphogen is initiated by selective nuclear transport in *Drosophila*. Cell 59, 1165-1177.

ZF-26   [zf]

Species: *Brachydanio rerio* (zebrafish)
Chromosomal location: -
Other names: -
Cognate genes: -
Type of homeobox: *Antp*-class
Accession number: M76382

■ Origin and description

This gene was isolated by PCR amplification with conserved primers targeted for the homeobox and first intron of *antennapedia*-class genes. PCR products were then ligated into the M13 vector. Twelve independent clones contained the identical 165 bp sequence of *ZF"26"*. Several vertebrate genes in the database are 70-80% similar to *ZF"26"*.

■ Remarks

It is not currently known what the cognate genes of this particular homeobox are. There are several sequences with 70-80% similarity identifying it as a homeobox, yet precise identification of homologues will only be made with longer sequence information.

■ References

Runstadler, J.A. and Kocher, T.D. (1991). A new *antennapedia*-class gene from the zebrafish. Nucleic Acids Res. 19, 5434.

zfh-1   [d]

Species: *Drosophila melanogaster*
Chromosomal location: chromosome III, cytological location 100A1-2
Other names: -
Cognate genes: -
Type of homeobox: Most closely related to the *C. elegans mec-3* homeobox
Accession number: M63449

■ Origin and description

The *zfh-1* gene was isolated by screening a *D. melanogaster* retinal cDNA expression library with an oligonucleotide probe containing the sequence CTAATTGAATT (Fortini et al., 1991). The gene spans ~25 kb of genomic DNA, contains four introns, and is predicted to encode a protein of 1060 amino acids with a molecular weight of 117 kDa. The protein sequence contains one homeodomain and nine C_2-H_2 zinc fingers. See also *zfh-2*.

■ Expression

The *zfh-1* gene is expressed in the mesoderm during early embryogenesis. During later embryonic development, expression is detected in mesodermal derivatives, including dorsal vessel cardioblasts, gonadal support cells, and adult muscle precursors, as well as in a restricted subset of CNS motorneurons (Lai et al., 1991).

■ References

Fortini, M.E., Lai, Z. and Rubin, G.M. (1991). The *Drosophila zfh*-1 and *zfh*-2 genes encode novel proteins containing both zinc-finger and homeodomain motifs. Mech. Dev. 34, 113-122.

Lai, Z., Fortini, M.E. and Rubin, G.M. (1991). The embryonic expression patterns of *zfh*-1 and *zfh*-2, two *Drosophila* genes encoding novel zinc-finger homeodomain proteins. Mech. Dev. 34, 123-134.

Species: *Drosophila melanogaster*
Chromosomal location: chromosome IV, cytological location 102C1-2
Other names: -
Cognate genes: -
Type of homeobox: Most closely related to the *C. elegans mec-3* homeobox
Accession number: M63450

■ Origin and description

The *zfh-2* gene was isolated by screening a *D. melanogaster* retinal cDNA expression library with an oligonucleotide probe containing the sequence CTAATTGAATT (Fortini et al., 1991). The gene spans ~30 kb of genomic DNA and is predicted to encode a protein of 3005 amino acids with a molecular weight of 332 kDa. The protein sequence contains three homeodomains and fifteen C_2-H_2 zinc fingers. See also *zfh*-1.

■ Expression

The *zfh*-2 gene is expressed in the CNS and hindgut during late stages of *Drosophila* embryogenesis (Lai et al., 1991).

■ References

Fortini, M.E., Lai, Z. and Rubin, G.M. (1991). The *Drosophila zfh*-1 and *zfh*-2 genes encode novel proteins containing both zinc-finger and homeodomain motifs. Mech. Dev. 34, 113-122.
Lai, Z., Fortini, M.E. and Rubin, G.M. (1991). The embryonic expression patterns of *zfh*-1 and *zfh*-2, two *Drosophila* genes encoding novel zinc-finger homeodomain proteins. Mech. Dev. 34, 123-134.

ZPOU-9 [zf]

Species: *Brachydanio rerio* (zebrafish)
Chromosomal location: -
Other names: -
Cognate genes: -
Type of homeobox: POU
Accession number: -

■ Origin and description

The gene was identified by PCR amplification with primers specific for POU domains in cDNA libraries prepared from either gastrula, neurula or adult zebrafish. The PCR product was used to screen the neurula library and a full-length clone encoding *zPOU-9* was isolated.

■ Expression

zPOU-9 is strongly expressed at the gastrula stage. Within the next 10 hours its expression drops dramatically to undetectable levels. At a not yet identified time point expression increases again and the transcript is found in the adult.

■ Reference

Gerster, T. unpublished.

ZPOU-12 [zf]

Species: *Brachydanio rerio* (zebrafish)
Chromosomal location: -
Other names: -
Cognate genes: -
Type of homeobox: POU, related to the *brn-1* class of POU domains
Accession number: -

■ Origin and description

The gene was identified by PCR amplification with primers specific for POU domains in genomic DNA and in cDNA libraries prepared from either neurula or 1 day old zebrafish embryos. The PCR product is currently used to screen the libraries.

■ Reference

Bornmann, C. and Gerster, T. unpublished.

ZPOU-23 [zf]

Species: *Brachydanio rerio* (zebrafish)
Chromosomal location: -
Other names: -
Cognate genes: -
Type of homeobox: POU, related to the *brn-1* class of POU domains
Accession number: -

■ Origin and description

The gene was identified by PCR amplification with primers specific for POU domains in genomic DNA and in cDNA libraries prepared from either neurula or 1-day old zebrafish embryos. The PCR product was used to screen the neurula library and two cDNAs presumably corresponding to alternatively processed mRNAs were isolated.

■ Reference

Spaniol, P. and Gerster, T. unpublished.

ZPOU-47 [zf]

Species: *Brachydanio rerio* (zebrafish)
Chromosomal location: -
Other names: -
Cognate genes: -
Type of homeobox: POU, related to the *brn-1* class of POU domains
Accession number: -

■ Origin and description

The gene was identified by PCR amplification with primers specific for POU domains in genomic DNA and in cDNA libraries prepared from either neurula or 1-day old zebrafish embryos. The PCR product was used to screen the 1-day library and a presumably full-length cDNA was isolated.

■ Expression

Expression of the gene starts between 10 and 12 hours of development. The precise expression pattern has not been analyzed yet.

■ Reference

Bornmann, C. and Gerster, T. unpublished.

ZPOU-50   [zf]

Species: *Brachydanio rerio* (zebrafish)
Chromosomal location: -
Other names: -
Cognate genes: -
Type of homeobox: POU, related to the *brn-1* class of POU domains
Accession number: -

■ Origin and description

The gene was identified by PCR amplification with primers specific for POU domains in genomic DNA and in cDNA libraries prepared from either neurula or 1-day old zebrafish embryos. The PCR product was used to screen the neurula library and a presumably full-length cDNA was isolated.

■ Expression

Expression of the gene starts between 10 and 12 hours of development. The precise expression pattern has not been analyzed yet.

■ Reference

Hauptmann, G. and Gerster, T. unpublished.

ZPOU-61   [zf]

Species: *Brachydanio rerio* (zebrafish)
Chromosomal location: -
Other names: -
Cognate genes: -
Type of homeobox: POU, related to the *brn-1* class of POU domains
Accession number: -

■ Origin and description

The gene was identified by PCR amplification with primers specific for POU domains in genomic DNA and in cDNA libraries prepared from either neurula or 1-day old zebrafish embryos. The PCR product is currently used to screen the libraries.

■ Reference

Bornmann, C. and Gerster, T. unpublished.

REFERENCES

Abbott, M.K. and Kaufman, T.C. (1986). The relationship between the functional complexity and the molecular organization of the *Antennapedia* locus of *Drosophila melanogaster*. Genetics 114, 919-942.

Acampora, D., D'Esposito, M., Faiella, A., Pannese, M., Migliaccio, E., Morelli, F., Stornaiuolo, A., Nigro, V., Simeone, A. and Boncinelli, E. (1989). The human HOX gene family. Nucleic Acids Res. 17, 10385-10402.

Acampora, D., Pannese, M., D'Esposito, M., Simeone, A. and Boncinelli, E. (1987). Human homeobox-containing genes in development. Hum. Reprod. 2, 407-414.

Acampora, D., Simeone, A. and Boncinelli, E. (1991). Human HOX homeobox genes. In: Oxford Surveys on Eukaryotic Genes (MacLean, N., ed.) Oxford University Press 7, 1-28.

Adams, M.D., Dubnick, M., Kerlavage, A.R., Moreno, R., Kelley, J.M., Utterback, T.R., Nagle, J.W., Fields, C. and Venter, J.C. (1992). Sequence identification of 2375 human brain genes. Nature 355, 632-634.

Affolter, M., Walldorf, U., Kloter, U., Schier, A. and Gehring, W.J. (1993). Regional repression of a *Drosophila* POU box gene in the endoderm involves interaction between germ layers. Development 117, 1199–1210.

Agarwal, V.R. and Sato, S.M. (1991). XLPOU 1 and XLPOU 2, two novel POU domain genes expressed in the dorsoanterior region of *Xenopus* embryos. Dev. Biol. 147, 363-373.

Akam, M.E. (1983). The location of the *Ultrabithorax* transcripts in *Drosophila* tissue sections. EMBO J. 2, 2075-2084.

Akam, M.E. and Martinez-Arias, A. (1985). The distribution of *Ultrabithorax* transcripts in *Drosophila* embryos. EMBO J. 4, 1689-1700.

Akam, M.E., Dawson, I. and Tear, G. (1988). Homeotic genes and the control of segment identity. Development 104, 123-133.

Ali, N. and Bienz, M. (1991). Functional dissection of *Drosophila Abdominal-B* protein. Mech. Dev. 35, 55-64.

Allen, J.D., Lints, T., Jenkins, N.A., Copeland, N.G., Strasser, A., Harvey, R.P. and Adams, J.M. (1991). Novel murine homeobox gene on chromosome 1 expressed in specific hematopoietic lineages and during embryogenesis. Genes Dev. 5, 509-520.

Andres, V., Nadal-Ginard, B. and Mahdavi, V. (1992). *Clox*, a mammalian homeobox gene related to *Drosophila cut* encodes DNA-binding regulatory proteins differentially expressed during development. Development 116, 321-334.

Angerer, L.M., Dolecki, G.J., Gagnon, M., Lum, R., Wang, G., Yang, Q., Humphreys, T. and Angerer, R.C. (1989). Progressively restricted expression of a homeobox gene within the aboral ectoderm of developing sea urchin embryos. Genes Dev. 3, 370-383.

Arcioni, L., Simeone, A., Guazzi, S., Zappavigna, V., Boncinelli, E. and Mavilio, F. (1992). The upstream region f the human homeobox gene HOX3D is a target for regulation by retinoic acid and HOX proteins. EMBO J. 11, 265-278.

Arndt, K.T., Styles, C. and Fink, G.R. (1987). Multiple global regulators control HIS4 transcription in yeast. Science 237, 874-880.

Auerbach, R. (1954). Analysis of the developmental effects of a lethal mutation in the house mouse. J. Exp. Zool. 127, 3905-329.

Averof, M. and Akam, M. (1993). Hom/Hox genes in a crustacean; implication for the origin of insect and crustacean body plans. Current biology 3, 73–78.

Awad, A.A.M., Gausz, J., Gyurkovics, H. and Parducz, A. (1981). A new homeotic mutation, SGA-62, in *Drosophila melanogaster*. Acta Biol. Acad. Sci. Hung. 32, 219-228.

Awgulewitsch, A. and Jacobs, D. (1990). Differential expression of Hox 3.1 protein in subregions of the embryonic and adult spinal cord. Development 108, 411-420.

Awgulewitsch, A., Bieberich, C., Bogarad, L., Shashikant, C. and Ruddle, F.H. (1990). Structural analysis of the Hox-3.1 transcription unit and the Hox-3.2-Hox-3.1 intergenic region. Proc. Natl. Acad. Sci. USA 87, 6428-6432.

Awgulewitsch, A., Utset, M.F., Hart, C.P., McGinnis, W. and Ruddle, F.H. (1986). Spatial restriction in expression of a mouse homeobox locus within the central nervous system. Nature 320, 328-335.

Bach, I., Galcheva-Gargova, Z., Mattei, M.-G., Simon-Chazottes, D., Guenet, J.L., Cereghini, S. and Yaniv, M. (1990). Cloning of human Hepatic Nuclear Factor 1 (HNF1) and chromosomal localization of its gene in man and mouse. Genomics 8, 155-164.

Bach, I., Mattei, M.-G., Cereghini, S. and Yaniv, M. (1991). Two members of an HNF1 homeoprotein family are expressed in human liver. Nucl. Acids Res. 13, 3553-3559.

Bachiller, D., Macias, A., Duboule, D. and Morata, G. (1994). Homeotic transformations in *Drosophila* produced by over-expression of vertebrate *AbdB*-related genes. EMBO J., in press.

Baker, B. (1987). Dros. Inf. Serv.

Baldwin, C.T., Hoth, C.F., Amos, J.A., da-Silva, E.O. and Milunski, A. (1992). An exonic mutation in the *HuP2* paired domain gene causes Waaredenburg's syndrome. Nature 355, 637-638.

Balkaschina, E.I. (1929). Ein Fall der Erbhomöosis (die Genoraviation "aristopedia") der *Drosophila melanogaster*. Arch. Entw. Mech. 115, 448-463.

Balling, R., Mutter, G., Gruss, P. and Kessel, M. (1989). Craniofacial abnormalities induced by ectopic expression of the homeobox gene *Hox-1.1* in transgenic mice. Cell 58, 337-347.

Baltzinger, M., Stiegler, P. and Remy, P. (1990). Cloning and sequencing of POU-boxes expressed in *Xenopus laevis* neurula embryos. Nucleic Acids Res. 18, 6131.

Banuett, F. (1992). *Ustilago maydis*, the delightful blight. Trends in Genet. 8, 174-180.

Barad, M., Erlebacher, A. and McGinnis, W. (1991). Despite expression in embryonic visceral mesoderm, *H2.0* is not essential for *Drosophila* visceral muscle morphogenesis. Dev. Genet. 12, 206-211.

Barad, M., Jack, T., Chadwick, R. and McGinnis, W. (1988). A novel, tissue-specific, *Drosophila* homeobox gene. EMBO J. 7, 2151-2161.

Barberis, A., Superti-Furga, G. and Busslinger, M. (1987). Mutually exclusive interaction of the CCAAT-binding factor and of a displacement protein with overlapping sequences of a histone gene promoter. Cell 50, 347-359.

Baron, A., Featherstone, M.S., Hill, R.E., Hall, A., Galliot, B. and Duboule, D. (1987). *Hox-1.6*: a mouse homeobox containing gene member of the *Hox-1* complex. EMBO J. 6, 2977-2986.

Basler, K., Yen, D., Tomlinson, A. and Hafen, E. (1990). Reprogramming cell fate in the developing *Drosophila retina*: transformation of R7 cells by ectopic expression of *rough*. Genes and Dev. 4, 728-739.

Bastian, H. and Gruss, P. (1990). A murine even-skipped homologue, Evx-1, is expressed during early embryogenesis and neurogenesis in a biphasic manner. EMBO J. 9, 1839-1852.

Bastian, H., Gruss, P., Duboule, D., Izpisúa-Belmonte, J.-C. (1992). The murine even-skipped-like gene *Evx-2* is closely linked to the *Hox-4* complex, but is transcribed in the opposite direction. Mammalian Genome 3, 241–243.

Baumgarten, M.J. and Macagno, E.R. (1986). Expression of homeobox-containing genes in the leech *Hirudo medicinalis*. Soc. Neurosci. Abstr. 12, 1122.

Baumgarten, M.J. Homeobox genes in the leech. Ph. D. Thesis, Columbia University (1990).

Baumgartner, S. (1988). Patterns of *paired* and *gooseberry* transcripts in wild-type and segmentation mutant embryos imply a combinatorial regulation of segmentation genes in *Drosophila*. Thesis, University of Basel.

Baumgartner, S. and Noll, M. (1990). Network of interactions among pair-rule genes regulating *paired* expression during primordial segmentation of *Drosophila*. Mech. Dev. 33, 1-18.

Baumgartner, S., Bopp, D., Burri, M. and Noll, M. (1987). Structure of two genes at the gooseberry locus related to the *paired* gene and their spatial expression during *Drosophila* embryogenesis. Genes Dev. 1, 1247-1267.

Baumhueter, S., Courtois, G. and Crabtree, G.R. (1988). A variant nuclear protein in dedifferentiated hepatoma cells binds to the same functional sequences in the b fibrinogen gene promoter as HNF1. EMBO J. 7, 2485-2493.

Baumhueter, S., Mendel, D.B., Conley, P.B., Kuo, C.J., Turk, C., Graves, M.K., Edwards, C.A., Courtois, G. and Crabtree, G.R. (1990). HNF-1 shares three sequence motifs with the POU domain proteins and is identical to LFB1 and APF. Genes Dev. 4, 372-379.

Beachy, P.A., Helfand, S.L. and Hogness, D.S. (1985). Segmental distribution of bithorax complex proteins during *Drosophila* development. Nature 313, 545-551.

Beachy, P.A., Krasnow, M.A., Gavis, E.R. and Hogness, D.S. (1988). An *Ultrabithorax* protein binds sequences near its own and the *Antennapedia* P1 promoter. Cell 55, 1069-1081.

Bedichek, S. (1934). Dros. Inf. Serv. 2, 9.

Bender, W., Akam, M., Karch, F., Beachy, P.A., Peifer, M., Spierer, P., Lewis, E.B. and Hogness, D.S. (1983). Molecular genetics of the bithorax complex in *Drosophila*. Science 221, 23-29.

Berben, G., Legrain, M. and Hilger, F. (1988). Studies on the structure, expression and function of the yeast regulatory gene PHO2. Gene 66, 307-312.

Berleth, T., Burri, M., Thoma, G., Bopp, D., Richstein, S., Frigerio, G., Noll, M. and Nüsslein-Volhard, C. (1988). The role of localization of *bicoid* RNA in organizing the anterior pattern of the *Drosophila* embryo. EMBO J. 7, 1749-1756.

Bermingham, Jr., J.R. and Scott, M.P. (1988). Developmentally regulated alternative splicing of transcripts from the *Drosophila* homeotic gene *Antennapedia* can produce four different proteins, EMBO J. 7, 3211-3222.

Bermingham, Jr., J.R., Martinez-Arias, A., Petitt, M.G. and Scott, M.P. (1990). Different patterns of transcription from the two *Antennapedia* promoters during *Drosophila* embryogenesis. Development 109, 553-566.

Bieberich, C.J., Ruddle, F.H. and Stenn, K.S. (1992). Differential expression of the Hox3.1 gene in adult mouse skin. Annals NY Acad. Sci. 642, 346-354.

Bieberich, C.J., Utset, M.F., Awgulewitsch, A. and Ruddle, F.H. (1990). Evidence for positive and negative regulation of the Hox-3.1 gene. Proc. Natl. Acad. Sci. USA 87, 8462-8466.

Bienz, M. and Tremml, G. (1988). Domain of *Ultrabithorax* expression in *Drosophila* visceral mesoderm from autoregulation and exclusion. Nature 33, 576-578.

Biggin, M. and Tjian, R. (1989). A purified *Drosophila* homeodomain protein represses transcription *in vitro*. Cell 58, 433-440.

Billin, A., Cockerill, K., and Poole, S. (1991). Isolation of a family of *Drosophila* POU domain genes expressed in early development. Mech. Dev. 34, 75-84.

Blochlinger, K., Bodmer, R., Jack, J., Jan, L.Y. and Jan, Y.N. (1988). Primary structure and expression of a product from *cut*, a *locus* involved in specifying sensory organ identity in *Drosophila*. Nature 333, 629-635.

Blochlinger, K., Bodmer, R., Jan, L.Y. and Jan, Y.N. (1990). Patterns of expression of *cut*, a protein required for external sensory organ development, in wild-type and *cut* mutant *Drosophila* embryos. Genes Dev.4, 1322-1331.

Blochlinger, K., Jan, L.Y. and Jan, Y.N. (1991). Transformation of sensory organ identity by ectopic expression of cut in Drosophila. Genes Dev. 5, 1124-1131.

Blum, M., Gaunt, S., Cho, K. W. Y., Steinbeisser, H., Bittner, D., Blumberg, B. and De Robertis, E. M. (1992) Gastrulation in the mouse: the role of the homeobox gene goosecoid. Cell 69, 1097-1106.

Blumberg, B., Wright, C.V.E., DeRobertis, E.M. and CHo, K.W.Y. (1991). Organizer-specific homeobox genes in *Xenopus laevis* embryos. Science 253, 194-196.

Blumenfeld, M., Maury, M., Chouard, T., Yaniv, M. and Condamine, H. (1991). Hepatic Nuclear Factor 1 (HNF1) shows a wider distribution than products of its known target genes. Development 113, 589-599.

Bodmer, R., Barbel, S., Shepherd, S., Jack, J.W., Jan, L.Y. and Jan, Y.N. (1987). Transformation of sensory organs by mutations of the *cut locus* of D. melanogaster. Cell 51, 293-307.

Bodmer, R., Jan, L.Y. and Jan, Y.N. (1990). A new homeobox-containing gene, *msh-2*, is transiently expressed early during mesoderm formation in *Drosophila*. Development 110, 661-669.

Boncinelli, E., Acampora, D., Pannese, M., D'Esposito, M., Somma, R., Gaudino, G., Stornaiuolo, A., Cafiero, M., Faiella, A. and Simeone, A. (1989). Organization of human class I homeobox genes. Genome 31, 745-756.

Boncinelli, E., Simeone, A., Acampora, D. and Malvilio, F. (1991). HOX gene activation by retinoic acid. Trends Genet. 7, 329-334.

Bopp, D., Burri, M, Baumgartner, S., Frigerio, G. and Noll, M. (1986). Conservation of a large protein domain in the segmentation gene *paired* and in functionally related genes of *Drosophila*. Cell 47, 1033-1040.

Bopp, D., Jamet, E., Baumgartner, S., Burri, M. and Noll, M. (1989). Isolation of two tissue-specific *Drosophila* paired box genes, *pox meso* and *pox neuro*. EMBO J. 8, 3447-3457.

Boulet, A., Lloyd, A. and Sakonju, S. (1991). Molecular definition of the morphogenetic and regulatory functions and the *cis*-regulatory elements of the *Drosophila Abd-B* homeotic gene. Development 111, 393-405.

Boulet, A.M. and Scott, M.P. (1988). Control elements of the P2 promoter of the *Antennapedia* gene. Genes Dev. 2, 1600-1614.

Braus, G., Mösch, H.-U., Vogel, K., Hinnen, A. and Hütter, R. (1989). Interpathway regulation of the TRP4 gene of yeast. EMBO J. 8, 939-945.

Breier, G., Bucan, M., Francke, U., Colberg-Poley, A.M. and Gruss, P. (1986). Sequential expression of murine homeobox genes during F9 EC cell differentiation. EMBO J. 5, 2209-2215.

Breier, G., Dressler, G.R. and Gruss, P. (1988). Primary structure and developmental expression pattern of *Hox-3.1*, a member of the murine Hox-3 homeobox gene cluster. EMBO J. 7, 1329-1336.

Brenner, S. (1974). The genetics of *Caenorhabditis elegans*. Genetics 77, 71-94.

Bridges, C.B. and Morgan, T.H. (1923). The third chromosome group of mutant characters of *Drosophila melanogaster*. Carnegie Inst. Wash. Pub. No. 327, 93.

Brown, R. and Brockes, J.P. (1991). Identification and expression of a regeneration-specific homeobox gene in the newt limb blastema. Development 111, 489-496.

Bucan, M., Yang-Feng, T., Colberg-Poley, A.M., Wolgemuth, D.J., Guenet, J.-L., Francke, U. and Lehrach, H. (1986). Genetic and cytogenetic localization of the homeobox containing genes on mouse chromosome 6 and human chromosome 7. EMBO J. 5, 2899-2905.

Bürglin, T.R., Wright, C.V.E. and De Robertis, E.M. (1987). Translational control in homeobox mRNAs? Nature 330, 701-702.

Bürglin, T.R. (1988). The yeast regulatory gene PHO2 encodes a homeodomain. Cell, 53, 339-340.

Bürglin, T.R. and Barnes, T.M. (1992). Introns in ESTs. Nature 367, 357.

Bürglin, T.R. and Ruvkun, G. (1992). A new motif in *PBX* genes. Nature Genet. 1, 319–320.

Bürglin, T.R., Finney, M., Coulson, A. and Ruvkun, G. (1989). *Caenorhabdiris elegans* has scores of homeobox-containing genes. Nature 341, 239-243.

Bürglin, T.R., Ruvkun, G., Coulson, A., Hawkins, N.C., McGhee, J.D., Schaller, D., Wittmann, C., Müller, F. and Waterston, R.H. (1991). Nematode homeobox cluster. Nature 351, 703.

Burri, M., Tromvoukis, Y., Bopp, D., Frigerio, G. and Noll, M. (1989). Conservation of the paired domain in metazoans and its structure in three isolated human genes. EMBO J. 8, 1183-1190.

Busturia, A., Casanova, J., Sanchez-Herrero, E., González, R. and Morata, G. (1989). Genetic structure of the *abd-A* gene of *Drosophila*. Development 107, 575-583.

Busturia, A., Vernos, I., Casanova, J. and Morata, G. (1990). Different forms of *Ultrabithorax* proteins generated by alternate splicing are functionally equivalent. EMBO J. 9, 3551-3555.

Cabrera, C., Botas, J. and Garcia-Bellido, A. (1985). Distribution of *Ultrabithorax* proteins in mutants of *Drosophila* bithorax complex and its transregulatory genes. Nature 318, 569-571.

Candia, A.F., Hu, J., Crosby, J., Lalley, P.A., Noden, D., Nadeau, J.H., Wright, C.V.E. (1992). *Mox-1* and *Mox-2* define a novel homeobox gene subfamily and are differentially expressed during early mesodermal patterning in mouse embryos. Development 116, 1123–1126.

Cannizzaro, L.A., Croce, C.M., Griffin, C.A., Simeone, A., Boncinelli, E. and Huebner, K. (1987). Human homeo box-containing genes located at chromosome regions 2q31Æ2q37 and 12q12Æ 12q13. Am J. Hum. Genet. 41, 1-15.

Caroll, S.B., Laymon, R.A., McCutcheon, M.A., Riley, P.D. and Scott, M.P. (1986). The localization and regulation of *Antennapedia* protein expression in *Drosophila* embryos. Cell 47, 113-122.

Carrasco, A.E. and Malacinski, G.M. (1987). Localization of *Xenopus* Homeobox gene transcripts during embryogenesis and in the adult Nervous System. Dev. Biol. 121, 69-81.

Carrasco, A.E., McGinnis, W., Gehring, W.J., and De Robertis, E.M. (1984). Cloning of a *Xenopus laevis* gene expressed during early embryogenesis that codes for a peptide region homologous to *Drosophila* homeotic genes. Cell 37, 409-414.

Carroll, S. and Scott, M. (1985). Localization of the *fushi tarazu* protein during *Drosophila* embryogenesis. Cell 43, 47-57.

Carroll, S.B., Laymon, R., McCuthcheon, M., Riley, P. and Scott, M.P. (1986). The localization and regulation of *Antennapedia* protein expression in *Drosophila* embryos. Cell 47, 113-122.

Casanova, J., Sanchez-Herrero, E. and Morata, G. (1986). Identification and characterization of a parasegment specific regulatory element of the *Abdominal-B* gene of D*rosophila*. Cell 47, 627-636.

Casanova, J., Sanchez-Herrero, E., and Morata, G. (1988). Developmental analysis of a hybrid gene composed of parts of the *Ubx* and *abd-A* genes of *Drosophila*. EMBO J. 7, 1097-1105.

Casselton, L. (1978). Dikaryon formation in higher basidiomycetes. In the filamentous fungi (ed. J.E. Smith and D.R. Berry), vol. 3, pp. 275-297.

Celniker, S.E. and Lewis, E.B. (1987). *Transabdominal*, a dominant mutant of the bithorax complex, produces a sexually dimorphic segmental transformation in *Drosophila*. Genes Dev. 1, 111-123.

Celniker, S.E., Keelan, D.J. and Lewis, E.B. (1989). The molecular genetics of the bithorax complex of *Drosophila*: Characterization of the products of the *Abdominal-B* domain. Genes Dev. 3, 1424-1435.

Celniker, S.E., Sharma S., Keelan, D.J. and Lewis, E.B. (1990). The molecular genetics of the bithorax complex of *Drosophila*: *cis*-regulation in the *Abdominal-B* domain. EMBO J. 9, 4277-4286.

Cereghini, S., Blumenfeld, M. and Yaniv, M. (1988). A liver-specific factor essential for albumin transcription differs between differentiated and dedifferentiated rat hepatoma cells. Genes Dev. 2, 957-974.

Cereghini, S., Yaniv, M. and Cortese, R. (1990). Hepatocyte dedifferentiation and extinction is accompanied by a block in the synthesis of mRNA coding for the transcription factor HNF1/LFB1. EMBO J. 9, 2257-2263.

Chadwick, R. and McGinnis, W. (1987). Temporal and spatial distribution of transcripts from the *Deformed* gene of *Drosophila*. EMBO J. 6, 779-789.

Chadwick, R., Jones, B., Jack, T. and McGinnis, W. (1990). Ectopic expression from the *Deformed* gene triggers a dominant defect in *Drosophila* adult head development. Dev. Biol. 141, 130-140.

Chalfie, M. and Sulston, J. (1981). Developmental genetics of mechanosensory neurons of *Caenorhabditis elegans* . Dev. Biol. 82, 358-370.

Chalfie, M., Horvitz, H.R. and Sulston, J. (1981). Mutations that lead to reiterations in the cell lineages of *C. elegans*. Cell 24, 59-69.

Chalfie, M., Sulston, J.E., White, J.G., Southgate, E., Thomson, J.N. and Brenner, S. (1985). The neural circuit for touch sensitivity in *Caenorhabditis elegans*. J. Neurosci. 5, 956-964.

Chavrier, P., Vesque, C., Galliot, B., Vigneron, M., Dollé, P., Duboule, D. and Charnay, P. (1990). The segment-specific gene Krox-20 encodes a transcription factor with binding sites in the promoter region of the Hox-1.4 gene. EMBO J. 9, 1209-1218.

Chisaka, O., Musci, T. S. and Capecchi, M. R. (1992). Developmental defects of the ear, cranial nerves and hindbrain resulting from targeted disruption of the mouse homeobox gene Hox-1.6. Nature 355, 516-521.

Chisholm, A. (1991). Control of cell fate in the tail region of *C. elegans* by the gene *eg*1-5. Development 111, 921-932.

Cho, K.W.Y. and De Robertis, E.M. (1990). Differential activation of *Xenopus* homeobox genes by mesoderm-inducing growth factors and retinoic acid. Genes Dev. 4, 1910-1916.

Cho, K.W.Y., Blumberg, B., Steinbeisser, H. and De Robertis, E.M. (1991). Molecular nature of Spemann's organizer: the role of the *Xenopus* homeobox gene *goosecoid*. Cell 67, 1111–1120.

Cho, K.W.Y., Goetz, J., Wright, C.V.E., Fritz, A., Hardwicke, J. and De Robertis, E.M. (1988). Differential utilization of the same reading frame in a *Xenopus* homeobox gene encodes two related proteins sharing the same DNA-binding specificity. EMBO J. 7, 2139-2149.

Cho, K.W.Y., Morita, E.A., Wright, C.V.E. and De Robertis, E.M. (1991). Overexpression of a homeodomain protein confers axis-forming activity to uncommitted *Xenopus* embryonic cells. Cell 65, 55-64.

Chouard, T., Blumenfeld, M., Bach, I., Vandekerckhove, J., Cereghini, S. and Yaniv, M. (1990). A distal dimerization domain is essential for DNA-binding by the atypical HNF1 homeodomain. Nucl. Acids Res. 18, 5853-5863.

Chuong, C-M., Oliver, G., Ting, S.A. and De Robertis, E.M. (1990). Gradients of homeoproteins in developing feather buds. Development, 110, 1021-1030.

Cianetti, L., Di Cristofaro, A., Zappavigna, V., Bottero, L., Boccoli, G., Testa, U., Russo, G., Boncinelli, E. and Peschle, C. (1990). Molecular mechanisms underlying the expression of the human HOX-5.1 gene. Nucleic Acids Res. 18, 4361-4368.

Civitareale, D., Lonigro, R., Sinclair, A.J. and Di Lauro, R. (1989). A thyroid specific nuclear protein essential for tissue-specific expression of the thyroglobulin promoter. EMBO J. 8, 2537-2542.

Clerc, R.G., Corcoran, L.M., LeBowitz, J.H., Baltimore, D. and Sharp, P.A. (1988). The B-cell-specific *Oct*-2 protein contains POU box- and homeo box-type domains. Genes Dev. 2, 1570-1581.

Coelho, C., Sumoy, L., Rodgers, B., Davidson, D., Hill, R., Upholt, W. and Kosher, R. (1991). Expression of the chicken homeobox-containing gene GHox 8 during embryonic chick limb development. Mech. Dev. 34, 143-154.

Cohen, B., McGuffin, M.E., Pfeifle, C., Segal, D. and Cohen, S.M. (1992). Apterous, a gene required for imaginal disc development in *Drosophila* encodes a member of the LIM family of developmental regulatory proteins. Genes Dev. 6, 715-729.

Cohen, B., Wimmer, E. and Cohen, S.M. (1991). Early development of the leg and wing primorida in the *Drosophila* embryo. Mech. Dev. 33, 229-240.

Cohen, S.M. (1990). Specification of limb development in the *Drosophila* embryo by positional cues from segmentation genes. Nature 343, 173-177.

Cohen, S.M. and Jürgens, G. (1989a). Proximal-distal pattern formation in *Drosophila*: cell autonomous requirement for Distal-less gene activity in limb development. EMBO J. 8, 2045-2055.

Cohen, S.M. and Jürgens, G. (1989b). Proximal-distal pattern formation in *Drosophila*: graded requirement for Distal-less gene activity during limb development in *Drosophila*. Roux's Archiv. Dev. Biol. 198, 157-169.

Cohen, S.M. and Jürgens, G. (1990). Mediation of *Drosophila* head development by gap-like segmentation genes. Nature 346, 482-485.

Cohen, S.M., Brönner, G., Küttner, F., Jürgens, G. and Jäckle, H. (1989). Distal-less encodes a homeodomain required for limb development in *Drosophila*. Nature 338, 432-434.

Colberg-Poley, A.M., Voss, S.D., Chowdhury, K. and Gruss, P. (1985). Clustered homeo boxes are differentially expressed during murine development. Cell 43, 39-45.

Colberg-Poley, M.A., Voss, S.D.K.C., Chowdhury, K. and Gruss, P. (1985b). Structural analysis of murine genes containing homeobox sequences and their expression in embryonal carcinoma cells. Nature 314, 713-718.

Colberg-Poley, M.A., Voss, S.D.K.C., Stewart, C.L., Wagner, E.F. and Gruss, P. (1985). Clustered homeoboxes are differentially expressed during murine development. Cell 43, 39-45.

Coletta, P.L., Shimeld, S.M., Chaudhuri, C., Muller, U., Clarke, J.P. and Sharpe, P.T. (1991a). Characterization of the murine *Hox*-3.3 gene and its promoter. Mech. Dev. 35, 129-142.

Coletta, P.L., Shimeld, S.M., Clarke, J.P. and Sharpe, P.T. (1991b). The murine *Hox*-3.3 gene. In: Development and regeneration of the nervous system. Chapmann and Hall, London.

Condie, B. G., and Harland, R. M. (1987). Posterior expression of a homeobox gene in early *Xenopus* embryos. Development 35, 206-211.

Condie, B. G., Hemmati Brivanlou, A. and Harland, R. M. (1990). Most of the homeobox-containing Xhox 36 transcripts in early *Xenopus* embryos cannot encode a homeodomain protein. Mol. Cell. Biol. 10, 3376-3385.

Condie, B.G. and Harland, R. (1987). Posterior expression of a homeobox gene in early *Xenopus* embryos. Development 101, 93-105.

Costa, M., Weir, M., Coulson, A., Sulston, J. and Kenyon, C. (1988). Posterior pattern formation in *C. elegans* involves position-specific expression of a gene containing a homeobox. Cell 55, 747-756.

Côté, S., Preiss, A., Haller, J., Schuh, R., Kienlin, A., Seifert, E. and Jäckle, H. (1987). The *gooseberry-zipper* region of *Drosophila*: five genes encode different spatially restricted transcripts in the embryo. EMBO J. 6, 2793-2801.

Cowing, D.W. and Kenyon, C. (1992). Expression of the homeotic gene *mab-5* during *Caenorhabditis elegans* embryogenesis. Development 116, 484–490.

Cribbs, D.L, Pattatucci, A.M., Pultz, M.A. and Kaufman, T.C. (1992b). Ectopic expression of the *Drosophila* homeotic gene *proboscipedia* under the control of the *Antennapedia* P1 promoter causes dominant thoracic defects. Genetics 132, 699-711.

Cribbs, D.L., Pultz, M.A., Johnson, D., Mazzulla, M. and Kaufman, T.C. (1992a). Structural complexity and evolutionary conservation of the *Drosophila* homeotic gene *proboscipedia*. EMBO J. 11, 1437-1449.

Cserjesi, P., Lilly, B., Bryson, L., Wang, Y., Sassoon, D.A. and Olson, E.N. (1992). MHox: a mesodermally restricted homeodomain protein that binds an essential site in the muscle creatine kinase enhancer. Development 115, 1087-1101.

D'Esposito, M., Morelli, F., Acampora, D., Migliaccio, E., Simeone, A. and Boncinelli, E. (1991). EVX2, a human homeobox gene homologous to the even-skipped segmentation gene, is localized at the 5' end of HOX4 locus on chromosome 2. Genomics 10, 43-50.

Dalton, D., Chadwick, R. and McGinnis, W. (1989). Expression and embryonic function of *empty spiracles*: A *Drosophila* homeobox gene with dual patterning functions on the anterior-posterior axis. Genes Dev. 3, 1940-1956.

Damante, G. and Di Lauro, R. (1991). Several regions of Antennapedia and thyroid transcription factor 1 homeodomains contribute to DNA binding specificity. Proc. Natl. Acad. Sci. USA 88, 5388-5392.

Davidson, D.R., Crawley, A., Hill, R.E. and Tickle, C. (1991). Position-dependent expression of two related homeobox genes in developing vertebrate limbs. Nature 352, 429-431.

Davidson, D.R, Graham, E., Sime, C. and Hill, R. (1988). A gene with sequence similarity to *Drosophila engrailed* is expressed during the development of the neural tube and vertebrae in the mouse. Development 104, 305-316.

Davis, C.A. and Joyner, A.L. (1988). Expression patterns of the homeo box-containing genes *En-1* and *En-2* and the proto-oncogene *int-1* diverge during mouse development. Genes and Development 2, 1736-1744.

Davis, C.A., Holmyard, D.P., Millen, K.J. and Joyner, A.L. (1991). Examining pattern formation in mouse, chicken and frog embryos with an *En*-specific antiserum. Development 111, 287-298.

Davis, C.A., Noble-Topham, S.E., Rossant, J., Joyner, A.L. (1988). Expression of the homeo box containing gene *En-2* delineates a specific region of the developing mouse brain. Genes and Development 2, 361-371.

De Robertis, E.M., Bürglin, T.R., Fritz, A., Wright, C.V.E., Jegalian, B., Schnegelsberg, P., Bittner, D., Morita, E., Oliver, G. and Cho, K.W.Y. (1989). Families of vertebrate homeodomain proteins.

In DNA-Protein Interactions in Transcription, Jay Gralla, ed. Alan R. Liss, Inc., New York, pp. 107-115.

De Robertis, E.M., Morita, E., and Cho, K.W.Y. (1991). Gradient-fields and homeobox genes. Development 112, 669-678.

De Robertis, E.M., Oliver, G., and Wright C.V.E. (1989). Determination of axial polarity in the vertebrate embryo: homeodomain proteins and homeogenetic induction. Cell 57, 189-191.

De Simone, V. and Cortese, R. (1991). Transcriptional regulation of liver-specific gene expression. Curr. Op. in Cell Biol. 3, 960-965.

De Simone, V., De Magistris, L., Lazzaro, D., Gerstner, J., Monaci, P., Nicosia, A. and Cortese, R. (1991). LFB3, a heterodimer-forming homeoprotein of the LFB1 family, is expressed in specialized epithelia. EMBO J. 10, 1435-1443.

Dearolf, C., Topol, J. and Parker, C. (1989). The *caudal* gene product is a direct activator of *fushi tarazu* transcription during *Drosophila* embryogenesis. Nature 341, 3430-343.

Deguchi, Y. and Kehrl, J.H. (1991). Nucleotide sequence of a novel diverged human homeobox gene encodes a DNA binding protein. Nucl. Acid Res. 19, 3742.

Deguchi, Y. and Kehrl, J.H. (1991). Selective expression of two homeobox genes in CD34 positive cells from human bone marrow. Blood 78, 323.

Deguchi, Y., Moroney, J.F., Wilson, G.L., Fox, C.H., Winter, H.S. and Kehrl, J.H. (1991). Cloning of a human homeobox gene that resembles a diverged *Drosophila* homeobox gene and is expressed in activated lymphocytes. New Biol. 3, 353.

DeLorenzi, M. and Bienz, M. (1990). Expression of *Abdominal-B* homeoproteins in *Drosophila* embryos. Development 108, 323-329.

DeLorenzi, M., Ali, N., Saari, G., Henry, C., Wilcox, M. and Bienz, M. (1988). Evidence that the *Abdominal-B* r element function is conferred by a *trans*-regulatory homeoprotein. EMBO J. 7, 3223-3231.

Deschamps, J., De-Laaf, R., Joosen, L., Meijlink, F. and Destree, O. (1987). Abundant expression of homeobox genes in mouse embryonal carcinoma cells correlates with chemically induced differentiation. Proc. Natl. Acad. Sci. U.S.A. 84, 1304-1308.

Deutsch, U. and Gruss, P. (1991). Murine paired domain proteins as regulatory factors of embryonic development. Sem. Dev. Biol. 2, 413–424.

Deutsch, U., Dressler, G.R. and Gruss, P. (1988). *Pax-1*, a member of a paired box homologous murine gene family, is expressed in segmented structures during development. Cell 53, 617-625.

Develin, C., Tice-Baldwin, K., Shore, D. and Arndt, K.T. (1991). RAP1 is required for BAS1/BAS2- and GCN4-dependent transcription of the yeast HIS4 gene. Mol. Cell. Biol. 11, 3642-3651.

Dick, T., Yang, X., Yeo, S. and Chia, W. (1991). Two closely linked *Drosophila* POU domain genes are expressed in neuroblasts and sensory elements. Proc. Natl. Acad. Sci. USA, 88, 7645-7649.

Diederich, R.J., Merrill, V.K.L., Pultz, M.A. and Kaufman, T.C. (1989). Isolation, structure and expression of labial, a homeotic gene of the *Antennapedia* complex involved in *Drosophila* head development. Genes Dev. 3, 399-414.

DiNardo, S., Kuner, J., Theis, J. and O'Farrell, P. (1985). Development of embryonic pattern in *D. melanogaster* as revealed by accumulation of the nuclear *engrailed* protein. Cell 43, 59-69.

Doe, C.Q., Hiromi, Y., Gehring, W.J. and Goodman, C.S. (1988). Expression and function of the segmentation gene *fushi tarazu* during *Drosophila* neurogenesis. Science 239, 170-175.

Doe, C.Q., Smouse, D. and Goodman, C.S. (1988). Control of neuronal fate by the *Drosophila* segmentation gene even-skipped. Nature 333, 376-378.

Dohrmann, C., Azpiazu, N. and Frasch, M. (1990). A new homeobox gene is expressed in mesodermal precursor cells of distinct muscles during embryogenesis. Genes Dev. 4, 2098-2111.

Dolecki, G.J. and Humphreys, T. (1988). An engrailed class homeobox gene in sea urchins. Gene 64, 21-31.

Dolecki, G.J., Wang, G. and Humphreys, T. (1988). Stage- and tissue-specific expression of two homeobox genes in sea urchin embryos and adults. Nucl. Acid Res. 16, 11543-11558.

Dolecki, G.J., Wannakrairoj, S., Lum, R., Wang, G., Riley, H.D., Carlos, R., Wang, A. and Humphreys, T. (1986). Stage-specific expression of a homeo box-containing gene in the non-segmented sea urchin embryo. EMBO J. 5, 925-930.

Dollé, P. and Duboule, D. (1989). Two gene members of the murine HOX-5 complex show regional and cell-type specific expression in developing limbs and gonads. EMBO J. 8, 1507-1515.

Dollé, P., Dierich, A., Schimmang, T., Schuhbaur, B., LeMeur, M., Chambon, P. and Duboule D. (1993). Disruption of the *Hoxd-13* gene induces localized heterochrony leading to mice with neotenic limbs. Cell 75, 431-441.

Dollé, P., Izpisúa-Belmonte, J-C., Boncinelli, E. and Duboule, D. (1991). The Hox-4.8 gene is localised at the 5' end of the HOX-4 complex and is expressed at the posterior extremity of the body during development. Mech. Dev. 36, 3-14.

Dollé, P., Izpisúa-Belmonte, J.-C., Falkenstein, H., Renucci, A. and Duboule, D. (1989). Coordinate expression of the murine Hox-5 complex homeobox-containing genes during limb pattern formation. Nature 342, 767-772.

Dollé, P., Izpisúa-Belmonte, J.-C., Tickle, C., Brown, J. and Duboule, D. (1991). Hox-4 genes and the morphogenesis of mammalian genitalia. Genes Dev. 5, 1767-1776.

Dollé, P., Lufkin, T., Krumlauf, R., Mark, M., Duboule, D. and Chambon, P. (1993). Local alterations of Krox-20 and Hox gene expression in the hindbrain of Hox-1.6 null embryos. Proc. Natl. Acad. Sci. U.S.A. 90, 7666–7670.

Dollé, P., Price, M. and Duboule, D. (1992). Expression of the mouse *Dlx-1* homeobox gene during facial, ocular and limb development. Differentiation 49, 93-100.

Dony, C. and Gruss, P. (1987). Specific expression of the Hox 1.3 homeo box gene in murine embryonic structures originating from or induced by the mesoderm. EMBO J. 6, 2965-2975.

Doyle, H., Harding, K., Hoey, T. and Levine, M. (1986). Transcripts encoded by a homeobox gene are restricted to dorsal tissue of *Drosophila* embryos. Nature 322, 76-79.

Doyle, H., Kraut, R. and Levine, M. (1989). Spatial regulation of *zerknüllt*: a dorsal-ventral patterning gene in *Drosophila*. Genes Dev. 3, 1518-1533.

Drees, B., Ali, Z., Soeller, W., Coleman, K., Poole, S. and Kornberg, T. (1987). The transcription unit of the *Drosophila* engrailed locus: an unusually small portion of a 70,000 bp gene. EMBO J. 6, 2803-2809.

Dressler, G.R. and Gruss, P. (1989). Anterior boundaries of Hox gene expression in mesoderm-derived structures correlate with the linear gene order along the chromosome. Differentiation 41, 193-201.

Dressler, G.R., Deutsch, U., Balling, R., Simon, D., Guénet, J.-L. and Gruss, P. (1988). Murine genes with homology to *Drosophila* segmentation genes. Development 104 (Supplement), 181-186.

Driever, W. and Nüsslein-Volhard, C. (1988a). A gradient of *bicoid* protein in *Drosophila* embryos. Cell 54, 83-93.

Driever, W. and Nüsslein-Volhard, C. (1988b). The *bicoid* protein determines position in the *Drosophila* embryo in a concentration dependent manner. Cell 54, 95-104.

Driever, W. and Nüsslein-Volhard, C. (1989). *bicoid* protein, a positive regulator of hunchback transcription in the early *Drosophila* embryo. Nature 337, 138-143.

Driever, W., Ma, J., Nüsslein-Volhard, C. and Ptashne, M. (1989b). Rescue of bicoid-mutant *Drosophila* embryos by *bicoid* fusion proteins containing heterologous activating sequences. Nature 342, 149-154.

Driever, W., Siegel, V. and Nüsslein-Volhard, C. (1990). Autonomous determination of anterior structures in the early *Drosophila* embryo by the *bicoid* morphogen. Development 109, 811-820.

Driever, W., Thoma, G. and Nüsslein-Volhard, C. (1989a). Determination of spatial domains of zygotic gene expression in the *Drosophila* embryo by the affinity of binding site for the bicoid morphogen. Nature 340, 363-367.

Duboule D., Boncinelli E., DeRobertis E., Featherstone M.S., Lonai P., Oliver G. and Ruddle, F. (1990). An update of mouse and human HOX genes nomenclature. Genomics 7, 458-459.

Duboule, D. and Dollé P. (1989). The structural and functional organization of the murine HOX gene family resembles that of *Drosophila* homeotic genes. EMBO J. 8, 1497-1505.

Duboule, D., Baron, A., Mähl, P. and Galliot, B. (1986). A new homeo-box is present in overlapping cosmid clones which define the mouse Hox-1 locus. EMBO J. 5, 1973-1980.

Duboule, D., Boncinelli, E., DeRobertis E., Featherstone, M.S., Lonai P., Oliver G. and Ruddle, F. (1990). An update of mouse and human HOX genes nomenclature. Genomics 7, 458-459.

Duncan, I. (1986). Control of bithorax complex functions by the segmentation gene *fushi tarazu* of *D. melanogaster.*, Cell 47, 297-309.

Duncan, I. (1987). The bithorax complex. Annu. Rev. Genet. 21, 285-319.

Duncan, I. and Lewis, E. (1982). Genetic control of body segment differentiation in *Drosophila* (eds. S. Subtelney and P.B. Green). Developmental order, its origin and regulation: 40th Symp. Dev. Biol. A. Liss, New York, pp. 533-554.

Duprey, P., Chowdhury, K., Dressler, G.R., Balling, R., Simon, D., Guenet, J.-L., and Gruss, P. (1988). A mouse gene homologous to the *Drosophila* gene *caudal* is expressed in epithelial cells from the embryonic intestine. Genes Dev. 2, 1647-1654.

Durston, A.J., Timmermans, J.P.M., Hage, W.J., Hendriks, H.F.J., de Vries, N.J., Heideveld, M. and Nieuwkoop, P.D. (1989). Retinoic acid causes an anteroposterior transformation in the developing central nervous system. Nature 340, 140-144.

Dush, M.K. and Martin, G.R. (1992). Analysis of mouse Evx genes: Evx-1 displays a graded expression in the primitive streak. Dev. Biol. 151, 273-287.

Eiken, H.G., Njølstad, P.R., Molven, A. and Fjose, A. (1987). A zebrafish homeobox-containing gene with embryonic transcription. Biochem. Biophys. Res. Commun. 149, 1165-1171.

Ekker, M., Wegner, J., Akimenko, M.A. and Westerfield, M. (1992). Coordinate embryonic expression of three zebrafish *engrailed* genes. Development 116, 1001-1010.

Engström, Y., Schneuwly, S. and Gehring, W.J. (1992). Spatial and temporal expression of an *Antennapedia/lacZ* gene construct integrated into the endogenous *Antennapedia* gene of *Drosophila melanogaster*. Roux's Arch. Dev. Biol. 201, 65-80.

Epstein, D.J., Vekemans, M. and Gros, P. (1991). splotch (Sp^{2H}), a mutation affecting development of the mouse neural tube, shows a deletion within the paired homeodomain of *Pax-3*. Cell 67, 767-774.

Ericson, J.U., Krauss, S. and Fjose, A. (1993). Genomic sequence and embryonic expression pattern of the zebrafish homeobox gene *hox*-3.4. Int. J. Dev. Biol. 37, 263–272.

Erselius, J.R., Goulding, M.D. and Gruss, P. (1990). Structure and expression pattern of the murine *Hox-3.2* gene. Development 110, 629-642.

Faiella, A., D'Esposito, M., Rambaldi, M., Acampora, D., Balsofiore, S., Stornaiuolo, A., Mallamaci, A., Migliaccio, E.,

Gulisano, M., Simeone, A. and Boncinelli, E. (1991). Isolation and mapping of EVX1, a human homeobox gene homologous to *even skipped*, localized at the 5' end of HOX 1 locus on chromosome 7. Nucl. Acids Res. 19, 6541-6545.

Fainsod, A. and Greunbaum, Y. (1989). A chicken homeobox gene with developmentally regulated expression. FEBS Lett. 250, 381-385.

Fainsod, A., Awgulewitsch, A. and Ruddle, F. H. (1987). Expression of the murine homeo box gene Hox 1.5 during embryogenesis. Dev. Biol. 124, 125-133.

Fainsod, A., Bogarad, L. D., Ruusala, T., Lubin, M., Crothers, D. M. and Ruddle, F. H. (1986). The homeo domain of a murine protein binds 5' to its own homeobox. Proc. Natl. Acad. Sci. U.S.A. 83, 9532-9536.

Fainsod, A., Margalit, Y.,Haffner, R. and Gruenbaum, Y. (1991). Non-immunological precipitation of protein-DNA complexes using glutathione-S-transferase fusion proteins. Nucl. Acids Res. 19, 4005.

Featherstone, M. S., Baron, A., Gaunt, S. J., Mattei, M.-G. and Duboule, D. (1988). Hox-5.1 defines a homeobox-containing gene locus on mouse chromosome 2. Proc. Natl. Acad. Sci. USA 85, 4760-4764.

Ferguson, E.L, Sternberg, P.W. and Horvitz, H.R. (1987). A genetic pathway for the specification of the vulval cell lineages of *Caenorhabditis elegans*. Nature 326, 259-267.

Ferguson, E.L. and Horvitz, H.R. (1985). Identification and characterization of 22 genes that affect the vulval cell lineages of the nematode Caenorhabditis elegans. Genetics 110, 17-72.

Fibi, M.B.Z., Zink, B., Kessel, M., Colberg-Poley, A.M., Labeit, S., Lehrach, H. and Gruss, P. (1988). Coding sequence and expression of the homeobox containing gene *Hox-1.3*. Development 102, 349-359.

Finkelstein, R. and Perrimon, N. (1990a). The orthodenticle gene is regulated by bicoid and torso and specifies *Drosophila* head development. Nature 346, 485-488.

Finkelstein, R., Smouse, D., Capaci, T.M., Spradling, A.C. and Perrimon, N. (1990a). The orthodenticle gene encodes a novel homeodomain protein involved in the development of the *Drosophila* nervous system and ocellar visual structures. Genes Dev. 4, 1516-1527.

Finney, M. (1990). The homeodomain of the transcription factor LFB1 has a 21 amino acid loop between helix 2 and helix 3. Cell 60, 5-6.

Finney, M. and Ruvkun, G. (1990). The *unc-86* gene product couples cell lineage and cell identity in *C. elegans*. Cell 63, 895-905.

Finney, M., Ruvkun, G. and Horvitz, H.R. (1988). The *C. elegans* cell lineage and differentiation gene *unc-86* encodes a protein with a homeodomain and extended similarity to transcription factors. Cell 55, 757-769.

Fitzpatrick V.D., Percival-Smith, A., Ingles, C.J. and Krause, H.M. (1992). Homeodomain-independent activity of the *fushi tarazu* polypeptide in *Drosophila* embryos. Nature 356, 610-612.

Fjose, A., Eiken, H.G., Njølstad, P.R., Molven, I. and Hordvik, I. (1988). A zebrafish *engrailed*-like sequence expressed during embryogenesis. FEBS Lett. 231, 355-360.

Fjose, A., McGinnis, W.J. and Gehring, W.J. (1985). Isolation of a homeobox containing gene from the engrailed region of *Drosophila* and the spatial distribution of its transcripts. Nature 313, 284-289.

Fjose, A., Njølstad, P.R., Nornes, S., Molven, A. and Krauss, S. (1992). Structure and early embryonic expression of the zebrafish *eng*-2 gene. Mech. Dev. 39, 51–62.

Fortini, M.E., Lai, Z. and Rubin, G.M. (1991). The *Drosophila zfh*-1 and *zfh*-2 genes encode novel proteins containing both zinc-finger and homeodomain motifs. Mech. Dev. 34, 113-122.

Frain, M., Swart, G., Monaci, P., Nicosia, A., Stampfli, S., Frank, R. and Cortese, R. (1989). The liver-specific transcription factor LFB1 contains a highly diverged homeobox DNA binding domain. Cell 59, 145-157.

Francis-Lang, H., Price, M., Polycarpou-Schwartz, M. and Di Lauro, R. (1992). Cell-type-specific expression of the rat thyroperoxidase promoter indicates common mechanisms for thyroid-specific gene expression. Mol. Cell. Biol. 12, 576-588.

Franz, T. (1990). Defective ensheatment of motoric nerves in the Splotch mutant mouse. Acta Anat. 138, 246-253.

Frasch, M.R.W. and Levine, M. (1987). Complementary patterns of even-skipped and fushi tarazu expression involve their differential regulation by a common set of segmentation genes in *Drosophila*. Genes Dev. 1, 981-995.

Frasch, M.R.W., Hoey, T., Rushlow, C., Doyle, H.J. and Levine, M. (1987). Characterization and localization of the even-skipped protein of *Drosophila*. EMBO J. 6, 749-759.

Frasch, M.R.W., Warrior, R., Tugwood, J. and Levine, M. (1988). Molecular analysis of *even-skipped* mutants in *Drosophila* development. Genes. Dev. 2, 1824-1838.

Freeling, M. and Hake, S. (1985). Developmental genetics of mutants that specify Knotted leaves in maize. Genetics 111, 617-634.

Freyd, G., Kim, S. and Horvitz, R.H. (1990). Novel cysteine-rich motif and homeodomain in the product of the *Caenorhabditis elegans* cell lineage gene *lin*-11. Nature 344, 876-879.

Frigerio, G., Burri, M., Bopp, D., Baumgartner, S. and Noll, M. (1986). Structure of the segmentation gene paired and the *Drosophila* PRD gene set as part of a gene network. Cell 47, 735-746.

Fritz, A.F. and De Robertis, E.M. (1988). *Xenopus* homeobox-containing cDNAs expressed in early development. Nucl. Acids Res. 16, 1453-1469.

Fritz, A.F., Cho, K.W.Y., Wright, C.V.E., Jegalian, B.G. and De Robertis, E.M. (1989). Duplicated homeobox genes in Xenopus. Dev. Biol.131, 584-588.

Frohman, M.A. and Martin, G.R. (1992). Isolation and spatial analysis of *Hox*-4.9, a new murine *labial*-like gene, reveals that *labial* subfamily members are expressed similarly only in early anteroposterior axis formation. Mech. Dev. 38, 55–68.

Frohman, M.A., Boyle, M. and Martin, G.R. (1990). Isolation of the mouse Hox 2.9 gene; analysis of embryonic expression suggests that positional information along the anterior-posterior axis is specified by mesoderm. Development 110, 589-607.

Frohman, M.A., Martin, G.R., Cordes, S.P. and Barsh, G.S. (1993). Altered rhombomere-specific gene expression in the mouse segmentation mutant, kreisler 1993. Development 117, 925–936.

Frohnhöfer, H.G. and Nüsslein-Volhard, C. (1986). Organization of anterior pattern in the *Drosophila* embryo by the maternal gene *bicoid*. Nature 324, 120-125.

Frohnhöfer, H.G. and Nüsslein-Volhard, C. (1987). Maternal genes required for the anterior localization of *bicoid* activity in the embryo of *Drosophila*. Genes Dev. 1, 880-890.

Frohnhöfer, H.G., Lehmann, R. and Nüsslein-Volhard, C. (1986). Manipulating the anterioposterior pattern of the *Drosophila* embryo. J. Embryol. Exp. Morphol. 97, 169-179.

Frumkin, A., Rangini, Z., Ben-Yehuda, A., Gruenbaum, Y. and Fainsod, A. (1991). A chicken *caudal* homologue, *CHox-cad*, is expressed in the epiblast with posterior localization and in the early endodermal lineage. Development 112, 207-219.

Furley, A.M., Morton, S.B., Manalo, D., Karagogeos, D., Dodd, J. and Jessell, T.M. (1990). The axonal glycoprotein TAG-1 is an immunoglobulin superfamily member with neurite outgrowth-promoting activity. Cell 61, 157-170.

Furukubo-Tokunaga, K., Müller, M., Affolter, M., Pick, L., Kloter, U. and Gehring, W.J. (1992). *In vivo* analysis of the helix-turn-helix motif of the *fushi tarazu* homeodomain of *Drosophila melanogaster*. Genes Dev. 6, 1082-1096.

Galliot, B., Dolle, P., Vigneron, M., Featherstone, M. S., Baron, A. and Duboule, D. (1989). The mouse Hox-1.4 gene: primary structure, evidence for promoter activity and expression during development. Development 107, 343-359.

Galliot, B., Dolle, P., Vigneron, M., Featherstone, M. S., Baron, A. and Duboule, D. (1989). The mouse Hox-1.4 gene: primary structure, evidence for promoter activity and expression during development. Development. 107, 343-359.

Garber, R.L., Kuroiwa, A. and Gehring, W.J. (1983). Genomic and cDNA clones of the homeotic locus *Antennapedia* in *Drosophila*. EMBO J. 2, 2027-2036.

Garbern, J., Odenwald, W.F., Tournier-Lasserve, E. and Lazzarini, R.A. (1989). Analysis of transcription of the murine homeobox gene Hox 1.3. In: Cell to Cell Signals in Mammalian Development, (S.W. de Laat, et al., eds.) NATO ASI Series, H26, 63-73, Springer-Verlag, Berlin.

Garcia-Bellido, A. (1977). Homeotic and atavic mutations in insects. Am. Zool. 17, 613-629.

Garcia-Fernandez, J., Baguñà, J. and Saló, E. (1991). Planarian homeobox genes: Cloning analysis and expression. Proc. Natl. Acad. Sci. USA 88, 7338-7342.

Gaunt, S.J. (1988). Mouse homeobox gene transcripts occupy different but overlapping domains in embryonic germ layers and organs: a comparison of Hox-3.1 and Hox-1.5. Development 103, 135-144.

Gaunt, S.J., Coletta, P.L, Pravtcheva, D. and Sharpe, P.T. (1990). Mouse *Hox-3.4*: homeobox sequence and embryonic expression patterns compared with other members of the Hox gene network. Devleopment 109, 329-339.

Gaunt, S.J., Krumlauf, R. and Duboule, D. (1989). Mouse homeogenes within a subfamily, Hox-1.4, -2.6 and -5.1 display similar anteroposterior domains of expression in the embryo, but show stage- and tissue-dependent differences in their regulation. Development 107, 131-141.

Gaunt, S.J., Sharpe, P.T., and Duboule, D. (1988). Spatially restricted domains of homeo-gene transcripts in mouse embryos: relation to a segmented body plan. Development, 104 (Supplement) 169-181.

Gaunt, S.J., Blum, M. and De Robertis, E. M.(1993) Expression of the mouse *goosecoid* gene during mid-embryogenesis may mark mesenchymal cell lineages in the developing head, limbs and ventral body wall. Development, 117, 769–778.

Geada, A.M.C., Gaunt, S.J., Azzawi, M., Shimeld, S.M., Pearce, J. and Sharpe, P.T. (1992). Sequence and expression of the mouse HOX-3.5 gene. Development 116, 497-506.

Gehring, W. (1966). Bildung eines vollständigen Mittelbeines mit Sternopleura in der Antennenregion bei der Mutante *Nasobemia* (*Ns*) von *Drosophila melanogaster*. Jul. Klaus. Arch. 41, 44-54.

Gehring, W.J. (1987). Homeo boxes and the study of development. Science 236, 1245-1252.

Gérard, M., Duboule, D. and Zákány, J. (1993). *In vivo* mapping of the mouse Hoxd-11 control elements. EMBO J. 12, 3539–3550.

German, M.S., Wang, J., Chadwick, R.B. and Rutter, W.J. (1992). Synergistic activation of the insulin gene by a LIM-homeodomain protein and a basic helix-loop-helix protein: building a functional insulin minienhancer complex. Genes Dev. 6, 2165-2176.

Gerster, T., Balmaceda, C.-G. and Roeder, R.G. (1990). The cell type-specific octamer transcription factor OTF-2 has two domains required for the activation of transcription. EMBO J. 9, 1635-1643.

Giampaolo, A., Acampora, D., Zappavigna, V., Pannese, M., D'Esposito, M., Carè, A., Faiella, A., Stornaiuolo, A., Russo, G., Simeone, A., Boncinelli, E. and Peschle, C. (1989). Differential expression of human HOX-2 genes along the anterior-posterior axis in embryonic central nervous system. Differentiation 40, 191-197.

Giasson, L., Specht, C.A., Milgrim, C., Novotny, C.P. and Ullrich, R.C. (1989). Cloning and comparison of Aa mating-type alleles of the Basidiomycete Schizophyllum commune. Mol. Gen. Genet. 218, 72-77.

Gibson, G. and Gehring, W.J. (1988). Head and thoracic transformations caused by ectopic expression of Antennapedia during Drosophila development. Development 102, 657-675.

Gibson, G., Schier, A., Le Motte, P. and Gehring W.J. (1990). The specification of sex combs reduced and Antennapedia are defined by a distinct portion of each protein that includes the homeodomain. Cell 62, 1087-1103.

Gillissen, B., Bergemann, J., Sandmann, C., Schroeer, B., Bolker, M. and Kahmann, R. (1992). A two-component regulatory system for self-nonself recognition in Ustilago maydis. Cell 66, 647-657.

Glicksman, M.A. and Brower, D.L. (1988). Expression of the Sex combs reduced protein in Drosophila larvae. Dev. Biol. 127, 113-118.

Golubovsky, M.D. and Kulakov, L.A. (1978). Regulator gene in Drosophila? Dros. Inf. Serv. 53, 133-134.

Goto, T., Macdonald, P. and Maniatis, T. (1989). Early and late periodic patterns of even-skipped expression are controlled by distinct regulatory elements that respond to different spatial cues. Cell 57, 413-422.

Gould, A.P., Brookman, J.J., Strutt, D.I. and White, A.H. (1990). Targets of homeotic gene control in Drosophila. Nature 348, 308-312.

Goulding, M.D., Chalepkis, G., Deutsch, U., Erselius, J.R. and Gruss, P. (1991). Pax-3, a novel murine DNA binding protein expressed during early neurogenesis. EMBO J. 10, 1135-1147.

Graham, A., Maden, M. and Krumlauf, R. (1991). The murine Hox-2 genes display dynamic dorsoventral patterns of expression during central nervous system development. Development 112, 255-264.

Graham, A., Papalopulu, N. and Krumlauf, R. (1989). The murine and Drosophila homeobox clusters have common features of organisation and expression. Cell 57, 367-378.

Graham, A., Papalopulu, N., Lorimer, J., McVey, J., Tuddenham, E. and Krumlauf, R. (1988). Characterization of a murine homeo box gene, Hox 2.6, related to the drosophila deformed gene. Genes Dev. 2, 1424-1438.

Green, J. B. and Smith, J. C. (1990). Graded changes in dose of a Xenopus activin A homologue elicit stepwise transitions in embryonic cell fate. Nature 347, 391-394.

Gruenberg, D.A., Natesan, S., Alexandre, C. and Gilman, M.Z. (1992). Human and Drosophila homeodomain proteins that enhance the DNA-binding activity of serum response factor. Science 257, 1089-1095.

Guazzi, S., Price, M., De Felice, M., Damante, G., Mattei, M.-G. and Di Lauro, R. (1990). Thyroid nuclear factor 1 (TTF-1) contains a homeodomain and displays a novel DNA binding specificity. EMBO J. 9, 3631-3639.

Guthrie, S., Muchamore, I., Kuriowa, A., Krumlauf, R. and Lumsden, A. (1991). Rhombomere transpositions in the chick embryo hindbrain reveal neuroectodermal autonomy of Hox-2.9 expression and segment phenotype. Nature 356, 157-159.

Gutjahr, T., Frei, E. and Noll, M. (1993). Complex regulation of early paired expression: Initial activation by gap genes and pattern modulation by pair-rule genes. Development, 117, 609–623.

Gyurkovics, H., Gausz, J., Kummer, J. and Karch, F. (1990). A new homeotic mutation in the Drosophila bithorax complex removes a boundary separating two domains of regulation. EMBO J. 9, 2579-2585.

Hafen, E., Kuroiwa, A. and Gehring, W.J. (1984). Spatial distribution of transcripts from the segmentation gene fushi tarazu during Drosophila embryonic development. Cell 37, 833-841.

Hafen, E., Levine, M. and Gehring, W.J. (1984). Regulation of Antennapedia transcript distribution by the bithorax complex in Drosophila. Nature 307, 287-289.

Hafen, E., Levine, M., Garber, R.L. and Gehring, W.J. (1983). An improved in situ hybridization method for the detection of cellular RNAs in Drosophila tissue sections and its application for localizing transcript of the homeotic Antennapedia gene complex. EMBO J. 2, 617-623.

Hake, S., Vollbrecht, E. and Freeling, M. (1989). Cloning Knotted, the dominant morphological mutant in maize using Ds2 as a transposon tag. EMBO J. 8, 15-22.

Hama, C., Ali, Z. and Kornberg, T.B. (1990). Region-specific recombination and expression are directed by portions of the Drosophila engrailed promoter. Genes Dev. 4, 1079-1093.

Hanes, S.D. and Brent, R. (1989). DNA specificity of the bicoid activator protein is determined by homeo domain recognition helix residue 9. Cell 57, 1275-1283.

Harding, K., Hoey, T., Warrior, R. and Levine, M. (1989). Autoregulatory and gap response elements of the even-skipped promoter of Drosophila. EMBO J. 8, 1205-1212.

Harding, K., Wedeen, C., McGinnis, W. and Levine, M. (1985). Spatially regulated expression of homeotic genes in Drosophila. Science 229, 1236-1242.

Hart, C., Awgulewitsch, A., Fainsod, A., McGinnis, W. and Ruddle, F. (1985). Homeobox gene complex on mouse chromosome 11: molecular cloning, expression in embryogenesis and homology to a human homeobox locus. Cell 43, 9-18.

Harvey, R. P., Tabin, C. J. and Melton, D. A. (1986). Embryonic expression and nuclear localization of Xenopus homeobox (Xhox) gene products. EMBO. J. 5, 1237-1244.

Harvey, R.P. and Melton, D.A. (1988). Microinjection of synthetic Xhox-1A homeobox mRNA disrupts somite formation in developing Xenopus embryos. Cell 53, 687-697.

Harvey, R.P., Tabin, C.J. and Melton, D.A. (1986). Embryoinic expression and nuclear localizationof Xenopus homeobox (Xhox) gene products. EMBO J. 5, 1237-1244.

Hasty, P., Ramirez-Solis, R., Krumlauf, R. and Bradley, A. (1991). Introduction of a subtle mutation into the Hox-2.6 locus in embryonic stem cells. Nature 350, 243-246.

Hatzopoulos, A.K., Stoykova, A.S., Erselius, J.R., Golding, M., Neuman, T. and Gruss, P. (1990). Structure and expression of the mouse Oct2a and Oct2b, two differentially spliced products of the same gene. Development 109, 349-362.

Hauser, C., Joyner, A., Klein, R., Learned, T., Martin, G. and Tjian, R. (1985). Expression of homologous homeobox containing genes in differentiated human teratocarcinoma cells and mouse embryos. Cell 43, 19-28.

Hawkins, N.C. and McGhee, J.D. (1990). Homeobox containing genes in the nematode Caenorhabditis elegans. (1990). Nucl. Acids Res. 18, 6101-6106.

Hayes, P.H., Sato, T. and Denell, R.E. (1984). Homeosis in Drosophila: The Ultrabithorax larval syndrome. Proc. Natl. Acad. Sci. U.S.A. 81, 545-549.

Hazelrigg, T. and Kaufman, T.C. (1983). Revertants of dominant mutations associated with the Antennapedia gene complex of Drosophila melanogaster. Cytology and genetics. Genetics 105, 581-600.

He, X., Gerrero, R., Simmons, D.M., Park, R.E., Lin, C.R., Swanson,

L.W. and Rosenfeld, M.G. (1991). Tst-1, a member of the POU domain gene family, binds the promoter of the gene encoding the cell adhesion molecule P_O. Molec. Cell. Biol. 11, 1739-1744.

He, X., Treacy, M.N., Simmons, D.M., Ingraham, H.A., Swanson, L.W. and Rosenfeld, M.G. (1989). Expression of a large family of POU domain regulatory genes in mammalian brain development. Nature 340, 35-42.

Heberlein, U., Mlodzik, M. and Rubin, G.M. (1991). Cell-fate determination in the developing *Drosophila* eye: role of the *rough* gene. Development 112, 703-712.

Hedgecock, E.M., Culotti, J.G., Hall, D.H. and Stern, B.D. (1987). Genetics of cell and axon migrations in *Caenorhabditis elegans*. Development 100, 365-382.

Hemmati-Brivanlou, A., de la Torre, J.R., Holt, C. and Harland, R.M. (1991). Cephalic expression and molecular characterization of *Xenopus En*-2. Development 111, 715-724.

Herr, W., Sturm, R.A., Clerc, R.G., Corcoran, L.M., Baltimore D., Sharp, P.A., Ingraham, H.A., Rosenfeld, M.G., Finney, M., Ruvkun, G. and Horvitz, H.R. (1988). The POU domain: a large conserved region in the mammalian *pit*-1, *oct*-1, *oct*-2 and *Caenorhabditis elegans unc*-86 gene products. Genes Dev. 2, 1513-1516.

Herskowitz, I. (1992). Yeast branches out. Nature 357, 190-191.

Higashijima, S.-i., Michiue, T., Emori, Y., and Saigo, K. (1992). Subtype determination of *Drosophila* embryonic external sensory organs by redundant homeo box genes *Bar H1* and *Bar H2*. Gene Dev. 6, 1005-1018.

Hill, R.E., Favor, J., Hogan, B.L.M., Ton, C.C.T., Saunders, G.F., Hanson, I.M., Prosser, J., Jordan, T., Hastie, N.D. and van Heyningen, V. (1991). Mouse *Small eye* results from mutations in a paired-like homeobox-containing gene. Nature 354, 522-525.

Hill, R.E., Jones, P.F., Rees, A.R., Sime, C.M., Justice, M.J., Copeland, N.G., Jenkins, N.A., Graham, E. and Davidson, D.R. (1989). A new family of mouse homeobox-containing genes: molecular structure, chromosomal location, and developmental expression of *Hox*-7.1. Genes Dev. 3, 26-37.

Hinnebusch, A.G. (1984). Evidence for translational regulation of the activator of general amino acid control in yeast. Proc. Natl. Acad. Sci. USA 81, 6442-6446.

Hiromi, Y. and Gehring, W.J. (1987). Regulation and function of the *Drosophila* segmentation gene *fushi tarazu*. Cell 50, 963-974.

Hiromi, Y., Kuroiwa, A. and Gehring, W.J. (1985). Control elements of the *Drosophila* segmentation gene *fushi tarazu*. Cell 43, 603-613.

Hirsch, M.R., Valarché, I., Deagostini-Bazin, H., Pernelle, C., Joliot, A. and Goridis, C. (1991). An upstream regulatory element of the NCAM promoter contains a binding site for homeodomains. FEBS Lett. 287, 197-202.

Hodgkin, J. (1983). Male phenotypes and mating efficiency in *Caenorhabditis elegans*. Genetics 103, 43-64.

Hodgkin, J., Horvitz, H.R. and Brenner, S. (1979). Nondisjunction mutants of the nematode *Caenorhabditis elegans*. Genetics 91, 67-94.

Hoey, T., Warrior, R., Manak, J. and Levine, M. (1988). DNA-binding activities of the *Drosophila melanogaster* even-skipped protein are mediated by its homeodomain and influenced by protein context. Mol. Cell. Biol. 8, 4598-4607.

Hogan, B.L.M., Hirst, E.M.A., Horsburgh, G. and Hetherington, C.M. (1988). *Small eye* (*Sey*): a mouse model for the genetic analysis of cranofacial abnormalities. Development 103 Supplement, 115-119.

Hogan, B.L.M., Horsburgh, G., Cohen, J., Hetherington, C.M., Fisher, G. and Lyon, M.F. (1986). *Small eyes* (*Sey*): a homozygous lethal mutation on chromosome 2 which affects the differentiation of both lens and nasal placodes in the mouse. J. Embryol. Exp. Morph. 97, 95-110.

Hogness, D.S., Lipshitz, H.D., Beachy, P.A., Peattie, D.A., Saint, R.B., Goldschmidt-Clermont, M., Harte, P.J., Gavis, E.R. and Helfand, S.L. (1985). Regulation and products of the *Ubx* domain of the bithorax complex. Cold Spring Harb. Symp. Quant. Biol. 50, 181-194.

Holland, P.W.H. (1991). Cloning and evolutionary analysis of *msh*-like homeobox genes from mouse, zebrafish and ascidian. Gene 98, 253-257.

Holland, P.W.H. and Hogan, B.L.M. (1988). Spatially restricted patterns of expression of the homeobox-containing gene Hox 2.1 during embryogenesis. Development 102, 159-174.

Holland, P.W.H. and Williams, N.A. (1990). Conservation of engrailed-like homeobox sequences during vertebrate evolution. FEBS Lett. 277, 250-252.

Holland, P.W.H., Williams, N.A. and Lanfear, J. (1991). Cloning of segment polarity gene homologues from the unsegmented brachiopod Terebratulina retusa (Linnaeus). FEBS Lett. 2, 211–213.

Hooper, J.E., Perez-Alonso, M., Bermingham, J.R., Prout, M., Rocklein, B.A., Wagenbach, M., Edström, J.E., de Frutos, R. and Scott, M.P. (1992). Evolutionary conservation of *Antennapedia* homeotic gene sequences. Genetics 132, 953–969.

Hunger, S.P., Galili, N., Carroll, A.J., Crist, W.M., Link, M.P. and Cleary, M.L. (1991). The t(1;19)(q23;p13) results in consistent fusion of E2A and PBX1 coding sequences in acute lymphoblastic leukemias. Blood 77, 687-693.

Hunt, P., Gulisano, M., Cook, M., Sham, M.-H., Faiella, A., Wilkinson, D., Boncinelli, E. and Krumlauf, R. (1991). A distinct *Hox* code for the branchial region of the head. Nature 353, 861-864.

Hunt, P., Whiting, J., Nonchev, S., Sham, M., Marshall, H., Graham, A., Cook, M., Allemann, R., Rigby, P., Gulisano, M., Faiella, A., Boncinelli, E. and Krumlauf, R. (1991b). The branchial *Hox* code and its implications for gene regulation, patterning of the nervous system and head evolution. Development 113 (Supplement 2), 63-77.

Hunt, P., Wilkinson, D. and Krumlauf, R. (1991b). Patterning the vertebrate head: murine Hox 2 genes mark distinct subpopulations of premigratory and migrating neural crest. Development 112, 43-51.

Immerglück, K., Lawrence, P.A. and Bienz, M. (1990). Induction across germ layers in *Drosophila* mediated by a genetic cascade. Cell 62, 261-268.

Ingham, P.W. and Martinez-Arias, A. (1986). The correct activation of *Antennapedia* and bithorax complex genes requires the *fushi tarazu* gene. Nature 324, 592-597.

Ingham, P.W., Baker, N.E. and Martinez-Arias, A. (1988). Regulation of segment polarity genes in the *Drosophila* blastoderm by fushi tarazu and even-skipped. Nature 331, 73-75.

Ip, Y., Kraut, R., Levine, M. and Rushlow, C. (1991). The *dorsal* morphogen is a sequence-specific DNA-binding protein that interacts with a long-range repression element in *Drosophila*. Cell 64, 439-446.

Irvine, K.D., Helfand, S.L. and Hogness, D.S. (1991). The large upstream control region of the homeotic gene *Ultrabithorax*. Development 111, 407-424.

Ishikiriyama, S., Tonoki, H., Shibuya, Y., Chin, S., Harada, N., Abe, K. and Niikawa, N. (1989). Waardenburg syndrome Type I in a child with de novo inversion (2) (q35q37.3). Am. J. Med. Genet 33, 505-507.

Izpisúa-Belmonte, J.-C., Falkenstein, H., Dollé, P., Renucci, A. and Duboule D. (1991). Murine genes related to the *Drosophila AbdB* homeotic gene are sequentially expressed during development of the posterior part of the body. EMBO J. 10, 2279-2289.

Izpisúa-Belmonte, J.C., Dollé, P., Renucci, A., Zappavigna, V., Falkenstein, H. and Duboule, D. (1990). Primary structure and embryonic expression pattern of the mouse Hox-4.3 homeobox gene. Development 110, 733-746.

Izpisùa-Belmonte, J.-C., De Robertis, E. M., Storey, K. G. and Stern, C. D. (1993). The homeobox gene *goosecoid* and the origin of organizer cells in the early chick blastoderm. Cell, 74, 645–660.

Jack, J.W. (1985). Molecular organization of the *cut locus* of *Drosophila melanogaster*. Cell 42, 869-876.

Jack, T., Regulski, M. and McGinnis, W. (1988). Pair-rule segmentation genes regulate the expression of the homeotic selector gene, *Deformed*. Genes Dev. 2, 635-651.

Jackson, I., Schofield, P. and Hogan, B. (1985). A mouse homeobox gene is expressed during embryogenesis and in adult kidney. Nature 317, 745-748.

Jagla, K., Georgel, Ph., Bellard, F., Dretzen, G. and Bellard, M. (1993). *nkch4*, a novel homeobox gene from the *Drosophila* 93E region. Gene 127, 167–171.

James, R. and Kazenwadel, J. (1991). Homeobox gene expression in the intestinal epithelium of adult mice. J. Biol. Chem. 266, 3246-3251.

Jegalian, B. G. and De Robertis, E. M. (1990). The *Xenopus laevis* Hox 2.1 homeodomain protein is expressed in a narrow band of the hindbrain. Int. J. Dev. Biol. 34, 453-456.

Johnson, W.A. and Hirsh, J. (1990). A *Drosophila* "POU Protein" binds to a sequence element regulating gene expression in specific dopaminergic neurons. Nature 343, 467-470.

Joly, J.-S., Maury, M., Joly, C., Duprey, P., Boulekbache, H. and Condamine, H. (1992). Expression of a zebrafish *caudal* homeobox gene correlates with the establishment of posterior cell lineages at gastrulation. Differentiation 50, 75-87.

Jones, F.S., Prediger, E.A., Bittner, D.A., De Robertis, E.M. and Edelman, G.M. (1992). Cell adhesion molecules as targets for Hox genes: N-CAM transcription is modulated by co-transfection of Hox 2.5 and 2.4. Proc. Natl. Acad. Sci. U.S.A. 89, 2086–2090.

Jorgensen, E.M. and Garber, R.L. (1987). Function and misfunction of the two promoters of the *Drosophila Antennapedia* gene. Genes Dev. 1, 544-555.

Jostes, B., Walther, C. and Gruss, P. (1991). The murine paired box gene, *Pax-7*, is expressed specifically during the development of the nervous and muscular system. Mech. Dev. 33, 27-38.

Joyner, A.L. and Hanks, M. (1991). The *engrailed* genes: Evolution of function. Sem. in Dev. Biol. 2, 435-445.

Joyner, A.L., Herrup, K., Auerbach, B.A., Davis, C.A. and Rossant, J. (1991). Subtle cerebellar phenotype in mice homozygous for a targeted deletion of the *En-2* homeobox- Science 251, 1239-1243.

Joyner, A.L., Kornberg, T., Coleman, K.G., Cox, D.R. and Martin, G.R. (1985). Expression during embryogenesis of a mouse gene with sequence homology to the *Drosophila engrailed* gene. Cell 43, 29-37.

Joyner, A.L. and Martin, G.R. (1987). *En-1* and *En-2*, two mouse genes with sequence homology to the *Drosophila engrailed* gene: expression during embryogenesis. Genes Dev. 1, 29-38.

Jürgens, G., Wieschaus, E., Nüsslein-Volhard, C. and Kluding, H. (1984). Mutations affecting the pattern of the larval cuticle in Drosophila melanogaster. 2. Zygotic loci on the third chromosome. Wilhelm Roux's Arch. Dev. Biol. 193, 283-295.

Kamb, A., Weir, M., Rudy, B., Varmus, H. and Kenyon, C. (1989). Identification of genes from pattern formation, tyrosine kinase and pottassium channel families by DNA amplification. Proc. Natl. Acad. Sci. U.S.A. 86, 4372-4376.

Kamps, M.P., Murre, C., Sun, X.-H. and Baltimore, D. (1990). A new homeobox gene contributes the DNA binding domain of the t (1;19) translocation protein in Pre-B ALL. Cell, 60, 547-555.

Karch, F., Bender, W, and Weiffenbach, B. (1990). *abdA* expression in *Drosophila* embryos. Genes Dev. 4, 1573-1587.

Karch, F., Weiffenbach, B., Peifer, M., Bender, W., Duncan, I., Celniker, S., Crosby, M. and Lewis, E.B. (1985). The abdominal region of the bithorax complex. Cell 43, 81-96.

Karr, T.L., Weir, M.J., Ali, Z. and Kornberg, T. (1989). Patterns of *engrailed* protein in early *Drosophila* embryos. Development 105, 605-612.

Kassis, J.A. (1990). Spatial and temporal control elements of the *Drosophila engrailed* gene. Genes Dev. 4, 433-443.

Kaufman, T.C. (1978). Cytogenetic analysis of chromosome 3 in *Drosophila melanogaster*. Isolation and characterization of four new alleles of the *proboscipedia* locus. Genetics 90, 579-596.

Kaufman, Lewis, R. and Wakimoto, B. (1980). Cytogenetic analysis of chromosome 3 in *Drosophila melanogaster*: the homeotic gene complex in polytene chromosome interval 84A-B. Genetics 94, 115-133.

Kaufman, T.C., Seeger, M.A. and Olsen, G. (1990). Molecular and genetic organization of the *Antennapedia* gene complex of *Drosophila melanogaster*. Adv. Genet. 27, 309-362.

Kelly, M., Burke, J., Smith, M., Klar, A. and Beach, D. (1988). Four mating-type genes control sexual differentiation in the fission yeast. EMBO J. 7, 1537-1547.

Kelsh, R.N. (1991). Ph.D. Thesis, University of Cambridge.

Kelsh, R.N., Dawson, I. and Akam, M.E. (1993). An analysis of Abdominal-B expression in the Locust *Schistocerca gregaria*. Development, 117, 293–302.

Kemler, I. and Schaffner, W. (1990). Octamer transcription factors and the cell type-specificity of immunoglobulin gene expression. FASEB J. 4, 1444-1449.

Kenyon, C. (1986). A gene involved in the development of the posterior body region of *C. elegans*. Cell 46, 477-487.

Kenyon, C. and Wang, B. (1991). A cluster of *Antennapedia*-class homeobox genes in a nonsegmented animal. Science 253, 516-517.

Kern, M.J., Witte, D.P., Valerius, M.T., Aronow, B.J. and Potter, S.S. (1992). A novel murine homeobox gene isolated by a tissue specific PCR cloning strategy. Nucl. Acid. Res. 20, 5189-5195.

Kerridge, S. and Morata, G. (1982). Developmental effects of some newly induced *Ultrabithorax* alleles of *Drosophila*. J. Embryol. Exp. Morph. 68, 211-234.

Kessel, M. and Gruss, P. (1991). Homeotic transformations of murine vertebrae and concomitant alteration of Hox codes induced by retinoic acid. Cell 67, 89-104.

Kessel, M., Balling, R. and Gruss, P. (1990). Variations of cervical vertebrae after expression of a *Hox-1.1* transgene in mice. Cell 61, 301-308.

Kessel, M., Schulze, F., Fibi, M. and Gruss, P. (1987). Primary structure and nuclear localization of a murine homeodomain protein. Proc. Natl. Acad. Sci. USA 84, 5306-5310.

Kilchherr, F., Baumgartner, S., Bopp, D., Frei, E. and Noll, M. (1986). Isolation of the *paired* gene of *Drosophila* and its spatial expression during early embryogenesis. Nature 321, 493-499.

Kim, Y. and Nierenberg, M. (1989). *Drosophila* NK-homeobox genes. Proc. Natl. Acad. Sci. U.S.A. 86, 7716-7720.

Kimmel, B.E., Heberlein, U. and Rubin, G.M. (1990). The homeo domain protein rough is expressed in a subset of cells in the developing *Drosophila* eye where it can specify photoreceptor cell subtype. Genes Dev. 4, 712-727.

Kissinger, C.R., Liu, B., Martin-Blanco, E., Kornberg, T.B. and Pabo, C.O. (1990). Structure of an *engrailed* homeodomain/DNA complex at 2.6Å resolution: a framework for understanding homeodomain/DNA interactions. Cell 63, 579-590.

Ko, H.-S., Fast, P., McBride, W. and Staudt, L.M. (1988). A human protein specific for the immunoglobulin octamer DNA motif contains a functional homeobox domain. Cell 55, 135-144.

Kohler, A., Logan, C., Joyner, A.L. and Muenki, M. (1993). Regional assignment of the human homeobox-containing gene *EN1* to chromosome 2q13-q21.2. Genomics, 15, 233–235.

Kongsuwan et al. (1988). Expression of multiple homeobox genes within diverse mammalian haemopoietic lineages. EMBO J. 7, 2131-2138.

Kornberg, T., Siden, I., O'Farrell, P. and Simon, M. (1985). The *engrailed* locus of *Drosophila*: *in situ* localization of transcripts reveals compartment-specific expression. Cell 40, 45-53.

Kornfeld, K., Saint, R.B., Beachy, P.A., Harte, P.J., Peattie, D.A. and Hogness, D.S. (1989). Structure and expression of a family of *Ultrabithorax* mRNAs generated by alternate splicing and polyadenylation in *Drosophila*. Genes Dev. 3, 243-258.

Korochkina, L.S. and Golubovsky, M.D. (1978). Cytogenetic analysis of induced mutations on the left end of the second chromosome of *D. melanogaster*. Dros. Inf. Serv. 53, 197-200.

Krasnow, M.A., Saffman, E.E., Kornfeld, K. and Hogness, D.S. (1989). Transcriptional activation and repression by *Ultrabithorax* proteins in cultured *Drosophila* cells. Cell 57, 1031-1043.

Krause, H.M. and Gehring, W.J. (1989). Stage-specific phosphorylation of the *fushi tarazu* protein during *Drosophila* development. EMBO J. 8, 1197-1204.

Krause, H.M., Klemenz, R. and Gehring W.J. (1988). Expression, modification and localization of the *fushi tarazu* protein in *Drosophila* embryos. Genes Dev. 2, 1021-1036.

Krauss, S., Johansen, T., Korzh, V. and Fjose, A. (1991a). Expression pattern of zebrafish *pax* genes suggests a role in early brain regionalization. Nature 353, 267-270.

Krauss, S., Johansen, T., Korzh, V., Moens, U., Ericson, J.U. and Fjose, A. (1991). Zebrafish *pax* [zf-a]: a paired box-containing gene expressed in the neural tube. EMBO J. 10, 3609-3619.

Kronstad, J.W. and Leong, S.A. (1990). The *b* mating type locus of *Ustilago maydis* contains variable and constant regions. Genes Dev. 4, 1384-1395.

Krumlauf, R., Holland, P.W.H., McVey, J.H. and Hogan, B.L.M. (1987). Developmental and spatial patterns of expression of the mouse homeobox gene, *Hox2.1*. Development 99, 603-617.

Kües, U. and Casselton, L. (1992). Homeodomains and regulation of sexual development in basidiomycetes. Trends Genet. 8, 154-155.

Kües, U., Richardson, W.V.J., Tymon, A.M., Mutasa, E.S., Göttgens, B., Gaubatz, S., Gregoriades, A. and Casselton, L.A. (1992). The combination of dissimilar alleles of the Aa and Ab gene complexes, whose proteins contain homeodomain motifs, determines sexual development in the mushroom *Coprinus cinereus*. Genes Dev. 6, 568-577.

Kuhn, D.T., Woods, D.F. and Andrew, D. (1981). Linkage analysis of the tumorous-head (*tuh-3*) gene in *Drosophila melanogaster*. Genetics 99, 99-107.

Kuhn, R., Monuki, E.S. and Lemke, G. (1991). The gene encoding the transcription factor SCIP has features of an expressed retroposon. Molec. Cell. Biol. 11, 4642-4650.

Kuner, J., Nakanishi, M., Ali, Z., Drees, B., Gustavson, E., Theis, J., Kauvar, L., Kornberg, T. and O'Farrell, P. (1985). Molecular cloning of *engrailed*, a gene involved in the development of pattern in *Drosophila* melanogaster. Cell 42, 309-316.

Kuo, C.J., Conley, P.B., Chen, L., Sladek, F.M., Darnell, J.E. Jr. and Crabtree, G.R. (1992). Inhibition by extinguishing loci of transcriptional hierarchy involved in cell-type specification. Nature 355, 457-461.

Kuo, C.J., Conley, P.B., Hsieh, C.L., Francke, U. and Crabtree, G.R. (1990). Molecular cloning, functional expression and chromosomal localization of mouse Hepatocyte Nuclear Factor 1. Proc. Natl. Acad. Sci. USA 87, 9838-9842.

Kuo, C.J., Mendel, D.B., Hansen, L.P. and Crabtree, G.R. (1991). Independent regulation of HNF1 alpha and HNF1 beta by retinoic acid in F9 teratocarcinoma cells. EMBO J. 10, 2231-2236.

Kuroiwa, A., Hafen, E. and Gehring, W.J. (1984). Cloning and transcriptional analysis of the segmentation gene *fushi tarazu* in *Drosophila*. Cell 37, 825-831.

Kuroiwa, A., Kloter, U., Baumgartner, P. and Gehring W.J. (1985). Cloning of the homeotic *Sex combs reduced* gene in *Drosophila* and *in situ* localization of its transcripts. EMBO J. 4, 3757-3764.

Kuziora, M.A. and McGinnis, W. (1988). Autoregulation of a *Drosophila* homeotic selector gene. Cell 55, 477-485.

Kuziora, M.A. and McGinnis, W. (1988). Different transcripts of the *Drosophila AbdB* gene correlate with distinct genetic sub-functions. EMBO J. 7, 3233-3244.

Lai, Z., Fortini, M.E. and Rubin, G.M. (1991). The embryonic expression patterns of *zfh*-1 and *zfh*-2, two *Drosophila* genes encoding novel zinc-finger homeodomain proteins. Mech. Dev. 34, 123-134.

LaRosa, G. J. and Gudas, L. J. (1988a). Early retinoic acid-induced F9 teratocarcinoma stem cell gene ERA-1: alternate splicing creates transcripts for a homeobox-containing protein and one lacking the homeobox. Mol. Cell. Biol. 8, 3906-3917.

Laughon, A and Scott, M.P. (1984). Sequence of a *Drosophila* segmentation gene: protein structure homology with DNA-binding proteins. Nature 310, 25-31

Laughon, A., Boulet, A.M., Bermingham, Jr. J.R., Laymon, R.A. and Scott, M.P. (1986). Structure of transcripts from the homeotic *Antennapedia* gene of *Drosophila melanogaster*: two promoters control the major protein-coding region. Mol. Cell. Biol. 6, 4676-4689.

Lawson, K. A., Meneses, J.J. and Pedersen, R. A. (1991). Clonal analysis of epiblast fate during germ layer formation in the mouse. Development 113, 891-911.

Lawson, K. A. and Pedersen, R. A. (1992). Clonal analysis of cell fate during gastrulation and early neurulation in the mouse. CIBA Foundation Symposium 165, 3-27.

Lazzaro, D., De Simone, V., De Magistris, L., Lehtonen, E. and Cortese, R. (1992). LFB1 and LFB3 homeoproteins are sequentially expressed during kidney development. Development, 114, 469–479.

Lazzaro, D., Price, M., De Felice, M. and Di Lauro, R. (1991). The transcription factor TTF-1 is expressed at the onset of thyroid and lung morphogenesis and in restricted regions of the foetal brain. Development 113, 1093-1104.

Le Calvez, J. (1948). In (3R) SSAr: Mutation *Aristapedia*, hétérozygote dominante, homozygote lethal chez *Drosophila melanogaster*. Bull. Biol. France Belg. 82, 97-113.

Le Mouellic, H., Condamine, H. and Brûlet, P. (1988). Pattern of transcription of the homeo gene Hox-3.1 in the mouse embryo. Genes Dev. 2, 125-135.

Le Mouellic, H., Lallemand, Y. and Brûlet, P. (1990). Targeted replacement of the homeobox gene Hox-3.1 by the Escherichia coli lacZ in mouse chimeric embryos. Proc. Natl. Acad. Sci. U.S.A. 87, 4712-4716.

Le Mouellic, H., Lallemand, Y. and Brûlet, P. (1992). Homeosis in the mouse induced by a null mutation in the Hox-3.1 gene. Cell 69, 251-264.

LeMotte, P., Kuroiwa, A., Fessler, L.I. and Gehring, W.J. (1989).

The homeotic gene *Sex combs reduced* of *Drosophila*: gene structure and embryonic expression. EMBO J. 8, 219-227.

Levine, M., Hafen, E., Garber, R.L. and Gehring, W.J. (1983). Spatial distribution of *Antennapedia* transcripts during *Drosophila* development. EMBO J. 2, 2037-2046.

Levine, M., Harding, K., Wedeen, C., Doyle, H., Hoey, T. and Radomska, H. (1985). Expression of the homeo box gene family in Drosophila. Cold Spring Harb. Symp Quant. Biol. 50, 209-222.

Lewis, E. B. (1951). Pseudoallelism and gene evolution. Cold Spring Harb. Symp. Quant. Biol. 16, 159-174.

Lewis, E.B. (1945). The relation of repeats to position effect in *Drosophila melanogaster*. Genetics 30, 137-166.

Lewis, E.B. (1955). Some aspects of position pseudoallelism. Amer. Nat. 89, 73-89.

Lewis, E.B. (1978). A gene complex controlling segmentation in *Drosophila*. Nature 276, 565-570.

Li, P.M., Reichert, J., Freyd, G., Horvitz, H.R. and Walsh, C.T. (1991). The LIM region of a presumptive *Caenorhabditis elegans* transcription factor is an iron-sulfur- and zinc-containing metallodomain. Proc. Natl. Acad. Sci. U.S.A. 88, 9210-9213.

Lindsley, D.L. and Grell, E.H. (1968). Genetic variation of *Drosophila melanogaster*. Carnegie Inst. Wash. Publ. 627.

Little, J., Byrd, C. and Brower, D. (1990). Effects of *abx, bx, pbx* mutations on expression of homeotic genes in *Drosophila* larvae. Genetics 124, 899-908.

Lloyd, A. and Sakonju, S. (1991). Characterization of two *Drosophila* POU domain genes, related to oct-1 and oct-2, and the regulation of their expression patterns. Mech. Dev. 36, 87-102.

Logan, C., Hanks, M.C, Noble-Topham, S., Nallainathan, D., Provart, N.J. and Joyner, A.L. (1992). Cloning and sequence comparison of the mouse, human and chicken *engrailed* genes reveal potential functional domains and regulatory regions. Developmental Genetics 13, 345-358.

Logan, C., Khoo, W., Cado, D. and Joyner, A.L. (1993). Two enhancer regions in the mouse *En-2* locus direct expression to the mid/hindbrain region and mandibular myoblasts. Development 117, 905–918.

Logan, C., Willard, H.F., Rommens, J.M. and Joyner, A.L. (1989). Chromosomal localization of the human homeo box-containing genes, EN1 and EN2. Genomics 4, 206-209.

Lonai, P., Arman, E., Czosnek, H., Ruddle, F. H. and Blatt, C. (1987). New murine homeoboxes: structure, chromosomal assignment, and differential expression in adult erythropoiesis. DNA 6, 409-418.

Lufkin, T., Dierich, A., LeMeur, M., Mark, M. and Chambon, P. (1991). Disruption of the Hox-1.6 homeobox gene results in defects in a region corresponding to its rostral domain of expression. Cell 66, 1105-1120.

Lufkin, T., Mark, M., Hart, C.P., Dollé, P., LeMeur, M. and Chambon, P. (1992). Homeotic transformation of the occipital bones of the skull by ectopic expression of a homeobox gene. Nature 359, 835-841.

MacDonald, P.M. and Struhl, G. (1986). A molecular gradient in early *Drosophila* embryos and its role in specifying the body pattern. Nature 324, 537-545.

MacDonald, P.M. and Struhl, G. (1988). *Cis*-acting sequences responsible for anterior localization of *bicoid* mRNA in *Drosophila* embryos. Nature 336, 595-598.

Macdonald, P.M. (1990). *bicoid* mRNA localization signal: phylogenetic conservation of function and RNA secondary structure. Development 110, 161-171.

Macdonald, P.M., Ingham, P. and Struhl, G. (1986). Isolation, structure and expression of even-skipped: a second pair-rule gene of *Drosophila* containing a homeobox. Cell 47, 721-734.

Macias, A., Casanova, J. and Morata, G. (1990). Expression and regulation of the *abd-A* gene of *Drosophila*. Development 110, 1197-1207.

Mackem, S. and Mahon, K. A. (1991). GHox-4.7: A chick homeobox gene expressed primarily in limb buds with limb-type differences in expression. Development 112, 791-806.

MacKenzie, A., Ferguson, M. W. J. and Sharpe, P. T. (1992). Expression patterns of the homeobox gene Hox-8 in the mouse embryo suggest a role in specifying tooth initiation and shape. Development 115, 403-420.

MacLeod, C.L., Fong, A.M., Seal, B.M., Walls, L. and Wilkinson, M.F. (1990). Isolation of novel cDNA clones from T-lymphoma cells: one encodes a putative multiple membrane spanning protein. Cell Growth & Diff. 1, 271-279.

Maden, M., Hunt, P. N., Eriksson, U., Kuroiwa, A., Krumlauf, R. and Summerbell, D. (1991). Retinoic acid-binding protein and homeobox expression in the rhombomeres of the chick embryo. Development 111, 35-44.

Magli, M.C., Barba, P., Celetti, A., De Vita G., Cillo, C. and Boncinelli, E. (1991). Coordinate regulation of HOX genes in human hematopoietic cells. Proc. Natl. Acad. Sci. U.S.A. 88, 6348-6352.

Mahaffey, J.W. and Kaufman, T.C. (1987a). Distribution of the *Sex combs reduced* gene products in *Drosophila melanogaster*. Genetics 117, 51-60.

Mahaffey, J.W. and Kaufman, T.C. (1987b). The homeotic genes of the *Antennapedia* complex and Bithorax complex of *Drosophila*.. Developmental genetics of higher organisms: a primer in developmental biology. Macmillan New York (ed. G.M. Malacinski) pp. 329-360.

Mahaffey, J.W., Diederich, R.J. and Kaufman, T.C. (1989). Novel patterns of homeotic protein accumulation in the head of the *Drosophila* embryo. Development 105, 167-174.

Mahon, K.A., Westphal, H. and Gruss, P. (1988). Expression of homeobox gene *Hox-1.1* during mouse embryogenesis. Development (Supplement) 102, 187-195.

Maier, D., Preiss, A. and Powell, J.R. (1990). Regulation of the segmentation gene *fushi tarazu* has been functionally conserved in *Drosophila*. EMBO J. 9, 3957-3966.

Malicki, J., Cianetti, L., Peschle, C. and McGinnis, W. (1992). A human *HOX4B* regulatory element provides head-specific expression in *Drosophila* embryos. Nature 358, 345-347.

Manoukian, A.S. and Krause, H.M. (1992). Concentration-dependent activities of the even-skipped protein in *Drosophila* embryos. Genes Dev. 6, 1740-1751.

Martin, G.R., Richman, M., Reinsch, S., Nadeau, J. and Joyner, A.L. (1990). Mapping of the two mouse *engrailed*-like genes: close linkage of *En-1* to *dominant hemimelia* (*Dh*) on Chromosome 1 and of *En-2* to *hemimelic extra-toes* (*Hx*) on Chromosome 5. Genomics 6, 302-308.

Martinez-Arias, A. (1986). The *Antennapedia* gene is required and expressed in parasegments 4 and 5 of the *Drosophila* embryo. EMBO J. 5, 135-141.

Martinez-Arias, A., Ingham, P.W., Scott, M.P. and Akam, M.E. (1987). The spatial and temporal deployment of *Dfd* and *Scr* transcripts during development of *Drosophila*. Development 100, 673-686.

Mavilio, F., Simeone, A., Boncinelli, E. and Andrews, P.W. (1988). Activation of four homeobox gene cluster in human embryonal carcinoma cells induced to differentiate by retinoic acid. Differentiation 37, 73-79.

Mavilio, F., Simeone, A., Giampaolo, A., Faiella, A., Zappavigna, V., Acampora, D., Poiana, G., Russo, G., Peschle, C. and Boncinelli, E. (1986). Differential and stage-related expression in embryonic tissues of a new human homeobox gene. Nature 324, 664-668.

McCarthy, B.J., Creasy, C.L. and Bergman, L.W. (1991). Molecular analysis of a temperature sensitive allele of the PHO2 gene of *Saccharomyces cerevisiae*. Nucl. Acids Res. 19, 3463.

McGinnis, N., Kuziora, M.A. and McGinnis, W. (1990). Human *Hox-4.2* and *Drosophila Deformed* encode similar regulatory specificities in *Drosophila* embryos and larvae. Cell 63, 969-976.

McGinnis, W., Garber, R.L., Wirz, J., Kuroiwa, A. and Gehring, W.J. (1984). A homologous protein-coding sequence in *Drosophila* homeotic genes and its conservation in other metazoans. Cell 37, 403-408.

McGinnis, W., Levine, M.S., Hafen, E., Kuroiwa, A. and Gehring, W. (1984). A conserved DNA sequence in homeotic genes of the *Drosophila Antennapedia* and bithorax complexes. Nature 308, 428-433.

McGinnis, W., Jack, T., Chadwick, R., Regulski, M., Bergson, C., McGinnis, N. and Kuziora, M.A. (1990). Establishment and maintenance of position-specific expression of the *Drosophila* homeotic selector gene *Deformed*. Adv. Genet. 27, 363-402.

McMahon, A.P., Joyner, A.L., Bradley, A. and McMahon, J.A. (1992). The mid-hindbrain phenotype of *wnt-1⁻/wnt-1⁻* mice results from stepwise deletion of *engrailed* expressing cells by 9.5 days *post-coitum*. Cell 69, 581-595.

Meijer, D., Graus, A., Kraay, R., Langeveld, A., Mulder, M.P. and Grosveld, G. (1990). The octamer binding factor *Oct*6: cDNA cloning and expression in early embryonic cells. Nucl. Acid Res. 18, 7357-7365.

Mendel, D.B. and Crabtree, G.R. (1991). HNF-1, a member of a novel dimerizing homeodomain protein. J. Biol. Chem. 266, 677-680.

Mendel, D.B., Hansen, L.P., Graves, M.K., Conley, P.B. and Crabtree, G.R. (1991). HNF1 alpha and HNF1 beta (vHNF1) share dimerization and homeo domains, but not activation domains, and form heterodimers *in vitro*. Genes Dev. 5, 1042-1056.

Mendel, D.B., Khavari, P.A., Conley, P.B., Graves, M.K., Hansen, L.P., Admon, A. and Crabtree, G.R. (1991). Characterization of a cofactor that regulates dimerization of a mammalian homeodomain protein. Science 254, 1762-1767.

Merrill, V.K.L., Diederich, R.J., Turner, F.R. and Kaufman, T.C. (1989). A genetic and developmental analysis of mutations in *labial*, a gene necessary for proper head formation in *Drosophila melanogaster*. Dev. Biol. 135, 376-391.

Merrill, V.K.L., Turner, F.R. and Kaufman, T.C. (1987). A genetic and developmental analysis of mutations in the *Deformed* locus in *Drosophila melanogaster*. Dev. Biol. 122, 379-395.

Miller, D.M., Shen, M.M., Shamu, C.E., Bürglin, T.R., Ruvkun, G., Dubois, M.L., Ghee, M. and Wilson, L. (1992). *C. elegans unc-4* gene encodes a homeodomain protein that determines the pattern of synaptic input to specific motor neurons. Nature 335, 841-845.

Mlodzik, M. and Gehring, W.J. (1987). Expression of the *caudal* gene in the germ line of Drosophila: formation of an RNA and protein gradient during early embryogenesis. Cell 48, 465-478.

Mlodzik, M., Fjose, A. and Gehring, W. J. (1988). Molecular structure and spatial expression of a homeobox gene from the labial region of the *Antennapedia*-complex. EMBO J. 7, 2569-2578.

Mlodzik, M., Fjose, A. and Gehring, W.J. (1985). Isolation of *caudal*, a *Drosophila* homeo box-containing gene with maternal expression, whose transcripts form a concentration gradient at the pre-blastoderm stage. EMBO J. 4, 2961-2969.

Mlodzik, M., Gibson, G. and Gehring, W.J. (1990). Effects of ectopic expression of *caudal* during *Drosophila* development. Development 109, 271-277.

Moase, C.E. and Trasler, D.G. (1989). Spinal ganglia reduction in the Splotch-delayed mouse neural tube defect mutant. Teratology 40, 67-75.

Mohler, J. and Pardue, M.-L. (1984). Mutational analysis of the region surrounding the 93D heat shock locus of *Drosophila melanogaster*. Genetics 106, 249-265.

Molven, A., Wright, C.V.E., Bremiller, R., De Robertis, E.M. and Kimmel, C.B. (1990). Expression of a homeobox gene product in normal and mutant zebrafish embryos: evolution of the tetrapod body plan. Development 109, 279-288.

Monaghan, A., Davidson, D., Sime, C., Graham, E., Baldock, R., Bhattacharya, S. and Hill, R. (1991). The msh-like homeobox genes define domains in the developing vertebrate eye. Development 112, 1053-1061.

Monaghan, A.P., Davidson, D.D., Sime, C., Graham, E., Baldock, R., Bhattacharya, S.S. and Hill, R.E. (1991). The *msh*-like homeobox genes define domains in the developing eye. Development 112, 1053-1061.

Monica, K., Galili, N., Nourse, J., Saltman, D. and Cleary, M.L. (1991). PBX2 and PBX3, new homeobox genes with extensive homology to the human proto-oncogene PBX1. Molec. Cell. Biol. 11, 6149-6157.

Monuki, E.S., Kuhn, R., Weinmaster, G., Trapp, B.D. and Lemke, G. (1990). Expression and activity of the POU transcription factor SCIP. Science 249, 1300-1303.

Monuki, E.S., Weinmaster, G., Kuhn, R. and Lemke, G. (1989). SCIP: A glial POU domain gene regulated by cAMP. Neuron 3, 783-793.

Morata, G. and Garcia-Bellido, A. (1976). Developmental analysis of some mutants of the bithorax system of *Drosophila*. Wilhem Roux's Arch. 179, 125-143.

Morata, G. and Kerridge, S. (1981). Sequential functions of the bithorax complex of *Drosophila*. Nature 290, 778-781.

Morata, G., Botas, J., Kerridge, S. and Struhl, G. (1983). Homeotic transformations of the abdominal segments of *Drosophila* caused by breaking or deleting a central portion of the bithorax complex. J. Embryol. Exp. Morph. 78, 319-341.

Morgan, B. A., Izpisúa-Belmonte, J.-C., Duboule, D. and Tabin, C. J. (1992). Targeted misexpression of Hox-4.6 in the avian limb bud causes apparent homeotic transformations. Nature 358, 236-239.

Morinaga, T., Yasuda, H., Hashimoto, T., Higashio, K. and Tamaoki, T. (1991). A human a-fetoprotein enhancer-binding protein, ATBF1, contains four homeodomains and seventeen zinc fingers. Mol. Cell. Biol. 11, 6041-6049.

Moriss-Kay, G.M., Murphy, P., Hill, R.E. and Davidson, D.R. (1991). Effects of retinoic acid excess on expression of *Hox-2.9* and *Krox*-20 and on morphological segmentation in the hindbrain of mouse embryos. EMBO J. 10, 2985-2995.

Muller, M., Carrasco, A.E., and De Robertis, E.M., (1984). Isolation of a maternally expressed *Xenopus* homeotic-like gene. Cell 39, 157-162.

Müller-Immerglück, M.M., Ruppert, S., Schaffner, W. and Matthias, P. (1988). A cloned transcription factor stimulates transcription from lymphoid-specific promoters in non-B cells. Nature 336, 544-551.

Müller-Immerglück, M.M., Schaffner, W. and Matthias, P. (1990). Transcription factor Oct-2A contains functionally redundant activation domains and works selectively from a promoter but not from a remote enhancer position in non-lymphoid (HeLa) cells. EMBO J. 9, 1625-1634.

Munke, M., Cox, D.R., Jackson, I.J., Hogan, B.L. and Francke, U. (1986). The murine Hox-2 cluster of homeo box containing genes maps distal on chromosome 11 near the tail-short (Ts) locus. Cytogenet. Cell. Genet. 42, 236-240.

Murphy, P. and Hill, R.E. (1991). Expression of the mouse *labial*-like homeobox-containing genes, *Hox*-2.9 and *Hox*1.6, during segmentation of the hindbrain. Development 111, 61-74.

Murphy, P., Davidson, D.R. and Hill, R.E. (1989). Segment-specific expression of a homeobox-containing gene in the mouse hindbrain. Nature 341, 156-159.

Murphy, S.P, Garbern, J., Odenwald, W.F., Lazzarini, E.A. and Linney, E. (1988). Differential eaxpression of the homeobox gene *Hox-1.3* in F9 embryonal carcinoma cells. Proc. Natl. Acad. Sci. U.S.A. 85, 5587-5591.

Murtha, M., Leckman, J.F. and Ruddle, F.H. (1991). Detection of homeobox genes in development and evolution. Proc. Natl. Acad. Sci. U.S.A. 88, 10711-10715.

Mutasa, E.S., Tymon, A.M., Gottgens, B., Mellon, F.M., Little, P.F.R. and Casselton, L. (1990). Molecular organization of an *A* mating type factor of the basidiomycete fugus *Coprinus cinereus*. Curr. Genet. 18, 223-229.

Nadeau, J.H., Berger, F.G., Cox, D.R., Crosby, J.L., Davisson, M.T., Ferrara, D., Fuchs, E., Hart, C., Hunihan, L. and Lalley, P.A. et al. (1989). A family of type I keratin genes and the homeobox-2 gene complex are closely linked to the rex locus on mouse chromosome 11. Genomics 5, 454-462.

Naito, M., Kohara, Y. and Kurosawa, Y. (1992). Identification of a homeobox-containing gene located between *lin-45* and *unc-24* on chromosome IV in the nematode *Caenorhabditis elegans*. Nucl. Acids Res. 20, 2967–2969.

Neufeld, E.J., Skalnik, D.G., Lievens, P.M.-J. and Orkin, S.H. (1992). Human CCAAT displacement protein is homologous to the *Drosophila* homeodomain protein, cut. Nature Genet. 1, 50-55.

Nicosia, A., Monaci, P., Tomei, L., De Francesco, R., Nuzzo, M., Stunnenberg, H. and Cortese, R. (1990). A myosin-like dimerization helix and an extra-large homeodomain are essential elements of the tripartite DNA binding structure of LFB1. Cell 61, 1225-1236.

Niehrs, C. and De Robertis, E.M. (1991). Ectopic expression of a homeobox gene changes cell fate in *Xenopus* embryos in a position-specific manner. EMBO J. 10, 3621-3629.

Niehrs, C., Cho, K.W.Y. and De Robertis, E.M. (1993). The Homeobox Genes *goosecoid* affects Gastrulation Movements and Dorsal Specification in *Xenopus* embryos. Cell 72, 491–503.

Njølstad, P.R,. Molven, A. and Fjose, A. (1988). A zebrafish homologue of the murine *Hox-2.1* gene. FEBS. Lett. 230, 25-30.

Njølstad, P.R. and Fjose, A. (1988). *In situ* hybridization patterns of zebrafish homeobox gene homologous to *Hox*-2.1 and *En*-2 of mouse. Biochem. Biophys. Res. Commun. 157, 426-432.

Njølstad, P.R., Molven, A. and Fjose, A. (1988a). A zebrafish homologue of the murine *Hox*-2.1 gene. FEBS Lett. 230, 25-30.

Njølstad, P.R., Molven, A., Apold, J. and Fjose, A. (1990). The zebrafish homeobox gene *hox*-2.2: transcription unit, potential regulatory regions and *in situ* localization of transcripts. EMBO J. 9, 515-524.

Njølstad, P.R., Molven, A., Hordvik, I., Apold, J. and Fjose, A. (1988b). Primary structure, developmentally regulated expression and potential duplication of the zebrafish homeobox gene ZF-21. Nucl. Acids Res. 19, 9097-9111.

Nourse, J., Mellentin, J.D., Galili, N., Wilkinson, J., Stanbridge, E., Smith, S.D. and Cleary, M.L. (1990). Chromosomal translocation t (1;19) results in synthesis of a homeobox fusion mRNA that codes for a potential chimeric transcription factor. Cell 60, 535-545.

Novotny, C.P., Stankis, M.M., Specht, C.A., Yang, H., Ullrich, R.C. and Giasson, L. (1991). The *Aa* mating type locus of *Schizophyllum commune*. In: More Gene Manipulations in Fungi; eds. Bennett, J.W. and Lasure, L.L. (Academic Press, New York) pp. 234-257.

Nüsslein-Volhard, C. and Wieschaus, E. (1980). Mutations affecting segment number and polarity in *Drosophila*. Nature 287, 795-801.

Nüsslein-Volhard, C., Kluding, H. and Jürgens, G. (1985). Genes affecting the segmental subdivision of the *Drosophila* embryo. Cold Spring Harb. Symp. Quant. Biol. 50, 145-154.

Nüsslein-Volhard, C., Wieschaus, E. and Kluding, H. (1984). Mutations affecting the pattern of the larval cuticle in *Drosophila melanogaster*. I. Zygotic loci on the second chromosome. Roux's Arch. Dev. Biol. 193, 267-282.

O'Connors, M.B., Binari, R., Perkins, L.A. and Bender, W. (1988). Alternative RNA products from the *Ultrabithorax* domain of the bithorax complex. EMBO J. 7, 435-445.

Odenwald, W.F., Garbern, J., Arnheiter, H., Tournier-Lasserve, E. and Lazzarini, R.A. (1989). The *Hox-1.3* homeobox protein is a sequence-specific DNA-binding phosphoprotein. Genes Dev. 3, 158-172.

Odenwald, W.F., Taylor, C.F., Palmer-Hill, F.J., Friedrich Jr. V., Tani, M. and Lazzarini, R.A. (1987). Expression of a homeo domain protein in noncontact-inhibited cultured cells and postmitotic neurons. Genes Dev. 1, 482-496.

Okamoto, K., Okazawa, H., Okuda, A., Sakai, M., Muramatsu, M. and Hamada, H. (1990). A novel octamer binding transcription factor is differentially expressed in mouse embryonic cells. Cell 60, 461-472.

Oliver, G., De Robertis, E.M., Wolpert, L. and Tickle, C. (1990). Expression of a homeobox gene in the chick wing bud following application of retinoic acid and grafts of polarizing region tissue. EMBO J. 9, 3093-3099.

Oliver, G., Sidell, N., Fiske, W., Heinjmann, C., Mohandas, T., Sparkes, R.S. and DeRobertis, E. M. (1989). Complementary homeoprotein gradients in developing limb buds. Genes Dev. 3, 641-650.

Oliver, G., Vispo, M., Mailhos, A., Martínez, C., Sosa-Pineda, B., Fielitz, W. and Ehrlich, R. (1992). Homeobox containing genes in flatworms. Gene 121, 337–342.

Oliver, G., Wright, C.V.E., Hardwicke, J. and De Robertis, E.M. (1988). Differential antero-posterior expression of two proteins encoded by a homeobox gene in *Xenopus* and mouse embryos. EMBO J. 7, 3199-3209.

Oliver, G., Wright, C.V.E., Hardwicke, J. and De Robertis, E.M. (1988b). A gradient of homeodomain protein in developing forelimbs of *Xenopus* and mouse embryos. Cell 55, 1017-1024.

Opstelten, D.-J.E., Vogels, R., Robert, B., Kalkhoven, E., Zwartkruis, F., de Laaf, L., Destrée, O.H., Deschamps, J., Lawson, K.A. and Meijlink, F. (1991). The mouse homeobox gene *S8* is expressed during embryogenesis predominantly in mesenchyma. Mech. Dev. 34, 29-42.

Oshima, Y. (1982). Regulation of phosphatases. Strathern, J.N., Jones, E.W. and Broach, J.R. (eds.) in Molecular Biology of the Yeast Saccharomyces. Cold Spring Harbor Laboratory Press, Cold Spring Harbor. 159-180.

Ott, M.-O., Rey-Campos, J., Cereghini, S. and Yaniv, M. (1991). vHNF1 is expressed in epithelial cells of distinct embryonic origin during development and precedes HNF1 expression. Mech. Dev. 36, 47-58.

Otting, G., Qian, Y.Q., Billeter, M., Müller, M., Affolter, M., Gehring, W.J. and Wüthrich, K. (1990). Protein-DNA contacts in the structure of a homeodomain-DNA complex determined by nuclear magnetic resonance spectroscopy in solution. EMBO J. 9, 3085-3092.

Oudejans, C.B.M., Pannese, M., Simeone, A., Meijer, C.I.L.M. and Boncinelli, E. (1990). The three most downstream genes of the HOX3 cluster are expressed in human extraembryonic tissues including trophoblast of androgenic origin. Development 108, 471-477.

Pannese, M., Stornaiuolo, A., Acampora, D., D'Esposito, M., Cafiero, M., Somma, R., Morelli, F., Menna, A., Faiella, A. and Boncinelli, E. (1989). Human class I homeobox gene expression in teratocarcinoma cells. In: Pathology of gene expression 131-149, Frati, L. and Aaronson, S.A. eds. Raven Press.

Papalopulu, N., Hunt, P., Wilkinson, D., Graham, A. and Krumlauf, R. (1990). Hox-2 homeobox genes and retinoic acid: potential roles in patterning the vertebrate nervous system. Advances in Neural Regeneration Research 291-307.

Papalopulu, N., Lovell-Badge, R., and Krumlauf, R. (1991). The expression of murine *Hox-2* genes is dependent on the differentiation pathway and displays collinear sensitivity to retinoic acid in F9 cells and *Xenopus* embryos. Nucl. Acids Res. 19, 5497-5506.

Patel, N.H. (1992). Evolution of insect pattern formation: a molecular analysis of short germband segmentation. In Spradling, A.C. (ed.): Evolutionary conservation of developmental mechanisms. New York: John-Wiley, pp. 85-110.

Patel, N.H., Ball, E.E. and Goodman, C.S. (1992). Changing role of even-skipped during the evolution of insect pattern formation. Nature 357, 339-342.

Patel, N.H., Martin-Blanco, E., Coleman, K.G., Poole, S.J., Ellis, M.C., Kornberg, T.B. and Goodman, C.S. (1989a). Expression of engrailed proteins in arthropods, annelids, and chordataes. Cell 58, 955-968.

Patel, N.H., Kornberg, T.B. and Goodman, C.S. (1989b). Expression of engrailed during segmentation in grasshopper and crayfish. Development 107, 201-212.

Patel, N.H., Schafer, B., Goodman, C.S. and Holmgren, R. (1989). The role of segment polarity genes during *Drosophila* neurogenesis. Genes Dev. 3, 890-904.

Peifer, M. and Bender, W. (1986). The anterobithorax and bithorax mutations of the bithorax complex. EMBO J. 5, 2293-2303.

Peifer, M., Karch, F. and Bender, W. (1987). The bithorax complex: control of segmental identity. Genes Dev. 1, 891-898.

Percival-Smith, A., Müller, M., Affolter, M. and Gehring, W.J. (1990). The interaction with DNA of wild-type and mutant *fushi tarazu* homeodomains. EMBO J. 9, 3967-3974.

Peterson, R.L, Jacobs, D.F. and Awgulewitsch, A. (1991). Hox-3.6: isolation and characterization of a new murine homeobox gene located in the 5' region of the *Hox*-3 cluster. Mech. Dev. 37, 151-166.

Peverali, A.F., D'Esposito, M., Acampora, D., Bunone, G., Negri, M., Faiella, A., Stornaiuolo, A., Pannese, M., Migliaccio, E., Simeone, A., Della Valle, G. and Boncinelli, E. (1990). Expression of HOX homeogenes in human neuroblastoma cell culture lines. Differentiation 45, 61-69.

Pokrywka, N.J. and Stephenson, E.C. (1991). Microtubules mediate the localization of *bicoid* RNA during *Drosophila* oogenesis. Development 113, 55-66.

Poole, S.J., Kauvar, L., Drees, B. and Kornberg, T. (1985). The *engrailed* locus of *Drosophila*: structural analysis of an embryonic transcript. Cell 40, 37-43.

Poole, S.J., Law, M.L., Kao, F.G. and Lau, Y.F. (1989). Isolation and chromosomal localization of the human *En-2* gene. Genomics 4, 225-231.

Pöpperl, H. and Featherstone, M.S. (1992). An autoregulatory element of the murine Hox-4.2 gene. EMBO J. 11, 3673-3680.

Pöpperl, H. and Featherstone, M.S. (1993). Identification of a retinoic acid response element upstream of the murine Hox-4.2 gene. Mol. Cell Biol. 13, 257–265.

Porteus, M.H., Brice, A.E.J, Bulfone, A., Usdin, T.B., Ciaranello, R.D. and Rubenstein, J.L.R. (1992). Isolation and characterization of a library of cDNA clones that are preferentially expressed in the embryonic telencephalon. Molecular Brain Research 12, 7-22.

Porteus, M.H., Bulfone, A., Ciaranello, R.D. and Rubenstein, J.L.R. (1991). Isolation and characterization of a novel cDNA encoding a homeodomain that is developmentally regulated in the ventral forebrain. Neuron 7, 221-229.

Pravcheva, M., Newman, M., Hunihan, L., Lonai, P. and Ruddle, F. (1989). Chromosome assignment of the murine *Hox*-4.1 gene. Genomics 5, 541-545.

Price, M., Lazzaro, D., Pohl, T., Mattei, M.-G., Rüther, U., Olivo, J.-C., Duboule, D. and DiLauro, R. (1992). Regional expression of the homeobox gene Nkx2.2 in the developing mammalian forebrain. Neuron 8, 241-255.

Price, M., Lemaistre, M., Pischetola, M., DiLauro, R. and Duboule D. (1991). A mouse distal-less related homeobox gene shows a restricted expression in the developing forebrain. Nature 351, 748-751.

Pultz, M.A., Diederich, R., Cribbs, D.L. and Kaufman, T.C. (1988). The *proboscipedia* locus of the *Antennapedia* complex: a molecular and genetic analysis. Genes Dev. 2, 901-920.

Püschel, A.W., Balling, R. and Gruss, P. (1990). Position-specific activity of the *Hox-1.1* promotor in transgenic mice. Development 108, 435-442.

Püschel, A.W., Balling, R. and Gruss, P. (1991). Separate elements cause lineage restriction and specify boundaries of *Hox-1.1* expression. Development 112, 279-287.

Qian, Y.-Q., Billeter, M., Otting, G., Müller, M., Gehring, W.J. and Wüthrich, K. (1989). The structure of the *Antennapedia* homeodomain determined by NMR spectroscopy in solution: Comparison with prokaryotic repressors. Cell 59, 573-580.

Rabin, M., Ferguson-Smith, A., Hart, C.P. and Ruddle, F.H. (1986). Cognate homeobox loci mapped on homologous human and mouse chromosomes. Proc. Natl. Acad. Sci. USA 83, 9104-9108.

Randazzo, F.M., Cribbs, D.L. and Kaufman, T.C. (1991). Rescue and regulation of *proboscipedia*: a homeotic gene of the *Antennapedia* complex. Development 113, 257-271.

Rangini, Z., Ben-Yehuda, A., Shapira, E., Gruenbaum, Y. and Fainsod, A. (1991). *CHox E* a chicken homeogene of the *H2.0* type exhibits dorso-ventral restriction in the proliferating region of the spinal cord. Mech. Dev. 35, 13-24.

Rangini, Z., Frumkin, A., Shani, G., Guttmann, M., Eyal-Giladi, H., Gruenbaum, Y. and Fainsod, A. (1989). The chicken homeobox genes Chox 1 and 3: Cloning, sequencing and expression during embryogenesis. Gene 76, 61-74.

Regulski, M., Harding, K., Kostriken, R., Karch, F., Levine, M. and McGinnis, W. (1985). Homeobox genes of the *Antennapedia* and bithorax complexes of *Drosophila*. Cell 43, 71-80.

Regulski, M., McGinnis, N., Chadwick, R. and McGinnis, W. (1987). Developmental and molecular analysis of deformed, a homeotic gene controlling *Drosophila* head development. EMBO J. 6, 767-777.

Renucci, A., Zappavigna, V., Zakany, J., Izpisúa-Belmonte, J-C., Bürki, K. and Duboule, D. (1992). Comparison of mouse and human Hox-4 complexes defines conserved promoter sequences involved in the regulation of the Hox-4.4 gene. EMBO J. 11, 1459-1468.

Reuter, R., Panganiban, G.E.F., Hoffmann, F.M. and Scott, M.P. (1990). Homeotic genes regulate the spatial expression of putative growth factors in the visceral mesoderm of *Drosophila* embryos. Development 101, 1031-1040.

Rey-Campos, J. and Yaniv, M. (1992). Regulation of albumin gene expression. In: Genetic intervention in diseases with unknown etiology. (T.O. Yoshida, ed.) Elsevier Science Publishers B.V. pp 1-14.

Rey-Campos, J., Chouard, T., Yaniv, M. and Cereghini, S. (1991). vHNF1 is a homeoprotein that activates transcription and forms heterodimers with HNF1. EMBO J. 10, 1445-1457.

Riley, P.D., Carroll, S.B. and Scott, M.P. (1987). The expression and regulation of *Sex combs reduced* protein in *Drosophila* embryos. Genes Dev. 1, 716-730.

Robert, B., Lyons, G., Simandl, B.K., Kuroiwa, A. and Buckingham, M. (1991). The apical ectodermal ridge regulates *Hox-7* and

Hox-8 gene expression in developing chick limb bud. Genes Dev. 5, 2363–2374.

Robert, B., Sassoon, D., Jacq, B., Gehring, W. and Buckingham, M. (1989). Hox 7, a mouse homeobox with a novel pattern of expression during embryogenesis. EMBO J. 8, 91-100.

Roberti, I., Sessa, G., Lucchetti, S. and Morelli, G. (1991). A novel class of plant proteins containing a homeodomain with a closely linked leucine zipper motif. EMBO J. 10, 1787-1791.

Roberts, R.C. (1967). *Small eyes*, a new dominant mutant in the mouse. Genet. Res. 9, 121-122.

Robinson, G.W., Wray, S. and Mahon, K.A. (1991). Spatially restricted expression of a member of a new family of murine Distal-less homeobox genes in the developing forebrain. The New Biologist 3, 1183-1194.

Rosa, F. (1989). *Mix.1*, a homeobox mRNA inducible by mesoderm inducers, is expressed in the presumptive endodermal cells of Xenopus embryos. Cell 57, 965-974.

Rosner, M.H., De Santo, R.J., Arnheiter, H. and Staudt, L.M. (1991). *Oct-3* is a maternal factor required for the first mouse embryonic division. Cell 64, 1103-1110.

Rosner, M.H., Vigano, M.A., Ozato, K., Timmons, P.M., Poirier, F., Rigby, P.W.J. and Staudt, L.M. (1990). A POU-domain transcription factor in early stem cells and germ cells of the mammalian embryo. Nature 345, 686-692.

Rowe, A. and Akam, M. 1988. The structure and expression of a hybrid homeotic gene. EMBO J. 7, 1107-1114.

Ruberti, I., Sessa, G., Lucchetti, S. and Morelli, G. (1991). A novel class of plant proteins containing a homeodomain with a closely linked leucine zipper motif. EMBO J. 7, 1787-1791.

Rubin, M.R., King, W., Toth, L.E., Sawcuk, I.S., Levine, M., D'Eustachio, P. and Nguyen-Huu, M.C. (1987). Murine Hox-1.7 Homeo-box gene: Cloning chromosomal location and expression. Mol. Cell. Biol. 7, 3836-3841, correction p. 5593.

Rubin, M.R., Toth, L.E., Patel, M.D., d'Eustachio, P. and Nguyen-Huu, M.C. (1986). A mouse homeobox gene is expressed in spermatocytes and embryos. Science 233, 663-667.

Rubock, M., Larin, Z., Cook, M., Papalopulu, N., Krumlauf, N. and Lehrach, H. (1990). A yeast artificial chromosome containing the mouse homeobox cluster Hox-2. Proc. Natl. Acad. Sci. U.S.A. 87, 4751-4755.

Ruiz i Altaba, A. (1990). Neural expression of the *Xenopus* homeobox gene Xhox3: evidence for a patterning neural signal that spreads through the ectoderm. Development 108, 595-604.

Ruiz i Altaba, A. and Jessell, T.M. (1991a). Retinoic acid modifies mesodermal patterning in early *Xenopus* embryos. Genes Dev. 5, 175-187.

Ruiz i Altaba, A. and Melton, D.A. (1989a). Bimodal and graded expression of the *Xenopus* homeobox gene *Xhox-3* during embryonic development. Development 106, 173-183.

Ruiz i Altaba, A. and Melton, D.A. (1989b). Involvement of the *Xenopus* homeobox gene Xhox3 in pattern formation along the anterior-posterior axis. Cell 57, 317-326.

Ruiz i Altaba, A. and Melton, D.A. (1989c). Interaction between peptide growth factors and homeobox genes in the establishment of anterior-posterior polarity in frog embryos. Nature 341, 33-38.

Ruiz i Altaba, A., Choi, T. and Melton, D.A. (1991). Expression of the Xhox3 homeobox protein in *Xenopus* embryos: blocking its early function suggests the requirement of Xhox3 for normal posterior development. Dev. Growth Diff. 33, 651-669.

Runstadler, J.A. and Kocher, T.D. (1991). A new *antennapedia*-class gene from the zebrafish. Nucl. Acids Res. 19, 5434.

Rushlow, C. and Levine, M. (1989). The role of the *zerknüllt* gene in dorsal-ventral pattern formation in *Drosophila*. In: Advances in Genetics, Genetic Regulatory Hierarchies in Development. T. Wright and J. Scandalios, eds. (San Diego: Academic Press, Inc.).

Rushlow, C. and Warrior, R. (1992). The rel family of proteins. Bioessays 14, 89-95.

Rushlow, C., Doyle, H., Hoey, T. and Levine, M. (1987). Molecular characterization of the *zerknüllt* region of the *Antennapedia* gene complex in *Drosophila*. Genes Dev. 1, 1268-1279.

Rushlow, C., Frasch, M., Doyle, H. and Levine, M. (1987). Maternal regulation of *zerknüllt*: a homeobox gene controlling differentiation of dorsal tissues in *Drosophila*. Nature 330, 583-586.

Rushlow, C., Han, K., Manley, J. and Levine, M. (1989). The graded distribution of the *dorsal* morphogen is initiated by selective nuclear transport in *Drosophila*. Cell 59, 1165-1177.

Sadoul, R. and Featherstone, M.S. (1991). Sequence analysis of the homeobox containing exon of the murine Hox-4.3 homeogene. Bioch.Biophys. Acta 1089, 259-261.

Saiga, H., Mizokami, A., Makabe, K.W., Satoh, N. and Mita, T. (1991). Molecular cloning and expression of a novel homeobox gene AHox1 of the ascidian, *Halocynthia roretzi*. Development 111, 821-828.

Saint, R., Kalionis, B., Lockett, T.J. and Elizur, A. (1988). Pattern formation in the developing eye of *Drosophila melanogaster* is regulated by the homeobox gene, *rough*. Nature 334, 151-154.

Salser, S.J. and Kenyon, C. (1992). Activation of a *C. elegans Antennapedia* homologue in migrating cells controls their direction of migration. Nature 355, 255-258.

Samson, M.L., Jackson-Grusby, L. and Brent, R. (1989). Gene activation and DNA binding by *Drosophila Ubx* and *abd-A* proteins. Cell 57, 1045-1052.

Sanchez-Herrero, E. (1991). Control of the expression of the bithorax complex *abdominal-A* and *Abdominal-B* by *cis*-regulatory regions in *Drosophila* embryos. Development 111, 437-449.

Sanchez-Herrero, E. and Crosby, M.A. (1988). The *Abdominal-B* gene of *Drosophila melanogaster*: overlapping transcripts exhibit two different spatial distributions. EMBO J. 7, 2163-2173.

Sanchez-Herrero, E., Vernos, I., Marco, R. and Morata, G. (1985). Genetic organization of the *Drosophila* bithorax complex. Nature 313, 108-113.

Sasaki, A., Doskow, J., MacLeod, C.L., Rogers, M., Gudas, L. and Wilkinson, M.F. (1991). The oncofetal gene *Pem* encodes a homeodomain and is regulated in embryonic and pre-muscle stem cells. Mech. Dev. 34, 155-164.

Sasaki, H. and Kuroiwa, A. (1990a). The nucleotide sequence of the cDNA encoding a chicken Deformed family homeobox gene, *Chox-Z*. Nucl. Acids Res. 18, 184.

Sasaki, H., Yamamoto, M. and Kuroiwa, A. (1992). Cell type dependent transcription regulation by chick homeodomain proteins. Mech. Dev. 37, 25–36.

Sasaki, H., Yokoyama, E. and Kuroiwa, A. (1990). Specific DNA binding of the two chicken Deformed family homeodomain proteins, *Chox*-1.4 and *Chox*-a. Nucl. Acids Res. 18, 1739-1747.

Sato, T., Hayes, P.H. and Denell, R.E. (1985). Homeosis in *Drosophila*: Roles and spatial patterns of expression of the *Antennapedia* and *Sex combs reduced* loci in embryogenesis. Dev. Biol. 111, 171-192.

Savard, P, Gates, P.B. and Brockes, J.P. (1988). Position dependent expression of a homeobox gene transcript in relation to amphibian limb regeneration. EMBO J. 7, 4275-4282.

Schaller, D., Wittmann, C., Spicher, A., Müller, F. and Tobler, H. (1990). Cloning and analysis of three new homeobox genes from the nematode *Caenorhabditis elegans*. Nucl. Acids Res. 18, 2033-2036.

Scheidereit, C., Cromlish, J.A., Gerster, T., Kawakami, K., Balmaceda, C.-G., Currie, R.A. and Roeder, R.G. (1988). A human lymphoid-specific transcription factor that activates immunoglobulin genes is a homeobox protein. Nature 336, 551-557.

Schena, M. and Davis, R. (1992). HD-Zip Proteins: Members of an *Arabidopsis* homeodomain protein superfamily. Proc. Natl. Acad. Sci. U.S.A. 89, 3894–3898.

Schier, A.F. and Gehring, W.J. (1992). Direct homeodomain-DNA interaction in the autoregulation of the *fushi tarazu* gene. Nature 356, 804-807.

Schierwater, B., Murtha, M., Dick, M., Ruddle, F.H., Buss, L.W. (1991). Homeoboxes in Cnidarians. J. Exp. Zool. 260, 413-416.

Schneitz, K., Spielmann, P. and Noll, M. (1993). Molecular genetics of *aristaless*, a *prd*-type homeobox gene involved in the morphogenesis of proximal and distal pattern elements in a subset of appendages in *Drosophila*. Genes Dev. 7, 114–129.

Schneuwly, S. and Gehring, W.J. (1985). Homeotic transformation of thorax into head: Developmental analysis of a new *Antennapedia* allele in *Drosophila melanogaster*. Dev. Biol. 108, 377-386.

Schneuwly, S., Klemenz, R. and Gehring, W.J. (1987a). Redesigning of the body plan of *Drosophila* by ectopic expression of the homeotic gene *Antennapedia*. Nature 325, 816-818.

Schneuwly, S., Kuroiwa, A. and Gehring, W.J. (1987b). Molecular analysis of the dominant homeotic *Antennapedia* phenotype. EMBO J. 6, 201-206.

Schneuwly, S., Kuroiwa, A., Baumgartner, P. and Gehring, W.J. (1986). Structural organization and sequence of the homeotic gene *Antennapedia* of *Drosophila melanogaster*. EMBO J. 5, 733-739.

Schöler, H.R. (1991). Octamania: The POU factors in murine development. Trends Genet. 7, 323-329.

Schöler, H.R., Dressler, G.R., Balling, R., Rohdewohld, H. and Gruss, P. (1990B). *Oct*-4: a germline-specific transcription factor mapping to the mouse t-complex. EMBO J. 9, 2185-2195.

Schöler, H.R., Hatzopoulos, A.K., Balling, R., Suzuki, N. and Gruss, P. (1989). A family of octamer-specific proteins present during mouse embryogenesis: evidence for germline-specific expression of an Oct-factor. EMBO J. 8, 2543-2550.

Schöler, H.R., Ruppert, S., Suzuki, N., Chowdhury, K. and Gruss. P. (1990). New type of POU domain in germline-specific protein *Oct*-4. Nature 344, 435-439.

Schreiber, E., Harshman, K., Kemler, I., Malipiero, U., Schaffner, W, Fontana, A. (1990). Astrocytes and glioblastoma cells express novel octamer-DNA binding proteins distinct from the ubiquitous *Oct*-1 and B cell type *Oct*-2 proteins. Nucl. Acids Res. 18, 5495-5503.

Schulz, B., Banuett, F., Dahl, M., Schlesinger, R., Schafer, W., Martin, T., Herskowitz, I. and Kahmann, R. (1990). The *b* alleles of *U. maydis*, whose combinations program pathogenic development, code for polypeptides containing a homeodomain-related motif. Cell 60, 295-306.

Schulze, F., Chowdhury, K., Zimmer, A., Drescher, U. and Gruss, P. (1987). The murine homeobox gene product, *Hox-1.1* protein, is growth controlled and associated with chromatin. Differentiation 36, 130-137.

Schummer, M., Scheurlen, I., Schaller, C. and Galliot, B. (1992). HOM/HOX homeobox genes are present in hydra (*Chlorohydra viridissima*) and are differentially expressed during regeneration. EMBO J. 11, 1815-1823.

Scott, M.P. and Weiner, A.J. (1984). Structural relationships among genes that control development: sequence homology between the *Antennapedia*, *Ultrabithorax*, and *fushi tarazu* loci of *Drosophila*. Proc. Natl. Acad. Sci. U.S.A. 81, 4115-4119.

Scott, M.P., Tamkun, J.W. and Hartzell III, G.W. (1989). The structure and function of the homeodomain. Biochim. Biophys. Acta 989, 25-48.

Scott, M.P., Weiner, A.J., Hazelrigg, T.I., Polisky, B.A., Pirotta, V., Scalenghe, F. and Kaufman, T.C. (1983). The molecular organization of the *antennapedia* locus of *Drosophila*. Cell 35, 763-776.

Sebastio, G., D'Esposito, M., Montanucci, M., Simeone, A., Auricchio, S. and Boncinelli, E. (1987). Modulated expression of human homeobox genes in differentiating intestinal cells. Biochem. Biophys. Res. Commun. 146, 751-756.

Seeger, M.A. and Kaufman, T.C. (1990). Molecular analysis of the *bicoid* gene from *Drosophila* pseudoobscura: identification of conserved domains within coding and noncoding regions of the *bicoid* mRNA. EMBO J. 9, 2977-2987.

Sengstag, C. and Hinnen, A. (1987). The sequence of the *Saccharomyces cerevisiae* gene PHO2 codes for a regulatory protein with unusual aminoacid composition. Nucl. Acids Res. 15, 233-246.

Sham, M.-H., Hunt, P., Nonchev, S., Papalopulu, N., Graham, A., Boncinelli, E. and Krumlauf, R. (1992). Analysis of the murine *Hox-2.7* gene: conserved alternative transcripts with differential distributions in the nervous system and the potential for shared regulatory regions. EMBO J. 11, 1825-1836.

Shapira, E., Yarus, S. and Fainsod, A. (1991). Genomic organization and expression during embryogenesis of the chicken repeat. Genomics 10, 931-939.

Sharpe, C.R. (1991). Retinoic acid can mimic endogenous signals involved in transformation of the *Xenopus* nervous system. Neuron 7, 239-247.

Sharpe, C.R. and Gurdon, J.B. (1990). The induction of anterior and posterior neural genes in *Xenopus laevis*. Development 109, 765-774.

Sharpe, C.R., Fritz, A., De Robertis, E.M. and Gurdon, J.B. (1987). A homeobox-containing marker of posterior neural differentiation shows the importance of predetermination in neural induction. Cell 50, 749-758.

Sharpe, P.T., Miller, J.R., Evans, E.P., Burtenshaw, M.D. and Gaunt, S.J. (1988). Isolation and expression of a new mouse homeobox gene. Development 102, 397-407.

Sherperd, J.C.W., McGinnis, E., Carrasco, A.E., DeRobertis, E.M. and Gehring, W.J. (1984). Fly and frog homeo domains show homology with yeast mating type regulatory proteins. Nature 310, 70-71.

Simeone, A., Acampora, D., Arcioni, L., Andrews, P. W., Boncinelli, E. and Mavilio, F. (1990). Sequential activation of HOX2 homeobox genes by retinoic acid in human embryonal carcinoma cells. Nature 346, 763-766.

Simeone, A., Acampora, D., D'Esposito, M., Faiella, A., Pannese, M., Scotto, L., Montanucci, M., D'Alessandro, G., Mavilio, F. and Boncinelli, E. (1989). Posttranscriptional control of human homeobox gene expression in induced NTERA-2 embryonal carcinoma cells. Mol. Rep. Dev. 1, 107-115.

Simeone, A., Acampora, D., D'Esposito, M., Faiella, A., Pannese, M., Scotto, L., Montanucci, M., D'Alessandro, G., Mavilio, F. and Boncinelli, E. (1989). Posttranscriptional control of human homeobox gene expression in induced NTERA-2 embryonal carcinoma cells. Mol. Rep. Dev. 1, 107-115.

Simeone, A., Acampora, D., Nigro, V., Faiella, A., D'Esposito, M., Stornaiuolo, A., Mavilio, F. and Boncinelli, E. (1991). Differential regulation by retinoic acid of the homeobox genes of the four HOX loci in human embryonal carcinoma cells. Mech. Dev. 33, 215-228.

Simeone, A., Mavilio, F., Acampora, D., Giampaolo, A., Faiella, A., Zappavigna, V., D'Esposito, M., Pannese, M., Russo, G., Boncinelli, E. and Peschle, C. (1987). Two human homeobox genes, c1 and c8: Structure analysis and expression in embryonic development. Proc. Natl. Acad. Sci. U.S.A. 84, 4914-4918.

Simeone, A., Mavilio, F., Bottero, L., Giampaolo, A., Russo, G., Faiella, A., Boncinelli, E. and Peschle, C. (1986). A human homeobox gene specifically expressed in spinal cord during embryonic development. Nature 320, 763-765.

Simeone, A., Pannese, M., Acampora, D., D'Esposito, M. and Boncinelli, E. (1988). At least three human homeoboxes on chromosome 12 belong to the same transcription unit. Nucl. Acids Res. 16, 5379-5387.

Simeone, A., Acampora, D., Gulisano, M., Stornaiuolo, A. and Boncinelli, E. (1992b). Nested expression domains of four homeobox genes in developing rostral brain. Nature 358, 687-690.

Simeone, A., Gulisano, M., Acampora, D., Stornaiuolo, A., Rambaldi, M.and Boncinelli, E. (1992a). 2 vertebrate homeobox genes related to the *Drosophila*-empty spiracles gene are expressed in the embryonic cerebral-cortex. EMBO J. 11, 2541-2550.

Sinha, N. and Hake, S. (1990). Mutant characters of Knotted maize leaves are determined in the innermost tissue layers. Dev. Biol. 141, 203-210.

Sive, H.L. and Cheng, P.F. (1991). Retinoic acid perturbs the expression of Xhox.lab genes and alters mesodermal determination in *Xenopus laevis*. Genes Dev. 5, 1321-1332.

Sive, H.L., Draper, B.W., Harland, R.M. and Weintraub, H. (1990). Identification of a retinoic acid-sensitive period during primary axis formation in *Xenopus laevis*. Genes Dev. 4, 932-942.

Skalnik, D.G., Strauss, E.C. and Orkin, S.H. (1991). CCAAT displacement protein as a repressor of the myelomonocytic-specific gp91-phox gene promoter. J. Biol. Chem. 266, 16736-16741.

Smith, D.P. and Old, R.W. (1990). Nucleotide sequence of *Xenopus laevis* Oct-1 cDNA. Nucl. Acids Res. 18, 369.

Specht, C.A., Stankis, M.M., Giasson, L. and Novotny, C.P. (1992). Functional analysis of the homeodomain-related proteins of the *Aa* locus of *Schizophyllum commune*. Proc. Natl. Acad. Sci. U.S.A. 89, 7174-7178.

Spicher, A., Schaller, D., Müller, F. and Tobler, H. (1988). Homeobox containing genes in nematodes. Experientia, 44, A27, abstract CMB 81.

St. Johnston, D., Driever, W., Berleth, T., Richstein, S. and Nüsslein-Volhard, C. (1989). Mutiple steps in the localization of *bicoid* mRNA to the anterior pole of the *Drosophila* oocyte. Development Supplement 108, 13-19.

Stankis, M.M., Specht, C.A., Yang, H., Giasson, L., Ullrich, R.C. and Novotny, C.P. (1992). The *Aa* mating locus of *Schizophyllum commune* encodes two dissimilar, multiallelic, homeodomain proteins. Proc. Natl. Acad. Sci. U.S.A .89, 7169-7173.

Stanojevic, D., Small, S. and Levine, M. (1991). Regulation of a segmentation stripe by overlapping activators and repressors in the *Drosophila* embryo. Science 254, 1385-1387.

Staudt, L.M., Clerc, R.G., Singh, H., LeBowitz, J.H., Sharp, P.A. and Baltimore, D. (1988). Cloning of a lymphoid-specific cDNA encoding a protein binding the regulatory octamer DNA motif. Science 241, 577-580.

Stern, C. and Bridges, C.B. (1926). The mutants of the extreme left end of the second chromosome of *Drosophila melanogaster*. Genetics 11, 503-530.

Stornaiuolo, A., Acampora, D., Pannese, M., D'Esposito, M., Morelli, F., Migliaccio, E., Rambaldi, M., Faiella, A., Nigro, V., Simeone, A. and Boncinelli, E. (1990). Human HOX genes are differentially activated by retinoic acid in embryonal carcinoma cells according to their position within the four loci. Cell Differ. Dev. 31, 119-127.

Stroeher, V.L., Gaiser, J.C. and Garber, R.L. (1988). Alternative RNA splicing that is spatially regulated: Generation of transcripts from the *Antennapedia* gene of *Drosophila melanogaster* with different protein-coding regions. Mol. Cell. Biol. 8, 4143-4154.

Stroeher, V.L., Jorgensen, E.M. and Garber, R.L. (1986). Multiple transcripts from the *Antennapedia* gene of *Drosophila melanogaster*. Mol. Cell. Biol. 6, 4667-4675.

Struhl, G. (1981). A homeotic mutation transforming leg to antenna in *Drosophila*. Nature 292, 635-638.

Struhl, G. (1985). Near-reciprocal phenotypes caused by inactivation or indiscriminate expression of the *Drosophila* segmentation gene *ftz*. Nature 318, 677.

Struhl, G. and White, A.H. (1985). Regulation of the *Ultrabithorax* gene of *Drosophila* by other bithorax complex genes. Cell 43, 507-519.

Struhl, G., Struhl, K. and Macdonald, P. (1989). The gradient morphogen *bicoid* is a concentration-dependent transcriptional activator. Cell 57, 1259-1273.

Stubbs, L., Poustka, A., Baron, A., Lehrach, H., Lonai, P. and Duboule, D. (1990). The murine genes *Hox*-5.1 and *Hox*-4.1 belong to the same complex on chromosome 2. Genomics 7, 422-427.

Sturm, R.A., Das, G. and Herr, W. (1988). The ubiquitous octamer binding protein Oct-1 contains a POU domain with a homeobox subdomain. Genes Dev. 2, 1582-1599.

Sundin, O. and Eichele, G. (1990). A homeo domain protein reveals the metameric nature of the developing chick hindbrain. Genes Dev. 4, 1267-1276.

Sundin, O. H., Busse, H. G., Rogers, M. B., Gudas, L. J. and Eichele, G. (1990). Region-specific expression in early chick and mouse embryos of Ghox-lab and Hox 1.6, vertebrate homeobox-containing genes related to *Drosophila* labial. Development. 108, 47-58.

Sunkel, C.E. and Whittle, J.R.S. (1987). Brista: a gene involved in the specification and differentiation of distal cephalic and thoracic structures in *Drosophila melanogaster*. Roux's Arch. Dev. Biol. 196, 124-132.

Superti-Furga, G., Barberis, A., Schreiber, E. and Busslinger, M. (1989). The protein CDP, but not CP1, footprints on the CCAAT region of the gamma-globin gene in unfractionated B-cell extracts. Biochim. Biophys. Acta 1007, 237-242.

Suzuki, N., Rohdewohld, H., Neuman, T., Gruss, P. and Schöler, H.R. (1990). *Oct-6*: a POU transcription factor expressed in embryonal stem cells and in the developing brain. EMBO J. 9, 3723-3732.

Taira, M., Jamrich, M., Good, P.J. and Dawid, I.B. (1992). The LIM domain-containing homeobox gene Xlim-1 is expressed specifically in the organizer region of *Xenopus* gastrula embryos. Genes Dev. 6, 356–366.

Takahashi, Y. and Le Douarin, N. (1990). cDNA cloning of a quail homeobox gene and its expression in neural crest-derived mesenchyme and lateral plate mesoderm. Proc. Natl. Acad. Sci. U.S.A. 87, 7482-7486.

Takahashi, Y., Bontoux, M. and Le Douarin, N. (1991). Epithelio-mesenchymal interactions are critical for Quox 7 expression and membrane bone differentiation in the neural crest derived mandibular mesenchyme. EMBO J. 10, 2387-2393.

Tanaka, M. and Herr, W. (1990). Differential transcriptional activation by oct-1 and oct-2: Interdependent activation domains induce oct-2 phosphorylation. Cell 60, 375-386.

Tanaka, M., Grossniklaus, U., Herr, W. and Hernandez, N. (1988). Activation of the U2 snRNA promoter by the octamer motif defines a new class of RNA polymerase II enhancer elements. Genes Dev. 2, 1764-1778.

Tani, M., Odenwald, W.F., Lazzarini, R.A. and Friedrich Jr. V.L. (1989). Progressive restriction in the distribution of Hox-1.3 homeodomain protein during embryogenesis. J. Neurosci. Res. 24, 457-469.

Tassabehji, M., Read, A.P., Newton, V.E., Harris, R., Balling, R., Gruss, P. and Strachan, T. (1992). Waardenburg's syndrome patients have mutations in the human homologue of the *Pax-3* paired box gene. Nature 355, 635-636.

Tear, G. (1990). D. Phil. thesis, University of Cambridge.

Tear, G., Akam, M. and Martinez-Arias, A. (1990). Isolation of an abdominal-A gene from the locust *Schistocerca gregaria* and its expression during embryogenesis. Development 110, 915–925.

Tearle, R. and Nüsslein-Volhard, C. (1987). Tübingen mutants and stock list. Dros. Inf. Serv. 66, 209-269.

Tetsuya Kojima., Satoshi Ishimaru, Shin-ichi Hiogashijima, Eiji Takayama, Hiroshi Akimaru, Masaki Sone, Yasufumi Emori and Kaoru Saigo. (1991). Identification of a different-type homeobox gene, BarH1, possibly causing Bar (B) and Om (Om) mutations in *Drosophila*. Proc. Natl. Acad. Sci. U.S.A. 88, 4343-4347.

Thali, M., Müller, M.M., DeLorenzi, M., Matthias, P. and Bienz, M. (1988). *Drosophila* homeotic genes encode transcriptional activators similar to mammalian OTF-2. Nature 336, 598-601.

Thireos, G., Penn, M.D. and Greer, H. (1984). 5′ untranslated sequences are required for the translational control of a yeast regulatory gene. Proc. Natl. Acad. Sci. U.S.A. 81, 5096-5100.

Thomas, P. Q and Rathjen, D. (1992). HES-1, a novel homeobox gene expressed by murine ES cells, identifies a new class of homeobox genes. Nucl. Acids Res. 20, 5840.

Tian, J.-M. and Schibler, U. (1991). Tissue-specific expression of the gene encoding hepatocyte Nuclear Factor 1 may involve hepatocyte Nuclear Factor 4. Genes Dev. 5, 2225-2234.

Tice-Baldwin, K., Fink, G.R. and Arndt, K.T. (1989). BAS1 has a Myb motif and activates HIS4 transcription only in combination with BAS2. Science 246, 931-935.

Tiong, S.Y.K., Bone, L.M. and Whittle, J.R.S. (1985). Recessive lethal mutations within the bithorax complex in *Drosophila*. Mol. Gen. Genet. 200, 335-342.

Tiong, S.Y.K., Gribbin, M.C. and Whittle, J.R.S. (1988). Mutational dissection of gene expression in the abdominal region of the bithorax complex of *Drosophila* in imaginal tissue. Roux's Arch. Dev. Biol. 197, 131-140.

Tokunaga, C. and Stern, C. (1969). Determination of bristle direction in *Drosophila*. Dev. Biol. 20, 411-425.

Tomlinson, A., Kimmel, B.E. and Rubin, G.M. (1988). *rough*, a *Drosophila* hoemobox gene required in photoreceptors R2 and R5 for inductive interactions in the developing eye. Cell 55, 771-784.

Ton, C.C.T., Hirvonen, H., Miwa, H., Weil, M.M., Monaghan, P., Jordan, T., van Heyningen, V., Hastie, N.D., Meijers-Heijboer, H., Drechsler, M., Royer-Pokora, B., Collins, F., Swaroop, A., Strong, L.C. and Saunders, G.F. (1991). Positional cloning of a paired box- and homeobox-containing gene from the Aniridia region. Cell 67, 1059-1074.

Toth, L.E., Slawin, K.L., Pintar, J.E. and Nguyen-Huu, C.M. (1987). Region-specific expression of mouse homeobox genes in the embryonic mesoderm and central nervous system. Proc. Natl. Acad. Sci. U.S.A. 84, 6790-6794.

Tournier-Lasserve, E., Odenwald, W.F., Garbern, J., Trojanowski, J. and Lazzarini, R.A. (1989). Remarkable intron and exon sequence conservation in human and mouse homeobox Hox 1.3 genes. Mol. Cell. Biol. 9, 2273-2278.

Treacy, M.N., He, X. and Rosenfeld, M.G. (1991). I-POU: a POU-domain protein that inhibits neuron-specific gene. Nature 350, 577-584.

Treacy, M.N., Nelson, L.I. et al. (1992). Twin of I-POU: A two amino acid difference in the I-POU homeodomain distinguishes an activator from an inhibitor of transcription. Cell 68, 491-505.

Treisman, J., Gönczy, P., Vashishtha, M., Harris, E. and Desplan, C.

(1989). A single amino acid can determine the DNA binding specificity of homeo domain proteins. Cell 59, 553-562.

Trent, C., Tsung, N. and Horvitz, H.R. (1983). Egg-laying defective mutants of the nematode *Caenorhabditis elegans*. Genetics 104, 619-647.

Tuggle, C.K., Zakany, J., Chianetti, L., Peschle, C. and Nguyen-Huu, M.C. (1990). Region-specific enhancers near two mammalian homeobox genes define adjacent rostrocaudal domains in the central nervous system. Genes Dev. 4, 180-189.

Tymon, A.M., Kues, U., Richardson, W.V.J. and Casselton, L. (1992). A fungal mating type protein that regulates sexual and asexual development contains a POU-related domain. EMBO J. 11, 1805-1813.

Utset, M.F., Awgulewitsch, A., Ruddle, F.H. and McGinnis, W. (1987). Region-specific expression of two mouse homeobox genes. Science 235, 1379-1382.

Veit, B., Vollbrecht, E., Mathern, J. and Hake, S. (1990). A tandem duplication causes the *Kn1-O* allele of Knotted, a dominant morphological mutant of maize. Genetics 125, 623-631.

Violette, S.M., Shashikant, C., Salbaum, M.J., Belting, H.G., Wang, J.C.H. and Ruddle, F.H. (1990). Repression of the b-amyloid gene in a Hox-3.1 producing cell line. Proc. Natl. Acad. Sci. U.S.A. 89, 3805-3809.

Vogel, K., Hörz, W. and Hinnen, A. (1989). The two positively acting regulatory proteins PHO2 and PHO4 physically interact with PHO5 upstream activation regions. Mol. Cell. Biol. 9, 2050-2057.

Vollbrecht, E., Veit, B., Sinha, N. and Hake, S. (1991). The developmental gene Knotted-1 is a member of a maize homeobox gene family. Nature 350, 241-243.

Wagner-Bernholz, J.T., Wilson, C., Gibson, G., Schuh, R. and Gehring, W.J. (1991). Identification of target genes of the homeotic gene *Antennapedia* by enhancer detection. Genes Dev. 5, 2467-2480.

Wakimoto, B.T. and Kaufman, T.C. (1981). Analysis of larval segmentation in lethal genotypes associated with the *Antennapedia* gene complex in *Drosophila melanogaster*. Dev. Biol. 81, 51-64.

Wakimoto, B.T., Turner, F.R. and Kaufman, T.C. (1984). Defects in embryogenesis in mutants associated with the *Antennapedia* gene complex of *Drosophila melanogaster*. Dev. Biol. 102, 147-172.

Wall, N., Jones, C., Hogan, B. and Wright, C. (1992). Expression and modification of *Hox-2.1* protein in mouse embryos. Mech. Dev. 37, 111-120.

Walldorf, U. and Gehring, W.J. (1992). *empty spiracles*, a gap gene containing a homeobox involved in *Drosophila* head development. EMBO J. 11, 2247-2259.

Walldorf, U., Fleig, R. and Gehring, W.J. (1989). Comparison of homeobox-containing genes of the honeybee and *Drosophila*. Proc. Natl. Acad. Sci. U.S.A. 86, 9971-9975.

Walther, C. and Gruss, P. (1991). *Pax-6*, a murine paired box gene, is expressed in the developing CNS. Development 113, 1435-1449.

Walther, C., Guénet, J.-L., Simon, D., Deutsch, U., Jostes, B., Goulding, M., Plachov, D., Balling, R. and Gruss, P. (1991). Pax: A murine multigene family of paired box containing genes. Genomics 11, 424-434.

Wang, G.V.L., Dolecki, G.J., Carlos, R. and Humphreys, T. (1990). Characterization and expression of two sea urchin homeobox gene sequences. Dev. Genet. 11, 77-87.

Waring, D.A. and Kenyon, C. (1990). Selective silencing of cell communication influences anteroposterior pattern formation in *C. elegans*. Cell 60, 123-131.

Waring, D.A. and Kenyon, C. (1991). Regulation of cellular

responsiveness to inductive signals in the developing *C. elegans* nervous system. Nature 350, 712-715.

Waterston, R., Martin, C., Craxton, M., Huynh, C., Coulson, A., Hillier, L., Durbin, R., Green, P., Shownkeen, R., Halloran, N., Metzstein, M., Hawkins, T., Wilson, R., Berks, M., Du, Z., Thomas, K., Thierry-Mieg, J. and Sulston, J. (1992). A survey of expressed genes in *Caenorhabditis elegans*. Nature Genet. 1, 114-123.

Way, J.C. and Chalfie, M. (1988). *mec*-3, a homeobox-containing gene that specifies differentiation of the touch receptor neurons in *C. elegans*. Cell 54, 5-16.

Way, J.C. and Chalfie, M. (1989). The *mec*-3 gene of *Caenorhabditis elegans* requires its own product for maintained expression and is expressed in three neuronal cell types. Genes Dev. 3, 1823-1833.

Way, J.C., Wang, L., Run, J.-Q. and Wang, A. (1991). The *mec*-3 gene contains *cis*-acting elements mediating positive and negative regulation in cells produced by asymmetric cell division in *Caenorhabditis elegans*. Genes Dev. 5, 2199-2211.

Webster, P.J. and Mansour, T.E. (1992). Conserved classes of homeodomains in *Schistosoma mansoni*, an early bilateral metazoan. Mech. Dev. 38, 25-32.

Wedeen, G., Harding, K. and Levine, M. (1986). Spatial regulation of *Antennapedia* and bithorax gene expression by the *Polycomb* locus in *Drosophila*. Cell 44, 739-748.

Weinzierl, R., Axton, J.M., Ghysen, A. and Akam, M. (1987).*Ultrabithorax* mutations in constant and variable regions of the protein coding sequence. Genes Dev. 1, 386-397.

Weir, M.P. and Kornberg, T. (1985). Patterns of *engrailed* and fushi tarazu transcripts reveal novel intermediate stages in *Drosophila* segmentation. Nature 318, 433-439.

White, J.G., Southgate, E. and Thomson, J.N. (1992). Mutations in the *Caenorhabditis elegans unc-4* gene alter the synaptic input to ventral cord motor neurons. Nature 355, 838-841.

White, R.A.H. and Akam, M.E. (1985). Contrabithorax mutations cause inappropriate expression of Ultrabithorax products in *Drosophila*. Nature 318, 567-569.

White, R.A.H. and Wilcox, M. (1984). Protein products of the bithorax complex in *Drosophila*. Cell 39, 163-171.

White, R.A.H. and Wilcox, M. (1985). Regulation of the distribution of *Ultrabithorax* proteins in *Drosophila*. Nature 318, 563-567.

Whiteley, M. and Armstrong, J.B. (1990). Isolation and characterization of a developmentally regulated homeobox sequence in the Mexican axolotl. Biochem. Cell. Biol. 68, 622-629.

Whiteley, M. and Armstrong, J.B. (1991a). Ectopic expression of a genomic fragment containing a homeobox causes neural defects in the axolotl. Biochem. Cell. Biol. 69, 366-374.

Whiteley, M. and Armstrong, J.B. (1991b). On the origin of the mesoderm in the Mexican axolotl, *Ambystoma mexicanum*. Can. J. Zool. 69, 1221-1225.

Wieschaus, E., Nusslein-Volhard, C. and Jurgens, G. (1984). Mutations affecting the pattern of the larval cuticle in *Drosophila melanogaster*. III. Zygotic loci on the X chromosome and the fourth chromosome. Roux Arch. Devl. Biol. 193, 296-307.

Wilde, C.D. and Akam, M. 1987. Conserved sequence elements in the 5' region of the *Ultrabithorax* transcription unit. EMBO J. 6, 1393-1401.

Wilkinson, D., Bhatt, S., Cook, M., Boncinelli, E. and Krumlauf, R. (1989). Segmental expression of hox 2 homeobox-containing genes in the developing mouse hindbrain. Nature 341, 405-409.

Wilkinson, M.F., Doskow, J., von Borstell, II., R.C., Fong, A.M. and MacLeod, C.L. (1991). The expression of several T cell specific and novel genes are repressed by trans-acting factors in immature T lymphoma clones. J. Exp. Med. 174, 269-280.

Wilkinson, M.F., Kleeman, J., Richards, J. and MacLeod, C.L. (1990). A novel onco-fetal gene expressed in a stage-specific manner in murine embryonic development. Dev. Biol. 141, 451-455.

Wirth, T., Priess, A., Annweiler, A., Zwillig, S. and Oeler, B. (1991). Multiple Oct2 isoforms are generated by alternative splicing. Nucl. Acids Res. 19, 43-51.

Wirz, J., Fessler, L.I. and Gehring, W.J. (1986). Localization of the *Antennapedia* protein in *Drosophila* embryos and imaginal discs. EMBO J. 5, 3327-3334.

Wolgemuth, D. J., Viviano, C.M., Gizang-Ginsberg, E., Frohman, M.A., Joyner, A. and Martin, G. R. (1987). Differential expression of the mouse homeobox-containing the Hox-1.4 during male germ cell differentiation and embryonic development. Proc. Natl. Acad. Sci. U.S.A. 84, 5813-5817.

Wolgemuth, D.J., Behringer, R.R:, Mostoller, M.P., Brinster, R.L. and Palmiter, R.D. (1989). Transgenic mice overexpressing the mouse homoeobox-containing gene Hox-1.4 exhibit abnormal gut development. Nature 337, 464-467.

Wolgemuth, D.J., Engelmyer, E., Duggel, R.N., Gizang-Ginsberg, E., Mutter, G.L., Ponzetto, C., Viviano, C. and Zakeri, Z.F. (1986). Isolation of a mouse cDNA coding for a devlopmentally regulated, testis-specific transcript containing homeo box homolgy. EMBO J. 5, 1229-1235.

Wright, C.V.E., Cho, K.W.Y., Hardwicke, J., Collins, R.H. and De Robertis, E.M. (1989). Interference with function of a homeobox gene in *Xenopus* embryos produces malformations of the anterior spinal cord. Cell 59, 81-93.

Wright, C.V.E., Cho, K.Y., Fritz, A., Burgling, T.R. and De Robertis, E.M. (1987). A *Xenopus laevis* gene encodes both homeobox-containing and homeobox-less transcripts. EMBO J. 6, 4083-4094.

Wright, C.V.E., Morita, E.A., Wilkin, D.J. and De Robertis, E.M. (1990). The *Xenopus* XlHbox 6 homeoprotein, a marker of posterior neural induction, is expressed in proliferating neurons. Development 109, 225-234.

Wright, C.V.E., Schnegelsberg, P. and De Robertis, E.M. (1989). Xlhbox 8: A novel *Xenopus* homeo protein restricted to a narrow band of endoderm. Development 104, 787-794.

Wysocka-Diller, J.W., Aisemberg, G.O., Baumgarten, M., Levine, M. and Macagno, E.R. (1989). Characterization of a homologue of bithorax-complex genes in the leech Hirudo medicinalis. Nature 341, 760-763.

Xue, D., Finney, M., Ruvkun, G. and Chalfie, M. (1992). Regulation of the *mec-3* gene by the *C. elegans* homeoproteins UNC-86 and MEC-3. EMBO J. 11, 4969-4979.

Yanopulos, G.D., Oltz, E.M., Rathbun, G., Berman, J.E., Smith, R.K., Lansford, R.D., Rothman, P., Okada, A., Lee, G., Morrow, M., Kaplan, K. and Alt, F. (1990). Coordinately regulated genes that are expressed in discrete stages of B-cell development. Proc. Natl. Acad. Sci. U.S.A. 87, 5759-5763.

Yeom, Y.I., Ha, H.-S., Balling, R., Schöler, H.R. and Artzt, K.(1991). Structure, expression and chromosomal location of the *Oct*-4 gene. Mech. Dev. 351, 171-180.

Yokouchi, Y., Sasaki, H. and Kuroiwa, A. (1991). Homeobox gene expression correlated with the bifurcation process of limb cartilage development. Nature 353, 443-445.

Zakany, J., Tuggle, C.K., Patel, M.D. and Nguyen-Huu, M.C. (1988). Spatial regulation of homeobox gene fusions in the embryonic central nervous system of transgenic mice. Neuron 1, 679-691.

Zappavigna, V., Renucci, A., Izpisúa-Belmonte, J.-C., Urier, G., Peschle, C. and Duboule, D. (1991). HOX4 genes encode transcription factors with potential auto- and cross-regulatory capacities. EMBO J. 10, 4177-4187.

Zaraiski, A. G., Lukyanov, S. A., Vasiliev, O. L., Smirnov, Y. V., Belyavsky, A. V. and Kasanskaya, O.V. (1992). A novel homeobox gene expressed in the anterior neural plate of the *Xenopus* embryo. Dev. Biol. 152, 373-382.

Zavortink, M. and Sakonju, S. (1989). The morphogenetic and regulatory functions of the *Drosophila Abdominal-B* gene are encoded in overlapping RNAs transcribed from separate promoters. Genes Dev. 3, 1969-1981.

Zhao, A.Z., Vansant, G., Bell, J., Humphreys, T. and Maxson, R. (1991). Activation of the L1 late H2B histone gene in blastula-stage sea urchin embryos by an *Antennapedia*-class homeoprotein. Mech. Dev. 34, 21-28.

Zhi-Gang Xue, Gehring, W.J. and Le Douarin, N. (1991). Quox-1, a quail homeobox gene expressed in the embryonic central nervous system, including the forebrain. Proc. Natl. Acad. Sci. USA, 88, 2427-2431.

Zimmer, A. and Gruss, P. (1989). Production of chimaeric mice containing embryonic stem (ES) cells carrying a homeobox Hox-1.1 allele mutated by homologous recombination. Nature 338, 150-153.

Indexes

Gene Index

1) The names under which the genes can be found in the entries are in bold.
2) Bold names are followed by a code [ ] indicating the species. The code is given below

Al	*Ascaris lumbricoides* (roundworm)
Am	*Apis mellis* (honeybee)
ar	*Artemia franciscana*
At	*Arabidopsis thaliana*
ax	*Ambystoma mexicanum* (axolotl)
c	*Gallus gallus* (chicken)
can	*Canis* (dog)
cb	*Caenorhabditis Briggsae* (roundworm)
Cc	*Coprinus cinereus*
ce	*Caenorhabditis elegans* (roundworm)
Ci	*Ciona intestinalis* (Ascidian)
Cv	*Caenorhabditis vulgarensis* (roundworm)
d	*Drosophila melanogaster*
dh	*Drosophila hydei*
Dj	*Dugesia japonica* (flatworm)
dO	*Drosophila melanogaster* (Oregon R)
Ds	*Drosophila subobscura*
Dt	*Dugesia tigrina* (flatworm)
Dv	*Drosophila virilis*
Ed	*Eleutheria dichotoma*
Eg	*Echinococcus granulosus* (flatworm)
h	*Homo sapiens*
Hal	*Halocynthia roretzi* (ascidian)
Hm	*Hirudo medicinalis* (leech)
Hs	*Hydractinia symbiolongicarpus* (cniderian)
Hc	*Clorohydra viridissima*
Lp	*Lampetra planeri* (brook lamprey)
m	*Mus musculus* (mouse)
Mg	*Myxine glutinosa* (hagfish)
Nv	*Notophthalmus viridescens* (newt)
qu	*Coturnix coturnix japonica* (quail)
r	*Rattus norvegicus*
Sa	*Schistocerca americana* (grasshopper)
Sac	*Saccharomyces cerevisiae*
Sc	*Schizophyllum commune* (fungus)
Sg	*Schistocerca gregaria* (grasshopper)
Sm	*Schistosoma mansoni*
Sp	*Schizosaccharomyces pombe*
Tg	*Tripneustes gratilla* (brachiopod)
Tr	*Terebratulina retusa* (sea urchin)
Um	*Ustilago maydis* (fungus)
x	*Xenopus laevis*
zf	*Brachydanio rerio* (zebrafish)
zm	*Zea mays*